A TREATISE ON
DIFFERENTIAL EQUATIONS

A. R. Forsyth

Sixth Edition

Dover Publications, Inc.
Mineola, New York

Published in Canada by General Publishing Company, Ltd., 30 Lesmill Road, Don Mills, Toronto, Ontario.

Published in the United Kingdom by Constable and Company, Ltd., 3 The Lanchesters, 162–164 Fulham Palace Road, London W6 9ER.

Bibliographical Note

This Dover edition, first published in 1996, is an unabridged republication of the 1956 printing of the sixth edition of the work originally published by Macmillan & Company, Ltd., London and New York, in 1929 (first edition: Macmillan, 1885).

Library of Congress Cataloging-in-Publication Data

Forsyth, Andrew Russell, 1858–1942.
    A treatise on differential equations / A.R. Forsyth. — 6th ed.
       p.   cm.
    Previously published: 6th ed. London : Macmillan ; New York : St. Martin's Press, 1929.
    Includes index.
    ISBN 0-486-69314-7 (pbk.)
    1. Differential equations.   I. Title.
QA371.F74   1996
515'.35—dc20   96-22069
                                                  CIP

Manufactured in the United States of America
Dover Publications, Inc., 31 East 2nd Street, Mineola, N.Y. 11501

PREFACE

In the present volume I have tried to make the discussion of the various parts of the subject, which are here given, as full as possible; and there will be found much which has hitherto not appeared except in mathematical journals. At the same time, the treatise does not profess to be complete. Among the parts omitted are the investigations by Fuchs on the integration of linear differential equations, those of Königsberger on the irreducibility of differential equations, the discussion of Pfaff's equation, the recent researches of Hermite and Halphen, and the geometrical applications of the hypergeometric series by Klein; only a very slight sketch of Jacobi's method for partial differential equations is attempted, and there is no indication of the methods of Cauchy, Lie and Mayer. These, and others here omitted, I hope to give in another volume at some future date.

While writing this volume I have consulted many authorities in the shape of treatises, memoirs and textbooks; and, though it is impossible to give in detail every reference, I wish in particular to mention, as having been of great use, Boole's Treatise and his Supplement, Moigno, Imschenetsky and Mansion; and I have used, to a slighter extent than these, Gregory's Examples, Serret and De Morgan. Many references to original memoirs will be found in various chapters.

There occur, scattered throughout the book, many examples, amounting in number to more than eight hundred. Most of these are taken from University and College Examination papers set in Cambridge at various times; some are new, and many of them are results extracted from memoirs which have been consulted. In the case of the last, the original authority is, I think, always indicated. I cannot hope that, among so many, all results given are correct and all equations set are soluble; and I shall be glad to receive corrections of any mistakes actually found.

In conclusion, I wish to express the very great obligations under which I lie to my friend and former tutor Mr H. M. Taylor, of Trinity College, Cambridge, for his kindness in the revision of the proof-sheets. He has caused the removal of many obscurities and has made many valuable suggestions of which I have continually availed myself. My thanks are also due to my friend Mr J. M. Dodds, of St Peter's College, Cambridge, for his kindness in reading some of the early sheets.

A. R. FORSYTH.

Trinity College, Cambridge.
September, 1885.

PREFACE TO THE SECOND EDITION

This edition will be found to differ very slightly from the first. In its preparation I have been much helped by the kindness of many friends and correspondents who have sent me notification of mistakes and misprints.

My thanks are specially due to Dr Hermann Maser of Berlin for the honour he has done me in translating my book into German.

A. R. F.

Trinity College, Cambridge.
September, 1888.

PREFACE TO THE THIRD EDITION

THE present edition will be found to differ substantially from the first two editions. Much of the book has been revised; and there are definite additions.

The principal changes in the portions that belong to the earlier editions occur in a modified treatment of Riccati's equation, in an amplified discussion of the condition of integrability of a total differential equation, and in a complete alteration of the proof of the main proposition relating to the integral of Lagrange's linear partial differential equation of the first order.

The principal additions consist of a brief sketch of Runge's method for the numerical solution of ordinary differential equations, here limited to equations of the first order: of an outline of the method devised by Frobenius for the integration of linear equations in series, the exposition being limited to equations of the second order and applied to Bessel's equations in particular: and of an introduction to Jacobi's theory of multipliers.

At the end of the book I have placed, as a group of general examples, a selection of between two and three hundred equations and questions chosen from examination papers which have been set within the University of Cambridge during the last fifteen years.

I must again acknowledge the kindness of friends and correspondents who have sent me notification of mistakes and misprints; and it is a pleasure to thank the Staff of the University Press for their unfailing assistance during the printing of the volume.

I would add my thanks to Dr Alfredo Arbicone for the honour he has done me by translating the book into Italian.

A. R. F.

TRINITY COLLEGE, CAMBRIDGE.
September, 1903.

PREFACE TO THE FOURTH EDITION

This edition, being the fourth, differs in several respects from the earlier editions. It is somewhat larger in size. But the general scope of the book remains unaltered. My purpose now, in accordance with my purpose in its original production, is to make the volume a practical working text-book on the subject, with (as I hope) sufficient explanation in developing the various methods that are adopted at the various stages. I have definitely, almost rigorously, avoided entering upon the general theory that covers all branches of the subject—such as existence-theorems, whether of ordinary equations (even of linear equations) or of partial equations, and the characteristics of integrals with their limitations. Such matters, of the utmost importance in themselves, I have discussed elsewhere in my *Theory of Differential Equations* to which many references are given in the text. But my opinion is that, in a treatise which aims at the exposition of practical methods, there is an advantage (even if only of conciseness) in assuming some results of theory, never however going beyond the results elsewhere established.

In the work which I have just quoted, the fundamental theory is developed in much detail; the present volume remains a practical text-book which, in all its stages, can be supplemented by the adequate exposition of the fuller theory. Accordingly, I have deliberately avoided dealing with many topics which usually or mainly do not lead to practical processes in the solution of differential equations.

Some portions of the book have been re-written, particularly the early part of the chapter on the hypergeometric series, and parts of the chapters on partial differential equations.

A few additions have been made, each of them brief in itself. In association with the note on the method of Frobenius for the integration of ordinary linear equations in series, I have given further notes on equations which have all or only some or even none of their integrals of the type called regular. The section dealing with total differential equations has been amplified so as to indicate the methods of obtaining an integral equivalent of Pfaffian equations when there are three variables and four variables respectively. In partial equations, the main changes are the insertion of a brief discussion of complete homogeneous linear systems of the first order, the use of these systems in the general construction of the intermediate integral of an equation of the second order when it possesses an intermediate integral, and a modified account of Ampère's method for equations of the second order.

In conclusion I desire, once again, to tender my cordial thanks to the ever-courteous and helpful Staff of the University Press.

A. R. F.

Imperial College of Science
and Technology, S.W.
May, 1914.

PREFACE TO THE FIFTH EDITION

THE present issue is almost an exact reprint of the fourth edition. The pagination is unaltered; and the changes mainly consist of corrections which were found to be necessary, when the volume of *Solutions of the Examples* (1918) was being prepared.

Once more, my thanks are tendered to the Staff of the Cambridge University Press, who maintain their high tradition of unfailing help.

A. R. F.

IMPERIAL COLLEGE OF SCIENCE
  AND TECHNOLOGY, S.W.
    September, 1921.

PREFACE TO THE SIXTH EDITION

THE only differences between the preceding edition and the present edition consist of occasional verbal modifications. No additions have been made: the pagination remains unaltered.

Again I would thank the Staff of the Cambridge University Press for their assistance and care during the printing of the volume.

A. R. F.

September, 1928.

CONTENTS

CHAPTER I.

INTRODUCTION.

CHAPTER II.

DIFFERENTIAL EQUATIONS OF THE FIRST ORDER.

CHAPTER III.

GENERAL LINEAR EQUATION WITH CONSTANT COEFFICIENTS.

CHAPTER IV.

MISCELLANEOUS METHODS.

CHAPTER V.

INTEGRATION IN SERIES.

CHAPTER VI.

THE HYPERGEOMETRIC SERIES.

CHAPTER VII.

SOLUTION BY DEFINITE INTEGRALS.

CHAPTER VIII.

ORDINARY EQUATIONS WITH MORE THAN TWO VARIABLES.

CHAPTER IX.

PARTIAL DIFFERENTIAL EQUATIONS OF THE FIRST ORDER.

CHAPTER X.

PARTIAL DIFFERENTIAL EQUATIONS OF THE SECOND AND HIGHER ORDERS.

CHAPTER I

INTRODUCTION

1. WHEN one variable quantity y is a function of another variable quantity x, the relation between the two may be exhibited by means of an equation such as

$$\phi(x, y) = 0.$$

In this equation constants may occur; let one of such constants be denoted by a. If the equation be solved for y in terms of x, this constant a will enter into the expression for y; and, by taking different values for a, there will in general be obtained a number of corresponding values for y. If it be desired to indicate in the fundamental relation the fact that the value of y depends on that of a, this may be done by writing the above equation in the form

$$\phi(x, y, a) = 0 \dots\dots\dots\dots\dots\dots\text{(i)}.$$

Now it is possible to derive from this equation another, which shall include all the values of y that can be obtained by assigning all the possible values to the constant a. The differential coefficient of y with regard to x is given by

$$\frac{\partial\phi}{\partial x} + \frac{\partial\phi}{\partial y}\frac{dy}{dx} = 0 \dots\dots\dots\dots\dots\text{(ii)},$$

in which $\dfrac{\partial}{\partial x}$ and $\dfrac{\partial}{\partial y}$ indicate partial differentiation with regard to x and y respectively. Equation (ii) will in general involve the constant a, which occurs in (i); and, if between these two equations the constant be eliminated, the result of the elimination will be of the form

$$f\left(x, y, \frac{dy}{dx}\right) = 0 \dots\dots\dots\dots\dots\text{(iii)},$$

where f is a definite function depending on the form of the function ϕ in equation (i). Now equation (iii) includes all the values of y, which can arise from (i); for, while it is derived from the two equations (i) and (ii), in each of which a occurs, yet of the particular value of this quantity no special account is taken and, were any other constant as a' substituted for a in all the steps of the elimination, the result would be the same, since the constant is made to disappear from the result.

In the same way, if y depended on two constants a and b in a manner defined by an equation

$$\Phi\,(x,\,y,\,a,\,b) = 0,$$

and if the equations which give the first and second differential coefficients of y with regard to x were written down, the two constants a and b could be eliminated, and the resulting equation would be of the form

$$F\left(x,\,y,\,\frac{dy}{dx},\,\frac{d^2y}{dx^2}\right) = 0 \,\ldots\ldots\ldots\ldots\text{(iii)}'.$$

In all cases, the functions f and F can be deduced (by any of the customary methods) when the forms ϕ and Φ are given.

In particular, if such a form be

$$\theta\,(x,\,y) = a,$$

from which a is to be eliminated, then, as the equation embracing all the values of y, we have at once

$$\frac{\partial\theta}{\partial x} + \frac{\partial\theta}{\partial y}\frac{dy}{dx} = 0,$$

no further elimination being needed.

Thus, for example, the equation

$$y^2 = 4ax$$

leads to the equation

$$y = 2x\frac{dy}{dx},$$

which is the general equation of all parabolas having the same axis and vertex.

2. Such relations as (iii) and (iii)$'$ are called *Differential Equations;* the equation (i), which is free from all differential

coefficients, is called a *solution* of (iii). As, in passing from (i)
to (iii), a single arbitrary constant was removed, so conversely,
in passing from (iii) to (i), it is just to expect that a single
arbitrary constant will be introduced; and since, in eliminating n
arbitrary constants, there are needed the equations giving the
first n differential coefficients in addition to the original equation,
so conversely, in passing from such a relation between differential
coefficients up to the nth inclusive to an equation free from them
and equivalent to this relation, it is just to expect that n arbitrary
constants will be introduced.

3. It is not difficult to see how these arbitrary quantities must
enter into the solution of the equation. For the sake of simplicity,
let us consider an equation such as

$$M + N\frac{dy}{dx} = 0,$$

in which M and N are functions of x and y. Let x and y represent
the Cartesian coordinates of a point P in a plane referred to two
rectangular axes. Then the equation (i) is the equation of a curve;
and $\frac{dy}{dx}$ is the trigonometrical tangent of the angle, which the
tangent to the curve at the point P makes with the axis of x,
so that the above differential equation gives the direction of a line
at every point in the plane. Let any point A be taken on the
axis of y, and let us proceed from A for a very short distance
in the direction given by the value which $\frac{dy}{dx}$ has at A; we shall
thus come to another point B. Let us proceed now from B
through a very short distance in the direction given by the value
which $\frac{dy}{dx}$ has at B; we shall thus come to another point C. If
this process be carried out for a number of directions in succession,
a figure will be traced in the plane; and, when each of the distances through which we suppose the tracing point to move
becomes indefinitely small, the figure will become a curve passing
through A. This curve will have a definite equation, which may
be exhibited in the form

$$F(x, y, y_0) = 0,$$

where y_0 is the ordinate of A. Had another initial point A' been chosen instead of A, then another curve would have been obtained, and into its equation the magnitude of the ordinate of A' would have entered. The same result would ensue from taking each point in succession on the axis of y, because usually one curve and only one passes through each such point. As each equation, or one single equation as the representative of all, may be considered a solution of the differential equation, it is evident that into the solution of the example we have been considering one arbitrary constant will enter; and therefore, if by any method we can obtain an equation free from differential coefficients, it must be expected that an arbitrary constant will be contained in that equation. But this arbitrary constant obtained by the latter method will not necessarily be the ordinate of the point, at which the curve, represented by the solution, and the axis of y intersect; an arbitrary element would have entered into the equation, had the tracing of the curve begun from a point in the plane not lying on one of the coordinate axes.

In the example considered, the equation giving $\dfrac{dy}{dx}$ had only a single root; when it is of the form

$$\left(\frac{dy}{dx}\right)^2 + P\frac{dy}{dx} + Q = 0,$$

then the integral equation may be expected to be of the form

$$A^2 + AP' + Q' = 0,$$

where A is an arbitrary constant. And it is not difficult to see that, if the differential equation be of the nth degree in $\dfrac{dy}{dx}$, then the corresponding integral equation may be expected to contain an arbitrary constant raised to the nth and lower powers.

4. From what has been said as to one of the methods by which differential equations can be constructed, it might be deemed an easy matter to return from the differential to the integral equation; but this is not so. The steps of an elimination cannot be retraced, and therefore some other method or methods must be adopted. The methods which are most effective for the solution of several different forms of differential equations will be discussed hereafter.

5. When we pass from a given integral function to the equivalent differential equation, the latter may prove to be of a form which is not included among those already known; so conversely, if we pass from a given differential equation, we must not expect to arrive necessarily at a function that will be included among those with the properties of which we are acquainted. It is therefore desirable to indicate what, in such a case, would be meant by the solution of the differential equation.

When, in algebra, we ask whether any particular equation can be solved, we thereby enquire whether the value of the variable, which occurs in it, can be expressed in terms of known functions. Thus, for instance, in the equation

$$ax = b,$$

the value of x can be obtained immediately by a process of division. But let the equation be

$$\xi^2 = K.$$

To solve this we have to introduce a function, which was not needed for the former equation; and, expressing ξ in the form

$$\xi = \pm K^{\frac{1}{2}},$$

we consider the equation solved. Now equations of the third and fourth degrees can be solved by means of functions strictly analogous to these—the cube root and the fourth root of quantities; but general equations of the fifth and higher degrees cannot be solved in terms of these functions or combinations of these with similar functions. It does not therefore follow that solutions of these equations do not exist; they can be solved only when functions, unused in the solution of equations of lower degrees, are introduced.

Similarly, in the case of a differential equation, when we say that it can be solved, we do not mean to imply that the solution must be expressible in terms of purely algebraic functions, of exponentials (including sines and cosines), and of logarithmic functions (including inverse circular functions). The equation

$$\frac{dy}{dx} = 2x$$

is equivalent to

$$y = x^2 + A.$$

But suppose that the properties of the logarithm were unknown, and that the differential equation

$$\frac{dy}{dx} = \frac{1}{x}$$

were proposed for solution. We should then have

$$y = A + \int \frac{dx}{x};$$

and, calling

$$\int \frac{dx}{x} = f(x),$$

we should prove the relation

$$f(x) + f(y) = f(xy),$$

and become acquainted with the properties of this new function so as to include it amongst known functions. But, had we not been able to deduce the properties of $f(x)$, the value of y given by

$$A + \int \frac{dx}{x}$$

would still have been considered a solution of the differential equation. In fact, *every differential equation is considered as solved, when the value of the dependent variable is expressed as a function of the independent variable by means either of known functions or of integrals, whether the integrations in the latter can or cannot be expressed in terms of functions already known.* Thus, for instance,

$$y = A + \int \frac{e^x}{x} \, dx$$

is a solution of

$$x \frac{dy}{dx} = e^x,$$

although the value of y cannot be expressed otherwise than in this form without the introduction of a new function the properties of which can be investigated. In this way the solution of differential equations is continually suggesting new functions to be added to the stock of those already known.

6. Before we proceed further, it is desirable to give definitions of some terms used in the subject.

Any equation, which expresses a relation between dependent variables, their differential coefficients of any order whatever, and the independent variables, is called a *differential equation*.

Differential equations are divided into two *species*, viz.:—

I. *Ordinary differential equations*, into which only a single independent variable enters, either explicitly or implicitly. All the differential coefficients have reference to this variable. Should there be several dependent variables, the number of equations necessary for their complete determination as functions of the independent variable is equal to the number of such variables. Thus, for instance, we might have

$$\frac{d^2x}{dt^2} + \mu x = 0,$$

in which x is a function of the only independent variable t; and

$$\left. \begin{aligned} (x^2 + y^2)^{\frac{3}{2}} \frac{d^2x}{dt^2} + \mu x = 0 \\ (x^2 + y^2)^{\frac{3}{2}} \frac{d^2y}{dt^2} + \mu y = 0 \end{aligned} \right\},$$

in which both x and y are functions of t.

II. *Partial differential equations*, into which two independent variables at least, and partial differential coefficients with regard to any or all of these variables, may enter. If several dependent variables be present, the number of separate equations must be not less than the number of the separate dependent variables; but the occurrence of such systems of equations is relatively rare. As examples of partial differential equations, we may consider

$$\frac{\partial^2 z}{\partial x^2} \frac{\partial^2 z}{\partial y^2} - \left(\frac{\partial^2 z}{\partial x \partial y} \right)^2 = 0,$$

and

$$\left. \begin{aligned} \frac{\partial \phi}{\partial x} &= \frac{\partial \psi}{\partial y} \\ \frac{\partial \phi}{\partial y} &= -\frac{\partial \psi}{\partial x} \end{aligned} \right\}.$$

The *order* of a differential equation is the same as the order of the highest differential coefficient it contains.

As a rule, only those equations are considered which are algebraic in the differential coefficients. The *degree* of the equation is the power to which the highest differential coefficient is raised, when the equation is rational and integral in the differential coefficients and is one-valued in the variables themselves.

The equation

$$y = x\frac{dy}{dx} + \frac{a}{\dfrac{dy}{dx}}$$

is of the first order and the second degree; the equation

$$\left\{1 + \left(\frac{dy}{dx}\right)^2\right\}^{\frac{3}{2}} = a\frac{d^2y}{dx^2}$$

is of the second order and the second degree.

If a differential equation be such that, when it is rationalised and freed from fractions, the differential coefficients and the dependent variable enter in the first power and there are no products of these, while the coefficients in the separate terms are either constants or functions of the independent variables, the equation is called *linear*. The following are examples of linear equations:—

$$(1 - x^2)\frac{d^2y}{dx^2} - 2x\frac{dy}{dx} + n(n+1)y = 0,$$

$$\frac{\partial^2 V}{\partial x^2} + \frac{\partial^2 V}{\partial y^2} + \frac{\partial^2 V}{\partial z^2} = 0,$$

$$x\frac{\partial z}{\partial x} + y\frac{\partial z}{\partial y} - z = 0.$$

The relation, which exists between the variables themselves without their differential coefficients and which is the most general one possible, is called sometimes the *general solution*, and sometimes the *primitive*, of the differential equation.

7. The process of constructing the primitive of a given differential equation will frequently be the deduction of a first integral of the differential equation, that is, an equation of an order lower by unity than that of the original equation and containing an arbitrary constant; then of a first integral of the latter which will be a second integral of the original equation; and so on, until differential coefficients cease to appear. The primitive will be found when the operation has been repeated the number of times equal to the order of the original differential equation. Now the form of the first integral may be affected by any transformation to which the equation may be subjected prior to integration; and, since a given equation may be transformed in a number of different ways, there might be a corresponding number of different first integrals. But these will not all be necessarily independent: that is, some among them may be obtained by combinations of the remainder, with appropriate relations among the arbitrary constants. As a matter of fact, *if the equation be of the nth order, it cannot have more than n independent first integrals.* For example, the differential equation

$$\frac{d^2y}{dx^2} + y = 0$$

has the following first integrals, viz.:—

$$\left(\frac{dy}{dx}\right)^2 + y^2 = A^2,$$

$$\frac{dy}{dx}\cos x + y\sin x = B,$$

$$-\frac{dy}{dx}\sin x + y\cos x = C,$$

$$\frac{dy}{dx} = y\cot(x+\alpha);$$

but they are not all independent, the four constants A, B, C, α, being connected by the equations

$$B = A\cos\alpha, \quad C = A\sin\alpha, \quad B^2 + C^2 = A^2.$$

When a system of first integrals has been so obtained in any case, it can be used as a simultaneous system, from which the highest differential coefficients can be eliminated; and if independent first integrals of the equation, equal in number to the order of the equation, have been obtained, all the differential coefficients can be eliminated from them so as to leave the primitive. Thus, from the second and the third integrals in the foregoing example, we might deduce

$$y = B\sin x + C\cos x,$$

and, from the first and the fourth,

$$y = A\sin(x+\alpha),$$

each being a primitive. These solutions are seen to coincide on
account of the relations between the constants.

8. We proceed now to give reasons for the statement made
in the last paragraph; but they must be regarded rather as
suggesting, than as proving, the inference. A rigorous proof
involves considerations connected with the general theory of
functions, and can be based upon existence-theorems which will
be found elsewhere*.

*A differential equation of the order n has n, and cannot have
more than n, independent first integrals.*

From what has already been said, it is clear that an integral
relation between y and x involving n arbitrary independent
constants would lead to a differential equation of the order n.
Let the given integral equation be differentiated $n-1$ times in
succession; the $n-1$ resulting equations will involve all the
differential coefficients up to the $(n-1)^{\text{th}}$ inclusive and there will,
with the original equation, be n equations in all. Now from n
equations, in which n quantities occur, all but one of these quantities can be eliminated. Let the n arbitrary constants be denoted
by $C_1, C_2, \ldots\ldots, C_n$; and from the n equations, which we have, let
us eliminate all the arbitrary constants except C_1. The resulting
equation will involve the variables and the derivatives of y up to
the $(n-1)^{\text{th}}$ inclusive and will also involve C_1; it will therefore be
a first integral of the differential equation of the order n which is
equivalent to the given integral relation. Now eliminate all the
arbitrary constants except C_2; the resulting equation will now
involve C_2 and, as before, derivatives of y up to the $(n-1)^{\text{th}}$ inclusive and will therefore be a first integral of the differential
equation; it will, moreover, be independent of the former, since C_2
is independent of C_1. Proceeding in this way with all the constants
in turn, we shall obtain n independent first integrals, each of which
arises from the elimination of all but one of the n independent
constants.

As there are not more than n independent constants occurring
in the general integral equation, any other constant, which could
appear in it, must depend on $C_1, C_2, \ldots\ldots\ldots, C_n$; let A be such

---

* See the author's *Theory of Differential Equations*, vol. II., chap. II.

a constant, and let the relation between them be denoted by the equation

$$\psi(A, C_1, C_2, \ldots\ldots\ldots, C_n) = 0.$$

Then between this, and the original integral equation, and the $n - 1$ equations obtained by differentiation, (forming $n + 1$ equations in all), the n constants C may be eliminated; the result will involve the differential coefficients up to the $(n - 1)^{\text{th}}$ inclusive and the constant A. This would be a first integral of the differential equation, but it is not independent of the n already obtained; for from these let the respective values of the quantities C in terms of the variables and the differential coefficients of y be derived from the separate equations, in which they occur singly and be substituted in the equation $\psi = 0$; this equation will then be one involving the differential coefficients up to the $(n - 1)^{\text{th}}$ and the constant A, and will therefore be the same as the foregoing. In fact, the two processes are merely different methods of obtaining the one result; and the second shews that the first integral so obtained is derivable from the other n first integrals.

Hence the differential equation of order n has not more than n independent first integrals.

9. It is convenient to add here two lemmas to which frequent reference will subsequently be made.

LEMMA I. Let $u_1, u_2, \ldots\ldots, u_n$, be n functions of the n variables $x_1, x_2, \ldots\ldots, x_n$, these variables being independent of one another; if among these functions any relation, which may be represented by

$$F(u_1, u_2, \ldots\ldots\ldots, u_n) = 0 \ \ldots\ldots\ldots\ldots\ldots\text{(i)},$$

be identically satisfied, so that $u_1, u_2, \ldots\ldots, u_n$, are not independent of one another, then the equation

$$\begin{vmatrix} \dfrac{\partial u_1}{\partial x_1}, & \dfrac{\partial u_1}{\partial x_2}, & \ldots\ldots\ldots\ldots\ldots, & \dfrac{\partial u_1}{\partial x_n} \\[2ex] \dfrac{\partial u_2}{\partial x_1}, & \dfrac{\partial u_2}{\partial x_2}, & \ldots\ldots\ldots\ldots\ldots, & \dfrac{\partial u_2}{\partial x_n} \\[2ex] \ldots\ldots\ldots\ldots\ldots\ldots\ldots\ldots\ldots \\[1ex] \dfrac{\partial u_n}{\partial x_1}, & \dfrac{\partial u_n}{\partial x_2}, & \ldots\ldots\ldots\ldots\ldots, & \dfrac{\partial u_n}{\partial x_n} \end{vmatrix} = 0 \ \ldots\ldots \text{(ii)}$$

is identically satisfied.

Since equation (i) is identically satisfied, when for $u_1, u_2, \ldots\ldots, u_n$, are substituted their values in terms of the independent variables, the partial differential coefficients of F of the first order with regard to each of these variables are separately zero. Thus we have

$$\frac{\partial F}{\partial u_1}\frac{\partial u_1}{\partial x_1} + \frac{\partial F}{\partial u_2}\frac{\partial u_2}{\partial x_1} + \ldots\ldots\ldots + \frac{\partial F}{\partial u_n}\frac{\partial u_n}{\partial x_1} = 0,$$

$$\frac{\partial F}{\partial u_1}\frac{\partial u_1}{\partial x_2} + \frac{\partial F}{\partial u_2}\frac{\partial u_2}{\partial x_2} + \ldots\ldots\ldots + \frac{\partial F}{\partial u_n}\frac{\partial u_n}{\partial x_2} = 0,$$

$$\ldots\ldots\ldots\ldots\ldots\ldots\ldots\ldots\ldots\ldots\ldots\ldots\ldots\ldots\ldots\ldots$$

$$\frac{\partial F}{\partial u_1}\frac{\partial u_1}{\partial x_n} + \frac{\partial F}{\partial u_2}\frac{\partial u_2}{\partial x_n} + \ldots\ldots\ldots + \frac{\partial F}{\partial u_n}\frac{\partial u_n}{\partial x_n} = 0.$$

Let the ratios of the n partial differential coefficients of F with regard to the u's be eliminated between these n equations, which are linear in these quantities; the result of the elimination is

$$\begin{vmatrix} \dfrac{\partial u_1}{\partial x_1}, & \dfrac{\partial u_2}{\partial x_1}, & \ldots\ldots\ldots\ldots\ldots, & \dfrac{\partial u_n}{\partial x_1} \\[2ex] \dfrac{\partial u_1}{\partial x_2}, & \dfrac{\partial u_2}{\partial x_2}, & \ldots\ldots\ldots\ldots\ldots, & \dfrac{\partial u_n}{\partial x_2} \\[2ex] \ldots & \ldots & \ldots\ldots\ldots\ldots & \ldots \\[2ex] \dfrac{\partial u_1}{\partial x_n}, & \dfrac{\partial u_2}{\partial x_n}, & \ldots\ldots\ldots\ldots\ldots, & \dfrac{\partial u_n}{\partial x_n} \end{vmatrix} = 0,$$

and this is identically satisfied. The value of a determinant is unaltered by the change of rows into columns and columns into rows; when these changes take place the above equation becomes equation (ii), which is therefore identically satisfied.

LEMMA II. The converse of this is also true: If $u_1, u_2, \ldots\ldots, u_n$, be n functions of n independent variables $x_1, x_2, \ldots\ldots\ldots, x_n$, and if the equation

$$\begin{vmatrix} \dfrac{\partial u_1}{\partial x_1}, & \dfrac{\partial u_1}{\partial x_2}, & \ldots\ldots\ldots\ldots\ldots, & \dfrac{\partial u_1}{\partial x_n} \\[2ex] \dfrac{\partial u_2}{\partial x_1}, & \dfrac{\partial u_2}{\partial x_2}, & \ldots\ldots\ldots\ldots\ldots, & \dfrac{\partial u_2}{\partial x_n} \\[2ex] \ldots & \ldots & \ldots\ldots\ldots\ldots & \ldots \\[2ex] \dfrac{\partial u_n}{\partial x_1}, & \dfrac{\partial u_n}{\partial x_2}, & \ldots\ldots\ldots\ldots\ldots, & \dfrac{\partial u_n}{\partial x_n} \end{vmatrix} = 0$$

be identically satisfied, then the functions $u_1, u_2, \ldots\ldots\ldots, u_n$, are not independent of one another, but are connected by a relation of the form

$$F(u_1, u_2, \ldots\ldots\ldots, u_n) = 0.$$

If the $n-1$ functions $u_1, u_2, \ldots\ldots\ldots, u_{n-1}$, be not independent of one another, then the proposition to be proved is at once granted; we may therefore suppose them independent of one another.

Between the n equations expressing the n functions u we can eliminate $n-1$ of the variables; if the remaining variable, say x_n, be not thereby eliminated, the result may be written in the form

$$u_n = \phi(u_1, u_2, \ldots, u_{n-1}, x_n).$$

If the equation of condition be written in the form

$$\frac{\partial(u_1, u_2, \ldots\ldots, u_n)}{\partial(x_1, x_2, \ldots\ldots, x_n)} = 0,$$

we may write the theorem for the multiplication of determinants in the form

$$\frac{\partial(u_1, u_2, \ldots\ldots, u_n)}{\partial(x_1, x_2, \ldots\ldots, x_n)} = \frac{\partial(u_1, u_2, \ldots\ldots, u_{n-1}, \phi)}{\partial(u_1, u_2, \ldots\ldots, u_{n-1}, x_n)} \times \frac{\partial(u_1, u_2, \ldots\ldots, u_{n-1}, x_n)}{\partial(x_1, x_2, \ldots\ldots, x_{n-1}, x_n)}.$$

The left-hand side is zero by hypothesis. Since the functions $u_1, u_2, \ldots\ldots, u_{n-1}$, are independent, the first factor on the right-hand side is $\dfrac{\partial\phi}{\partial x_n}$, and the second is $\dfrac{\partial(u_1, u_2, \ldots, u_{n-1})}{\partial(x_1, x_2, \ldots, x_{n-1})}$. One of these must therefore vanish.

If it be the former, then ϕ is explicitly independent of x_n, so that u_n is a function of $u_1, u_2, \ldots, u_{n-1}$ only; and there is thus a relation between the original n functions.

If it be the latter, we have

$$\frac{\partial(u_1, u_2, \ldots\ldots, u_{n-1})}{\partial(x_1, x_2, \ldots\ldots, x_{n-1})} = 0,$$

an equation, which corresponds to the given equation of condition but in which there are only $n-1$ functions of $n-1$ variables, since for the differentiations that now occur x_n may be considered a constant. This equation is treated in the same manner as before; and we should find either that there is a relation between $u_1, u_2, \ldots, u_{n-1}$, considered as functions of $x_1, x_2, \ldots\ldots, x_{n-1}$, or that a new equation involving $n-2$ functions of $n-2$ variables would hold. If the relation between $u_1, u_2, \ldots, u_{n-1}$ exist, it will be of the form

$$\psi(u_1, u_2, \ldots\ldots, u_{n-1}, x_n) = 0;$$

which will involve x_n since we have assumed that $u_1, u_2, \ldots, u_{n-1}$, are independent of one another. Between $\psi = 0$ and $u_n = \phi$ we can eliminate x_n, and obtain a relation between $u_1, u_2, \ldots, u_n$.

Proceeding in this manner and diminishing by unity each time the number of functions, which enter into the equation of condition, we can prove that one of the two necessary inferences at each reduction is the statement contained in the proposition. And when the reduction has been repeated $n - 1$ times, the only alternative to the proposition is that any one of the n functions u, say u_r, selected at will, should be such as to satisfy the relation $\frac{\partial u_r}{\partial x_s} = 0$, where x_s is any one of the n variables x. This is manifestly not the case: for each of the functions u involves some of the variables.

Hence the proposition follows.

10. As a particular instance of the general lemmas, we have the following. Let U and V be two functions of two independent variables x and y; if V can be expressed as a function of U alone, we must have

$$\frac{\partial U}{\partial x}\frac{\partial V}{\partial y} - \frac{\partial U}{\partial y}\frac{\partial V}{\partial x} = 0;$$

and conversely, if this equation be satisfied identically, there is a relation between U and V, satisfied for all values whatever of x and y, such that

$$V = f(U).$$

Ex. 1. Are the functions

$$x + 2y + z, \quad x - 2y + 3z, \quad 2xy - xz + 4yz - 2z^2$$

independent of one another?

The equation of condition is

$$\begin{vmatrix} 1 & , & 1 & , & 2y - z \\ 2 & , & -2 & , & 2x + 4z \\ 1 & , & 3 & , & -x + 4y - 4z \end{vmatrix} = 0,$$

which is evidently satisfied identically; therefore the three functions are dependent.

To find the relation between them, if we call them u_1, u_2, u_3, we have

$$2x = u_1 + u_2 - 4z, \quad 4y = u_1 - u_2 + 2z;$$

and therefore

$$4u_3 = u_1{}^2 - u_2{}^2,$$

on substituting these values.

Ex. 2. Prove that the functions $ax^2 + by^2 + cz^2$, $Ax + By + Cz$, and $a^2x^2(B^2c + C^2b) + b^2y^2(C^2a + A^2c) + c^2z^2(A^2b + B^2a) - 2abc(BCyz + CAzx + ABxy)$, are not independent; and find the relation between them.

Now consider two coexistent equations $U=0$ and $V=0$, involving two variables x and y; and suppose that, for simultaneous values of x and y which satisfy them, they determine a common value of $\frac{dy}{dx}$. Then, as this common value is given by the relations

$$\frac{\partial U}{\partial x}+\frac{\partial U}{\partial y}\frac{dy}{dx}=0, \quad \frac{\partial V}{\partial x}+\frac{\partial V}{\partial y}\frac{dy}{dx}=0,$$

we must have

$$J=\frac{\partial U}{\partial x}\frac{\partial V}{\partial y}-\frac{\partial U}{\partial y}\frac{\partial V}{\partial x}=0.$$

If this equation $J=0$ is satisfied identically, the quantities U and V are not independent of one another; in that event, the two coexistent equations, $U=0$ and $V=0$, are not independent of one another.

If the equation $J=0$ is not satisfied identically, but only in virtue of $U=0$, or $V=0$, or $U=0$ and $V=0$, then we cannot infer the relative dependence of the two equations. Neither of them can be satisfied solely by means of the other.

Ex. 3. As an example of the first case, let

$$U=y\,(1-x^2)^{\frac{1}{2}}+x\,(1-y^2)^{\frac{1}{2}}, \quad V=(1-x^2)^{\frac{1}{2}}\,(1-y^2)^{\frac{1}{2}}-xy.$$

It is easy to prove that the quantity

$$\frac{\partial U}{\partial x}\frac{\partial V}{\partial y}-\frac{\partial U}{\partial y}\frac{\partial V}{\partial x}$$

vanishes identically. The quantities U and V are then connected by some relation; the relation is

$$U^2+V^2=1.$$

The equations $U-a=0$, $V-b=0$, are not independent; we must have $a^2+b^2=1$ as a condition of coexistence, and then the two equations are effectively equivalent to a single equation.

Ex. 4. As an example of the alternative case, let

$$U=x^2+y^2-1, \quad V=x\cos\alpha+y\sin\alpha-1,$$

where α is a constant. Then

$$J=\frac{\partial U}{\partial x}\frac{\partial V}{\partial y}-\frac{\partial V}{\partial x}\frac{\partial U}{\partial y}=2\,(x\sin\alpha-y\cos\alpha),$$

which manifestly does not vanish identically. The quantities U and V are independent of one another.

The quantity J vanishes when $U=0$ and $V=0$, for

$$\tfrac{1}{4}J^2=U-V^2-2V.$$

That is, when the two equations $U=0$ and $V=0$ are postulated, J vanishes; but it does not vanish identically. The two equations are independent of one another; they have the property merely of leading to the same value of $\frac{dy}{dx}$ for particular values of y and x which satisfy both of them.

The same kind of result is true, when we have any number n of coexistent equations

$$u_1 = 0, \quad u_2 = 0, \quad \ldots, \quad u_n = 0,$$

involving n variables $x_1, x_2, \ldots, x_n$. Let J denote their Jacobian, so that

$$J = \frac{\partial (u_1, u_2, \ldots, u_n)}{\partial (x_1, x_2, \ldots, x_n)}.$$

When J vanishes identically, the n equations are not independent of one another.

When J vanishes, not identically but only in virtue of some or of all the equations, we cannot infer any relative dependence among the equations. No one of them can be satisfied solely by means of the rest.

When J does not vanish, there is no question of the relative dependence of the equations.

CHAPTER II

DIFFERENTIAL EQUATIONS OF THE FIRST ORDER

11. THE general differential equation of the first order may be represented by

$$F\left(x, y, \frac{dy}{dx}\right) = 0,$$

where F is a rational and integral function so far as the differential coefficient is concerned. In this general form, the equation cannot be integrated; but there are certain particular forms, to one or other of which many equations can be reduced, and which admit of immediate solution. These forms are called *standard* forms.

12. Before considering them in detail, we will prove a proposition, which is merely a particular case of the general theorem indicated in § 8, viz., that a differential equation expressible in the form

$$M \frac{dy}{dx} = N,$$

where M and N are one-valued functions of x and y, *can have only one independent primitive.*

Suppose that, if it be possible, two primitives

$$\phi_1 (x, y) = a,$$
$$\phi_2 (x, y) = b,$$

have been obtained. From the first of these, the value of $\frac{dy}{dx}$ is given by

$$\frac{\partial \phi_1}{\partial x} + \frac{\partial \phi_1}{\partial y} \frac{dy}{dx} = 0,$$

and therefore

$$M\frac{\partial\phi_1}{\partial x} + N\frac{\partial\phi_1}{\partial y} = 0.$$

Treating the second primitive in the same way, we should obtain the equation

$$M\frac{\partial\phi_2}{\partial x} + N\frac{\partial\phi_2}{\partial y} = 0.$$

The elimination of the one-valued functions M and N between these two equations gives

$$\frac{\partial\phi_1}{\partial x}\frac{\partial\phi_2}{\partial y} - \frac{\partial\phi_1}{\partial y}\frac{\partial\phi_2}{\partial x} = 0.$$

This equation must be satisfied identically: for it does not contain a or b, and therefore it cannot be satisfied in virtue of $\phi_1 = a$ or $\phi_2 = b$. Consequently (§ 10) ϕ_2 is some function of ϕ_1. Hence the two primitives are not independent; and the second can be expressed in the form

$$F(\phi_1) = b,$$

which is resoluble into equations of the form

$$\phi_1 = a,$$

each of which is only a repetition of the first of the primitives.

If therefore in solving such a differential equation *any* primitive has been obtained, this may be looked upon as the general solution of the equation; for, from it, all other primitives can be derived.

13. Standard I.

The equation $Mdy = Ndx$ can always be solved when the variables can be separated. For, in this case, the equation may be changed to the form

$$Ydy = Xdx,$$

where Y is a function of y alone, and X a function of x alone; and the equation can be integrated in the form

$$\int Ydy = \int Xdx + A,$$

A being an arbitrary constant.

Ex. 1. Solve
$$\frac{dy}{dx} + \left(\frac{1-y^2}{1-x^2}\right)^{\frac{1}{2}} = 0.$$

The variables can be separated and the equation becomes
$$\frac{dy}{(1-y^2)^{\frac{1}{2}}} + \frac{dx}{(1-x^2)^{\frac{1}{2}}} = 0,$$
one integral of which is
$$\text{arcsin } y + \text{arcsin } x = c.$$

But the equation may be written
$$(1-x^2)^{\frac{1}{2}} dy + (1-y^2)^{\frac{1}{2}} dx = 0,$$
which, after integration by parts, gives
$$y(1-x^2)^{\frac{1}{2}} + \int \frac{xy\,dx}{(1-x^2)^{\frac{1}{2}}} + x(1-y^2)^{\frac{1}{2}} + \int \frac{xy\,dy}{(1-y^2)^{\frac{1}{2}}} = C.$$

But
$$\frac{xy\,dx}{(1-x^2)^{\frac{1}{2}}} + \frac{xy\,dy}{(1-y^2)^{\frac{1}{2}}} = 0;$$
and therefore an integral is
$$y(1-x^2)^{\frac{1}{2}} + x(1-y^2)^{\frac{1}{2}} = C.$$

This affords an illustration of the proposition in the preceding paragraph; for the latter primitive can be derived from the former by taking the sine of both members, and the relation between the constants is
$$C = \sin c.$$

Ex. 2. Solve
$$(x - y^2)\,dx + 2xy\,dy = 0.$$

The variables, though not immediately separable, become so after substitution. Write $y^2 = v$; the equation is
$$x\,dx + x\,dv - v\,dx = 0,$$
so that
$$\frac{dx}{x} + d\left(\frac{v}{x}\right) = 0,$$
and therefore
$$\log x + \frac{v}{x} = \text{constant},$$
or
$$xe^{\frac{y^2}{x}} = A.$$

Ex. 3. Solve the equations

(i) $x(1+y^2)^{\frac{1}{2}} + y(1+x^2)^{\frac{1}{2}} \dfrac{dy}{dx} = 0$;

(ii) $\sec^2 x \tan y\,dx + \sec^2 y \tan x\,dy = 0$;

(iii) $(x+y)^2 \dfrac{dy}{dx} = a^2$;

(iv) $(1+y^2)\,dx - \{y + (1+y)^{\frac{1}{2}}\}(1+x)^{\frac{3}{2}}\,dy = 0$;

(v) $(y-x)(1+x^2)^{\frac{1}{2}} \dfrac{dy}{dx} = n(1+y^2)^{\frac{3}{2}}$.

14. STANDARD II. *Linear Form.*

When the equation of the first order is linear, it may be written in the form

$$\frac{dy}{dx} + Py = Q,$$

where P and Q are functions of x and do not involve y.

If the function Q were zero, the integral of the equation could be obtained by the method of § 13, and would be found to be

$$y = ue^{-\int P dx},$$

where u is a constant.

When Q is different from zero, we assume the same form for y, but we do not restrict u to be constant. Substituting this value of y in the original equation, we find

$$\frac{dy}{dx} = \left(\frac{du}{dx} - Pu\right)e^{-\int P dx},$$

and therefore

$$\frac{du}{dx}e^{-\int P dx} = Q,$$

so that

$$\frac{du}{dx} = Qe^{\int P dx},$$

and therefore

$$u = C + \int Qe^{\int P dx},$$

where C is an arbitrary constant. Thus

$$y = ue^{-\int P dx}$$
$$= Ce^{-\int P dx} + e^{-\int P dx}\int Qe^{\int P dx}\, dx$$

is the primitive of the differential equation.

Note 1. This method of obtaining an integral of the equation when Q is zero, and then regarding the parameter u in that integral as actually variable when Q is not zero, is sometimes called the method of the *variation of parameters.* It is of frequent use for linear equations of the second and higher orders.

Note 2. The preceding analysis shews that

$$\frac{dy}{dx} + Py = \frac{du}{dx}e^{-\int P dx},$$

and therefore

$$e^{\int Pdx}\left(\frac{dy}{dx}+Py\right)$$

is a perfect differential. Such a quantity as $e^{\int Pdx}$, multiplying a quantity $\frac{dy}{dx}+Py$ so that it becomes a perfect differential, is often called an *integrating factor*.

Ex. 1. Solve $\qquad \frac{dy}{dx}+\frac{x}{1+x^2}y=\frac{1}{x(1+x^2)}.$

By the result in the general case, we have

$$ye^{\int\frac{xdx}{1+x^2}}=C+\int\frac{dx}{x(1+x^2)}e^{\int\frac{xdx}{1+x^2}};$$

hence $\qquad y(1+x^2)^{\frac{1}{2}}=C+\int\frac{dx}{x(1+x^2)^{\frac{1}{2}}}$

$$=C+\log\frac{x}{1+(1+x^2)^{\frac{1}{2}}}.$$

Ex. 2. Solve

(i) $x(1-x^2)\frac{dy}{dx}+(2x^2-1)y=ax^3$;

(ii) $\frac{dy}{dx}+y\cos x=\frac{1}{2}\sin 2x$;

(iii) $y\frac{dy}{dx}+cy^2=a\cos(x+\beta)$;

(iv) $\frac{dy}{dx}+y\frac{d\phi}{dx}=\phi(x)\frac{d\phi}{dx}$.

Ex. 3. Shew that the solution of the general equation may be exhibited in the form

$$y=\frac{Q}{P}-e^{-\int Pdx}\left[C+\int e^{\int Pdx}d\,\frac{Q}{P}\right].$$

15. An important associated form, which can be solved by the same method, is

$$\frac{dy}{dx}+Py=Qy^n,$$

where P and Q are functions of x alone, and n is neither 0 nor 1.

Divide by y^n; the equation then is

$$-\frac{1}{n-1}\frac{d}{dx}\left(\frac{1}{y^{n-1}}\right)+P\frac{1}{y^{n-1}}=Q,$$

or $\qquad \frac{d}{dx}\left(\frac{1}{y^{n-1}}\right)-(n-1)P\frac{1}{y^{n-1}}=-Q(n-1),$

which is the standard form; and the general solution is

$$\frac{1}{y^{n-1}} e^{-(n-1)\int P dx} = C - (n-1) \int Q e^{-(n-1)\int P dx} dx.$$

Ex. 4. Solve $\qquad x\frac{dy}{dx} + y = y^2 \log x.$

This becomes, after a transformation similar to the above,

$$\frac{d}{dx}\left(\frac{1}{y}\right) - \frac{1}{y}\frac{1}{x} = -\frac{1}{x}\log x,$$

the primitive of which is

$$\frac{1}{y} e^{-\int \frac{dx}{x}} = C - \int \frac{dx \log x}{x} e^{-\int \frac{dx}{x}}.$$

This is
$$\frac{1}{xy} = C - \int \frac{dx \log x}{x^2}$$

$$= C + \frac{\log x}{x} + \frac{1}{x},$$

whence
$$\frac{1}{y} = 1 + Cx + \log x.$$

Ex. 5. Solve $\qquad$ (i) $\dfrac{dz}{dx} + 2xz = 2ax^3z^3$;

$\qquad$ (ii) $(1-x^2)\dfrac{dz}{dx} - xz = axz^2$;

$\qquad$ (iii) $\dfrac{dy}{dx} + xy = y^2 \sin x$;

$\qquad$ (iv) $\dfrac{dy}{dx}(x^2y^3 + xy) = 1.$

Ex. 6. Shew that the four equations of the first order in § 7 lead to the same primitive.

16. STANDARD III. *Homogeneous Equations.*

(i) The equation, being of the first degree and expressed in the form

$$M\frac{dy}{dx} = N,$$

is said to be homogeneous, when M and N are homogeneous functions of x and y of the same degree. In this case, we can write

$$M = x^r \phi\left(\frac{y}{x}\right),$$

$$N = x^r \psi\left(\frac{y}{x}\right),$$

r being the common degree of M and N. On the substitution of

$$y = vx,$$

so that v may be considered a new dependent variable, the equation becomes

$$\left(v + x\frac{dv}{dx}\right)\phi(v) = \psi(v),$$

or

$$\frac{dx}{x} + \frac{\phi(v)\,dv}{v\phi(v) - \psi(v)} = 0,$$

in which the variables are separated; the integral is

$$\log x + \int\frac{\phi(v)\,dv}{v\phi(v) - \psi(v)} = A.$$

The primitive will be given by the substitution of $\dfrac{y}{x}$ for v after the integration has been performed.

Note. This result furnishes another instance of an *integrating factor*, of which one instance was indicated in § 14, *Note* 2. The equation can be taken in the form

$$M\,dy - N\,dx = 0,$$

so that

$$x^r\,\phi(v)(v\,dx + x\,dv) - x^r\,\psi(v)\,dx = 0,$$

and therefore

$$x^r\{v\phi(v) - \psi(v)\}\,dx + x^{r+1}\phi(v)\,dv = 0.$$

If $v\phi(v) - \psi(v)$ vanishes identically, then x^{-r-1} is an integrating factor: for it changes the equation into

$$\phi(v)\,dv = 0,$$

which is exactly integrable.

If $v\phi(v) - \psi(v)$ does not vanish identically, then the equation becomes exactly integrable in the form

$$\frac{dx}{x} + \frac{\phi(v)}{v\phi(v) - \psi(v)}\,dv = 0,$$

on division by the quantity

$$x^{r+1}\{v\phi(v) - \psi(v)\}.$$

But

$$My - Nx = x^r y\,\phi\left(\frac{y}{x}\right) - x^{r+1}\psi\left(\frac{y}{x}\right)$$

$$= x^{r+1}\{v\phi(v) - \psi(v)\};$$

and therefore the integrating factor is

$$\frac{1}{My - Nx},$$

provided $My - Nx$ be not identically zero.

(ii) If the equation however be not of the first degree but still be homogeneous in x and y, it may be written in the form

$$F\left\{\frac{y}{x}, \frac{dy}{dx}\right\} = 0.$$

There are now two methods of proceeding. The first method is to resolve the equation, considering it as an equation in $\frac{dy}{dx}$; let the result be expressed by

$$\frac{dy}{dx} = f\left(\frac{y}{x}\right).$$

This is the case already discussed.

The second method is to resolve the equation considered as an equation in $\frac{y}{x}$; then we should have

$$\frac{y}{x} = f_1\left(\frac{dy}{dx}\right) = f_1(p),$$

or
$$y = xf_1(p),$$

where p is written for $\frac{dy}{dx}$. Differentiating this with respect to x, we have

$$p = f_1(p) + xf_1'(p)\frac{dp}{dx},$$

and therefore

$$\frac{dx}{x} = \frac{f_1'(p)\,dp}{p - f_1(p)}.$$

This gives on integration

$$\log x = C + \int\frac{f_1'(p)\,dp}{p - f_1(p)}$$
$$= C + \psi(p),$$

say; the elimination of p between the last equation and

$$y = xf_1(p)$$

will give the primitive. But it is not always desirable to eliminate

p; it may be retained as the parameter of a point on the corresponding curve, in which case its use would be similar to that of the eccentric angle of a point on an ellipse.

(iii) It may happen that, in the case of an equation

$$F\left\{\frac{y}{x}, \frac{dy}{dx}\right\} = 0,$$

it is more convenient, or more easily possible, to express $\frac{y}{x}$ and $\frac{dy}{dx}$ in terms of a new quantity u, than to solve the equation either for $\frac{dy}{dx}$ or for $\frac{y}{x}$. We then should have relations of the form

$$\frac{y}{x} = f(u), \quad \frac{dy}{dx} = g(u).$$

The former gives

$$y = xf(u),$$

and therefore

$$\frac{dy}{dx} = f(u) + xf'(u)\frac{du}{dx},$$

so that

$$g(u) - f(u) = xf'(u)\frac{du}{dx}.$$

The variables are separable; and we have

$$\frac{dx}{x} = \frac{f'(u)}{g(u) - f(u)} du,$$

and therefore

$$\log x = A + \int \frac{f'(u)}{g(u) - f(u)} du,$$

where A is an arbitrary constant. This equation, combined with $y = xf(u)$, gives the primitive of the original differential equation.

Ex. 1. Solve $\qquad x + y\frac{dy}{dx} = 2y.$

When we write $y = vx$, the equation becomes

$$\frac{v\,dv}{(1-v)^2} + \frac{dx}{x} = 0,$$

whence

$$\frac{1}{1-v} + \log(1-v) + \log x = A,$$

or

$$(x-y)\,e^{\frac{x}{x-y}} = e^A = C.$$

Ex. 2. Solve

$$\text{(i)} \quad x + y \frac{dy}{dx} = my;$$

$$\text{(ii)} \quad y^2 + x^2 \frac{dy}{dx} = xy \frac{dy}{dx}.$$

Ex. 3. Solve

$$(ax + by + c) \frac{dy}{dx} = Ax + By + C.$$

Let $x = h + \xi$ and $y = k + \eta$, and suppose h and k so chosen that

$$ah + bk + c = 0, \quad Ah + Bk + C = 0;$$

then the equation becomes

$$(a\xi + b\eta) \frac{d\eta}{d\xi} = A\xi + B\eta,$$

which is homogeneous.

If however $\dfrac{A}{a} = \dfrac{B}{b}$, but $\dfrac{C}{c}$ differs from each of these fractions, then the equations giving h and k are inconsistent. Let each of the equal ratios be equal to m; then

$$(ax + by + c) \frac{dy}{dx} = m(ax + by) + C.$$

Substitute

$$ax + by = v;$$

then

$$a + b\, \frac{mv + C}{v + c} = \frac{dv}{dx},$$

and the variables are separable.

If $\dfrac{A}{a} = \dfrac{B}{b} = \dfrac{C}{c} = n$, the equation is

$$\frac{dy}{dx} = n,$$

so that

$$y = nx + E.$$

Ex. 4. Solve

$$\text{(i)} \quad 3y - 7x + 7 = (3x - 7y - 3)\frac{dy}{dx};$$

$$\text{(ii)} \quad (2x + 4y + 3)\frac{dy}{dx} = 2y + x + 1;$$

$$\text{(iii)} \quad (3x + 5y + 6)\frac{dy}{dx} = 7y + x + 2.$$

Ex. 5. Shew that the equation

$$(P + Qx)\frac{dy}{dx} = R + Qy,$$

in which P, Q, and R, are homogeneous functions of x and y, P and R being of the same degree, may be solved by the substitution $y = vx$.

Ex. 6. Solve

$$(Ax^2 + Bxy + \alpha x + \beta y + \gamma)\frac{dy}{dx} = Axy + By^2 + \alpha' x + \beta' y + \gamma'.$$

17. Let now the curves, whose equations are the complete primitives of the homogeneous equation, be traced; they form a system of similar curves. For let there be drawn through the origin any radius vector cutting all these curves and making an angle θ with the axis of x; the inclination to the axis of x of the tangent to one of the curves, at the point where this radius vector meets it, is given by

$$\tan \phi = \frac{dy}{dx} = f_1\left(\frac{y}{x}\right) = f_1(\tan \theta),$$

and therefore all the tangents at points lying on this line are parallel. Consequently all the curves are similar and similarly situated.

18. STANDARD IV.

Equations arise in which one of the two variables does not explicitly occur.

Consider, first, that class from which the independent variable is absent. The equation is of the form

$$\phi\left(y, \frac{dy}{dx}\right) = 0.$$

As in the general equation under Standard III, there are two methods of proceeding. If it be possible, we may resolve for $\frac{dy}{dx}$ so that

$$\frac{dy}{dx} = f(y),$$

in which the variables are separable; the primitive is

$$\int \frac{dy}{f(y)} = x + A.$$

Or, if it be possible, we may resolve for y; suppose the result to be given by

$$y = f_1\left(\frac{dy}{dx}\right) = f_1(p).$$

Differentiating with respect to x, we have

$$p = f_1'(p) \frac{dp}{dx},$$

in which the variables are separable : and the integral is

$$x = \int \frac{f_1'(p)}{p} \, dp + A,$$

which, when combined with

$$y = f_1(p)$$

for the elimination of p, will furnish the primitive. It may be more convenient to leave p uneliminated.

Let us now consider the class from which the dependent variable is absent. The equation is of the form

$$\phi_1 \left(x, \frac{dy}{dx} \right) = 0.$$

Since

$$\frac{dy}{dx} \frac{dx}{dy} = 1,$$

the equation may be written

$$\phi_1 \left(x, \frac{1}{\frac{dx}{dy}} \right) = 0,$$

or

$$\Phi \left(x, \frac{dx}{dy} \right) = 0,$$

an equation of the former class, and soluble by the methods thereto applying. These methods however may be applied to the equation without making it undergo this transformation. Resolving the equation if possible for $\frac{dy}{dx}$, we shall have

$$\frac{dy}{dx} = F(x),$$

and the primitive is therefore

$$y = \int F(x) \, dx + A.$$

Or resolving for x in terms of $\frac{dy}{dx}$, when this is possible, we shall obtain

$$x = F_1 \left(\frac{dy}{dx} \right) = F_1(p).$$

Differentiating with respect to y, the absent variable, we have

$$\frac{1}{p} = F_1'(p) \frac{dp}{dy},$$

the integral of which is

$$y = \int p F_1' (p) \, dp + C.$$

This equation, combined with

$$x = F_1 (p),$$

constitutes the primitive.

It may happen that, in the case of an equation

$$\phi (y, p) = 0,$$

it is more convenient, or more easily possible, to express y and p in terms of a new quantity u, than to resolve the equation either for p or for y. We then should have relations of the form

$$y = f(u), \quad p = g(u).$$

The first of these gives

$$p = f'(u) \frac{du}{dx},$$

so that

$$dx = \frac{f'(u)}{g(u)} \, du,$$

and therefore

$$x - C = \int \frac{f'(u)}{g(u)} \, du.$$

This equation, combined with

$$y = f(u),$$

constitutes the primitive.

Ex. 1. Shew how to obtain the primitive of the equation

$$\psi(x, p) = 0,$$

when it is more convenient to express x and p in terms of a variable u than to solve the equation for p.

Ex. 2. Solve $\quad$ (i) $y = a \dfrac{dy}{dx} + b \left(\dfrac{dy}{dx} \right)^2$;

$\qquad\qquad\qquad$ (ii) $1 + \left(\dfrac{dy}{dx} \right)^2 = \dfrac{(x+a)^2}{x^2 + 2ax}.$

19. STANDARD V.

When the equation of the first order is of the nth degree, suppose it arranged in descending powers of the differential coefficient, so that it may be written

$$\left(\frac{dy}{dx} \right)^n + P_1 \left(\frac{dy}{dx} \right)^{n-1} + P_2 \left(\frac{dy}{dx} \right)^{n-2} + \dots + P_{n-1} \frac{dy}{dx} + P_n = 0,$$

in which $P_1, P_2, \ldots\ldots, P_n$, denote functions of x and y. If we look upon this as an algebraical equation in $\dfrac{dy}{dx}$, which has n roots $p_1, p_2, \ldots\ldots, p_n$, these being functions of x and y, the equation becomes

$$\left(\frac{dy}{dx} - p_1\right)\left(\frac{dy}{dx} - p_2\right)\ldots\ldots\ldots\left(\frac{dy}{dx} - p_n\right) = 0.$$

This can be true only if one or more of the factors on the left-hand side vanish; and therefore any relation between x and y, which makes a factor vanish, will be a solution of the original equation, while no relation which does not make some factor vanish can be a solution. Suppose then that the primitives of the equations

$$\frac{dy}{dx} - p_1 = 0, \quad \frac{dy}{dx} - p_2 = 0, \quad \ldots\ldots\ldots, \quad \frac{dy}{dx} - p_n = 0,$$

(deduced by means of one or other of the preceding methods) are

$$\phi_1(x, y, C_1) = 0, \quad \phi_2(x, y, C_2) = 0, \quad \ldots\ldots\ldots, \quad \phi_n(x, y, C_n) = 0,$$

respectively. All possible solutions of the given equation will be contained in

$$\phi_1(x, y, C_1)\, \phi_2(x, y, C_2) \ldots\ldots\ldots \phi_n(x, y, C_n) = 0.$$

But the generality of this integral will still be maintained, if all the constants $C_1, C_2, \ldots\ldots, C_n$, be made the same, say C; for in order to find a value of y, we must equate to zero some factor on the left-hand side of the new form, and this would give an equation of the form

$$\phi_r(x, y, C) = 0.$$

Now C is an arbitrary constant; if then all possible numerical values be given to it, there must be included in the series of consequent equations all the integrals, which can be derived similarly from the corresponding factor of the first product. Hence we have as the general complete primitive of the original differential equation

$$\phi_1(x, y, C)\, \phi_2(x, y, C) \ldots\ldots\ldots \phi_n(x, y, C) = 0.$$

Ex. 1. $x^2 p^2 - 2xyp + y^2 = x^2 y^2 + x^4.$

Then $xp - y = \pm x (x^2 + y^2)^{\frac{1}{2}},$

which, by the substitution $y = xz$, becomes

$$\frac{dz}{(1 + z^2)^{\frac{1}{2}}} = \pm\, dx.$$

When the positive sign is taken, the solution is

$$z = \sinh(x+c).$$

The negative sign gives

$$z = \sinh(c-x);$$

hence the general solution is

$$[y - x \sinh(x+c)][y - x \sinh(c-x)] = 0.$$

Ex. 2. Solve   (i) $\left(\dfrac{dy}{dx}\right)^2 - \dfrac{a}{x} = 0$;

               (ii) $\left(\dfrac{dy}{dx}\right)^2 + 2\dfrac{y}{x}\dfrac{dy}{dx} = 1$.

Ex. 3. Solve

     (i) $x^2 p^2 + 3xyp + 2y^2 = 0$;

     (ii) $x^2 p^2 + 3xyp + 3y^2 = 0$;

     (iii) $p(p+y) = x(x+y)$;

     (iv) $p^3 - (x^2 + xy + y^2)p^2 + (x^3 y + x^2 y^2 + xy^3)p - x^3 y^3 = 0$;

     (v) $(a^2 - x^2)p^3 + bx(a^2 - x^2)p^2 - p - bx = 0$;

     (vi) $\left(1 - y^2 - \dfrac{y^4}{x^2}\right)p^2 - 2\dfrac{y}{x}p + \dfrac{y^2}{x^2} = 0$;

     (vii) $p^2 + \left(x + y - 2\dfrac{y}{x}\right)p + xy + \dfrac{y^2}{x^2} - y - \dfrac{y^2}{x} = 0$.

Ex. 4. Shew that, if the general equation be homogeneous in x and y, it can be solved by the substitutions

$$y = tx, \qquad x\frac{dt}{dx} = z.$$

Hence solve

$$p^3 - \tfrac{1}{2}p^2 + \frac{x}{y}p - \frac{x}{2y} = 0.$$

20. STANDARD VI. *Clairaut's Form.*

The equation, to which this name is usually applied, is

$$y = px + f(p),$$

in which p stands for $\dfrac{dy}{dx}$.

Differentiate the equation with regard to x: then

$$p = p + [x + f'(p)]\frac{dp}{dx},$$

so that either

$$\frac{dp}{dx} = 0,$$

or
$$x + f'(p) = 0.$$

Taking the first of these, we have $p = c$, an arbitrary constant; hence the primitive is

$$y = cx + f(c).$$

The second equation expresses x as a function of p; therefore, if p be eliminated between this equation and

$$y = px + f(p),$$

a relation between y and x will be obtained.

Of these, the former is evidently a solution of the equation, and from it the differential equation can be deduced at once; for, on differentiating, we obtain

$$p = c,$$

and eliminating c we have

$$y = px + f(p).$$

If now we turn to the other relation between x and y, which will be that derived from the elimination* of p between

$$\begin{aligned} y &= px + f(p) \\ 0 &= x + f'(p) \end{aligned} \Bigg\} ,$$

we see at once that it contains no arbitrary constant and therefore is not a *general* solution. Yet it may be a solution of the equation; for differentiating the first equation, we have

$$\frac{dy}{dx} = p + [x + f'(p)]\frac{dp}{dx}$$
$$= p,$$

by the second equation unless $\frac{dp}{dx}$ be infinite; eliminating p from the equations $y = px + f(p)$ and $\frac{dy}{dx} = p$, we obtain

$$y = x\frac{dy}{dx} + f\left(\frac{dy}{dx}\right),$$

which is the original equation.

---

* It should be noticed that, for purposes of elimination, p is merely a quantity likely to depend upon y and x; it is not now necessarily $\frac{dy}{dx}$.

Note 1. The relation between the two solutions, when both exist, is easily indicated by geometrical considerations. The first solution

$$y = cx + f(c)$$

represents a family of straight lines; if they have an envelope, it is found by differentiating the equation with respect to c (in fact, this is equivalent to giving c a pair of equal values for the same values of x and y). Then we have

$$0 = x + f'(c).$$

The result of the elimination of c between these two equations is the same as that of eliminating p between the two equations

$$y = px + f(p),$$
$$0 = x + f'(p);$$

and therefore the curve represented by the latter is the envelope of the family of lines represented by the first solution, should these lines have an envelope.

Such a solution of the equation, which is not included in the primitive (but which may be derived from it in the above manner), is called a *Singular Solution*. We shall shortly return to a more detailed discussion of singular solutions.

Note 2. That the two solutions are distinct can be seen as follows. Let

$$U = y - cx - f(c), \quad V = y - px - f(p),$$

where, in V, the value of p is given by the equation $x + f'(p) = 0$. Now

$$\frac{\partial U}{\partial x} = -c, \quad \frac{\partial U}{\partial y} = 1, \quad \frac{\partial V}{\partial y} = 1,$$

$$\frac{\partial V}{\partial x} = -p - \{x + f'(p)\}\frac{dp}{dx} = -p.$$

Hence
$$\frac{\partial U}{\partial x}\frac{\partial V}{\partial y} - \frac{\partial U}{\partial y}\frac{\partial V}{\partial x} = p - c,$$

which does not vanish identically; it does vanish in virtue of the two equations $U = 0$ and $V = 0$, when we take these equations simultaneously.

Hence (§ 10) the two equations $U = 0$ and $V = 0$ are independent of one another; and the two solutions are therefore distinct from one another.

Ex. 1. Solve $$y=xp+\frac{a}{p}.$$

The first solution is

$$y=cx+\frac{a}{c}.$$

The second is given by the elimination of p between

$$0=x-\frac{a}{p^2}$$

and the original equation; eliminating p, we have

$$y^2=4ax.$$

The latter is a solution, and so it is the singular solution; the curve represented is touched by all the lines included in the primitive.

Ex. 2. Solve   (i) $y=px+(1+p^2)^{\frac{1}{2}}$;

(ii) $y=px+p-p^2$;

(iii) $ayp^2+(2x-b)\,p=y$;

(iv) $x^2\,(y-xp)=yp^2$;

(v) $y=2xp+y^2p^3$.

21. There is an extended form of the equation, which can be solved in a similar manner, viz.:

$$y=xf(p)+\phi(p).$$

To solve this, let the equation be differentiated with regard to x; then

$$p=f(p)+[xf'(p)+\phi'(p)]\frac{dp}{dx},$$

or $$\frac{dx}{dp}+x\,\frac{f'(p)}{f(p)-p}=\frac{\phi'(p)}{p-f(p)},$$

which is linear in x and comes under Standard II.

Let the integral be

$$F(x,\,p,\,c)=0.$$

The result of eliminating p between this equation and the original equation is the primitive.

Ex. 1. Solve $x+yp=ap^2$,

or $$y=ap-\frac{x}{p}.$$

Differentiating with regard to x, we have

$$p=a\frac{dp}{dx}-\frac{1}{p}+\frac{x}{p^2}\frac{dp}{dx},$$

and therefore
$$\frac{dx}{dp} - \frac{x}{p(1+p^2)} = \frac{ap}{1+p^2},$$
the integral of which is
$$\frac{x(1+p^2)^{\frac{1}{2}}}{p} = C + a \log \{p + (1+p^2)^{\frac{1}{2}}\}.$$

This equation, combined with the original equation, gives the primitive.

The equation could also have been solved by differentiating with regard to y.

Ex. 2. Solve

(i) $x = yp + ap^2$;

(ii) $y = xp + ax(1+p^2)^{\frac{1}{2}}$;

(iii) $y = xmp + n(1+p^3)^{\frac{1}{3}}$;

(iv) $y = yp^2 + 2px$;

(v) $y(1+p^2)^{\frac{1}{2}} = n(x+yp)$.

SINGULAR SOLUTIONS.

22. From the investigation in § 20, it is clear that a solution of a differential equation can sometimes be found which, when it exists, is usually distinct from the primitive; such a solution does not involve in its expression any arbitrary constant. It may, however, be included in the primitive, by assigning a particular value to the arbitrary constant; such a solution is then regarded as being both a singular solution and a particular case of the primitive*.

We proceed now to consider the theory of these Singular Solutions of the general differential equation of the first order which will be written
$$\phi(x, y, p) = 0.$$

If the differential equation either be linear or be resoluble into a set of rational linear equations (as in the case of Standard V.), then it has no singular solution; any solution of it apparently of this nature is merely a particular solution, derived from the primitive by giving a particular value to the arbitrary constant therein contained. For the present purpose, therefore, the equation in p may be considered irresoluble: if it can be resolved into factors which are not linear and not resoluble into linear factors, then we should consider in turn each of these irresoluble factors.

---

* An example is given below, Ex. 6, § 29.

We may thus consider $\phi = 0$ as a rational and irresoluble equation of degree n. Moreover, we shall assume that ϕ is a one-valued function, and that it contains no factor which is independent of p; such a factor, if it were retained and equated to zero, would satisfy the equation, but would not involve the differential coefficient. If in any case these factors occurred, we should suppose them removed.

23. The considerations adduced in the Introduction furnish the inference that, if x and y be the coordinates of a point in a plane, the differential equation determines a system of curves in that plane, which depend upon a single independent variable parameter; and as the differential equation determines at any point a direction through that point, there will be n directions, given by the values of p there, and therefore n curves will pass through any point in the plane. To represent this system algebraically, we need an equation of the form

$$f(x, y, c_1, c_2, \ldots\ldots, c_m) = 0,$$

the constants in which are involved rationally and algebraically; but as only a single independent parameter is needed, there must be $m - 1$ algebraical relations among these m constants. Further, this function f will be one-valued; and any factor, involving x and y (or either of them) but none of the constants, would be rejected for the same reason as led to the rejection of similar factors from the differential equation. As the differential equation cannot be resolved into simpler equations of a lower degree, the algebraical equation is not so resoluble; if it were, to each algebraical equation of lower degree there would be a corresponding differential equation of lower degree—a result excluded by hypothesis. And the reason that m constants connected by $m - 1$ relations are inserted instead of a single constant is this; the equation in the latter case would be the same as that derived from the former with all the constants eliminated except one, and as this elimination would usually imply operations (such as squaring) which introduce equations other than that wanted, the result would be that the final equation would represent more than the single equation desired. For example, suppose that by any process an integral is obtained in the form

$$\{x^2 + y^2 - a(x \cos \alpha + y \sin \alpha)\}^2 = a^2(x^2 + y^2),$$

or, changing to algebraical constants,
$$\{x^2 + y^2 - a\,(lx + my)\}^2 = a^2\,(x^2 + y^2),$$
with the condition
$$l^2 + m^2 = 1\;;$$
then the equivalent equation containing one of these constants, as m, alone would represent not only this equation but also
$$\{x^2 + y^2 - a\,(-lx + my)\}^2 = a^2\,(x^2 + y^2),$$
with the same limiting condition, and therefore would not be equivalent solely to the first of these.

Further, there are n curves passing through every point in the plane; hence the equation $f = 0$, with the $m - 1$ equations between the constants, must give at every point n sets of values for these constants. Let the aggregate of the constants be denoted by c, so that for any point in the plane c will have n values.

24. Consider now the formation of the differential equation from the primitive
$$f(x, y, c) = 0.$$
It is obtained by eliminating the constants between the $m - 1$ relations, this equation, and the equation
$$\frac{\partial f}{\partial x} + \frac{\partial f}{\partial y}\frac{dy}{dx} = 0.$$
But suppose the quantities c replaced by functions of x; the deduction of the differential equation will proceed as before, except that for the last equation we must substitute
$$\frac{\partial f}{\partial x} + \frac{\partial f}{\partial y}\frac{dy}{dx} + \frac{\partial f}{\partial c}\frac{dc}{dx} = 0.$$
The result will be actually the same as before, if
$$\frac{\partial f}{\partial c}\frac{dc}{dx} = 0.$$
To satisfy this equation we must have either $\dfrac{dc}{dx}$ zero, which leaves c constant; or c must be determined by
$$\frac{\partial f}{\partial c} = 0.$$
Let the value of c so determined be substituted in the function f. We may thus in general, as a possible solution of the same

differential equation, equate to zero the discriminant of f with regard to c; let this be written

$$\text{Disct}_c f(x, y, c) = 0.$$

25. This locus is the locus of all points in the plane at which the parametric constants c have two or more equal values. In it there will therefore be included the following:—

(i) the locus of all the *nodal points* (double, treble, etc.) of the system of curves; for at such a point there are as many values of c equal to each other as there are branches through the point, since the branches belong to the same curve;

(ii) the locus of all the *cusps* of the system, for similar reasons;

(iii) the *envelope* of the system of curves, which may be either a single curve or several curves; for any point on the envelope may be considered as belonging to two separate but consecutive curves of the system, the constants of these consecutive curves being ulti-mately equal. [In the case, when the envelope can be decomposed into several curves, it may happen that one of these is merely a particular curve of the system $f(x, y, c) = 0$; its equation gives part of the envelope and a particular curve of the primitive.]

Let these three respectively be called the *nodal* locus, the *cuspidal* locus, and the *envelope* locus.

Note. The occurrence of the node-locus and the cusp-locus in the equation

$$\text{Disct}_c f(x, y, c) = 0$$

may be explained as follows.

The coordinates of a node or a cusp on the curve $f = 0$ satisfy the equations

$$\frac{\partial f}{\partial x} = 0, \quad \frac{\partial f}{\partial y} = 0, \quad f = 0;$$

let these coordinates be denoted by

$$X = g(c), \quad Y = h(c).$$

On the consecutive curve $f(x, y, c + dc) = 0$, the coordinates of the node or the cusp consecutive to X and Y may be denoted by $X + dX$, $Y + dY$; and then, among the three equations to be

satisfied, we have
$$f(X + dX,\ Y + dY,\ c + dc) = 0,$$
that is, when we include only quantities of the first order,
$$\frac{\partial f}{\partial X}\,dX + \frac{\partial f}{\partial Y}\,dY + \frac{\partial f}{\partial c}\,dc = 0.$$
Taking account of the equations
$$\frac{\partial f}{\partial X} = 0,\ \ \frac{\partial f}{\partial Y} = 0,$$
which are satisfied by X and Y, we infer that the equation
$$\frac{\partial f}{\partial c} = 0$$
holds along the node-locus or the cusp-locus. Hence the locus of all the nodes and the locus of all the cusps of the curves $f = 0$, for different values of c, may be expected to occur in the eliminant of
$$f = 0 \quad \frac{\partial f}{\partial c} = 0,$$
that is, in the discriminant.

26. If we now consider the differential equation
$$\phi\,(x,\,y,\,p) = 0$$
in connection with the system of curves, whose equation constitutes its general solution, it is evident that *the envelope of the system is a solution of the equation;* for at any point on the envelope (which is a point on two consecutive curves) the direction of the tangent is the same as that of the tangent to either of these curves at that point; and since the differential equation is satisfied by the quantities, which are connected with the element of the system of curves, it must be satisfied by these (unaltered) quantities, which are connected with the element of the envelope.

But the *nodal locus* is not a solution of the equation; if it were, the differential equation would, for the values of x and y at any node, be satisfied by the corresponding value of p at this point on the nodal locus. Remembering that the nodal locus is formed by a series of points on our system of curves, we know that the values of p at any such point which satisfy the differential equation are those given by that curve of the system which passes through the point. But as the tangent to the nodal locus at such a point will not in general be a tangent to any of the branches of the curve of the system at the point, it follows that the value of p for the nodal locus differs from those values of p for the curve of the system, which satisfy the equation when substituted in it with the coordinates of the point. And it would only be by accident that the value of p for the nodal locus could coincide with any of the remaining values of p, which do not belong to the curve on which the node lies, but are furnished by other curves of the system through that point. Hence the value of p for the nodal locus at the point will usually be such as not to satisfy the differential equation; and the *nodal locus will therefore, in general, not be a solution of the differential equation.*

Exactly similar considerations, applied to the cuspidal locus, lead to a similar conclusion:—*the cuspidal locus, in general, is not a solution of the differential equation.*

27. Now the envelope of the system can be derived from a knowledge of the differential equation alone, i.e. without a knowledge of the primitive. At any point on the envelope, at least two of the branches of the different curves coincide in direction; and therefore for such a point we shall have equal values of p belonging to *different* but consecutive curves.

If now we express the condition that two values of p shall be equal, by means of the equation

$$\frac{\partial \phi}{\partial p} = 0,$$

and eliminate p between this and the original differential equation (in fact, if we equate the discriminant of ϕ to zero), then the locus

$$\text{Disct}_p \, \phi \, (x, y, p) = 0$$

will be one at points along which two values of p will be equal, and will obviously include the envelope.

But, besides including the envelope, this equation will also give the locus of all points

(i) at which two branches of the *same* curve touch, i.e. will give all the cusps; this locus, as before, is the cuspidal locus;

(ii) at which two curves which are different but not consecutive touch; this locus is called a *tac-locus.* Thus, for instance, if we have two infinite series of concentric circles one round each of two points, the straight line joining the centres (and produced both ways) is the locus of points of contact of two circles, one belonging to each system.

As before, the cuspidal locus is rejected, not being a solution in general; and reasoning, exactly similar to that which led to the rejection of the nodal locus, indicates that the tac-locus is not a solution in general.

Hence, of all these loci, the only solution of the differential equation is the envelope-locus; and this, and this alone, we call the "*Singular Solution*" of the differential equation. Either method of obtaining the envelope-locus may introduce some of the other loci which have just been shewn not to be solutions; and therefore in any particular case, unless the equation derived obviously represents the envelope and nothing but the envelope, it is necessary to try whether the result satisfies the differential equation. Should it not do so, it may happen that the integral equation can be resolved into others that are simpler, and one or more than one of them may satisfy the equation; these will then constitute the Singular Solution. And those resolved parts which do not satisfy the differential equation will be found to be loci, which according to the principles above explained ought to be rejected.

It may be added that the locus, obtained by equating the c-discriminant to zero, contains the envelope-locus as a factor once, the node-locus twice, and the cusp-locus thrice; and that the locus, obtained by equating the p-discriminant to zero, contains the envelope-locus as a factor once, the cusp-locus once, and the tac-locus twice[*]. The results are often expressed in the forms

$$\text{Disct}_c f\,(x,\,y,\,c) = EN^2C^3,$$
$$\text{Disct}_p \phi\,(x,\,y,\,p) = ET^2C.$$

28. We can assign an analytical condition that the relation, obtained by the elimination of p between

$$\phi = 0,\quad \frac{\partial\phi}{\partial p} = 0,$$

---

[*] Cavley, *Mess. of Math.*, vol. II. (1873), pp. 11, 12, vol. XII. (1883), p. 3, *Coll. Math. Papers*, vol. VIII. pp. 533, 534; M. J. M. Hill, *Proc. Lond. Math. Soc.*, vol. XIX. (1888), pp. 561—589. Cayley inferred the results from geometrical considerations; an analytical proof is given by Hill.

should provide a solution of the equation, which of course will be a singular solution.

Imagine the second equation solved, so as to express p in terms of x and y; and let the value so obtained be substituted in ϕ, so that it becomes ϕ'. The equation then is

$$\phi' = 0.$$

If this equation provides a solution of the differential equation, it must lead to the appropriate value of p which, in that case, will be the value that was substituted in ϕ. Now the value of p arising through ϕ' is given by

$$\frac{\partial \phi'}{\partial x} + \frac{\partial \phi'}{\partial y} p = 0;$$

moreover,

$$\phi' = \phi,$$

so that

$$\frac{\partial \phi'}{\partial x} + p \frac{\partial \phi'}{\partial y} = \frac{\partial \phi}{\partial x} + p \frac{\partial \phi}{\partial y} + \frac{dp}{dx} \frac{\partial \phi}{\partial p}$$

$$= \frac{\partial \phi}{\partial x} + p \frac{\partial \phi}{\partial y},$$

because $\dfrac{\partial \phi}{\partial p} = 0$. Hence p is given by

$$\frac{\partial \phi}{\partial x} + p \frac{\partial \phi}{\partial y} = 0;$$

and this value of p is to agree with the value deduced from $\dfrac{\partial \phi}{\partial p} = 0$ and substituted in $\phi = 0$, which by hypothesis then provided a solution of the differential equation. Accordingly, we infer that, *if a singular solution of the equation $\phi (x, y, p) = 0$ exists, it must simultaneously satisfy the equations*

$$\phi = 0, \quad \frac{\partial \phi}{\partial p} = 0, \quad \frac{\partial \phi}{\partial x} + p \frac{\partial \phi}{\partial y} = 0.$$

If, however, the value of p be infinite, the last equation requires that $\dfrac{\partial \phi}{\partial y}$ should be zero. In that case, the best method of testing the character of the equation resulting from the elimination of p is actually to try whether it does provide a solution of the differential equation.

29. It is to be understood that an irreducible differential equation has not necessarily a singular solution. Thus let the discriminant with regard to p of

$$\phi(x, y, p) = 0$$

be denoted by U, where U is a function of the variable coefficients of p in this equation, and suppose that U cannot be resolved into simple factors.

If the equation $U = 0$ be a solution of the differential equation, then the value of p is given by

$$\frac{\partial U}{\partial x} + \frac{\partial U}{\partial y} p = 0,$$

and we must have the equation

$$\phi\left(x, y, -\frac{\dfrac{\partial U}{\partial x}}{\dfrac{\partial U}{\partial y}}\right) = 0$$

satisfied for values of x and y connected by $U = 0$. In other words, there must be a relation between the coefficients of p in ϕ and their differential coefficients with regard to x and y; but this will not in general be the case.

If we consider in particular the equation of the second degree in the form

$$Lp^2 + 2Mp + N = 0,$$

then the singular solution, when one exists, is $S = 0$, where S is either $LN - M^2$ or a factor of this. In general, $LN - M^2$ cannot be resolved into factors; and it is not itself a solution, unless

$$L\left(\frac{\partial S}{\partial x}\right)^2 - 2M\frac{\partial S}{\partial x}\frac{\partial S}{\partial y} + N\left(\frac{\partial S}{\partial y}\right)^2 = 0,$$

where $LN = M^2$; and these in general would be two independent simultaneous equations determining x and y as constant quantities. Yet, from what we have seen, the primitive of the differential equation is of the form

$$L'c^2 + 2M'c + N' = 0.$$

This equation may possess a general envelope; if it does, the envelope is contained in

$$L'N' - M'^2 = 0,$$

and it is a singular solution of the differential equation. If the integral equation does not possess an envelope, (and it does not always possess such an envelope even when it is an algebraic equation*), then there is no singular solution.

In fact, the exceptions in the first case—when the differential equation has a singular solution—are the exceptions in the second case—when the integral equation represents a family of curves with a genuine envelope.

The theory of singular solutions of differential equations of the first order, as at present accepted, was first given by Cayley, *Mess. of Math.*, vol. II. (1873), pp. 6—12, *Coll. Math. Papers*, vol. VIII. pp. 529—534; and an important memoir was independently published by Darboux about the same time, *Bull. des Sc. Math.*, vol. IV. (1873), pp. 158—176. Many memoirs dealing with the subject have since been written; it will be sufficient to mention those by Workman, *Quart. Journ.*, vol. XXII. (1887), pp. 175—198, 308—324; M. J. M. Hill, *Lond. Math. Soc. Proc.*, vol. XIX. (1888), pp. 561—589; Chrystal, *Trans. R. S. E.*, vol. XXXVIII. (1897), pp. 803—824; and to refer to author's *Theory of Differential Equations*, vol. II., ch. VIII., for an analytical discussion of the subject and, in particular, for an exposition of Hamburger's method of treatment.

We now proceed to consider some examples of the general theory.

In the case of each example, the corresponding figure should be drawn.

Ex. 1. Solve $p^2 y + p(x - y) - x = 0.$

The condition that p should have equal values is
$$(x - y)^2 + 4xy = 0,$$
i.e. $$(x + y)^2 = 0,$$
or $$y = -x,$$
which is not a solution. Now the equation may be written
$$(p - 1)(py + x) = 0,$$
the solutions of which are
$$y - x = c \text{ and } y^2 + x^2 = c.$$

The different curves represented are obvious.

This is an example of the remark (§ 22) that, if the equation be reducible to linear and rational factors, it has no singular solution.

Ex. 2. Solve $p^2 y^2 \cos^2\alpha - 2pxy \sin^2\alpha + y^2 - x^2 \sin^2\alpha = 0.$

The condition that p should have equal values is
$$x^2 y^2 \sin^4\alpha = y^2 \cos^2\alpha (y^2 - x^2 \sin^2\alpha),$$

---

* As to this result, see the memoir by Chrystal, pp. 819, 820, quoted below.

that is, $(x^2 \sin^2 a - y^2 \cos^2 a) y^2 = 0,$

so that $y = 0,$

and $y = \pm x \tan a.$

The primitive is

$$x^2 + y^2 - 2cx + c^2 \cos^2 a = 0;$$

and the condition that c should have equal values is

$$x^2 = (x^2 + y^2) \cos^2 a,$$

or $y = \pm x \tan a.$

The curves represented are a series of circles; their envelope is the two straight lines $y = \pm x \tan a$, which constitute the singular solution.

The line $y = 0$ is a tac-locus.

It is easy to verify that the lines $y = \pm x \tan a$ satisfy the tests in § 28, so that they constitute a solution of the equation.

Ex. 3. Solve $4p^2 x (x - a) (x - b) = \{3x^2 - 2x (a + b) + ab\}^2.$

The condition that p should have equal values is

$$x (x - a) (x - b) \{3x^2 - 2x (a + b) + ab\}^2 = 0.$$

The primitive is

$$(y + c)^2 = x (x - a) (x - b);$$

and the condition that c shall have equal values is

$$x (x - a) (x - b) = 0.$$

The differential equation is satisfied by $x = 0$, $x = a$, $x = b$ (and the corresponding infinite values of p); and these are singular solutions. The remaining factor in the p-discriminant gives

$$3x = a + b \pm (a^2 - ab + b^2)^{\frac{1}{2}};$$

and these lines are tac-loci.

The cubic curve $y^2 = x (x - a) (x - b),$

$(0 < a < b)$ consists of an oval cutting the axis of x at the origin and at a distance a, and of a curve like a parabola cutting the axis of x at a distance b; the tangents at all these points are parallel to the axis of y. The system of curves is obtained by moving the cubic curve parallel to the axis of y. The straight lines $x = 0$, $x = a$, $x = b$, are envelopes of the system; the line $3x = a + b - (a^2 - ab + b^2)^{\frac{1}{2}}$ is a tac-locus of real points of contact; the line $3x = a + b + (a^2 - ab + b^2)^{\frac{1}{2}}$ is a tac-locus of imaginary points of contact.

Ex. 4. In the foregoing, make $a = b$; and remove (see § 22) the factor $(x - a)^2$. The differential equation is

$$4xp^2 = (3x - a)^2;$$

the condition that p should have equal values is

$$x (3x - a)^2 = 0.$$

The integral equation is

$$(y+c)^2 = x(x-a)^2,$$

and the condition that c should have equal values is

$$x(x-a)^2 = 0.$$

Common to these we have $x=0$, which (with the corresponding infinite value of p) is a solution of the equation, and therefore a singular solution. Every curve of the system has a double point; the locus of these is $x=a$, which is a nodal locus; the line $x=\frac{1}{3}a$ is a tac-locus.

Ex. 5. In the foregoing, let $a=0$ and remove the factor x. The differential equation is

$$4p^2 = 9x;$$

the condition that p should have equal values is

$$x=0.$$

The primitive is

$$(y+c)^2 = x^3,$$

and the condition that c should have equal values is

$$x^3 = 0.$$

The differential equation is not satisfied by $x=0$ (with the corresponding infinite value of p).

The curve $y^2 = x^3$ is the semi-cubical parabola having a cusp at the origin; and the system is obtained by moving the curve parallel to the axis of y, so that $x=0$ is the locus of cusps, and therefore is not a singular solution.

Ex. 6.
$$p^3 - 4xyp + 8y^2 = 0;$$
the condition that p shall have equal values is

$$y^4 - \tfrac{4}{27}x^3y^3 = 0.$$

The primitive is

$$y = c(x-c)^2,$$

and the condition that c shall have equal values is obtained by eliminating c between this and

$$(x-c)(x-3c) = 0,$$

so that either

$$y=0 \text{ or } y = \tfrac{4}{27}x^3,$$

agreeing with the former. Both of these satisfy the differential equation. The first of them is a particular solution (corresponding to $c=0$), and it therefore is both a singular solution and a particular case of the primitive; the latter is only a singular solution.

Ex. 7. Obtain the primitives and the singular solutions (where these exist) of the following equations; and specify the nature of the loci, which are not solutions but which are obtained with the singular solution.

(a)
$$xp^2 - 2yp + 4x = 0;$$

Primitive $\qquad x^2 = c(y-c);$

Singular solutions $\qquad y = \pm 2x.$

(β) $(x^2 - a^2) p^2 - 2xyp - x^2 = 0;$

    Primitive $c^2 + 2cy + a^2 = x^2;$

    Singular solution $x^2 + y^2 = a^2;$

    Tac-locus $x = 0.$

(γ) $p^4 = 4y\,(xp - 2y)^2;$

    Primitive $y = c^2\,(x - c)^2;$

    Singular solution $x^4 - 16y = 0;$

    Singular solution, also particular,

$$y = 0.$$

(δ) $xyp^2 + (x^2 - y^2 - b^2)\,p - xy = 0;$

(ϵ) $(1 - y^2)\,p^2 = 1;$

(ζ) $p^2\,(1 - x^2) = 1 - y^2;$

(η) $(bx - ay)^2\,(b^2 + a^2 p^2) = c^2\,(b + ap)^2.$

Further examples occur in the paper by Cayley, *Mess. of Math.*, vol. VI. (1877), pp. 23—27, *Coll. Math. Papers*, vol. X. pp. 19—23: and in one by J. W. L. Glaisher, *Mess. of Math.*, vol. XII. (1882), pp. 1—14.

PRINCIPLE OF DUALITY.

30. There exists in ordinary differential equations a duality, in virtue of which each equation is connected with some other equation of the same order by relations of a perfectly reciprocal character. We shall consider here only equations of the first order.

We take a new dependent variable Y given by

$$Y = px - y,$$

and we have

$$dY = pdx + xdp - dy$$
$$= xdp.$$

We take p as our new independent variable, and we write it X for symmetry, so that

$$X = p;$$

and then we have

$$x = \frac{dY}{dp} = \frac{dY}{dX} = P.$$

With these relations, we have

$$y = px - y$$
$$= PX - Y,$$

so that the relations between the variables are reciprocal.

If now we have an equation of the form

$$\phi(x, y, p) = 0,$$

the above relations transform it to

$$\phi(P, PX - Y, X) = 0.$$

When the integral of either of these is known, the integral of the other can generally be deduced by a process of algebraical elimination. Thus let an integral of the second be given in the form

$$f(X, Y) = 0.$$

Then we have

$$\frac{\partial f}{\partial X} + P \frac{\partial f}{\partial Y} = 0,$$

that is,

$$x \frac{\partial f}{\partial Y} + \frac{\partial f}{\partial X} = 0;$$

and

$$-y \frac{\partial f}{\partial Y} = (Y - px) \frac{\partial f}{\partial Y}$$

$$= Y \frac{\partial f}{\partial Y} + X \frac{\partial f}{\partial X}.$$

The elimination of X and Y between these three equations will leave an equation between x and y, which will be a solution of

$$\phi(x, y, p) = 0.$$

Note. The preceding process of constructing one differential equation from another is the analytical equivalent of the geometrical construction of a polar reciprocal of a curve.

Let a curve C be drawn; a tangent to it at a point x, y is

$$\eta - y = p(\xi - x),$$

where ξ and η are current coordinates. Let the pole of this tangent, taken with respect to the parabola $\xi^2 - 2\eta = 0$, be X, Y; then its equation is

$$X\xi - \eta - Y = 0.$$

As the two equations are the same, we have

$$X = p, \quad Y = px - y.$$

The locus of X, Y is a curve C', the polar reciprocal of C; and C is known to be the polar reciprocal of C', that is, the locus of the poles of the tangents to C', taken with respect to the parabola.

Thus the point x, y on C is the pole of the tangent to C' at the point X, Y; by a similar process, we should find

$$x = P, \quad y = PX - Y.$$

These are the relations[*] used in the analysis connected with the differential equation.

Ex. 1. Solve the equation

$$(y - px)\, x = y.$$

Effecting the transformation determined by

$$px - y = Y, \quad p = X, \quad x = P,$$

we have the equation in the form

$$-YP = PX - Y,$$

that is,

$$P = \frac{Y}{Y + X}.$$

Writing $Y = VX$, we have

$$X\frac{dV}{dX} = \frac{V}{V+1} - V = \frac{-V^2}{V+1},$$

and therefore

$$\left(\frac{1}{V} + \frac{1}{V^2}\right) dV + \frac{dX}{X} = 0.$$

Hence

$$\log V - \frac{1}{V} + \log X = \text{const.,}$$

that is,

$$f(X, Y) = \log Y - \frac{X}{Y} = \text{const.}$$

Thus we have

$$\frac{\partial f}{\partial X} = -\frac{1}{Y}, \quad \frac{\partial f}{\partial Y} = \frac{X + Y}{Y^2};$$

and therefore

$$x = \frac{Y}{X + Y}, \quad y = -\frac{Y^2}{X + Y}.$$

Thus

$$Y = -\frac{y}{x}, \quad \frac{X}{Y} = \frac{1}{x} - 1;$$

so that, as

$$\log Y - \frac{X}{Y} = \text{const.,}$$

we have

$$\log\left(-\frac{y}{x}\right) - \frac{1}{x} + 1 = \text{const.;}$$

and therefore

$$y = ax\, e^x$$

is the primitive of the original equation.

---

[*] These relations are the simplest instance of what are called *contact* (or *tangential*) transformations.

Ex. 2. Integrate the equations

$$\text{(i)} \quad (y - px)\, x = cy,$$

where c is a constant ·

$$\text{(ii)} \quad (y - px)(py + y - px) = p.$$

MISCELLANEOUS EXAMPLES.

1. Solve the equations:

(i) $y - xp = x + yp$;

(ii) $a(xp + 2y) = xyp$;

(iii) $x^2 + y = p^2$;

(iv) $p^2 + 2xp = y$;

(v) $my - nxp = yp^2$;

(vi) $p^3 = y^4(y + xp)$;

(vii) $p^3 + x^3 = axp$;

(viii) $x^3 p^2 + x^2 yp + a^3 = 0$;

(ix) $ax^2 y^n p + y = 2xp$;

(x) $p^2 + 2yp \cot x = y^2$;

(xi) $y - 2xp = f(xp^2)$;

(xii) $x^2 - \dfrac{xy}{p} = f(y^2 - xyp)$;

(xiii) $(1 - p)^2 - e^{-2y} = p^2 e^{-2x}$;

(xiv) $(nx + yp)^2 = (1 + p^2)(y^2 + nx^2)$;

(xv) $(1 + 6y^2 - 3x^2 y)\, p = 3xy^2 - x^2$;

(xvi) $y = x\{ p + (1 + p^2)^{\frac{1}{2}} \}$;

(xvii) $ay + bxp = x^m y^n (cy + exp)$;

(xviii) $yp(x^2 + y^2 + a^2) + x(x^2 + y^2 - a^2) = 0$;

(xix) $(xp - y)^2 = p^2 - 2\dfrac{y}{x} p + 1$;

(xx) $(xp - y)^2 = a(1 + p^2)(x^2 + y^2)^{\frac{3}{2}}$;

(xxi) $(a^2 + x^2)^{\frac{1}{2}} p + y = (a^2 + x^2)^{\frac{1}{2}} - x$;

(xxii) $y = px + (1 + p^2)^{\frac{1}{2}} \phi(x^2 + y^2)$;

(xxiii) $\left(x \cos \dfrac{y}{x} + y \sin \dfrac{y}{x} \right) y = \left(y \sin \dfrac{y}{x} - x \cos \dfrac{y}{x} \right) xp$;

(xxiv) $(x^3 y^3 + x^2 y^2 + xy + 1)\, y + (x^3 y^3 - x^2 y^2 - xy + 1)\, xp = 0$;

(xxv) $\{ (x^2 - y^2) \sin \alpha + 2xy \cos \alpha - y(x^2 + y^2)^{\frac{1}{2}} \}\, p$
$$= 2xy \sin \alpha - (x^2 - y^2) \cos \alpha + x(x^2 + y^2).$$

2. Shew that, if

$$u = 1 + A_1 x + \frac{1}{2!} A_2 x^2 + \frac{1}{3!} A_3 x^3 + \dots,$$

where the quantities A are connected by the relation

$$A_m = m A_{m-1} - \tfrac{1}{2}(m - 1)(m - 2) A_{m-3},$$

then
$$\log \{ u(1 - x)^{\frac{1}{2}} \} = \tfrac{1}{2} x + \tfrac{1}{4} x^2.$$

3. Integrate the equation

$$\cos \theta (\cos \theta - \sin \alpha \sin \phi)\, d\theta + \cos \phi (\cos \phi - \sin \alpha \sin \theta)\, d\phi = 0.$$

Shew that, if the arbitrary constant be determined by the condition that the equation must be satisfied by the values $(0, \alpha)$ of (θ, ϕ), the equation is satisfied by putting $\theta + \phi = \alpha$.

4. Prove that, if the differential equation
$$cy\,dx - (y+a+bx)\,dy - nx\,(x\,dy - y\,dx) = 0$$
be transformed into an equation between u and x by the substitution
$$u\,(y+a+bx+nx^2) = y\,(c+nx),$$
then the variables are separable; and reduce the equation to the form
$$\frac{dv}{\phi\,(v)} = \frac{dx}{\phi\,(x)}$$
by the further substitution $v = \alpha u + \beta$, α and β being suitably determined.

(Euler.)

5. Reduce the equation
$$axyp^2 + (x^2 - ay^2 - b)\,p - xy = 0$$
to Clairaut's form; and hence solve the equation.

Solve the equation
$$\alpha\,\frac{x}{dx} + \beta\,\frac{y}{dy} + \gamma\,\frac{x+y-1}{dx+dy} = 0,$$
where $\alpha + \beta + \gamma = 0$.

6. Shew that, if y_1 and y_2 be solutions of the equation
$$\frac{dy}{dx} + Py = Q,$$
where P and Q are functions of x alone, and $y_2 = y_1 z$, then
$$z = 1 + ae^{-\int \frac{Q}{y_1} dx},$$
where a is an arbitrary constant.

7. Prove that the variables in the equation
$$\{x\,(x+y) + a^2\}\,\frac{dy}{dx} = y\,(x+y) + b^2$$
may be separated by the substitution $x = u + v$ and $y = ku - v$, provided k be properly chosen; and integrate the equation.

8. Shew that the equations
$$y - xp = a\,(y^2 + p) \quad \text{and} \quad y - xp = b\,(1 + x^2 p)$$
are derivable from a common primitive, and determine it.

Are the pair
$$x + p\,(1+p^2)^{-\frac{1}{2}} = a \quad \text{and} \quad y - (1+p^2)^{-\frac{1}{2}} = b$$
so derivable? Also the pair
$$yp = ax \quad \text{and} \quad y^2\,(1-p^2) = b?$$

9. Integrate the differential equation
$$x\,\{ay^3 + (ay+bx)^3\} + y\,\frac{dy}{dx}\,\{bx^3 + (ay+bx)^3\} = 0.$$

A tangent to a curve at any point P cuts the tangent and the normal at a fixed point O in the points M and N, and the rectangle $OMP'N$ is completed.

Find the curve which is such that the triangle formed by the tangents at any three points P, Q, R is equal to the triangle formed by the corresponding points P', Q', R'.

10. Determine the system of curves which satisfy the differential equation

$$dx\{(1+x^2)^{\frac{1}{2}}+ny\}+dy\{(1+y^2)^{\frac{1}{2}}+nx\}=0;$$

and shew that the curve, which passes through the point $x=0$ and $y=n$, contains as part of itself the conic

$$x^2+y^2+2xy(1+n^2)^{\frac{1}{2}}=n^2.$$

11. Integrate the equation

$$\frac{x^2}{a}+\frac{y^2}{b}=\frac{a-b}{a+b}\frac{x-yp}{x+yp},$$

and examine the nature of the solution

$$\frac{x^2}{a}+\frac{y^2}{b}=1.$$

12. Discuss the question whether $y=0$ is a particular solution or a singular solution of the equation

$$2\left(x\frac{dy}{dx}+y\right)^3=y\frac{dy}{dx}.$$

13. Obtain, and interpret, the primitive and the singular solution (if there be one) of each of the equations

(i) $p^3+\mu p^2=a(y+\mu x);$　　(ii) $xp^2-2yp+x+2y=0;$

(iii) $y(1+p^2)=2xp;$　　(iv) $p^2=(4y+1)(p-y).$

14. Prove that, if a locus of points of inflexion can be obtained from the integral family of curves of the equation $\phi(x, y, p)=0$, it will be included in the result obtained by the elimination of p between the equations

$$\phi=0,\quad \frac{\partial\phi}{\partial x}+p\frac{\partial\phi}{\partial y}=0.$$

Discuss the solution of the equation

$$(4p+2x+x^3)^2=(1+x^2)(16y+4x^2+x^4).$$

(Darboux.)

15. Obtain the primitive of the differential equation

$$2y=xp+\frac{a}{p}.$$

Shew that exactly the same equation is obtained by expressing the condition that p should have equal values in the differential equation as by expressing the condition that c (the arbitrary constant) should have equal values in the primitive; and determine the geometrical meaning of this equation. Is it a singular solution?

16. The primitive of the differential equation

$$(2x^2+1)\,p^2+(x^2+2xy+y^2+2)\,p+2y^2+1=0$$

is $c^2+c\,(x+y)+1-xy=0$. Verify this; and obtain the singular solution both from the equation in p and from the equation in c, explaining the geometrical significance of the irrelevant factors that present themselves.

17. Shew that the solution of the equation

$$a^2yp^2-4xp+y=0$$

is
$$c^2+2cx\,(3a^2y^2-8x^2)-3x^2a^4y^4+a^6y^6=0.$$

Is $2x=\pm ay$ a singular solution?

Trace the curve and the locus given by the equation independent of an arbitrary constant. (Woolsey Johnson.)

SUPPLEMENTARY NOTE:

RUNGE'S METHOD FOR THE NUMERICAL SOLUTION OF DIFFERENTIAL EQUATIONS.

It is not always possible to obtain the explicit expression for a quantity y as defined by a differential equation even of the first order. Cases arise in which the quadratures required do not belong to known forms; cases occur in which reduction to quadratures is not possible, that is to say, the equation cannot be solved analytically in simple forms.

There is, however, a convenience, particularly for numerical problems, in being able to assign a numerical solution; and a method has been devised by Runge* which is effective for this purpose. The following account of Runge's method relates only to insoluble differential equations of the first order; reference should be made to his memoir for applications to equations of higher order. Briefly stated, the question may be propounded in the form:

A quantity y is defined by the differential equation

$$\frac{dy}{dx}=f(x,\,y),$$

and it is subject to the condition that $y=b$, when $x=a$; what is the value of y, when $x=c$?

* *Mathematische Annalen*, t. XLVI. (1895), pp. 167—178.

For the purpose of the question, it is assumed that a, b, c, are real; that c is greater than a; and that $f(x, y)$ remains finite throughout. As a matter of fact, we take $f(x, y)$ to be not greater than 1 numerically: if at any stage the value of $f(x, y)$ is greater than 1, we deal with the equation

$$\frac{dx}{dy} = \frac{1}{f(x, y)} = g(x, y),$$

which satisfies the assumption made.

The range between a and c is divided into portions, not necessarily equal to one another; if h denote one of these portions, the governing consideration is that the approximation shall be accurate to quantities of the order h inclusive. According to the degree of accuracy required, we shall have an indication of the extent of the portions.

For brevity, write

$$f(x, y), \ \frac{\partial}{\partial x}f(x, y), \ \frac{\partial^2}{\partial x^2}f(x, y), \ \frac{\partial}{\partial y}f(x, y), \ \frac{\partial^2}{\partial y^2}f(x, y), \ \frac{\partial^2}{\partial x \partial y}f(x, y),$$

$$= f_0, \ f_1, \ f_{11}, \ f_2, \ f_{22}, \ f_{12},$$

respectively, when x is made equal to a and y to b after the differentiations are affected. Let

$$k_1 = f(a + \tfrac{1}{2}h, \ b + \tfrac{1}{2}f_0h)\, h;$$

and let

$$k_2 = \tfrac{1}{2}(k' + k'''),$$

where

$$k' = f_0 h,$$

$$k'' = f(a + h, \ b + k')\, h,$$

$$k''' = f(a + h, \ b + k'')\, h.$$

Then, expanding and rejecting powers of h beyond h^3, we have

$$k_1 = f_0 h + \tfrac{1}{2}(f_1 + f_0 f_2)\, h^2 + \tfrac{1}{8}(f_{11} + 2f_0 f_{12} + f_0^2 f_{22})\, h^3 + \ldots,$$

$$k_2 = f_0 h + \tfrac{1}{2}(f_1 + f_0 f_2)h^2 + \tfrac{1}{4}\left\{f_{11} + 2f_0 f_{12} + f_0^2 f_{22} + 2f_2(f_1 + f_0 f_2)\right\} h^3 + \ldots.$$

Let the value of y when $x = a + h$ be denoted by $b + k$; then, if

$$k = a_1 h + \tfrac{1}{2}a_2 h^2 + \tfrac{1}{6}a_3 h^3 + \ldots,$$

we must have

$$a_1 + a_2 h + \tfrac{1}{2} a_3 h^2 + \dots$$
$$= f(a+h, b+k)$$
$$= f_0 + h f_1 + k f_2 + \tfrac{1}{2}(h^2 f_{11} + 2hk f_{12} + k^2 f_{22}) + \dots$$
$$= f_0 + h(f_1 + a_1 f_2) + \tfrac{1}{2} h^2 (a_2 f_2 + f_{11} + 2a_1 f_{12} + a_1^2 f_{22}) + \dots.$$

Consequently

$$a_1 = f_0,$$
$$a_2 = f_1 + f_0 f_2,$$
$$a_3 = f_{11} + 2f_0 f_{12} + f_0^2 f_{22} + f_2(f_1 + f_0 f_2),$$

which are the initial coefficients in the accurate value* of k. From the expressions for k_1 and k_2, we have

$$k = \tfrac{2}{3} k_1 + \tfrac{1}{3} k_2$$
$$= k_1 + \tfrac{1}{3}(k_2 - k_1),$$

accurately as far as quantities of the order h^3 inclusive.

When we write k in the form

$$\tfrac{1}{6}(k' + 4k_1 + k'''),$$

the approximation takes the form stated in Simpson's rule†.

Having now obtained $b + k$ as the value of y, which is associated with $a + h$ as the value of x to the required degree of approximation, we can take $a + h$ and $b + k$ as initial values of the variables for the next portion of the range of variation of x: and obtain a final value $b + k + k'$ of y which is associated with the final value $a + h + h'$ of x, to the same degree of approximation as before. We thus proceed from portion to portion of the range until the ultimate value of x is attained.

Ex. 1. A solution of the equation

$$\frac{dy}{dx} = \frac{y^2 - 2x}{y^2 + x}$$

is to satisfy the condition that $y = 1$ when $x = 0$; find its value when $x = \tfrac{1}{2}$.

For the sake of illustration, the range from $x = 0$ to $x = \cdot5$ will be divided into two portions: from $x = 0$ to $x = \cdot2$, and from $x = \cdot2$ to $x = \cdot5$.

For the first portion of the range, we have

$$a = 0, \quad b = 1, \quad h = \cdot2, \quad f_0 = 1.$$

---

* It would, of course, be possible to use this expansion to calculate k; but the calculations are long, and might be intricate. They are avoided by the adoption of Runge's result.

† Lamb, *Infinitesimal Calculus* (2nd ed.), pp. 260, 277; in this connection, see Runge, p. 168 of the memoir quoted on p. 53.

Thus
$$k_1 = f(\cdot1, 1\cdot1) \times \cdot2$$
$$= \cdot2 \times \frac{1\cdot01}{1\cdot31} = \cdot154.$$

Also
$$k' = \cdot2\ ;$$
$$k'' = f(\cdot2, 1\cdot2) \times \cdot2 = \cdot2 \times \frac{1\cdot44 - \cdot4}{1\cdot44 + \cdot2} = \cdot127\ ;$$

$$f(a+h,\ b+k'') = f(\cdot2, 1\cdot127) = \frac{1\cdot270 - \cdot4}{1\cdot270 + \cdot2} = \cdot592\ ;$$
$$k_2 = \tfrac{1}{2}(k' + k''')$$
$$= \tfrac{1}{2}h\{f_0 + f(a+h,\ b+k'')\}$$
$$= \cdot1 \times 1\cdot592 = \cdot159.$$

Here
$$\tfrac{1}{3}(k_2 - k_1) = \cdot002,$$
and therefore
$$k = k_1 + \tfrac{1}{3}(k_2 - k_1) = \cdot156.$$
Consequently
$$y = 1\cdot156,\quad x = \cdot2,$$
are simultaneous values for the solution determined by
$$y = 1,\quad x = 0.$$

The calculations in the second stage are similar to those which precede: only the results will be stated. We have
$$a = \cdot2,\quad b = 1\cdot156,\quad h = \cdot3\ ;$$
$$k_1 = \cdot135\ ;$$
$$k' = \cdot183,$$
$$k'' = \cdot104,$$
$$k''' = \cdot085,$$
$$k_2 = \cdot134.$$
Hence we have
$$k = \cdot135\ ;$$
and therefore the value of y, when $x = \cdot5$, is given by
$$y = 1\cdot156 + \cdot135$$
$$= 1\cdot291,$$
with a possible error of one unit in the third place of decimals.

Ex. 2. Apply the process to the equation
$$\frac{dy}{dx} = \frac{y-x}{y+x}$$
for a solution which is 1 when $x = 0$; and prove that $y = 1\cdot499$ when $x = 1$. (It is convenient to divide the range into three portions 0 to $\cdot2$, $\cdot2$ to $\cdot5$, and $\cdot5$ to 1.)

Integrate the equation; and compare the result obtained by the numerical process with the accurate value.   (Runge.)

CHAPTER III

THE GENERAL LINEAR DIFFERENTIAL EQUATION WITH CONSTANT COEFFICIENTS

Preliminary Formulæ.

31. Before proceeding to the discussion of the linear equation of the nth order with constant coefficients, it is convenient to formulate and prove certain theorems in differentiation and integration, which will be required in that discussion.

Let D stand for $\dfrac{d}{dx}$; D^2 for $\dfrac{d^2}{dx^2}$; and so on. Then this symbol D obviously is subject to the fundamental laws of algebra; for evidently

$$(D^r + D^n)\,u = (D^n + D^r)\,u\,;$$
$$D^r \cdot D^n u = D^n \cdot D^r u = D^{n+r}u\,;$$
$$D\,(u+v) = Du + Dv.$$

It is necessary to deal with negative indices; thus if we have

$$Du = v,$$

and if, after the algebraical analogy, we write

$$u = D^{-1}v,$$

we have

$$v = Du = D \cdot D^{-1}v,$$

so that

$$D \cdot D^{-1} = 1.$$

Thus D^{-1} represents such an operation on any quantity that, if the operation represented by D be subsequently performed, the quantity is left unaltered. It at once follows that these symbols with negative indices also follow the laws of algebra; and an operation with a negative index is equivalent to an integration.

But it is important to point out that the special object of these inverse operations is to find an integral but not the complete integral; and the arbitrary constant which arises in integration is therefore omitted.

In what follows, ψ denotes a functional symbol; and $\psi(x)$ everywhere denotes a rational function of x which can be expanded in ascending or descending integral powers (or both) of the variable.

32. *Theorem I.*

$$\psi(D)\,e^{ax} = \psi(a)\,e^{ax}.$$

For, since D stands for $\dfrac{d}{dx}$,

$$De^{ax} = ae^{ax}.$$

When each side is operated on with D^{-1}, the equation becomes

$$D^{-1}.\,De^{ax} = aD^{-1}e^{ax};$$

or transposing the sides of the equation and dividing by a, we have

$$D^{-1}e^{ax} = a^{-1}e^{ax}.$$

Repeating these operations, we obtain the equations

$$D^n e^{ax} = a^n e^{ax},$$
$$D^{-m} e^{ax} = a^{-m} e^{ax}.$$

Now as ψ is a rational function which can be expanded in powers, we may write

$$\psi(D)\,e^{ax} = [A_0 + A_1 D + \ldots + A_r D^r + \ldots + B_1 D^{-1} + B_2 D^{-2} + \ldots\,]\,e^{ax}$$
$$= [A_0 + A_1 a + \ldots + A_r a^r + \ldots + B_1 a^{-1} + B_2 a^{-2} + \ldots\,]\,e^{ax}$$
$$= \psi(a)\,e^{ax}.$$

33. *Theorem II.* If X denote any function whatever of x, then

$$\psi(D)\,\{e^{ax}X\} = e^{ax}\psi(D+a)\,X.$$

A single operation with D gives

$$D\,\{e^{ax}X\} = e^{ax}(D+a)\,X,$$

from which, if both sides be multiplied by e^{-ax}, we have

$$(e^{-ax}De^{ax})\,X = (D+a)\,X,$$

so that the effect of operating on X with $e^{-ax}De^{ax}$ is to give $D+a$ operating on X. Let the operation be repeated; then

$$(e^{-ax}De^{ax})\,(e^{-ax}De^{ax})\,X = (D+a)\,(D+a)\,X,$$

or $$(e^{-ax} D^2 e^{ax}) X = (D+a)^2 X.$$

Operate again with $e^{-ax} D e^{ax}$: then

$$(e^{-ax} D e^{ax})(e^{-ax} D^2 e^{ax}) X = (D+a)(D+a)^2 X,$$

or $$(e^{-ax} D^3 e^{ax}) X = (D+a)^3 X,$$

and so on. If the operation be performed n times, the resulting equation will be

$$(e^{-ax} D^n e^{ax}) X = (D+a)^n X,$$

which, multiplied by e^{ax}, gives

$$D^n \{e^{ax} X\} = e^{ax} (D+a)^n X,$$

where n denotes a positive integer.

Consider now the case of negative indices; write

$$(D+a)^n X = X_1,$$

so that $$X = (D+a)^{-n} X_1.$$

Then the result just obtained may be written

$$D^n e^{ax} (D+a)^{-n} X_1 = e^{ax} X_1.$$

Operate on each side with D^{-n}: the result is

$$e^{ax} (D+a)^{-n} X_1 = D^{-n} e^{ax} X_1.$$

Now no limitations were assigned to the form of X and there are therefore none on that of X_1, which can thus represent any function of x; denoting it therefore by X, we have

$$D^{-n} \{e^{ax} X\} = e^{ax} (D+a)^{-n} X.$$

Let $\psi(D)$ be expanded in integral powers of D, positive and negative (if necessary); and let $e^{ax} X$ be operated on by these integral powers in succession, the equivalent values derived from the foregoing equations being substituted and the terms collected as before; then the result is

$$\psi(D) \{e^{ax} X\} = e^{ax} \psi(D+a) X.$$

Corollary. If we write

$$e^{ax} X = Y,$$

so that Y is a function of x, then

$$\psi(D) Y = e^{ax} \psi(D+a) \{Y e^{-ax}\},$$

a theorem which is useful. For example, let it be required to find a particular value of y to satisfy the equation

$$\frac{dy}{dx} + ky = V.$$

With the notation adopted, this will be

$$y = \frac{1}{D+k} V$$

$$= e^{ax} \frac{1}{D+k+a} V e^{-ax};$$

or choosing a so that $a + k = 0$, this is

$$y = e^{-kx} \frac{1}{D} V e^{kx}$$

$$= e^{-kx} \int V e^{kx} dx.$$

34. *Theorem III.* If $\psi(x)$ be an even function of x, say $\phi(x^2)$, then

$$\psi(D) \sin(ax + \alpha) = \phi(-a^2) \sin(ax + \alpha).$$

For $\qquad D^2 \sin(ax + \alpha) = (-a^2) \sin(ax + \alpha);$

and the theorem follows as before.

Corollary. If $\psi(x)$ be not an even function of x, it can be expressed in the form

$$\phi(x^2) + x\chi(x^2)$$

where ϕ and χ are even functions of x; in this case,

$$\psi(D) \sin(ax + \alpha) = \{\phi(D^2) + D\chi(D^2)\} \sin(ax + \alpha)$$

$$= \phi(-a^2) \sin(ax + \alpha) + a\chi(-a^2) \cos(ax + \alpha).$$

If the function to be operated upon be the cosine instead of the sine, the corresponding changes are obvious.

35. *Theorem IV.* This is really an extension of Leibnitz's theorem for the successive differentiation of the product of two quantities whose differential coefficients are known.

If $\psi(x)$ as before denote any rational function expansible in integral powers of x, and $\psi'(x)$, $\psi''(x)$, $\psi'''(x)$, ... denote its first, second, third, ... differential coefficients with regard to x, then the extended theorem is

$$\psi(D) uv$$

$$= u\psi(D) v + Du \, \psi'(D) v + \frac{D^2 u}{2!} \psi''(D) v + \frac{D^3 u}{3!} \psi'''(D) v + \dots.$$

The proof depends on Leibnitz's theorem, and is similar to that of the preceding propositions.

The advantage of this theorem arises in cases where one of the two quantities u and v is a power of x, or is the sum of powers of x. If, for instance, $u = x^{m-1}$, the series on the right-hand side need only be written as far as the mth term; and such inverse operations as are to be carried out will be performed on a single quantity v.

Ex. Shew that, if
$$(D+k)^2 y = x^2 V,$$
where V is a function of x only, y is given by
$$e^{-kx} \left(x^2 \iint e^{kx} V dx^2 - 4x \iiint e^{kx} V dx^3 + 6 \iiiint e^{kx} V dx^4 \right).$$

36. Another important operator which sometimes occurs is $x \dfrac{d}{dx}$ or, with the previous notation, xD; and similar theorems concerning this can be enunciated.

Let $F(z)$ denote a rational function of z expansible in powers of z; then in $F(xD)$ we shall have terms of the form $(xD)^n$ which means, not $x^n \dfrac{d^n}{dx^n}$, but $x \dfrac{d}{dx} \cdot x \dfrac{d}{dx} \dots$ operating n times. The relation between these two operators will be obtained immediately (§ 37).

Theorem I. $F(xD) x^m = F(m) x^m.$

For $(xD) x^m = m x^m,$
$$(xD)^2 x^m = (xD) m x^m = m^2 x^m,$$

and so for all integral powers positive and negative. Hence the theorem.

Ex. Prove that, if U be a function of x of the form
$$A + Bx + Cx^2 + Dx^3 + \dots,$$
then
$$\frac{1}{F(xD)} U = \frac{A}{F(0)} + \frac{B}{F(1)} x + \frac{C}{F(2)} x^2 + \frac{D}{F(3)} x^3 + \dots.$$

Theorem II. $F(xD) x^m V = x^m F(xD + m) V.$

We have $xD(x^m V) = x^m (xD + m) V,$
or $(x^{-m} \cdot xD \cdot x^m) V = (xD + m) V,$

so that the operators $x^{-m} \cdot xD \cdot x^m$ and $xD + m$ are equivalent. The course of proof follows lines exactly similar to those for the corresponding theorem with $F(D)$; and the result is in the enunciated form.

37. The relation between the operators D^n and xD is given by the formula

$$x^n D^n = xD (xD - 1) (xD - 2) \ldots (xD - n + 1).$$

The theorem can be established directly; for if u the subject of operation be expanded in a series of terms of the form $A_m x^m$, the result of operating on this with D^n and multiplying by x^n is zero if $m < n$, and is

$$m (m - 1) (m - 2) \ldots (m - n + 1) A_m x^m,$$

if $m \geqq n$; but this is also the result of operating with the right-hand side. Hence the operators are equivalent for each term of u and so for the sum of all the terms of u, i.e. for u itself.

The theorem can also be established by induction; for suppose

$$x^n D^n u = xD (xD - 1) (xD - 2) \ldots (xD - n + 1) u,$$

and write $\qquad u = (xD - n) v;$

then $\qquad D^n u = xD^{n+1} v,$

and so $\quad x^{n+1} D^{n+1} v = xD (xD - 1) (xD - 2) \ldots (xD - n) v.$

Now u is any general function; hence v is also a general function. The theorem, if true for n, is thus true for $n + 1$; it is obviously true for the values 1 and 2, and so it is true generally.

Some Properties of the General Linear Differential Equation.

38. The general type of linear differential equation of the nth order is

$$\frac{d^n y}{dx^n} + X_1 \frac{d^{n-1} y}{dx^{n-1}} + X_2 \frac{d^{n-2} y}{dx^{n-2}} + \ldots + X_{n-1} \frac{dy}{dx} + X_n y = V,$$

in which $X_1, X_2, \ldots, X_n, V$, are functions of x (or constants) but do not contain y; for the sake of shortness, let it be written

$$\Phi (D) y = V.$$

If this equation be integrated step by step so that each integration reduces the order of the equation by unity, every time such a reduction is effected an arbitrary constant enters, and therefore, when ultimately the integral equation is obtained, n arbitrary constants in all will have entered; or we shall expect the primitive of a given linear differential equation to contain a number of arbitrary constants equal to the order of the equation.

There are certain properties appertaining to all linear equations in common which simplify to some extent their integration; the most important of these are the following.

39. I. Let η be any particular value of y, which satisfies the equation; and let

$$y = \eta + Y.$$

Then substituting this value of y in the equation we have

$$\Phi(D) Y + \Phi(D) \eta = V.$$

But, since η is some solution of

$$\Phi(D) y = V,$$

the equation now becomes

$$\Phi(D) Y = 0,$$

so that to solve the original equation we must solve generally this equation, which is the same as the original equation except that the right-hand side is now zero. When the primitive of this modified equation, which will contain n arbitrary constants because it is of the nth order, has been obtained, it must be added to η; and the result equated to y will be the primitive of the given equation. The primitive then consists of two parts:

First, the quantity η, which is called the *Particular Integral* and is any solution whatever (the simpler the better) of the original equation;

Second, the quantity Y, which is called the *Complementary Function* and is the primitive of the equation when the right-hand side is made zero.

The sum of these two parts is the primitive of the general equation. If in any particular case the right-hand side should already be zero, the former of these parts will not occur.

The various methods available for the deduction of the Particular Integral occur later in § 46. The remaining properties are useful in the investigation of the Complementary Function.

40. II. If $Y = Y_1$ be a solution of the equation

$$\Phi(D) Y = 0,$$

then $Y = C_1 Y_1$ is also a solution, where C_1 is a constant; and if $Y_1, Y_2, \ldots\ldots, Y_n$, be particular solutions, then

$$Y = C_1 Y_1 + C_2 Y_2 + \ldots\ldots + C_n Y_n$$

is also a solution, where $C_1, C_2, \ldots\ldots, C_n$, are constants.

For $\quad \Phi(D) Y = \Phi(D) C_1 Y_1 + \Phi(D) C_2 Y_2 + \ldots\ldots,$

and each term on the right-hand side is zero. No restriction whatever has been laid on the values of the constants C, and they therefore are completely arbitrary; the above value of Y is thus the primitive of the simpler equation

$$\Phi(D) Y = 0,$$

and so it is the complementary function in the integral of the equation

$$\Phi(D) y = V.$$

Hence the determination of the complementary function is reduced to that of particular solutions of the simpler equation.

41. III. If a single particular solution of the simpler equation be known, the order of the given differential equation can be lowered by unity.

Let Y_1 be a solution of

$$\Phi(D) Y = 0,$$

and let the substitution of the value $Y_1 z$ be made in the equation

$$\Phi(D) y = V;$$

then, by § 35, the left-hand side becomes

$$z\Phi(D) Y_1 + Dz \frac{\partial \Phi}{\partial D} Y_1 + \frac{D^2 z}{2!} \frac{\partial^2 \Phi}{\partial D^2} Y_1 + \ldots + \frac{D^n z}{n!} \frac{\partial^n \Phi}{\partial D^n} Y_1,$$

in which the operations $\dfrac{\partial \Phi}{\partial D}, \ldots$ are derived from Φ by temporarily considering D as a magnitude and obtaining the partial differential coefficients with regard to D. But

$$\frac{\partial^n \Phi}{\partial D^n} = n!,$$

$$\frac{\partial^{n-1} \Phi}{\partial D^{n-1}} = \frac{n!}{1!} D + (n-1)! X_1,$$

$$\frac{\partial^{n-2} \Phi}{\partial D^{n-2}} = \frac{n!}{2!} D^2 + \frac{(n-1)!}{1!} X_1 D + (n-2)! X_2,$$

and so on; hence, re-writing the equation, we obtain

$$Y_1 D^n z + (X_1 Y_1 + nD Y_1) D^{n-1} z + \ldots\ldots + Dz \frac{\partial \Phi}{\partial D} Y_1 + z\Phi (D) Y_1 = V.$$

But by hypothesis

$$\Phi (D) Y_1 = 0,$$

so that the last term on the left-hand side disappears; the quantity Y_1 is supposed known, and therefore all the functions of it on the left-hand side may be considered known. Let Z be written for Dz; then the equation becomes

$$Y_1 D^{n-1} Z + (X_1 Y_1 + nD Y_1) D^{n-2} Z + \ldots + Z \frac{\partial \Phi}{\partial D} Y_1 = V,$$

an equation of order $n - 1$.

Ex. As a corollary prove that, if m particular solutions of the simpler equation be known, the order of the original differential equation can be reduced by m.

42. IV. The given equation may be transformed into an equation, from which the second term (i.e. the term involving the differential coefficient of order one less than the order of the equation) is absent.

The substitution of $Y_1 z$ for y gives, for the coefficient of $D^{n-1} z$,

$$X_1 Y_1 + nD Y_1,$$

(and up to this point in the last section the assumed value of Y_1 was not used, so that the equation there was perfectly general). Since the term in $D^{n-1} z$ is to be absent, we now have

$$X_1 Y_1 + nD Y_1 = 0,$$

and therefore

$$\log Y_1 = -\frac{1}{n} \int X_1 dx,$$

or

$$Y_1 = e^{-\frac{1}{n} \int X_1 dx},$$

no arbitrary constant being inserted as the differential equation remains linear and of the nth order. If this value of Y_1 be substituted the differential equation in z is freed from the term in $D^{n-1} z$.

Of these properties, I. and II. will be immediately useful.

GENERAL LINEAR EQUATION WITH CONSTANT COEFFICIENTS.

43. If, in the general linear equation, the coefficients of y and of its differential coefficients be constants, it may be written

$$\frac{d^n y}{dx^n} + a_1 \frac{d^{n-1} y}{dx^{n-1}} + \ldots + a_{n-1} \frac{dy}{dx} + a_n y = V,$$

or say
$$f(D) y = V,$$

in which $f(D)$ is a polynomial function of D alone, and V is any function of x. It has already been proved that the solution of the equation consists of two parts which can be obtained separately; these will be taken in turn.

44. *To find the Complementary Function.*

The complementary function is the primitive of

$$f(D) y = 0.$$

Now it has been proved that

$$f(D) e^{ax} = f(a) e^{ax},$$

so that $y = e^{ax}$ will be a particular solution of the equation, if a be such as to make

$$f(a) = 0.$$

But $f(z)$ is a polynomial function of degree n, and therefore there are n roots of the equation

$$f(z) = 0.$$

Let these n roots be $\alpha, \beta, \ldots, \lambda$; then $e^{ax}, e^{\beta x}, \ldots, e^{\lambda x}$, are n particular solutions of the equation

$$f(D) y = 0;$$

and the primitive is therefore

$$y = A e^{ax} + B e^{\beta x} + \ldots + L e^{\lambda x},$$

in which $A, B, \ldots, L$, are n arbitrary constants. This value of y is the complementary function of the original equation; and, if the roots be all real and different from one another, it is complete.

If however *two roots be equal* to one another, say α and β, then the value of y becomes

$$y = (A + B) e^{ax} + C e^{\gamma x} + \ldots + L e^{\lambda x}$$
$$= A_1 e^{ax} + C e^{\gamma x} + \ldots + L e^{\lambda x},$$

A_1 being a single arbitrary constant (equal to the sum of two arbitrary constants). There are now only $n-1$ arbitrary constants in y; and the expression therefore ceases to be the primitive. In order to obtain the primitive, we may suppose that the roots are not equal but differ by some quantity h which will ultimately be made zero; the part depending on the roots α and β will then be

$$A e^{\alpha x} + B e^{(\alpha+h) x}$$

$$= e^{\alpha x} \left\{ A + B \left(1 + hx + \frac{h^2 x^2}{2!} + \dots \right) \right\}$$

$$= e^{\alpha x} \left\{ (A+B) + Bhx + Bh \frac{h}{2!} x^2 + \dots \right\}.$$

As the equations A and B are arbitrary, we may assume them infinite in such a way that, as h approaches zero, Bh is finite and equal to B_1, while A and B are of opposite sign and their numerical difference (or algebraical sum) is finite and equal to A_1; thus the sum of the two terms $A e^{\alpha x} + B e^{\beta x}$ becomes

$$e^{\alpha x} \left\{ A_1 + B_1 \left(x + \frac{h}{2!} x^2 + \frac{h^2}{3!} x^3 + \dots \right) \right\} = (A_1 + B_1 x) e^{\alpha x}$$

ultimately, when h is made zero.

Similarly, if r *roots be equal*, the corresponding r terms in the complementary function will apparently coalesce into a single term. It is easy to shew, by reasoning similar to that adopted for the case of two equal roots, that the r terms will be transformed to

$$e^{\alpha x} [A_1 + A_2 x + A_3 x^2 + \dots + A_r x^{r-1}],$$

α denoting the common value of the r equal roots; and the complementary function will then be

$$y = e^{\alpha x} [A_1 + A_2 x + \dots + A_r x^{r-1}] + \dots + L e^{\lambda x}.$$

Again, if the roots be *not all real*, those which are imaginary must occur in pairs, when (as is usual) all the coefficients a in $f(z)$ are real; let such a pair be $\theta + \phi i$*. The corresponding terms of the complementary function are

$$e^{\theta x} [A' e^{x \phi i} + B' e^{-x \phi i}],$$

which it is sometimes necessary to express in a form free from

---

* Throughout the book, i will be used to denote $\sqrt{-1}$, in accordance with custom.

imaginary quantities. If cosines and sines be substituted for the exponentials, this expression becomes

$$e^{\theta x}\{(A'+B')\cos\phi x+i(A'-B')\sin\phi x\}.$$

Since A' and B' are arbitrary constants, we may write

$$A'+B'=F,$$
$$i(A'-B')=G,$$

and we then have F and G arbitrary; the corresponding terms in the complementary function therefore become

$$e^{\theta x}(F\cos\phi x+G\sin\phi x).$$

Lastly, *if an imaginary root be repeated,* the conjugate imaginary root will also be repeated; and the corresponding terms in y will be

$$e^{x(\theta+\phi i)}(A'+A''x)+e^{x(\theta-\phi i)}(B'+B''x).$$

Using the same method as before, and writing

$$A'+B'=F, \qquad\qquad A''+B''=F',$$
$$i(A'-B')=G, \qquad\qquad i(A''-B'')=G',$$

we obtain as the corresponding part of the complementary function

$$e^{\theta x}\{(F+F'x)\cos\phi x+(G+G'x)\sin\phi x\}.$$

Results, analogous to those in the case of multiple repetition of real roots, are obtained in the case of multiple repetition of imaginary roots.

45. In some cases of the general linear equation, when the coefficients are not constants but are some functions of x, a method somewhat similar to this will apply. Thus, it might happen that, when for y in the equation

$$(D^n+X_1 D^{n-1}+\ldots+X_{n-1}D+X_n)y=0$$

there is substituted $\psi(m, x)$, where ψ is a function of definite form, the resulting equation had a factor independent of x such as $\phi(m)$; if this were so, the factor would usually be of the degree n, and so equated to zero would satisfy the differential equation and would furnish n values of m which may be denoted by m_1, $m_2, \ldots, m_n$; the primitive would then be

$$y=A_1\psi(m_1, x)+A_2\psi(m_2, x)+\ldots+A_n\psi(m_n, x).$$

If two roots were equal, as m_1 and m_2, then writing $m_2 = m_1 + h$ we have for the corresponding part of y

$$(A_1 + A_2)\, \psi\,(m_1,\, x) + hA_2 \left[\frac{\partial \psi\,(m_1,\, x)}{\partial m_1} + \frac{h}{2\,!} \frac{\partial^2 \psi\,(m_1,\, x)}{\partial m_1{}^2} + \dots \right],$$

or

$$A'\psi\,(m_1,\, x) + B' \frac{\partial}{\partial m_1} \psi\,(m_1,\, x),$$

on changing the constants and making h ultimately zero as before. A similar process holds for the case of a multiple repetition of any root m_1. In the case of imaginary roots, the corresponding parts of y should usually have the constants changed in the modified expression, so as to leave the latter free from imaginary symbols.

This process was adopted in the case of constant coefficients, the special form of ψ used being e^{mx}. When the equation is homogeneous (§ 55), that is, when it takes the form

$$x^n \frac{d^n y}{dx^n} + A_1 x^{n-1} \frac{d^{n-1} y}{dx^{n-1}} + \dots + A_{n-1} x \frac{dy}{dx} + A_n y = 0,$$

in which the quantities A are constants, the proper form of ψ (see § 36) to be substituted is x^m. Occasionally, by a suitable change of variable, a given equation can be reduced to the above shape.

Ex. 1. Solve $\dfrac{d^2 y}{dx^2} + 3 \dfrac{dy}{dx} + 2y = 0.$

When we substitute $y = e^{mx}$, the equation for m is

$$(m+1)\,(m+2) = 0,$$

so that $y = A e^{-x} + B e^{-2x}.$

Ex. 2. Solve $\dfrac{d^2 y}{dx^2} - 2\lambda \dfrac{dy}{dx} + (\lambda^2 + \mu^2)\, y = 0.$

The equation for m is

$$(m - \lambda)^2 + \mu^2 = 0,$$

so that $y = e^{\lambda x} C \cos\,(\mu x + \alpha),$

or $y = e^{\lambda x} (A \cos \mu x + B \sin \mu x).$

Cor. The solution of

$$\frac{d^2 y}{dx^2} + \mu^2 y = 0$$

is $y = A \cos \mu x + B \sin \mu x.$

Ex. 3. Solve $\dfrac{d^2y}{dx^2} - 2\dfrac{dy}{dx} + y = 0.$

The equation for m is $(m-1)^2 = 0,$

and therefore $y = e^x(A + Bx).$

Ex. 4. Solve $\dfrac{d^4y}{dx^4} + 2n^2\dfrac{d^2y}{dx^2} + n^4y = 0.$

The equation for m is

$$(m^2 + n^2)^2 = 0,$$

and the value of y is

$$(A + Bx)\cos nx + (C + Dx)\sin nx.$$

Ex. 5. Solve $x^2\dfrac{d^2y}{dx^2} + x\dfrac{dy}{dx} - y = 0.$

When we substitute x^m for y, the equation for m is

$$m(m-1) + m - 1 = 0,$$

so that $m = +1$ or -1; the value of y is therefore

$$Ax + \frac{B}{x}.$$

Ex. 6. Solve $x^3\dfrac{d^3y}{dx^3} - 3x^2\dfrac{d^2y}{dx^2} + 7x\dfrac{dy}{dx} - 8y = 0.$

With the same substitution as in Ex. 5, the equation for m is

$$m(m-1)(m-2) - 3m(m-1) + 7m - 8 = 0,$$

or $m^3 - 6m^2 + 12m - 8 = 0,$

giving $m = 2$ thrice. Hence the value of y is

$$Ax^m + B\frac{\partial}{\partial m}x^m + C\frac{\partial^2}{\partial m^2}x^m,$$

m being put equal to 2 after differentiation; and thus the integral is

$$x^2\{A + B\log x + C(\log x)^2\}.$$

Ex. 7. Solve $(a + bx)^2\dfrac{d^2y}{dx^2} + A(a + bx)\dfrac{dy}{dx} + By = 0.$

Let $a + bx = z$; the equation will then be similar in form to the last two.

Ex. 8. Solve

(i) $(D^4 + 5D^2 + 6)y = 0;$

(ii) $(D^4 + a^4)y = 0;$

(iii) $(D^6 - a^6)y = 0;$

(iv) $\dfrac{d^8y}{dx^8} = y;$

(v) $x^2\dfrac{d^2y}{dx^2} + 3x\dfrac{dy}{dx} + y = 0;$

(vi) $(1+x)^3\dfrac{d^3y}{dx^3} + (1+x)^2\dfrac{d^2y}{dx^2} + 3(1+x)\dfrac{dy}{dx} - 8y = 0.$

46. Returning now to the linear equation, in which the co-efficients of the differential coefficients of y are constants, it is necessary to find *a Particular Integral* of the equation

$$f(D)\,y = V,$$

in which V is a function of x. Solving by the method of symbolical operators, we have

$$y = \frac{1}{f(D)}\,V;$$

the evaluation of the right-hand side will furnish a satisfactory value of y.

In some particular cases the form of V renders evaluation easy; we will proceed to mention some of these which occur most frequently.

I. Let V be a polynomial in x; suppose the highest power of x in V to be the nth. To find the particular integral, $\dfrac{1}{f(D)}$ must be expanded in ascending powers of D; and, because D^{n+1} and operators of a higher order would reduce to zero all the terms of V, the terms in this expansion beyond D^n may be omitted. Further, if the lowest power of D in $f(D)$ be D^k, then the expansion will begin with D^{-k} and it does not need to be carried on beyond D^n, i.e. $D^{-k+(k+n)}$; hence in $f(D)$ all terms of order higher than D^{n+k} may in this case at once be omitted before expansion.

Ex. 1. Solve　　　　　$(D^2 - 4D + 4)\,y = x^2$.

We have　　　　　$y = \dfrac{1}{(2 - D)^2}\,x^2$

$$= \frac{1}{4}\left[1 + 2\frac{D}{2} + 3\frac{D^2}{2^2}\right]x^2 = \frac{x^2}{4} + \frac{x}{2} + \frac{3}{8},$$

as the particular integral. The complementary function is $e^{2x}(A + Bx$; hence the primitive is

$$y = e^{2x}(A + Bx) + \tfrac{1}{8}(2x^2 + 4x + 3).$$

Ex. 2. Solve　　　　　$(D^4 - a^4)\,y = x^3$.

The primitive is evidently

$$y = -\frac{x^3}{a^4} + Ae^{ax} + Be^{-ax} + C\cos(ax + a).$$

Ex. 3. Solve $(D^4 - 2D^3 + D^2) y = x^3$.

We have
$$y = \frac{1}{D^2(1-D)^2} x^3$$

$$= \frac{1}{D^2}(1 + 2D + 3D^2 + 4D^3 + 5D^4 + 6D^5)\, x^3,$$

terms up to the fifth being retained (§ 46). Now $1 + 2D + \ldots$ and $\frac{1}{D^2}$ may be considered separate operators; operating with the former first and remembering that only a particular value is wanted, so that constants need not be inserted with $\frac{1}{D^2}$, the value for y is

$$\frac{x^5}{20} + \frac{x^4}{2} + 3x^3 + 12x^2.$$

Now if $\frac{1}{D^2}$ had operated first (or if the second operator had been taken distributively, each term with $\frac{1}{D^2}$, so as to be

$$\frac{1}{D^2} + \frac{2}{D} + 3 + 4D + 5D^2 + 6D^3),$$

then the value for y would have become

$$\frac{x^5}{20} + \frac{x^4}{2} + 3x^3 + 12x^2 + 30x + 36.$$

The primitive is

$$y = (A' + B'x)\, e^x + A + Bx + \frac{x^5}{20} + \frac{x^4}{2} + 3x^3 + 12x^2.$$

The apparently additional part of the particular integral obtained, when the operators are taken in the second order, is seen to be included in the complementary function, since A and B are arbitrary constants.

Note 1. It is easy to see that in general not merely may the terms of an order higher than D^{n+k} be at once removed from $f(D)$, but also that in the development of the expanded factor of $f(D)$ all powers of D of an order higher than D^n can be omitted, whatever be the magnitude of k.

Note 2. In particular, if X be a constant, only the lowest power in $f(D)$ need be retained.

Ex. 4. Solve

  (i) $(D^4 + 2D^3 + 3D^2 + 2D + 1) y = 1 + x + x^2$;

  (ii) $(D^3 + D^2 - D + 15) y = x^2$.

II. This method may be applied to evaluate y, when V is an exponential, and to simplify the process (by rendering the evaluation more proximate) when V contains an exponential factor. In either case we may write

$$V = e^{ax}\, X,$$

and then

$$y = \frac{1}{f(D)} V = \frac{1}{f(D)} e^{ax} X$$

$$= e^{ax} \frac{1}{f(D+a)} X.$$

If X be a constant, the value of y is now at once obtainable by the preceding method. The quantity a may or may not be a root of $f(z) = 0$. Suppose it to be a root occurring r times, so that for a single root $r = 1$. If a be not a root, $r = 0$. Then expanding $f(D+a)$, we have

$$f(D+a) = \frac{D^r}{r!} f^{(r)}(a) + \frac{D^{r+1}}{(r+1)!} f^{(r+1)}(a) + \dots,$$

in which $f^{(\mu)}(a)$ means the μth differential coefficient of $f(z)$ with respect to z, when a is substituted for z; for y, we have (by attending to the remark at the end of Ex. 3 on p. 72)

$$y = e^{ax} \frac{1}{\dfrac{f^{(r)}(a)}{r!} D^r} C$$

$$= C \frac{e^{ax} x^r}{f^{(r)}(a)}.$$

In particular, if $r = 0$, then

$$y = \frac{C}{f(a)} e^{ax}.$$

Ex. 5. Solve $(D^2 + D + 1) y = e^{2x}$.

Here 2 is not a root of $z^2 + z + 1 = 0$, and therefore

$$y = \frac{e^{2x}}{2^2 + 2 + 1} = \frac{1}{7} e^{2x};$$

and the primitive is

$$y = e^{-\frac{1}{2}x} \left(A \cos \frac{3^{\frac{1}{2}} x}{2} + B \sin \frac{3^{\frac{1}{2}} x}{2} \right) + \frac{1}{7} e^{2x}.$$

Ex. 6. Solve $(D^2 - 4D + 3) y = 2e^{3x}$.

Here
$$y = \frac{1}{(D-1)(D-3)} 2e^{3x}$$

$$= e^{3x} \frac{1}{D(D+2)} 2$$

$$= e^{3x} \frac{1}{2D} 2 = x e^{3x};$$

and the primitive is

$$y = A e^x + B e^{3x} + x e^{3x}.$$

Ex. 7. Solve $\qquad$ (i) $(D-a)^n y = e^{ax}$;

$\qquad\qquad\qquad\qquad$ (ii) $(D^2-6D+8)y = e^x + e^{2x}$.

Ex. 8. The roots of the equation $f(z)=0$ are n in number, being $a_1, a_2, \ldots, a_n$; obtain the particular integral of the equation

$$f(D)y = e^{a_1 x} + e^{a_2 x} + \ldots + e^{a_n x}.$$

Discuss the case when two of the roots (a_1 and a_2) are equal.

If X be a polynomial in x, then the quantity

$$\frac{1}{f(D+a)}X$$

must be evaluated as before in I.

Ex. 9. Solve $\qquad$ $(D^2-2D+1)y = x^2 e^{3x}$.

Here $\qquad\qquad$ $y = \dfrac{1}{(D-1)^2} x^2 e^{3x}$

$$= e^{3x} \frac{1}{(D+2)^2} x^2$$

$$= e^{3x}\left(\frac{x^2}{4} - \frac{x}{2} + \frac{3}{8}\right);$$

and the primitive is

$$y = (A+Bx)e^x + e^{3x}\left(\frac{x^2}{4} - \frac{x}{2} + \frac{3}{8}\right).$$

Ex. 10. Solve $\qquad$ $(D-2)^3 y = x^2 e^{2x}$.

Here $\qquad\qquad$ $y = \dfrac{1}{(D-2)^3} x^2 e^{2x}$

$$= e^{2x} \frac{1}{D^3} x^2 = \tfrac{1}{60} x^5 e^{2x};$$

and the primitive is

$$y = e^{2x}(A + Bx + Cx^2 + \tfrac{1}{60}x^5).$$

Ex. 11. Solve $\qquad$ (i) $(D^2 + D + 1)^2 y = xe^x$;

$\qquad\qquad\qquad\qquad$ (ii) $(D^4 - 1)^2 y = x^4 e^x$.

III. Suppose that V contains a sine or a cosine as a factor, so that

$$V = X \cos(nx + a),$$

in which n and a are constants. Then we have to evaluate

$$y = \frac{1}{f(D)} X \cos(nx + a).$$

Let
$$y_1 = \frac{1}{f(D)} X \sin(nx + \alpha);$$

then
$$y + iy_1 = \frac{1}{f(D)} X e^{i(nx+a)}$$

$$= e^{i(nx+a)} \frac{1}{f(D + in)} X.$$

It now remains to evaluate

$$\frac{1}{f(D + in)} X,$$

which may come under one or other of the given rules. Let its value be $u + iv$; then, equating real and imaginary parts, we have

$$y = u \cos(nx + \alpha) - v \sin(nx + \alpha).$$

In the case when X is a constant and $\cos nx$ is not part of the complementary function, so that in is not a root of $f(z) = 0$, the evaluation is immediate; for then

$$\frac{1}{f(D + in)} X = \frac{1}{f(in)} C.$$

If, however, $\cos nx$ be a part of the complementary function, so that in is a root occurring r times, then since

$$f(D + in) = \frac{D^r}{r!} f^{(r)}(in) + \frac{D^{r+1}}{(r+1)!} f^{(r+1)}(in) + \ldots\ldots,$$

we have

$$\frac{1}{f(D + in)} C = \frac{C x^r}{f^{(r)}(in)};$$

we must separate and equate the real and imaginary parts as before.

Ex. 12. Solve $\dfrac{d^2 y}{dx^2} + n^2 y = x \cos ax.$

Then
$$y = \frac{1}{D^2 + n^2} x \cos ax$$

$$= \text{real part of } e^{axi} \frac{1}{(D + ai)^2 + n^2} x$$

$$= \ldots\ldots\ldots\ldots\ e^{axi} \frac{1}{n^2 - a^2} \left(1 - \frac{2ai}{n^2 - a^2} D\right) x$$

$$= \ldots\ldots\ldots\ldots\ e^{axi} \left\{\frac{x}{n^2 - a^2} - \frac{2ai}{(n^2 - a^2)^2}\right\}$$

$$= \frac{x \cos ax}{n^2 - a^2} + \frac{2a \sin ax}{(n^2 - a^2)^2}.$$

Ex. 13. Solve $\qquad\qquad \dfrac{d^2y}{dx^2}+y=\cos x.$

Then $\qquad\qquad\qquad y=\dfrac{1}{D^2+1}\cos x$

$$=\text{real part of } e^{xi}\,\dfrac{1}{(D+i)^2+1}\,1$$

$$=\ldots\ldots\ldots\ldots\ e^{xi}\,\dfrac{1}{2iD+D^2}\,1$$

$$=\ldots\ldots\ldots\ldots\ e^{xi}\,\dfrac{x}{2i}$$

$$=\dfrac{x}{2}\sin x\,;$$

and the primitive is

$$y=A\cos x+B\sin x+\tfrac{1}{2}x\sin x.$$

Ex. 14. Solve $\qquad\qquad \phi(D)y=\cos nx,$

$\cos nx$ not being a part of the complementary function.

Let $\qquad\qquad\qquad \phi(D)=\phi_1(D^2)+D\phi_2(D^2)\,;$

then $\qquad\quad y=\dfrac{1}{\phi_1(D^2)+D\phi_2(D^2)}\cos nx$

$$=\dfrac{1}{\phi_1(-n^2)+D\phi_2(-n^2)}\cos nx$$

$$=\dfrac{\phi_1(-n^2)-D\phi_2(-n^2)}{\{\phi_1(-n^2)+D\phi_2(-n^2)\}\{\phi_1(-n^2)-D\phi_2(-n^2)\}}\cos nx$$

$$=\dfrac{\phi_1(-n^2)\cos nx+n\phi_2(-n^2)\sin nx}{\{\phi_1(-n^2)\}^2+n^2\{\phi_2(-n^2)\}^2}\,.$$

If, however, $\cos nx$ be a part of the complementary function, then the denominator will vanish and apparently render the particular integral infinite. But it is merely a part of the complementary function, multiplied by an infinite constant, which may be absorbed into the arbitrary constant. To evaluate the particular integral, it would be sufficient to evaluate

$$\dfrac{1}{\phi_1(D^2)+D\phi_2(D^2)}\cos(n+h)\,x,$$

assigning the infinite part (when h is made zero) to the complementary function, and retaining the finite part as the particular integral. It is however better in such cases to use the former method; in fact, this method is preferable only in the case of examples like that just treated.

Ex. 15. Solve

(i) $\dfrac{d^2y}{dx^2}+y=\sin nx$ (both when n is, and when it is not, unity);

(ii) $\dfrac{d^2y}{dx^2}+\dfrac{dy}{dx}+y=\sin 2x\,;$

(iii) $\dfrac{d^2y}{dx^2}+\dfrac{dy}{dx}+y=e^{-\frac{1}{2}x}\sin\dfrac{3^{\frac{1}{2}}}{2}x$;

(iv) $\dfrac{d^4y}{dx^4}+2\dfrac{d^2y}{dx^2}+y=x^2\cos ax$ (when a is, and when it is not, unity);

(v) $\dfrac{d^2y}{dx^2}+4y=x\sin^2x$;

(vi) $(D^2+m^2)^r\,y=(1-x)^2\cos mx$;

(vii) $(D^2-2D+4)^2\,y=xe^x\cos(3^{\frac{1}{2}}x+a)$;

(viii) $\dfrac{d^3y}{dx^3}-2\dfrac{dy}{dx}+4y=e^x\cos x$;

(ix) $\dfrac{d^4y}{dx^4}+n^4y=\sin\lambda x+\rho^{\mu x}+x^5$;

(x) $\{D^4+(m^2+n^2)\,D^2+m^2n^2\}\,y=\cos\tfrac{1}{2}(m+n)\,x\cos\tfrac{1}{2}(m-n)\,x$;

(xi) $\dfrac{d^6y}{dx^6}+y=\sin\tfrac{3}{2}x\sin\tfrac{1}{2}x$.

Ex. 16. The equation

$$\frac{d^2y}{dx^2}-2a\frac{dy}{dx}+by=ce^{px}\sin(qx+a),$$

where a, b, c, p, q, a, are constants, is of great importance in dynamics. The form of the primitive depends, to some extent, upon relations (of equality or inequality) among the constants. The results, for the various cases, are as follows.

I. Let $b>a^2$, and take $b=a^2+\kappa^2$.

(i) If p is not equal to a, and q^2 is not equal to κ^2; or if $p=a$ while q^2 is not equal to κ^2; or if $q^2=\kappa^2$ while p is not equal to a; the primitive is

$$y=e^{ax}(A\cos\kappa x+B\sin\kappa x)+e^{px}(E\cos qx+F\sin qx),$$

where A and B are arbitrary constants, and where E and F are definite constants given by the equations

$$\left.\begin{array}{l}E\{(p-a)^2+\kappa^2-q^2\}+2Fq\,(p-a)=c\sin a\\[4pt]-2Eq\,(p-a)+F\{(p-a)^2+\kappa^2-q^2\}=c\cos a\end{array}\right\}.$$

(ii) If $p=a$ and $q=\kappa$, so that the equation is

$$\frac{d^2y}{dx^2}-2a\frac{dy}{dx}+(a^2+\kappa^2)\,y=ce^{ax}\sin(\kappa x+a),$$

the primitive is

$$y=e^{ax}(A\cos\kappa x+B\sin\kappa x)-\frac{c}{2\kappa}\,xe^{ax}\cos(\kappa x+a),$$

where A and B are arbitrary constants.

II. Let $b=a^2$, so that the equation is

$$\frac{d^2y}{dx^2}-2a\frac{dy}{dx}+a^2y=ce^{px}\sin(qx+a).$$

(i) If p is not equal to a and q is not zero; or if $p=a$ while q is not zero; or if p is not equal to a while q is zero; then the primitive is

$$y=(A+Bx)\,e^{ax}+e^{px}\,(E\cos qx+F\sin qx),$$

where A and B are arbitrary constants, and where E and F are definite constants given by the equations

$$\left.\begin{aligned}E\{(p-a)^2-q^2\}+2Fq\,(p-a)&=c\sin\alpha\\-2Eq\,(p-a)+F\{(p-a)^2-q^2\}&=c\cos\alpha\end{aligned}\right\}.$$

(ii) If $p=a$ and $q=0$, so that the equation is

$$\frac{d^2y}{dx^2}-2a\frac{dy}{dx}+a^2y=ce^{ax}\sin\alpha,$$

the primitive is

$$y=(A+Bx+\tfrac{1}{2}cx^2\sin\alpha)\,e^{ax},$$

where A and B are arbitrary constants.

III. Let $b<a^2$, and take $b=a^2-\mu^2$.

(i) When q is not zero, the primitive is

$$y=Ae^{(a+\mu)x}+Be^{(a-\mu)x}+e^{px}\,(G\cos qx+H\sin qx),$$

where A and B are arbitrary constants, and where G and H are definite constants given by the equations

$$\left.\begin{aligned}E\{(p-a)^2-\mu^2-q^2\}+2Fq\,(p-a)&=c\sin\alpha\\-2Eq\,(p-a)+F\{(p-a)^2-\mu^2-q^2\}&=c\cos\alpha\end{aligned}\right\},$$

the form being effective for all values of p.

(ii) When q is zero, the preceding form is effective unless $(p-a)^2=\mu^2$.

(iii) When q is zero and $(p-a)^2=\mu^2$, the equation may be taken in the form

$$\frac{d^2y}{dx^2}-2a\frac{dy}{dx}+(a^2-\mu^2)\,y=ce^{(a+\mu)x}+ke^{(a-\mu)x},$$

covering both the exceptional values of p. The primitive of the equation is

$$y=Ae^{(a+\mu)x}+Be^{(a-\mu)x}+\frac{x}{2\mu}\{ce^{(a+\mu)x}-ke^{(a-\mu)x}\},$$

where A and B are arbitrary constants.

IV. If V contain a power of x as a factor, so that we may write

$$V=x^m\,T,$$

then for the determination of the particular integral we may use the extended form (§ 35) of Leibnitz's theorem.

Thus

$$y=\frac{1}{f(D)}x^m\,T$$

$$=x^m\frac{1}{f(D)}T+mx^{m-1}\left\{\frac{d}{dD}\frac{1}{f(D)}\right\}T+\frac{m\,(m-1)}{2\,!}x^{m-2}\left\{\frac{d^2}{dD^2}\frac{1}{f(D)}\right\}T+\dots,$$

where the series must be carried to the $(m+1)^{\text{th}}$ term; each of these terms still leaves a quantity to be evaluated which may be effected by the methods applied to one of the preceding cases; if it may not, the quantity may be obtained by the next method, which is of universal application. The success of this general method depends solely on the solution of an equation (the solution being requisite to obtain the complementary function) and on the integration of resulting expressions.

V. Suppose that all the factors, which occur in V and can be dealt with by one or other of the foregoing methods, have been taken outside the operator and that the quantity remaining comes under none of these heads, so that we have to evaluate expressions of the form

$$\frac{1}{\psi(D)}\,U.$$

Let $\dfrac{1}{\psi(D)}$ be expressed in partial fractions, each having for its denominator a linear factor or a power of a linear factor of $\psi(D)$, the constant quantities occurring not being necessarily real; then the fractions will be of the form

$$\frac{A_n}{(D-\alpha)^n},$$

where n is an integer, A_n and α constants, and α a root of $\psi(z)=0$. Hence

$$\frac{1}{\psi(D)}\,U = \Sigma\,\frac{A_n}{(D-\alpha)^n}\,U$$

$$= \Sigma e^{\alpha x}\frac{A_n}{D^n}\,e^{-\alpha x}\,U$$

$$= \Sigma A_n e^{\alpha x}\iint\ldots\ldots e^{-\alpha x}\,U dx^n.$$

If imaginary quantities enter into any expression, the conjugate imaginary quantities will enter into some other; such a pair of expressions must in general be combined so as to leave no imaginary quantity in the explicit expression of the particular integral.

Ex. 17. Solve $\qquad (D^2-5D+6)\,y=\log x.$

We have $\qquad\dfrac{1}{D^2-5D+6}=\dfrac{1}{D-3}-\dfrac{1}{D-2}.$

Hence the particular integral is

$$\frac{1}{D-3}\log x - \frac{1}{D-2}\log x = e^{3x}\int e^{-3x}\log x\, dx - e^{2x}\int e^{-2x}\log x\, dx\, ;$$

and the complementary function is

$$Ae^{2x} + Be^{3x}.$$

Ex. 18. Let the right-hand side in the preceding example be $x\log x$ instead of $\log x$; then we may either integrate by parts or use the extension of Leibnitz's theorem. The latter gives

$$y = x\frac{1}{D-3}\log x - \frac{1}{(D-3)^2}\log x - x\frac{1}{D-2}\log x + \frac{1}{(D-2)^2}\log x$$

$$= xe^{3x}\int e^{-3x}\log x\, dx - e^{3x}\iint e^{-3x}\log x\, dx^2 - xe^{2x}\int e^{-2x}\log x\, dx + e^{2x}\iint e^{-2x}\log x\, dx^2.$$

Ex. 19. Solve

$$\frac{d^2y}{dx^2} + n^2y = U,$$

where U is a function of x. We have

$$y = \frac{1}{D^2+n^2}U$$

$$= \frac{1}{2in}\left\{\frac{1}{D-in}U - \frac{1}{D+in}U\right\}$$

$$= \frac{1}{2in}\left\{e^{inx}\int Ue^{-inx}dx - e^{-inx}\int Ue^{inx}dx\right\};$$

or, changing the variable under the sign of integration,

$$y = \frac{1}{2in}\int^x U_\xi\{e^{in(x-\xi)} - e^{-in(x-\xi)}\}d\xi$$

$$= \frac{1}{n}\int^x U_\xi \sin n(x-\xi)\, d\xi,$$

in which U_ξ is the same function of ξ as U is of x.

Note. There is another method of integrating this equation which proceeds on different lines. Multiply throughout by $\sin nx$; then

$$\frac{d}{dx}\left(\frac{dy}{dx}\sin nx - ny\cos nx\right) = U\sin nx,$$

and therefore

$$\frac{dy}{dx}\sin nx - ny\cos nx = -An + \int^x U_\xi \sin n\xi\, d\xi.$$

Similarly, multiplying by $\cos nx$ and writing the equation in the corresponding form, we find an integral

$$\frac{dy}{dx}\cos nx + ny\sin nx = Bn + \int^x U_\xi \cos n\xi\, d\xi.$$

Eliminating $\dfrac{dy}{dx}$ between these, we obtain

$$y = A \cos nx + B \sin nx + \frac{1}{n} \int^x U_\xi \sin n\,(x - \xi)\,d\xi,$$

agreeing with the former result.

Or we may proceed thus. Multiply throughout by e^{nxi}; then

$$\frac{d}{dx}\left(\frac{dy}{dx}\,e^{nxi} - iny\,e^{nxi}\right) = U e^{nxi},$$

so that

$$\frac{dy}{dx} - iny = A' e^{-nxi} + \int^x U_\xi e^{ni(\xi - x)}\,d\xi.$$

Similarly

$$\frac{dy}{dx} + iny = B' e^{nxi} + \int^x U_\xi e^{-ni(\xi - x)}\,d\xi.$$

Subtracting the one equation from the other and changing the constants of integration, we again obtain the original result.

Ex. 20. Solve
(i) $\dfrac{d^2 y}{dx^2_1} + n^2 y = x^2 \cos ax,$

when $n \gtrless a$ and when $n = a$;

(ii) $\dfrac{d^2 y}{dx^2} - n^2 y = U,$

where U is any function of x;

(iii) $\dfrac{d^2 y}{dx^2} - 2y = 4x^2 e^{x^2}.$

Ex. 21. By means of (iii) in Ex. 20, prove that

$$\sqrt{2}e^{x\sqrt{2}} \int^x_{-\frac{1}{\sqrt{2}}} x^2 e^{x^2 - x\sqrt{2}}\,dx - \sqrt{2}e^{-x\sqrt{2}} \int^x_{\frac{1}{\sqrt{2}}} x^2 e^{x^2 + x\sqrt{2}}\,dx = e^{x^2}.$$

47. Owing to the close similarity between the linear equation with constant coefficients and the homogeneous linear equation, the latter may be dealt with here; it may be written in the form

$$x^n \frac{d^n y}{dx^n} + A_1 x^{n-1} \frac{d^{n-1} y}{dx^{n-1}} + \ldots + A_{n-1} x \frac{dy}{dx} + A_n y = V,$$

where V is a function of x alone and may be a constant C. In the latter case, the particular integral is at once obtainable; it is evidently

$$\frac{1}{A_n}\,C.$$

If the operator $x \dfrac{d}{dx}$ be denoted by ϑ, then (§ 37)

$$x^m \frac{d^m}{dx^m} = \vartheta\,(\vartheta - 1)\,\ldots\ldots(\vartheta - m + 1);$$

and the differential equation may be written

$$F(\vartheta)\, y = V.$$

Consider the two points of the primitive separately. The complementary function is the primitive of

$$F(\vartheta)\, y = 0.$$

Now we have already seen that

$$F(\vartheta)\, x^p = F(p)\, x^p.$$

Hence, if p be so chosen that

$$F(p) = 0,$$

then x^p is a solution of the equation; and if $p_1, p_2, \ldots, p_n$, be the roots of $F(z) = 0$, the complementary function is

$$y = A_1 x^{p_1} + A_2 x^{p_2} + \ldots\ldots + A_n x^{p_n}.$$

The case of equal roots has been discussed already (§ 45); if two roots be imaginary and conjugate, say p_1 and p_2, so that

$$p_1 = \alpha + i\beta, \quad p_2 = \alpha - i\beta,$$

then the corresponding part of y will be

$$x^\alpha \{A_1' \cos(\beta \log x) + A_2' \sin(\beta \log x)\},$$

the arbitrary constants having been changed.

Ex. If the imaginary roots $\alpha \pm i\beta$ are of multiplicity r, the corresponding part of the complementary function will be

$$x^\alpha [\{A_1' + A_2' \log x + A_3'(\log x)^2 + \ldots + A_r'(\log x)^{r-1}\} \cos(\beta \log x)$$
$$+ \{B_1' + B_2' \log x + B_3'(\log x)^2 + \ldots + B_r'(\log x)^{r-1}\} \sin(\beta \log x)].$$

48. The particular integral is the value of

$$\frac{1}{F(\vartheta)}\, V.$$

The evaluation may be effected in two ways, which are really equivalent save for the difference in operators employed.

If V either be a power of x, or contain as a factor a power of x, say x^m, then

$$y = \frac{1}{F(\vartheta)}\, x^m T$$

$$= x^m \frac{1}{F(\vartheta + m)}\, T.$$

In the case when T is a constant, the evaluation is easy. If m be not a root of $F(z) = 0$, then we may expand $\{F(\vartheta + m)\}^{-1}$ in ascending powers of ϑ and neglect all but the first term, which is independent of ϑ and in fact gives

$$y = \frac{Cx^m}{F(m)}.$$

The same method (of expansion) will apply when T is a rational integral function of $\log x$; and since

$$\vartheta \log x = 1,$$

the expansion does not need to be carried beyond ϑ^n, where n is the index of the highest power of $\log x$ in T.

If however m be a root of multiplicity r for $F(z) = 0$, then

$$F(\vartheta + m) = \frac{\vartheta^r}{r!} F^{(r)}(m) + \frac{\vartheta^{r+1}}{(r+1)!} F^{(r+1)}(m) + \dots;$$

and we have to evaluate

$$y = \frac{1}{\dfrac{\vartheta^r}{r!} F^{(r)}(m) + \dots} T.$$

If T be a constant C, then since

$$\frac{1}{\vartheta} 1 = \log x,$$

the value of y is

$$\frac{C(\log x)^r}{F^{(r)}(m)}.$$

If it be a function of $\log x$ as before, the operator should be expanded in ascending powers of ϑ up to ϑ^n (ϑ^r being retained in the denominator), and the value of y will be given as the sum of a number of terms of the form

$$\frac{1}{\vartheta^r} (\log x)^s,$$

that is, of a number of terms of the form

$$\frac{s!}{(s+r)!} (\log x)^{s+r}.$$

A general expression can be given for the particular integral in the case when V takes none of these forms. Let $\dfrac{1}{F(\vartheta)}$ be expanded in partial fractions, and suppose some term to be

$$\frac{A}{\vartheta - \alpha}:$$

then y will be the sum of terms of the form

$$\frac{A}{\vartheta - \alpha} V,$$

which is equivalent to

$$A x^\alpha \frac{1}{\vartheta} V x^{-\alpha}, \text{ or } A x^\alpha \int V x^{-\alpha-1} dx.$$

Note. Another method of proceeding is to change the independent variable from x to z, where x is e^z; this changes ϑ into $\dfrac{d}{dz}$ or D, and all the methods of § 46 will now apply. It is easy to see that all the cases indicated for ϑ are strict analogues of cases indicated for D.

Ex. Solve

(i) $\quad x^2 \dfrac{d^2 y}{dx^2} - 4x \dfrac{dy}{dx} + 6y = x;$

(ii) $\quad x^2 \dfrac{d^2 y}{dx^2} - x \dfrac{dy}{dx} + 2y = x \log x;$

(iii) $\quad x^3 \dfrac{d^3 y}{dx^3} - x^2 \dfrac{d^2 y}{dx^2} + 2x \dfrac{dy}{dx} - 2y = x^3 + 3x;$

(iv) $\quad x^4 \dfrac{d^4 y}{dx^4} + 6x^3 \dfrac{d^3 y}{dx^3} + 9x^2 \dfrac{d^2 y}{dx^2} + 3x \dfrac{dy}{dx} + y = (1 + \log x)^2;$

(v) $\quad x^2 \dfrac{d^2 y}{dx^2} - 3x \dfrac{dy}{dx} + 4y = 2x^2;$

(vi) $\quad x^2 \dfrac{d^2 y}{dx^2} - 2y = x^2 + \dfrac{1}{x};$

(vii) $\quad x^2 \dfrac{d^2 y}{dx^2} - (2m-1) x \dfrac{dy}{dx} + (m^2 + n^2) y = n^2 x^m \log x.$

MISCELLANEOUS EXAMPLES.

1. If there be two linear equations of orders m and n $(n>m)$ satisfied by the same dependent variable, a third linear equation of order $n-m$ can without any integration be derived from the first two; and the equations of orders m and $n-m$ (when integrated) will suffice to furnish the integral of the equation of order n.   (Liouville.)

2. Solve the equations

$$(a) \quad \frac{d^2y}{dx^2} + \frac{2}{x}\frac{dy}{dx} = n^2 y;$$

$$(\beta) \quad \frac{d^2y}{dx^2} + \frac{2}{x}\frac{dy}{dx} = \left(n^2 + \frac{2}{x^2}\right) y;$$

$$(\gamma) \quad \frac{d^2y}{dx^2} + y = \sin x \sin 2x.$$

3. Prove that the solution of

$$(D+c)^n y = \cos ax$$

is $y = e^{-cx}(A_1 + A_2 x + \ldots + A_n x^{n-1}) + (c^2 + a^2)^{-\frac{n}{2}} \cos\left(ax - n \arctan \cot \frac{c}{a}\right).$

4. Obtain the general solution of the equation

$$\frac{d^2y}{dt^2} + \kappa \frac{dy}{dt} + n^2 y = U$$

in the form

$$y = e^{-\frac{1}{2}\kappa t}(A \cos n't + B \sin n't) + \frac{1}{n'}\int_0^t e^{-\frac{1}{2}\kappa(t-t')} \sin n'(t-t')\, U'dt',$$

where U' is the same function of t' as U is of t, and n' is given by

$$n'^2 = n^2 - \tfrac{1}{4}\kappa^2.$$

(See also the results in Ex. 16, § 46.)

5. Solve the equations

$$(i) \quad x^4\frac{d^4y}{dx^4} + 6x^3\frac{d^3y}{dx^3} + 4x^2\frac{d^2y}{dx^2} - 2x\frac{dy}{dx} - 4y = x^2 + 2\cos(\log x);$$

$$(ii) \quad \frac{d^4y}{dx^4} + 12\frac{d^2y}{dx^2} + 12y = 16x^4 e^{x^2};$$

$$(iii) \quad \frac{d^4y}{dx^4} + 32\frac{dy}{dx} + 48y = xe^{-2x} + e^{2x}\cos 2^{\frac{3}{2}}x;$$

$$(iv) \quad \frac{d^3y}{dx^3} - 3\frac{d^2y}{dx^2} + 4\frac{dy}{dx} - 2y = e^x + \cos x;$$

$$(v) \quad \frac{d^4y}{dx^4} + \frac{d^2y}{dx^2} + y = ax^2 + be^{-x}\sin 2x;$$

$$(vi) \quad \left(\frac{d}{dx} + 1\right)^3 y = e^{-x} + x^2 + x^{-1}.$$

6. Obtain the complementary function of the equation

$$\frac{d^{2n}y}{dx^{2n}} - a^{2n}y = f(x)$$

in the form

$$y = Ce^{ax} + De^{-ax} + \sum_{r=1}^{r=n-1} e^{ax \cos \frac{r\pi}{n}} \left\{ A_r \cos \left(ax \sin \frac{r\pi}{n} \right) + B_r \sin \left(ax \sin \frac{r\pi}{n} \right) \right\};$$

and shew that the part of the particular integral, corresponding to the typical terms under the summation sign, is

$$\frac{1}{na^{2n-1}} \int^x e^{a(x-\xi)\cos\frac{r\pi}{n}} \cos \left\{ \frac{r\pi}{n} + a(x-\xi)\sin\frac{r\pi}{n} \right\} f(\xi)\, d\xi.$$

7. Prove that the solution of the equation

$$\left(\cos \frac{d}{dx} \right) y = \cos x$$

is
$$y = \sum_{n=0}^{n=\infty} \left\{ A_n e^{(n+\frac{1}{2})\pi x} + B_n e^{-(n+\frac{1}{2})\pi x} \right\} + \frac{2\cos x}{e + e^{-1}}.$$

8. Prove that

$$f\left(\frac{d}{dx} - \frac{p}{x+a} \right)(x+a)^p\, \phi(x) = (x+a)^p f\left(\frac{d}{dx} \right) \phi(x).$$

9. Prove that

(i) $2^{2n+1} D^n x^{n+\frac{1}{2}} D^{n+1} e^{x^{\frac{1}{2}}} = e^{x^{\frac{1}{2}}}$;

(ii) $D^m x^{m+r} D^r x^{-m} D^{n-r} \phi(x) = x^r D^{m+n} \phi(x)$;

(iii) $D^n (\vartheta - n)^r y = \vartheta^r D^n y$.

10. Prove that, if ϑ denote $x\dfrac{d}{dx}$,

$$\frac{(\vartheta - n)!}{\vartheta!} 0 = A_0 + A_1 x + \ldots + A_{n-1} x^{n-1},$$

where $A_0, A_1, \ldots, A_{n-1}$, are arbitrary constants.

11. If P, Q, R, be commutative symbols of operation, the solution of $P \cdot Q \cdot R \cdot u = 0$ is

$$u = P^{-1} \cdot 0 + Q^{-1} \cdot 0 + R^{-1} \cdot 0.$$

Apply this result to obtain the complementary function of the equation

$$f(D) y = 0,$$

(i) when the equation $f(\theta) = 0$ has no equal roots, (ii) when the equation $f(\theta) = 0$ has some equal roots.

CHAPTER IV

MISCELLANEOUS METHODS

49. BEFORE we discuss the linear equation of the second order with variable coefficients, there are several miscellaneous methods which it is advisable to consider; they apply to systems of equations which admit either of complete solution or of approach to a solution in the shape of a first integral. It is to be understood that the equations hereafter given are typical and not merely isolated equations which can be integrated; it is frequently possible to include others under some one of the following classes, by means of well-selected substitutions for either the dependent or the independent variable. Such substitutions point out however the limits within which the methods are for the most part effective, so that it must be borne in mind that the methods are not of general application to all linear equations of the second order.

50. The simplest case of all is that in which the equation is of the form

$$\frac{d^n y}{dx^n} = X,$$

where X is a function of x alone. It is immediately integrable; the result of integration is

$$\frac{d^{n-1}y}{dx^{n-1}} = \int X \, dx + A_1,$$

A_1 denoting an arbitrary constant. A second integration gives

$$\frac{d^{n-2}y}{dx^{n-2}} = \int dx \int X \, dx + A_1 x + A_2,$$

A_2 being another arbitrary constant. Proceeding in this way, we shall have after n integrations as the general solution

$$y = \int\!\!\int \ldots \ldots X (dx)^n + B_1 x^{n-1} + B_2 x^{n-2} + \ldots \ldots + B_{n-1} x + B_n,$$

in which B_r replaces $\dfrac{A_r}{(n-r)!}$ and is therefore an arbitrary constant.

Ex. Shew that the Particular Integral can be expressed in the form

$$\frac{1}{(n-1)!}\int_0^x T(x-t)^{n-1}dt,$$

where T is the same function of t as X is of x.   (Jordan.)

51. Another very simple equation to be considered is

$$\frac{d^n y}{dx^n} = Y,$$

in which Y is a function of y alone; but in general it is integrable in simple terms, only when n is either 1 or 2.

In the case when n is 2, let the equation be multiplied by $2\dfrac{dy}{dx}$; then each side may be integrated, and we have

$$\left(\frac{dy}{dx}\right)^2 = 2\int Y\,dy + A$$

$$= \psi(y) + A,$$

suppose. The variables can be separated; the general solution of the equation

$$\frac{d^2 y}{dx^2} = Y$$

is

$$\int \frac{dy}{\{\psi(y) + A\}^{\frac{1}{2}}} = x + B.$$

Ex. Solve $\dfrac{d^2 y}{dx^2} + a^2 y = 0.$

A first integral is

$$\left(\frac{dy}{dx}\right)^2 + a^2 y^2 = A = a^2 c^2,$$

where c is an arbitrary constant; separation of the variables gives

$$\frac{dy}{(c^2 - y^2)^{\frac{1}{2}}} = a\,dx,$$

and therefore

$$\text{arc} \sin \frac{y}{c} = ax + \alpha,$$

or $y = c \sin(ax + \alpha).$

52. Any differential equation which merely expresses a relation between two differential coefficients, whose orders differ by either 1 or 2, admits of solution. As a type of the differential equation, when the orders differ by 1, we may take

$$\frac{d^n y}{dx^n} = F\left(\frac{d^{n-1}y}{dx^{n-1}}\right).$$

Let $\dfrac{d^{n-1}y}{dx^{n-1}} = Y$; then the equation becomes

$$\frac{dY}{dx} = F(Y),$$

the integral of which is

$$\psi(Y) = \int \frac{dY}{F(Y)} = x + A.$$

Suppose this equation can be solved for Y and that the solution is

$$Y = \phi(x + A),$$

that is,

$$\frac{d^{n-1}y}{dx^{n-1}} = \phi(x + A).$$

Then this is one of the cases already discussed (§ 50), and the general integral can be obtained.

Or, after obtaining the equation $\psi(Y) = x + A$, we may proceed thus. Since

$$\frac{d}{dx}\left(\frac{d^{n-2}y}{dx^{n-2}}\right) = Y,$$

we have

$$\frac{d^{n-2}y}{dx^{n-2}} = \int Y dx = \int \frac{Y dY}{F(Y)}.$$

Similarly

$$\frac{d^{n-3}y}{dx^{n-3}} = \int dx \int \frac{Y dY}{F(Y)}$$

$$= \int \frac{dY}{F(Y)} \int \frac{Y dY}{F(Y)},$$

and so on, until

$$y = \int \frac{dY}{F(Y)} \int \frac{dY}{F(Y)} \cdots\cdots \int \frac{Y dY}{F(Y)},$$

an arbitrary constant being introduced after every one of the integrations, which must be taken in order from right to left. Then we have two equations between x, y, Y, from which Y is to be eliminated. The eliminant will be the primitive.

It is evident that the equation

$$f\left(\frac{d^n y}{dx^n}, \frac{d^{n-1} y}{dx^{n-1}}\right) = 0$$

can be solved by this method.

Ex. 1. Solve $\qquad a\dfrac{d^3 y}{dx^3} = \dfrac{d^2 y}{dx^2}.$

Let $\qquad\qquad\qquad \dfrac{d^2 y}{dx^2} = \xi;$

then $\qquad\qquad\qquad a\dfrac{d\xi}{dx} = \xi,$

of which the integral is

$$\xi = A' e^{\frac{x}{a}}.$$

Therefore $\qquad\qquad y = A e^{\frac{x}{a}} + Bx + C,$

where A, B, C, are arbitrary constants.

Ex. 2. Integrate

$$\text{(i)}\qquad a\frac{d^2 y}{dx^2} = \left\{1 + \left(\frac{dy}{dx}\right)^2\right\}^{\frac{1}{2}};$$

$$\text{(ii)}\quad -a\frac{d^2 y}{dx^2} = \left\{1 + \left(\frac{dy}{dx}\right)^2\right\}^{\frac{3}{2}};$$

$$\text{(iii)}\quad a^3 \frac{d^3 y}{dx^3}\frac{d^2 y}{dx^2} = \left\{1 + c^2\left(\frac{d^2 y}{dx^2}\right)^2\right\}^{\frac{1}{2}}.$$

53. As a type of the differential equations which connect differential coefficients, whose orders differ by 2, we may write

$$\frac{d^n y}{dx^n} = f\left(\frac{d^{n-2} y}{dx^{n-2}}\right).$$

Let $\dfrac{d^{n-2} y}{dx^{n-2}} = z$; then the equation becomes

$$\frac{d^2 z}{dx^2} = f(z),$$

the solution of which has been obtained in the form

$$\int \frac{dz}{\{A + 2\int f(z)\,dz\}^{\frac{1}{2}}} = x + B.$$

If, after the integrations have been carried out, the equation can be resolved for z in terms of x, say

$$z = \theta(x),$$

where the function $\theta(x)$ will involve the constants A and B, then $n - 2$ direct integrations will furnish the primitive. But if it should be impossible to effect this resolution, then we have

$$\frac{dz}{dx} = \{A + 2\int f(z)\,dz\}^{\frac{1}{2}}.$$

Hence $$\frac{d^{n-3}y}{dx^{n-3}} = \int z\,dx = \int \frac{z\,dz}{\{A + 2\int f(z)\,dz\}^{\frac{1}{2}}}.$$

$$\frac{d^{n-4}y}{dx^{n-4}} = \int \frac{dz}{\{A + 2\int f(z)\,dz\}^{\frac{1}{2}}} \int \frac{z\,dz}{\{A + 2\int f(z)\,dz\}^{\frac{1}{2}}},$$

and so on. Ultimately we shall obtain y as a function of z; and the primitive will be the eliminant with regard to z of the equation between y and z and the equation between x and z.

Ex. 1. Solve $$a^2 \frac{d^4y}{dx^4} = \frac{d^2y}{dx^2}.$$

When we write z for $\frac{d^2y}{dx^2}$, the equation becomes

$$a^2 \frac{d^2z}{dx^2} = z,$$

so that $$z = c_1 e^{\frac{x}{a}} + c_2 e^{-\frac{x}{a}},$$

and therefore $$y = A e^{\frac{x}{a}} + B e^{-\frac{x}{a}} + Cx + D,$$

in which A and B replace $c_1 a^2$ and $c_2 a^2$ respectively.

Ex. 2. Solve

$$\text{(i)} \quad x^2 \frac{d^4y}{dx^4} = \lambda \frac{d^2y}{dx^2};$$

$$\text{(ii)} \quad c^3 \frac{d^3y}{dx^3} = \frac{dy}{dx} \left\{ 1 + \left(\frac{dy}{dx} \right)^2 \right\}^2.$$

54. In some particular cases, the general differential equation of the second order can, by substitution, be depressed so as to become a differential equation of the first order. Such cases occur when one of the variables is explicitly absent from the equation.

First, consider an equation in which x does not occur, so that it may be written in the form

$$\psi \left(y, \frac{dy}{dx}, \frac{d^2y}{dx^2} \right) = 0.$$

Let $\frac{dy}{dx} = p$, and then $\frac{d^2y}{dx^2} = p\frac{dp}{dy}$; the equation thus becomes

$$\psi \left(y, p, p\frac{dp}{dy} \right) = 0,$$

a differential equation of the first order to find p in terms of y. Let the solution be

$$p = f(y)$$

in which $f(y)$ will include an arbitrary constant. Then the variables are separable, since we may write

$$\frac{dy}{f(y)} = dx;$$

and integration of this equation will lead to the primitive.

Next, consider an equation in which y does not occur, so that it may be written in the form

$$\phi \left(x, \frac{dy}{dx}, \frac{d^2y}{dx^2} \right) = 0.$$

Let $\frac{dy}{dx} = p$; then $\frac{d^2y}{dx^2} = \frac{dp}{dx}$; the equation is transformed into

$$\phi \left(x, p, \frac{dp}{dx} \right) = 0,$$

an equation of the first order to find p in terms of x. Let the solution be

$$p = F(x),$$

where F includes an arbitrary constant. Integrating this, we obtain as the primitive

$$y = A + \int F(x)\, dx.$$

Note. The foregoing explanations relate to an equation of the second order; but they apply to similar equations of any order. When either of the variables is explicitly absent from an equation, whatever the order of the equation may be, the same substitution can be applied; the order of the transformed equation is less by unity than the order of the original equation.

Ex. 1. Solve
$$2(2a-y)\frac{d^2y}{dx^2}=1+\left(\frac{dy}{dx}\right)^2.$$

When we write $\frac{dy}{dx}=p$, the equation is transformed into
$$\frac{2p\frac{dp}{dy}}{1+p^2}=\frac{1}{2a-y},$$

the integral of which is
$$(1+p^2)(2a-y)=\mu,$$

where μ is an arbitrary constant; the primitive is given by the evaluation of
$$\int dy\left\{\frac{2a-y}{\mu-2a+y}\right\}^{\frac{1}{2}}=x+B.$$

Ex. 2. Solve
$$a^2\frac{d^2y}{dx^2}=2x\left\{1+\left(\frac{dy}{dx}\right)^2\right\}^{\frac{3}{2}}.$$

The substitution $\frac{dy}{dx}=p$ transforms the equation into
$$\frac{a^2}{(1+p^2)^{\frac{3}{2}}}\frac{dp}{dx}=2x;$$

on integration, this gives
$$\frac{a^2p}{(1+p^2)^{\frac{1}{2}}}=x^2+A,$$

and therefore
$$p^2=\frac{(x^2+A)^2}{a^4-(x^2+A)^2},$$

so that the primitive is
$$y=\int p\,dx=B+\int\frac{x^2+A}{\{a^4-(x^2+A)^2\}^{\frac{1}{2}}}\,dx.$$

Ex. 3. Integrate

(i) $1+\left(\frac{dy}{dx}\right)^2+x\frac{dy}{dx}\frac{d^2y}{dx^2}=a\frac{d^2y}{dx^2}\left\{1+\left(\frac{dy}{dx}\right)^2\right\}^{\frac{1}{2}};$

(ii) $ab\frac{d^2y}{dx^2}=\left\{y^2+a^2\left(\frac{dy}{dx}\right)^2\right\}^{\frac{1}{2}};$

(iii) $y\frac{d^2y}{dx^2}+\left\{\left(\frac{dy}{dx}\right)^2+a^2\left(\frac{d^2y}{dx^2}\right)^2\right\}^{\frac{1}{2}}=\left(\frac{dy}{dx}\right)^2;$

(iv)　$(1+x^2)\dfrac{d^2y}{dx^2}+1+\left(\dfrac{dy}{dx}\right)^2=0$;

(v)　$\dfrac{dy}{dx}+a^2\left(\dfrac{d^2y}{dx^2}\right)^2=2ax\dfrac{d^2y}{dx^2}$;

(vi)　$(1-x^2)\dfrac{d^2y}{dx^2}-x\dfrac{dy}{dx}=2$;

(vii)　$y\dfrac{d^2y}{dx^2}-\left(\dfrac{dy}{dx}\right)^2=y^2\log y$;

(viii)　$y(1-\log y)\dfrac{d^2y}{dx^2}+(1+\log y)\left(\dfrac{dy}{dx}\right)^2=0$.

Homogeneous Equations.

55. There are certain classes of differential equations in which a kind of homogeneity subsists; and the solution of these can, by suitable transformations, be made to depend upon that of equations of lower orders. The homogeneity is constituted as follows: if y be considered to be of n dimensions, while x is of one dimension, then $\dfrac{dy}{dx}$, since it is the limit of $\dfrac{\Delta y}{\Delta x}$, is of $n-1$ dimensions; $\dfrac{d^2y}{dx^2}$, being the limit of $\dfrac{\Delta p}{\Delta x}$, is of $n-2$ dimensions, and so on; and the equation is said to be homogeneous when, if these dimensions be assigned to the corresponding quantities, all the terms in the equation are of the same dimensions. The simplest case is, of course, that in which n is unity.

First, let n be unity so that x and y may both be considered of one dimension. Let $y=xz$ and $x=e^\theta$; then

$$\frac{dy}{dx}=\frac{dz}{d\theta}+z,$$

$$\frac{d^2y}{dx^2}=\left(\frac{d^2z}{d\theta^2}+\frac{dz}{d\theta}\right)e^{-\theta},$$

and so on; the resulting differential equation becomes one between z and θ. Now it will be noticed that the coefficient of θ in the index of the exponential, wherever it occurs in any differential coefficient, is the number representing the dimensions of that differential coefficient; and therefore, when substitution takes place in the differential equation, supposed homogeneous, the index of θ in the exponential will be the same for each term of the equation, and this exponential will therefore be a factor which may be removed. The new independent variable θ will no longer occur explicitly in the equation, which will therefore be of the class already discussed in § 54 and can have its order depressed.

Ex. 1. Solve
$$x^2\frac{d^2y}{dx^2}=\left\{mx^2\left(\frac{dy}{dx}\right)^2+ny^2\right\}^{\frac{1}{2}}.$$

Making the substitutions of § 55, we have
$$\frac{d^2z}{d\theta^2}+\frac{dz}{d\theta}=\left\{m\left(\frac{dz}{d\theta}+z\right)^2+nz^2\right\}^{\frac{1}{2}}.$$

When we write $\dfrac{dz}{d\theta}=v$, the equation becomes
$$\frac{dv}{d\theta}+v=\{m\,(v+z)^2+nz^2\}^{\frac{1}{2}};$$

or, if $v=zs$,
$$s^2+\frac{ds}{d\theta}+s=\{m\,(1+s)^2+n\}^{\frac{1}{2}},$$

and therefore
$$\frac{ds}{\{m\,(1+s)^2+n\}^{\frac{1}{2}}-s^2-s}=a\theta.$$

The variables are separated; and the equation can be integrated.

Ex. 2. Solve

(i) $n\dfrac{d^2y}{dx^2}(x^2+y^2)^{\frac{1}{2}}=\left\{1+\left(\dfrac{dy}{dx}\right)^2\right\}^{\frac{3}{2}};$

(ii) $nx^3\dfrac{d^2y}{dx^2}=\left(y-x\dfrac{dy}{dx}\right)^2;$

(iii) $x^4\dfrac{d^2y}{dx^2}=\left(y-x\dfrac{dy}{dx}\right)^3.$

Passing now to the case in which homogeneity is constituted on the assumption of n dimensions for y, we write
$$x=e^\theta,\quad y=x^nz=ze^{n\theta}.$$
We now have
$$\frac{dy}{dx}=\left(\frac{dz}{d\theta}+nz\right)e^{(n-1)\,\theta},$$
$$\frac{d^2y}{dx^2}=\left\{\frac{d^2z}{d\theta^2}+(2n-1)\frac{dz}{d\theta}+n\,(n-1)z\right\}e^{(n-2)\,\theta},$$

and so on. It is obvious that the coefficient of θ in the index of the exponential, which occurs in the expression of every differential coefficient, exactly measures the dimensions of that differential coefficient. As in the former case, the exponential will disappear when substitution takes place; the differential equation, having been thus transformed into one from which the independent variable is explicitly absent, can have its order lowered by unity.

Ex. 1. Solve $\qquad x^4\dfrac{d^2y}{dx^2}=(x^3+2xy)\dfrac{dy}{dx}-4y^2.$

This is homogeneous if y be considered to be of two dimensions while x is of one. Hence we substitute

$$x=e^\theta,\quad y=x^2z=ze^{2\theta};$$

and the equation becomes

$$\frac{d^2z}{d\theta^2}+3\frac{dz}{d\theta}+2z=(1+2z)\left(\frac{dz}{d\theta}+2z\right)-4z^2,$$

or $\qquad\qquad \dfrac{d^2z}{d\theta^2}+2\,(1-z)\dfrac{dz}{d\theta}=0.$

A first integral is given by

$$\frac{dz}{d\theta}-(1-z)^2=A,$$

in this, the variables can be separated in the form

$$\frac{dz}{A+(1-z)^2}=d\theta,$$

the integral of which will vary (being either an inverse circular function or a logarithm) according to the sign of A.

Ex. 2. Solve

(i) $\quad x^4\dfrac{d^2y}{dx^2}-x^3\dfrac{dy}{dx}=x^2\left(\dfrac{dy}{dx}\right)^2-4y^2;$

(ii) $\quad x\dfrac{d^2y}{dx^2}+2\dfrac{dy}{dx}=x^2\left(\dfrac{dy}{dx}\right)^2-y^2;$

(iii) $\quad \dfrac{1}{x^2}\dfrac{d^2y}{dx^2}+4\dfrac{y^2}{x^2}-6\dfrac{y}{x^4}=\left(\dfrac{dy}{dx}\right)^2.$

A particular set of cases arises when n is made infinite; all the quantities $y,\dfrac{dy}{dx},\ldots\ldots$ have then the same dimensions. The simplest method of solution is to adopt the substitution

$$y=e^{\int u\,dx};$$

the resulting equation between u and x is of an order lower by unity than the given equation.

Ex. 3. Solve

$$\text{(i)}\quad ay\frac{d^2y}{dx^2}+b\left(\frac{dy}{dx}\right)^2=y\frac{dy}{dx}(c^2+x^2)^{-\frac{1}{2}};$$

$$\text{(ii)}\quad xy\frac{d^2y}{dx^2}-x\left(\frac{dy}{dx}\right)^2=y\frac{dy}{dx}+bx\left(\frac{dy}{dx}\right)^2(a^2-x^2)^{-\frac{1}{2}}.$$

EXACT DIFFERENTIAL EQUATIONS.

56. A differential equation of the form

$$f\left(\frac{d^ny}{dx^n},\,\frac{d^{n-1}y}{dx^{n-1}},\,\ldots\ldots,\,\frac{dy}{dx},\,y,\,x\right)=0$$

is said to be exact when, on representing the left-hand member by V, the expression $V\,dx$ is the exact differential of some function U, which is necessarily of the form

$$f_1\left(\frac{d^{n-1}y}{dx^{n-1}},\,\ldots\ldots,\,\frac{dy}{dx},\,y,\,x\right).$$

Consider first a linear exact differential equation, which may be represented by

$$P_n\frac{d^ny}{dx^n}+P_{n-1}\frac{d^{n-1}y}{dx^{n-1}}+\ldots\ldots+P_1\frac{dy}{dx}+P_0y=P,$$

where all the coefficients are functions of x. An equation of this form will not in general be an exact differential equation; but we proceed to shew that, if a certain relation be satisfied by these quantities P, the equation can be integrated once.

Indicating for convenience differentiation with regard to x by means of dashes, we have on direct integration

$$\int P_0y\,dx \;=\; \int P_0y\,dx,$$

$$\int P_1\frac{dy}{dx}\,dx = -\int P_1{'}y\,dx \;+P_1y,$$

$$\int P_2\frac{d^2y}{dx^2}\,dx = \;\int P_2{''}y\,dx \;-P_2{'}y+P_2y',$$

$$\int P_3\frac{d^3y}{dx^3}\,dx = -\int P_3{'''}y\,dx+P_3{''}y-P_3{'}y'+P_3y'',$$

$\ldots\ldots\ldots\ldots\ldots\ldots\ldots\ldots\ldots\ldots\ldots\ldots\ldots\ldots\ldots\ldots\ldots\ldots\ldots\ldots\ldots\ldots$

and therefore

$$\int P dx = \int (P_0 - P_1' + P_2'' - P_3''' + \ldots\ldots) y \, dx$$
$$+ (P_1 - P_2' + P_3'' - \ldots\ldots) y$$
$$+ (P_2 - P_3' + P_4'' - \ldots\ldots) y'$$
$$+ (P_3 - P_4' + P_5'' - \ldots\ldots) y'' + \ldots\ldots$$
$$= \int Q_0 y \, dx + Q_1 y + Q_2 y' + \ldots\ldots + Q_n \frac{d^{n-1} y}{dx^{n-1}},$$

where the law of formation of the successive coefficients $Q_0, Q_1, Q_2, \ldots$ is the same for all and, in particular,

$$Q_n = P_n,$$
$$Q_{n-1} = P_{n-1} - P_n'.$$

Now the condition of integrability evidently is that there shall be no term remaining which involves an integral of y; and so the *necessary and sufficient condition* is that

$$Q_0 = 0,$$

that is,

$$P_0 - \frac{dP_1}{dx} + \frac{d^2 P_2}{dx^2} - \ldots\ldots + (-1)^n \frac{d^n P_n}{dx^n} = 0.$$

When this condition is satisfied, the first integral of the equation is

$$Q_n \frac{d^{n-1} y}{dx^{n-1}} + Q_{n-1} \frac{d^{n-2} y}{dx^{n-2}} + \ldots\ldots + Q_1 y = \int P dx + A_1,$$

where A_1 is an arbitrary constant.

If now the coefficients Q satisfy the corresponding condition

$$Q_1 - \frac{dQ_2}{dx} + \frac{d^2 Q_3}{dx^2} - \ldots\ldots + (-1)^{n-1} \frac{d^{n-1} Q_n}{dx^{n-1}} = 0,$$

the equation is again integrable; and the process can be continued so long as the coefficients of each successive equation thus derived satisfy the condition of integrability.

Ex. 1. Express the successive conditions in terms of the coefficients P only.

Ex. 2. The equation

$$(ax^2 - bx) \frac{d^3 y}{dx^3} + (cx - e) \frac{d^2 y}{dx^2} + x \frac{dy}{dx} + y = x$$

is an exact equation; for we have

$$P_0 = 1, \quad P_1' = 1, \quad P_2'' = 0, \quad P_3''' = 0;$$

and so the condition is satisfied. Integrating each side, we have

$$(ax^2 - bx)\frac{d^2y}{dx^2} - \{(2a-c)\,x + e - b\}\,\frac{dy}{dx} + (2a - c + x)\,y = \tfrac{1}{2}x^2 + A.$$

In practice, it is sometimes easy to see that a given equation is integrable. In many cases the quantities P are either of the form ax^m or sums of expressions of this form; and $x^m \dfrac{d^n y}{dx^n}$ is a perfect differential coefficient, if m be less than n; for integrating it by parts we have

$$x^m \frac{d^{n-1}y}{dx^{n-1}} - m x^{m-1} \frac{d^{n-2}y}{dx^{n-2}} + m\,(m-1)\,x^{m-2} \frac{d^{n-3}y}{dx^{n-3}} - \ldots + (-1)^m\,m\,!\,\frac{d^{n-m-1}y}{dx^{n-m-1}}.$$

If $n = m + 1$, the last term is $(-1)^m m\,!\,y$.

When we apply this lemma to the present example, the terms involving $\dfrac{d^3y}{dx^3}$, $\dfrac{d^2y}{dx^2}$, are seen to be perfect differential coefficients; and $x\dfrac{dy}{dx} + y$ is $\dfrac{d}{dx}(xy)$; so that the left-hand side is a perfect differential coefficient and the equation is therefore exact.

Ex. 3. Prove that the equation in Ex. 2 cannot be further integrated by the foregoing method.

Ex. 4. Obtain a first integral of the equations

$$\text{(i)}\quad x\frac{d^3y}{dx^3} + (x^2 - 3)\frac{d^2y}{dx^2} + 4x\frac{dy}{dx} + 2y = 0;$$

$$\text{(ii)}\quad 2x + y^2 + 2xy\frac{dy}{dx} + x\frac{d^2y}{dx^2} + x^2\frac{d^3y}{dx^3} = \frac{dy}{dx};$$

and shew that the equation

$$x^2\frac{d^3y}{dx^3} + 4x\frac{d^2y}{dx^2} + (x^2 + 2)\frac{dy}{dx} + 3xy = 2$$

becomes integrable on being multiplied by some power of x. Obtain its integral.

57. The method, used for integrating exact equations which are not linear, may be illustrated by considering an example.

Ex. 1. Solve

$$y + 3x\frac{dy}{dx} + 2y\left(\frac{dy}{dx}\right)^3 + \left(x^2 + 2y^2\frac{dy}{dx}\right)\frac{d^2y}{dx^2} = 0.$$

On the supposition that this is an exact differential equation, we may write

$$dU = (y + 3xp + 2yp^3)\,dx + (x^2 + 2y^2p)\,dp,$$

where p stands for $\dfrac{dy}{dx}$. Let U_1 denote what would be the value of U if p alone were variable, so that

$$U_1 = x^2p + y^2p^2.$$

Let the restriction now be removed, so that

$$dU_1 = (2xp + 2yp^3)\, dx + (x^2 + 2y^2 p)\, dp,$$

and therefore

$$dU - dU_1 = (y + xp)\, dx = d\,(xy),$$

which gives on integration

$$U - U_1 = xy + A,$$

that is,

$$U = x^2 \frac{dy}{dx} + y^2 \left(\frac{dy}{dx}\right)^2 + xy + A\,;$$

and therefore the first integral is

$$x^2 \frac{dy}{dx} + y^2 \left(\frac{dy}{dx}\right)^2 + xy = C.$$

The preceding method will be seen to lead to the following general rule for the integration of an exact differential equation of the nth order. The equation, being derivable from one of order $n-1$ by direct differentiation, will contain $\dfrac{d^n y}{dx^n}$ only in the first degree; if this condition be not satisfied, the equation is not exact.

Let the equation be written in the form $V = 0$, and integrate $V\,dx$ as if $\dfrac{d^{n-1}y}{dx^{n-1}}$ were the only variable occurring in V and $\dfrac{d^n y}{dx^n}$ its differential coefficient; let the result be U_1. Then $V\,dx - dU_1$ involves differential coefficients of y of the order $n-1$ at the utmost; as it is an exact differential, the highest differential coefficient of y which occurs can enter only in the first degree. Repeating the process as often as necessary, we shall ultimately have

$$V\,dx - dU_1 - dU_2 - \ldots = 0.$$

Then a first integral of the given equation is

$$U_1 + U_2 + \ldots = C.$$

Ex. 2. Obtain a first integral of the equations

$$\text{(i)} \quad \frac{dy}{dx}\frac{d^2 y}{dx^2} - vx^2 \frac{dy}{dx} = xy^2\,;$$

$$\text{(ii)} \quad x^2 \frac{d^3 y}{dx^3} + x \frac{d^2 y}{dx^2} + (2xy - 1)\frac{dy}{dx} + y^2 = 0.$$

Ex. 3. Shew that the equation

$$\frac{d^2 y}{dx^2} + \frac{a^2 y}{(y^2 + x^2)^2} = 0$$

becomes integrable on multiplication by the factor $2x^2\dfrac{dy}{dx} - 2xy$. Hence deduce a first integral and the primitive.

Ex. 4. Obtain a first integral of the equation

$$\frac{d^2y}{dx^2} + \frac{\alpha y}{(\beta y^2 + \gamma + 2\delta x + \epsilon x^2)^2} = 0$$

having given that there is an integrating factor of the form $X_1\dfrac{dy}{dx} + X_2 y$.

(Euler.)

Linear Equation of the Second Order.

58. We shall here prove some of the leading properties of the linear equation of the second order; but the present investigation will not for the most part anticipate the discussion of the general linear equation, for the properties here established belong solely to the equation of the second order.

The general form of the equation is

$$\frac{d^2y}{dx^2} + P\frac{dy}{dx} + Qy = R,$$

in which $P, Q,$ and $R,$ are functions of x; they may in special cases be merely constant quantities.

Substitute in the equation for y a value vw, where v and w are both functions of x; as yet, the only limitation on them is that their product must be equal to y. We then have

$$w\frac{d^2v}{dx^2} + \left(2\frac{dw}{dx} + Pw\right)\frac{dv}{dx} + \left(\frac{d^2w}{dx^2} + P\frac{dw}{dx} + Qw\right)v = R.$$

As we may choose a relation arbitrarily between v and w or make either of them satisfy some condition, we will suppose it possible to determine w so that the coefficient of v may vanish, that is,

$$\frac{d^2w}{dx^2} + P\frac{dw}{dx} + Qw = 0,$$

which, it will be noticed, is the same as the original equation with the right-hand side equated to zero. The quantity w being now considered known, the modified equation becomes

$$\frac{d^2v}{dx^2} + \left(\frac{2}{w}\frac{dw}{dx} + P\right)\frac{dv}{dx} = \frac{R}{w},$$

so that

$$w^2\frac{dv}{dx}e^{\int Pdx} = A + \int wRe^{\int Pdx}\,dx;$$

and therefore

$$v = B + A\int\frac{dx}{w^2}e^{-\int Pdx} + \int\frac{dx}{w^2}e^{-\int Pdx}\int wRe^{\int Pdx}\,dx.$$

It therefore follows that, *if any solution whatever of the original equation with the right-hand side equated to zero can be found, the complete primitive of the original equation in its general form can also be found.* The problem of deducing this complete primitive is therefore resolved into that of finding some single solution of the simpler equation. This, in the most general case of P and Q unrestricted to particular functions of x, has not yet been effected; but in special instances it is possible to determine such a solution as is desired, sometimes by inspection, sometimes by means of a converging series, sometimes by means of a definite integral; but in the two latter cases (which are usually closely connected) the explicit evaluation of the form obtained for v is difficult or impossible, though this form (§ 5) still remains the solution.

Ex. 1. Solve

$$\frac{d^2y}{dx^2} - x^2\frac{dy}{dx} + xy = x^{m+1}.$$

A particular solution of

$$\frac{d^2y}{dx^2} - x^2\frac{dy}{dx} + xy = 0$$

is evidently $y = x$. Writing $y = xv$ in the original equation, we get

$$x\frac{d^2v}{dx^2} + 2\frac{dv}{dx} - x^2\left(x\frac{dv}{dx} + v\right) + x^2v = x^{m+1},$$

or

$$\frac{d^2v}{dx^2} + \left(\frac{2}{x} - x^2\right)\frac{dv}{dx} = x^m.$$

Hence

$$\frac{dv}{dx}x^2e^{-\frac{1}{3}x^3} = A + \int x^{m+2}e^{-\frac{1}{3}x^3}\,dx;$$

and therefore

$$v = B + \int \frac{A\,dx}{x^2} e^{\frac{1}{3}x^3} + \int \frac{dx}{x^2} e^{\frac{1}{3}x^3} \int x^{m+2} e^{-\frac{1}{3}x^3}\,dx.$$

If $m=0$, or if m be any positive multiple of 3, this can be simplified.

Ex. 2. Solve

$$\text{(i)} \quad \frac{d^2y}{dx^2} - x\frac{dy}{dx} + (x-1)y = X;$$

$$\text{(ii)} \quad (ax - bx^2)\frac{d^2y}{dx^2} + 2a\frac{dy}{dx} + 2by = x^{n-1};$$

$$\text{(iii)} \quad (1+x^2)\frac{d^2y}{dx^2} - 2x\frac{dy}{dx} + 2y = X.$$

59. If, however, a solution of the equation when R has been put zero cannot be obtained, then it is sometimes useful to remove from the transformed differential equation the term involving $\frac{dv}{dx}$. In that case, w must satisfy

$$2\frac{dw}{dx} + Pw = 0,$$

from which we find

$$w = e^{-\frac{1}{2}\int P\,dx};$$

there is no necessity for adding a constant in the integration, as it will afterwards disappear. Insert this value of w in the equation, and write

$$I = Q - \tfrac{1}{2}\frac{dP}{dx} - \tfrac{1}{4}P^2;$$

then the equation becomes

$$\frac{d^2v}{dx^2} + Iv = Re^{\frac{1}{2}\int P\,dx}$$

In some particular instances, this equation admits of immediate solution; but they occur much less frequently than those to which the preceding method applies. The advantage of the new form, which will be indicated shortly, lies in an altogether different direction. Now we know that, if a solution of this equation with the right-hand side equated to zero can be obtained, the primitive of the general equation is obtainable; and we may therefore quote the equation in the form

$$\frac{d^2v}{dx^2} + Iv = 0.$$

Ex. 1. Solve

$$\frac{d^2y}{dx^2} - \frac{1}{x^{\frac{1}{2}}} \frac{dy}{dx} + \frac{y}{4x^2}(-8 + x^{\frac{1}{2}} + x) = 0.$$

Hence $P = -\frac{1}{x^{\frac{1}{2}}}$, and therefore $w = e^{-\frac{1}{2}\int P dx} = e^{x^{\frac{1}{2}}}$. Also

$$I = \frac{-2}{x^2} + \frac{1}{4x^{\frac{3}{2}}} + \frac{1}{4x} - \frac{1}{4x^{\frac{3}{2}}} - \frac{1}{4x} = -\frac{2}{x^2},$$

so that the equation giving v is

$$\frac{d^2v}{dx^2} - \frac{2}{x^2}v = 0.$$

The solution of this is

$$v = Ax^2 + \frac{B}{x};$$

and therefore the general integral of the first equation is

$$y = (Ax^2 + Bx^{-1})e^{x^{\frac{1}{2}}}.$$

Ex. 2. Solve

    (i) $\dfrac{d^2y}{dx^2} - 2bx\dfrac{dy}{dx} + b^2x^2y = x$;

    (ii) $\dfrac{d^2y}{dx^2} - \dfrac{2}{x}\dfrac{dy}{dx} + \left(a^2 + \dfrac{2}{x^2}\right)y = 0$;

    (iii) $\dfrac{d^2y}{dx^2} - 4x\dfrac{dy}{dx} + (4x^2 - 3)y = e^{x^2}$.

60. The advantage of using the form

$$\frac{d^2v}{dx^2} + Iv = 0,$$

instead of

$$\frac{d^2y}{dx^2} + P\frac{dy}{dx} + Qy = 0,$$

as typical of the linear differential equation of the second order, lies in the fact that, for all substitutions such as $zf(x)$ for y in the latter equation, I is a function of P and Q of such a form that, when the new equation

$$\frac{d^2z}{dx^2} + P_1\frac{dz}{dx} + Q_1z = 0$$

has its second term removed by the substitution

$$z = we^{-\frac{1}{2}\int P_1 dx},$$

it takes the form

$$\frac{d^2w}{dx^2} + Iw = 0.$$

Thus I is exactly the same function of P_1 and Q_1 as it is of P and Q; and we may therefore call I an *invariant* of the coefficients of the differential equation*. The equation so reduced may be said to be in its *normal form*. Any two linear equations, such as the equations in y and z, can be transformed into one another if the normal form of each be the same.

If it be known that two given equations are so transformable and the equation of substitution between the dependent variables be desired, this can easily be obtained by using the normal form as an intermediate transformed equation. Thus, in the general example, the equation in y becomes transformed to that in v by writing

$$ye^{\frac{1}{2}\int P dx} = v,$$

and the equation in v passes into that in z by writing

$$v = ze^{\frac{1}{2}\int P_1 dx};$$

and therefore the relation which transforms directly the y-equation into the z-equation is

$$ye^{\frac{1}{2}\int P dx} = ze^{\frac{1}{2}\int P_1 dx}.$$

Ex. 1. Prove that the equations

$$(1-x^2)\frac{d^2z}{dx^2} + (1-3x)\frac{dz}{dx} + kz = 0,$$

and

$$(1-x^2)\frac{d^2\zeta}{dx^2} - (1+x)\frac{d\zeta}{dx} + (k+1)\zeta = 0,$$

can be transformed into one another; and find the relation between z, ζ, and x.

Ex. 2. Find the value of Q which is such that the equation

$$\frac{d^2y}{dx^2} + a\frac{dy}{dx} + Qy = 0$$

may be transformed, by a substitution $y = zf(x)$, into

$$\frac{d^2z}{dx^2} + \frac{1}{x}\frac{dz}{dx} + \left(1 - \frac{n^2}{x^2}\right)z = 0.$$

Obtain the value of $f(x)$.

61. Let y_1 and y_2 be two particular integrals of the equation

$$\frac{d^2y}{dx^2} + P\frac{dy}{dx} + Qy = 0,$$

* Malet, *Phil. Trans.* (1882), p. 751.

and v_1 and v_2 the corresponding particular integrals of

$$\frac{d^2v}{dx^2} + Iv = 0;$$

then

$$v_1 = y_1 e^{\frac{1}{2}\int P dx} \text{ and } v_2 = y_2 e^{\frac{1}{2}\int P dx},$$

and therefore

$$\frac{y_1}{y_2} = \frac{v_1}{v_2} = s,$$

so that s is the quotient of two different solutions of either differential equation. We now proceed to find the equation which is satisfied by s. Since each of the quantities y (or v) may consist of two terms each containing an arbitrary constant factor, the quotient of one by the other may contain three arbitrary constants (not four, since without altering the value or generality of such a quotient any of the four constants may be made unity); therefore the differential equation satisfied by s, a function involving three arbitrary constants, must be of the third order.

Indicating differentiation with regard to x by dashes, we may write

$$v_1'' + Iv_1 = 0,$$
$$v_2'' + Iv_2 = 0.$$

Substituting sv_2 for v_1 in the former equation, we have

$$s''v_2 + 2s'v_2' + sv_2'' + Isv_2 = 0,$$

that is,

$$s''v_2 + 2s'v_2' = 0,$$

so that

$$\frac{s''}{s'} = -2\frac{v_2'}{v_2}.$$

Differentiating this, we have

$$\frac{s'''}{s'} - \left(\frac{s''}{s'}\right)^2 = -2\frac{v_2''}{v_2} + 2\left(\frac{v_2'}{v_2}\right)^2$$

$$= 2I + \tfrac{1}{2}\left(\frac{s''}{s'}\right)^2;$$

and the transposition of the last term gives

$$\frac{s'''}{s'} - \tfrac{3}{2}\left(\frac{s''}{s'}\right)^2 = 2I.$$

This is the differential equation satisfied by s; and it is of the third order, as was indicated.

The function of the differential coefficients of s with regard to x, which occurs on the left-hand side of the equation, has been called by Cayley the Schwarzian Derivative* and is denoted by him by $\{s, x\}$; it is so called because its properties are discussed, and it is of fundamental importance, in a memoir by Schwarz in *Crelle's Journal* (t. LXXV), though the function is not originally due to him †.

62. If now *any* solution of this equation can be obtained, then a solution of the original differential equation can be immediately deduced. For let such a solution of the new equation be denoted by s; then since

$$\frac{v_2'}{v_2} = -\tfrac{1}{2}\frac{s''}{s'},$$

we have, on integration,

$$v_2 = Cs'^{-\frac{1}{2}},$$

where C is arbitrary. This is one solution; another is

$$v_1 = v_2 s = Cs'^{-\frac{1}{2}} s.$$

From these, the corresponding solutions of the equation in y are derived by inserting the exponential factor. When any one solution of a linear equation of the second order is known, we can obtain the general solution; and hence any particular value of s satisfying its differential equation will lead to the complete solution of the first of the differential equations.

This theorem holds in regard to the general linear equation of the second order; one of its chief applications arises when the linear equation is that satisfied by the hypergeometric series, to be discussed in Chapter VI.

---

* Cayley, *Camb. Phil. Trans.* (1880), vol. XIII. p. 5; *Coll. Math. Papers*, vol. XI. p. 148.

† It occurs implicitly in Lagrange's memoir "Sur la construction des cartes géographiques," *Œuvres*, vol. IV. p. 651 (this reference is due to Schwarz), and Jacobi's *Fundamenta Nova*; and explicitly for the first time in Kummer's memoir on the hypergeometric series in *Crelle*, t. XV., which is referred to in Chapter VI.; see also Cayley, *l.c.*

Ex. 1. Prove that, if

$$s(ax+b)=cx+d,$$

the Schwarzian derivative of s vanishes.

Ex. 2. Find the general value of s when

$$x^2\{s,\ x\}+a=0,$$

where a is a constant.

Ex. 3. Prove that

(i) $\{s,\ x\}=-\left(\dfrac{ds}{dx}\right)^2\{x,\ s\};$

(ii) $\left\{\dfrac{as+b}{cs+d},\ x\right\}=\{s,\ x\};$

(iii) $\{s,\ x\}=\left(\dfrac{dy}{dx}\right)^2[\{s,\ y\}-\{x,\ y\}];$

(iv) $\{s,\ x\}=\left(\dfrac{dy}{dx}\right)^2\{s,\ y\}-\left(\dfrac{dv}{dx}\right)^2\{x,\ v\}+\left(\dfrac{dv}{dx}\right)^2\{y,\ v\}.$

(Cayley.)

63. Another method which is sometimes effective is that of *changing the independent variable.*

Take z as the new independent variable; then

$$\frac{dy}{dx}=\frac{dy}{dz}\frac{dz}{dx},$$

$$\frac{d^2y}{dx^2}=\frac{d^2y}{dz^2}\left(\frac{dz}{dx}\right)^2+\frac{dy}{dz}\frac{d^2z}{dx^2};$$

and the original equation becomes

$$\frac{d^2y}{dz^2}\left(\frac{dz}{dx}\right)^2+\frac{dy}{dz}\left(\frac{d^2z}{dx^2}+P\frac{dz}{dx}\right)+Qy=0.$$

As yet, z is quite arbitrary; it may therefore be chosen to satisfy any assignable condition. Thus we may choose to make the coefficient of $\dfrac{dy}{dz}$ vanish; then

$$\frac{d^2z}{dx^2}+P\frac{dz}{dx}=0,$$

and therefore z is given in terms of x by the equation

$$z=\int dx\,e^{-\int Pdx}.$$

The eliminant of this relation between z and x and the transformed equation may furnish a differential equation which proves integrable.

One integrable case occurs when the value of z, thus obtained, is such as to satisfy the relation

$$\mu \left(\frac{dz}{dx}\right)^2 = Qz^2,$$

where μ is a constant; and then the equation takes the form

$$z^2 \frac{d^2y}{dz^2} + \mu y = 0,$$

of which the integral is

$$y = A z^\alpha + B z^\beta,$$

α and β being the roots of

$$m(m-1) + \mu = 0.$$

It is not difficult to prove that the relation, which must exist between P and Q in order that this may be the case, is

$$\frac{Q}{\mu^{\frac{1}{2}}} + \frac{d}{dx}(Q^{\frac{1}{2}}) + PQ^{\frac{1}{2}} = 0.$$

Another integrable case would be furnished by

$$\mu \left(\frac{dz}{dx}\right)^2 = Q,$$

and so for other cases. It will be noticed that in each instance the equation is reduced to what may be called a known form, that is, one of which the primitive can be obtained.

Ex. 1. Solve

$$(1-x^2)\frac{d^2y}{dx^2} - x\frac{dy}{dx} = c^2y.$$

Hence

$$P(1-x^2) = -x,$$

so that

$$\frac{dz}{dx} = e^{-\int P dx} = e^{\int \frac{x dx}{1-x^2}}$$

$$= (1-x^2)^{-\frac{1}{2}},$$

and

$$z = \arcsin x.$$

When the independent variable is changed to z, the equation becomes

$$\frac{d^2y}{dz^2} = c^2 y,$$

and therefore

$$y = A e^{c \, \text{arc sin} \, x} + B e^{c \, \text{arc cos} \, x}.$$

Ex. 2. Solve (i) $(x^2 - 1)\dfrac{d^2y}{dx^2} + x\dfrac{dy}{dx} = c^2 y$;

(ii) $\dfrac{d^2y}{dx^2} + \dfrac{dy}{dx}\tan x + y\cos^2 x = 0$;

(iii) $\dfrac{d^2y}{dx^2} - \dfrac{3x+1}{x^2-1}\dfrac{dy}{dx} + y\left\{\dfrac{6(x+1)}{(x-1)(3x+5)}\right\}^2 = 0$;

(iv) $\dfrac{d^2y}{dx^2} + \dfrac{2}{x}\dfrac{dy}{dx} + \dfrac{a^2}{x^4}y = 0$;

(v) $(1+x^2)\dfrac{d^2y}{dx^2} + x\dfrac{dy}{dx} + 2y = 0$.

64. The property used in § 60 to obtain the relations between the dependent variables in two equations, which are transformable into one another—viz. that the equations have the same normal form—can be used to obtain the relations between the dependent variables in two equations, the independent variables in which are different, on the hypothesis that the equations ultimately determine the same function. The process adopted will be similar to the former one, as both equations will be reduced to their normal forms in the same variable and these, being assumed identical, will give the conditions necessary for the justification of the hypothesis.

Let the two equations, which are to be thus transformable into one another by changing both the dependent and the independent variables, be

$$\frac{d^2y}{dx^2} + 2P\frac{dy}{dx} + Qy = 0 \quad\dots\dots\dots\dots\dots(\text{i}),$$

and

$$\frac{d^2v}{dz^2} + 2R\frac{dv}{dz} + Sv = 0 \quad\dots\dots\dots\dots(\text{ii}),$$

in which P and Q are functions of x, and R and S functions of z.

Writing in (i)

$$y e^{\int P dx} = y_1,$$

and

$$I = Q - \frac{dP}{dx} - P^2,$$

we have

$$\frac{d^2 y_1}{dx^2} + I y_1 = 0 \quad \dotfill \text{(iii)}.$$

Writing in (ii)

$$v e^{\int R dz} = v_1,$$

and

$$J = S - \frac{dR}{dz} - R^2,$$

we have

$$\frac{d^2 v_1}{dz^2} + J v_1 = 0 \quad \dotfill \text{(iv)}.$$

In (iii), changing the independent variable from x to z, we obtain

$$\frac{d^2 y_1}{dz^2} \left(\frac{dz}{dx} \right)^2 + \frac{dy_1}{dz} \frac{d^2 z}{dx^2} + I y_1 = 0,$$

or

$$\frac{d^2 y_1}{dz^2} + \frac{dy_1}{dz} \frac{z''}{z'^2} + \frac{I}{z'^2} y_1 = 0,$$

in which dashes indicate differentiation with regard to x. To reduce this to its normal form, we write

$$y_1 e^{\frac{1}{2} \int \frac{z''}{z'^2} dz} = y_2,$$

or, on the evaluation of the integral in the exponent,

$$y_1 z'^{\frac{1}{2}} = y_2 ;$$

the equation then becomes

$$\frac{d^2 y_2}{dz^2} + G y_2 = 0 \quad \dotfill \text{(v)},$$

where

$$G = \frac{I}{z'^2} - \tfrac{1}{4} \left(\frac{z''}{z'^2} \right)^2 - \tfrac{1}{2} \frac{d}{dz} \left(\frac{z''}{z'^2} \right)$$

$$= \frac{I}{z'^2} - \tfrac{1}{4} \frac{z''^2}{z'^4} - \tfrac{1}{2} \left(\frac{z'''}{z'^2} - 2 \frac{z''}{z'^3} \right)^2 \frac{1}{z'}$$

$$= \frac{I}{z'^2} - \tfrac{1}{2} \frac{\{z, x\}}{z'^2},$$

and $\{z, x\}$ is the Schwarzian derivative of z.

If, then, the equations be transformable into one another, the normal forms will be the same when expressed in terms of the same independent variable; hence. comparing (iv) and (v), which are the normal forms, we have

$$y_2 = v_1,$$

and $$G = J.$$

Substituting for G in the latter equation, we have

$$I - \tfrac{1}{2}\{z, x\} = Jz'^2,$$

or $$\tfrac{1}{2}\{z, x\} + \left(\frac{dz}{dx}\right)^2 \left(S - \frac{dR}{dz} - R^2\right) - \left(Q - \frac{dP}{dx} - P^2\right) = 0;$$

and substituting their values for y_2 and v_1 in the former equation, we have

$$y \left(\frac{dz}{dx}\right)^{\frac{1}{2}} e^{\int P dx} = v e^{\int R dz}.$$

These two equations are the conditions that the differential equations (i) and (ii) should be transformable into one another. The first of them gives the relation which must exist between the independent variables. When the first is satisfied, the second gives the relation which must exist between the dependent variables.

The foregoing equations enable us to obtain the general form of all differential equations into which (i) is transformable, and also to obtain the connection between two given related equations. Thus, for instance, the equation in a given independent variable z equivalent to (i) would have as its normal form

$$\frac{d^2 v_1}{dz^2} + v_1 J = 0,$$

where

$$v_1 = y \left(\frac{dz}{dx}\right)^{\frac{1}{2}} e^{\int P dx},$$

and

$$J = \frac{I}{z'^2} - \tfrac{1}{2}\frac{\{z, x\}}{z'^2};$$

and since z and I are known in terms of x, J is also known in terms of x and can therefore be expressed in terms of z. Every differential equation, which is equivalent to (i) and has z for its independent variable, must have the foregoing equation in v_1 for its normal form.

Ex. 1. Prove that the equations

$$(1-x^2)\frac{d^2z}{dx^2} - 2x\frac{dz}{dx} + n(n+1)z = 0$$

and

$$(1-k^2)\frac{d^2v}{dk^2} + \frac{1-3k^2}{k}\frac{dv}{dk} = \left\{1 + \frac{(2n+1)^2}{1-k^2}\right\}v$$

are transformable into one another by the relation

$$x(1-k^2) = 1+k^2;$$

and find the relation between z and v.

(G. H. Stuart.)

Ex. 2. Prove that the equations

$$\frac{d^2v}{dz^2}(1-z)^2 + 2(B-1+z)\frac{dv}{dz} + k(1-k)v = 0$$

and

$$\frac{d^2y}{dx^2} + \frac{2k}{x}\frac{dy}{dx} - B^2y = 0$$

are transformable into one another by the relation

$$x-1 = xz;$$

and find the relation between y and v.

METHOD OF VARIATION OF PARAMETERS.

65. It was proved (§ 58) that, if a solution of the equation

$$\frac{d^2y}{dx^2} + P\frac{dy}{dx} + Qy = 0$$

be known, the primitive of the equation

$$\frac{d^2y}{dx^2} + P\frac{dy}{dx} + Qy = R$$

can be obtained. The following method, which is different from the methods already indicated, is effective in giving for this equation, and for other linear equations, the Particular Integral; and it can be applied where the earlier methods cease to be applicable.

Let y_1 be a solution of the equation

$$\frac{d^2y}{dx^2} + P\frac{dy}{dx} + Qy = 0,$$

so that

$$\frac{d^2y_1}{dx^2} + P\frac{dy_1}{dx} + Qy_1 = 0.$$

Eliminating Q, we have

$$y_1 \frac{d^2y}{dx^2} - y \frac{d^2y_1}{dx^2} + P \left(y_1 \frac{dy}{dx} - y \frac{dy_1}{dx} \right) = 0;$$

and therefore

$$y_1 \frac{dy}{dx} - y \frac{dy_1}{dx} = A e^{-\int P dx},$$

where the constant A clearly is not zero when y denotes the most general solution of the equation

$$\frac{d^2y}{dx^2} + P \frac{dy}{dx} + Qy = 0.$$

We now have a relation between y and y_1 of the first order; its integral is

$$y = B y_1 + A y_1 \int \frac{dx}{y_1^2} e^{-\int P dx}.$$

Let y_2 stand for the quantity of which A is the coefficient, so that the primitive is

$$y = B y_1 + A y_2,$$

and y_2 is a particular solution of the differential equation. Then the preceding analysis shews that any two particular solutions y_1 and y_2 are connected by the equation

$$y_1 \frac{dy_2}{dx} - y_2 \frac{dy_1}{dx} = C e^{-\int P dx},$$

where the value of C is no longer arbitrary but depends on the forms of y_1 and y_2, the two particular solutions of the equation.

66. Let us now take the above value of y and substitute it in the equation

$$\frac{d^2y}{dx^2} + P \frac{dy}{dx} + Qy = R,$$

on the supposition that A and B are no longer constants but functions of x to be so chosen that the equation shall be satisfied. Thus the form of y is the same for the two equations, but the constants which occur in the former case are changed in the latter into functions of the independent variable; to this process is applied the name *Variation of Parameters* (§ 14, *Note* 1).

We have now two unknown quantities A and B, in terms of which y, a single unknown, is expressed; and we are therefore at liberty to choose any relation between them that may be most convenient for our purpose. When we differentiate y, we obtain

$$\frac{dy}{dx} = B\frac{dy_1}{dx} + A\frac{dy_2}{dx} + y_1\frac{dB}{dx} + y_2\frac{dA}{dx}$$

$$= B\frac{dy_1}{dx} + A\frac{dy_2}{dx},$$

provided

$$y_1\frac{dB}{dx} + y_2\frac{dA}{dx} = 0;$$

we shall take this last equation as the chosen relation between A and B. Again, if we differentiate $\frac{dy}{dx}$, so that

$$\frac{d^2y}{dx^2} = B\frac{d^2y_1}{dx^2} + A\frac{d^2y_2}{dx^2} + \frac{dB}{dx}\frac{dy_1}{dx} + \frac{dA}{dx}\frac{dy_2}{dx},$$

and substitute these values in the original equation, then, since y_1 and y_2 are particular solutions of the equation when $R = 0$, we have as the result

$$\frac{dB}{dx}\frac{dy_1}{dx} + \frac{dA}{dx}\frac{dy_2}{dx} = R.$$

Thus

$$\frac{\frac{dA}{dx}}{y_1} = \frac{\frac{dB}{dx}}{-y_2} = \frac{R}{y_1\frac{dy_2}{dx} - y_2\frac{dy_1}{dx}} = \frac{R}{C}e^{\int Pdx};$$

and therefore

$$A = E + \frac{1}{C}\int Ry_1 e^{\int Pdx}\, dx,$$

$$B = F - \frac{1}{C}\int Ry_2 e^{\int Pdx}\, dx,$$

where E and F are arbitrary constants, and C is a specific constant depending upon the forms of y_1 and y_2.

If now, in the differential equation, we write $\phi(x)$ for P and $\psi(x)$ for R; $f_1(x)$ for y_1 and $f_2(x)$ for y_2; then *the general solution of*

$$\frac{d^2y}{dx^2} + \phi(x)\frac{dy}{dx} + Qy = \psi(x)$$

is

$$y = E f_2(x) + F f_1(x) + \frac{1}{C} \int^x \psi(\xi) e^{\int^\xi \phi(z)\,dz} \{f_2(x)f_1(\xi) - f_1(x)f_2(\xi)\}\,d\xi,$$

where $f_1(x)$ and $f_2(x)$ are particular solutions of

$$\frac{d^2y}{dx^2} + \phi(x)\frac{dy}{dx} + Qy = 0,$$

and are therefore connected by the relation

$$f_1\frac{df_2}{dx} - f_2\frac{df_1}{dx} = Ce^{-\int^x \phi(z)\,dz}.$$

It may be noticed that we may make C unity without loss of generality; for, if it be not unity, we may substitute for $f_2(x)$ the quantity $\frac{1}{C}f_2(x)$ which, while still a particular solution, will render the constant unity.

Ex. 1. Solve

$$x\frac{dy}{dx} - y = (x-1)\left(\frac{d^2y}{dx^2} - x + 1\right).$$

Arranged in the ordinary form, this is

$$\frac{d^2y}{dx^2} - \frac{x}{x-1}\frac{dy}{dx} + \frac{1}{x-1}y = x-1.$$

Particular solutions of the equation, without the right-hand member, are x and e^x; hence, if we take

$$f_1(x) = x, \quad f_2(x) = e^x,$$

we may proceed as above, and have as the primitive

$$y = Ae^x + Bx.$$

As in the general case, A and B are connected by the relation

$$\frac{dA}{dx}e^x + \frac{dB}{dx}x = 0,$$

while

$$\frac{dA}{dx}e^x + \frac{dB}{dx} = x-1.$$

Thus

$$\frac{dA}{dx} = xe^{-x} \text{ and } \frac{dB}{dx} = -1;$$

and therefore

$$A = E + \int^x \xi e^{-\xi}\,d\xi$$

$$= E - e^{-x}(x+1),$$

and

$$B = F - x.$$

The primitive is therefore

$$y = Ee^x + Fx - (x^2 + x + 1).$$

Ex. 2. Integrate by this method the equation

$$\frac{dy}{dx} + Qy = Ry^2,$$

where Q and R are functions of x alone.

Ex. 3. Solve

(i) $\dfrac{d^2y}{dx^2} + n^2y = \sec nx$;

(ii) $(1 - x^2)\dfrac{d^2y}{dx^2} - 4x\dfrac{dy}{dx} - (1 + x^2)y = x.$

67. The method of variation of parameters may be applied in a manner, different in regard to the terms neglected, to obtain a subsidiary integral, the constants in which are subsequently made variable parameters. Thus consider the equation

$$\frac{d^2y}{dx^2} + \left(\frac{dy}{dx}\right)^2 f(y) + F(y) = 0.$$

Neglect the term involving $F(y)$ in order to obtain a subsidiary integral; this integral satisfies

$$\frac{d^2y}{dx^2} + \left(\frac{dy}{dx}\right)^2 f(y) = 0,$$

and therefore is

$$\frac{dy}{dx} e^{\int f(y)\, dy} = C.$$

Suppose now that C, instead of being a constant, is a function of x and let this relation be differentiated; then

$$\left\{\frac{d^2y}{dx^2} + f(y)\left(\frac{dy}{dx}\right)^2\right\} e^{\int f(y)\, dy} = \frac{dC}{dx},$$

or

$$-F(y) e^{\int f(y)\, dy} = \frac{dC}{dx}.$$

Therefore

$$C\frac{dC}{dx} = -F(y) e^{2\int f(y).dy} \frac{dy}{dx},$$

and so

$$C^2 = A - 2\int dy\, F(y) e^{2\int f(y)\, dy}.$$

A first integral of the original equation therefore is

$$\frac{dy}{dx} e^{\int f(y)\,dy} = \{A - 2 \int dy\, F(y)\, e^{2\int f(y)\,dy}\}^{\frac{1}{2}}.$$

This can be again integrated since the variables are separable.

Ex. 1. Solve in this manner the equation

$$\frac{d^2 y}{dx^2} + \frac{dy}{dx} f(x) + \left(\frac{dy}{dx}\right)^2 \phi(x) = 0.$$

Shew also that the integral of this equation may be derived by the method of § 54.

By changing the independent variable in this example from x to y, obtain the integral of the equation

$$\frac{d^2 y}{dx^2} + f(y)\frac{dy}{dx} + \phi(y)\left(\frac{dy}{dx}\right)^2 = 0.$$

Ex. 2. Integrate the general equation

$$\frac{d^2 y}{dx^2} + f(x)\frac{dy}{dx} + F(y)\left(\frac{dy}{dx}\right)^2 = 0,$$

firstly, by neglecting the last term to obtain a subsidiary equation and then varying the parameters;

secondly, by applying the same method to the integral derived from neglecting the second term;

thirdly, by multiplying by $\left(\frac{dy}{dx}\right)^{-1}$ and then integrating each term.

It thus appears from these examples that

$$\frac{d^2 y}{dx^2} + P\frac{dy}{dx} + Q\left(\frac{dy}{dx}\right)^2 = 0$$

is integrable in the cases :—

      (α) when both P and Q are functions of x,

      (β) when both P and Q are functions of y,

      (γ) when P is a function of x and Q a function of y.

Two Particular Methods.

68. When, in the equation $\frac{d^2 v}{dx^2} + Iv = 0$, the quantity I is a rational function of a fractional form such that the denominator is of a higher degree in the variable than the numerator, the following method is sometimes of use.

Let a quantity

$$ze^{\int P_1\,dx}$$

be substituted for v; then the equation becomes

$$\frac{d^2z}{dx^2} + 2P_1\frac{dz}{dx} + P_2z = 0,$$

where

$$P_2 = I + P_1{}^2 + \frac{dP_1}{dx}.$$

On integrating the equation as if the left-hand side were a perfect differential, we have

$$\frac{dz}{dx} + 2P_1z + \int z\left(P_2 - 2\frac{dP_1}{dx}\right)dx = A.$$

Since the quantities P_1 and P_2 are connected as yet by only a single relation, we may assign as a further condition to determine them

$$P_2 = 2\frac{dP_1}{dx}.$$

This gives as the equation for P_1

$$\frac{dP_1}{dx} - P_1{}^2 = I;$$

while, if *any* value of P_1 satisfying this be obtained, an integral of the original equation is obtained in the shape

$$\frac{dz}{dx} + 2P_1z = A.$$

It should be pointed out that the possible utility of this method depends on the form of the equation which gives P_1; this would be lost by the substitution

$$P_1 = -\frac{1}{w}\frac{dw}{dx},$$

for then the equation giving P_1 becomes changed to the original.

With the assumption which was made as to the form of I, we may write

$$I = \frac{V}{T^2U} = \frac{VU}{T^2U^2} = \frac{UV}{\phi^2},$$

say, where T, U, and V, are rational integral functions of x. Then we may assume

$$P_1 = \frac{f(x)}{\phi},$$

leaving the constants in $f(x)$ as the quantities to be determined from the equation; but in general there are not sufficient disposable constants arising in f to allow the equation to be satisfied. Hence this method, like the other methods which have been proposed for the solution of the linear equation of the second order, is not one of universal application, but is effective only in particular cases.

Ex. 1. Solve $\qquad x(1-x)^2 \dfrac{d^2v}{dx^2} = 2v.$

Here the equation for P_1 is

$$\frac{dP_1}{dx} - P_1^2 = -\frac{1}{x(1-x)^2}.$$

Let $P_1 = \dfrac{E}{x} + \dfrac{F}{1-x}$ and substitute; the equation is satisfied by $E = F = -1$, and therefore a first integral is

$$\frac{dz}{dx} - \frac{2}{x(1-x)} z = A,$$

where $\qquad\qquad \log\dfrac{v}{z} = -\displaystyle\int\dfrac{dx}{x} - \int\dfrac{dx}{1-x},$

or $\qquad\qquad vx = z(1-x).$

The primitive can easily be deduced, for the first integral equation in z is linear of the first order.

Ex. 2. Solve

(i) $\quad (1-x^2)^2 \dfrac{d^2v}{dx^2} + v = 0;$

(ii) $\quad (2x+1)^2(x^2+x+1)\dfrac{d^2v}{dx^2} = 18v;$

(iii) $\quad \dfrac{d^2y}{dx^2} = 4y\,\dfrac{\sin 3x}{\sin^3 x}.$

If a term involving $\dfrac{dy}{dx}$ should occur in the equation, this term should be removed before applying the above method.

Ex. 3. Solve

(i) $\dfrac{d^2y}{dx^2} + \dfrac{\gamma - (\alpha+2)x}{x(1-x)} \dfrac{dy}{dx} - \dfrac{\alpha y}{x(1-x)} = 0$;

(ii) $\dfrac{d^2y}{dx^2} + \dfrac{\alpha - (\alpha+\beta+1)x}{x(1-x)} \dfrac{dy}{dx} - \dfrac{\alpha\beta y}{x(1-x)} = 0$;

(iii) $\dfrac{d^2y}{dx^2} + \dfrac{\alpha+1 - (\alpha+\beta+1)x}{x(1-x)} \dfrac{dy}{dx} - \dfrac{\alpha\beta}{x(1-x)} y = 0$.

Ex. 4. Shew that this method will apply to the equation

$$\frac{d^2v}{dx^2} = \frac{A'x^2 + 2B'x + C'}{(x^2 + 2Ax + B)^2} v,$$

provided there be a single relation between A', B', and C'; and find this relation.

69. A certain class of linear differential equations can be solved by the *resolution of the operator on y into the product of operators.* Thus consider the equation

$$u \frac{d^2y}{dx^2} + v \frac{dy}{dx} + wy = 0,$$

in which u, v and w are functions of x; then, if the operator

$$u \frac{d^2}{dx^2} + v \frac{d}{dx} + w$$

be resoluble into the product

$$\left(p \frac{d}{dx} + q \right) \left(r \frac{d}{dx} + s \right),$$

p, q, r and s being functions of x, the equation can be integrated. For, if we write

$$\left(r \frac{d}{dx} + s \right) y = z,$$

we have

$$p \frac{dz}{dx} + qz = 0,$$

and therefore

$$z = A e^{-\int \frac{q}{p} dx};$$

and we must now integrate

$$r \frac{dy}{dx} + sy = A e^{-\int \frac{q}{p} dx},$$

which is linear of the first order. In order that this resolution may take place, we have the three equations

$$pr = u,$$

$$qr + p\left(\frac{dr}{dx} + s\right) = v,$$

$$qs + p\frac{ds}{dx} = w,$$

to determine four quantities p, q, r and s; but we may consider p and r as known factors of u, and use the two remaining equations to determine q and s.

But these cannot be solved in general; and again therefore the method will apply only in particular cases.

Ex. 1. Solve

$$(x^2 + x - 2)\frac{d^2y}{dx^2} + (x^2 - x)\frac{dy}{dx} - (6x^2 + 7x)\,y = 0.$$

Here we may write $p = x + 2$ and $r = x - 1$.

If $q = Ex + F$ and $s = E'x + F''$, we have

$$\left.\begin{aligned} E + E' &= 1 \\ -E + F + 2E' + F'' &= -2 \\ F - 2F'' &= 2 \end{aligned}\right\}, \qquad \left.\begin{aligned} EE' &= -6 \\ E'F + EF'' + E' &= -7 \\ FF'' + 2E' &= 0 \end{aligned}\right\},$$

which are satisfied by

$$E = 3; \quad E' = -2; \quad F = 4; \quad F'' = 1.$$

Hence the equation may be written

$$\left\{(x+2)\frac{d}{dx} + 3x + 4\right\}\left\{(x-1)\frac{d}{dx} - (2x-1)\right\}y = 0.$$

A first integral is

$$(x-1)\frac{dy}{dx} - (2x-1)\,y = A\,(x+2)^2 e^{-3x},$$

and the primitive is

$$y = (x-1)\,e^{2x}\left\{B + A\int\left(\frac{x+2}{x-1}\right)^2 e^{-5x}dx\right\}.$$

Ex. 2. Solve

(i) $\quad ax\dfrac{d^2y}{dx^2} + (3a + bx)\dfrac{dy}{dx} + 3by = 0;$

(ii) $\quad (x-1)(x-2)\dfrac{d^2y}{dx^2} - (2x-3)\dfrac{dy}{dx} + 2y = 0;$

(iii) $\quad (2x-1)\dfrac{d^2y}{dx^2} - (3x-4)\dfrac{dy}{dx} + (x-3)\,y = 0;$

(iv) $(x^2+3x+2)\dfrac{d^2y}{dx^2}+(5x^2+\tfrac{2}{2}x+4)\dfrac{dy}{dx}+(6x^2+\tfrac{17}{2}x+4)y=0;$

(v) $(x^2-1)\dfrac{d^2y}{dx^2}-(3x+1)\dfrac{dy}{dx}-(x^2-x)y=0;$

(vi) $x^2(a-bx)\dfrac{d^2y}{dx^2}-2x(2a-bx)\dfrac{dy}{dx}+2(3a-bx)y=6a^2.$

70. There is a particular form into which the ordinary linear differential equation of the second order may be changed; multiplying

$$\frac{d^2y}{dx^2}+P\frac{dy}{dx}+Qy=0$$

throughout by $e^{\int P\,dx}$, we may write it

$$\frac{d}{dx}\left\{e^{\int P\,dx}\frac{dy}{dx}\right\}+Qe^{\int P\,dx}\,y=0.$$

Let a new independent variable z be taken such that

$$dz=Qe^{\int P\,dx}\,dx;$$

then the equation becomes

$$\frac{d}{dz}\left\{Qe^{2\int P\,dx}\frac{dy}{dz}\right\}+y=0.$$

Now $Qe^{2\int P\,dx}$ is a definite function of x and therefore of z; let it be denoted by $\dfrac{1}{U}$, where U is a function of z. Then the equation is

$$\frac{d}{dz}\left\{\frac{1}{U}\frac{dy}{dz}\right\}+y=0,$$

which is the form referred to.

Sir William Thomson (afterwards Lord Kelvin) indicated a method of approximating to a solution of this equation by mechanical means*.

Ex. Express $P\dfrac{d^2u}{dx^2}+Q\dfrac{du}{dx}+Ru=0$ in the form $\dfrac{d^2v}{dx^2}+\mu v=0$. Prove that $v=S_0-S_1+S_2-\ldots$, where

$$S_0=C+C'x,\qquad S_{n+1}=\int_0^x dx\int_0^x \mu S_n\,dx,$$

expresses the solution of this in a series necessarily converging for all values of x, provided μ remains finite.

Work out the case when $\mu=x^n$.

* See *Proc. Roy. Soc.* vol. xxiv. (1876), p. 269.

GENERAL LINEAR EQUATION.

71. The general linear equation with variable coefficients is of the form

$$X_0\frac{d^ny}{dx^n} + X_1\frac{d^{n-1}y}{dx^{n-1}} + X_2\frac{d^{n-2}y}{dx^{n-2}} + \ldots\ldots + X_{n-1}\frac{dy}{dx} + X_ny = V \quad \ldots(\text{i}),$$

in which X_0, X_1, X_2, ..., X_n, and V, are functions of x alone; equations, in which the coefficients of the differential coefficients of y are constants, have already been considered. The coefficients X_0, X_1, ..., X_n, may be taken to be integral functions of x; if in any equation they were not actually so, the equation could be transformed so that its coefficients would be integral functions of x by multiplication throughout by the least common multiple of the denominators of such fractions as occurred in the given form.

The primitive of the differential equation consists, as before, of two parts :—

First. The *Particular Integral*, which is any value of y (the simpler the better) satisfying the equation ;

Second. The *Complementary Function*, which is the general solution of the equation without the second member, that is, of the equation

$$X_0\frac{d^ny}{dx^n} + X_1\frac{d^{n-1}y}{dx^{n-1}} + \ldots\ldots + X_{n-1}\frac{dy}{dx} + X_ny = 0 \quad \ldots(\text{ii}).$$

The equation (ii), being of the nth order, will have in its general solution n arbitrary constants—the necessary number for the primitive of (i), which is the sum of these two parts.

72. If y_1 be a solution of (ii), then A_1y_1 is also a solution since the equation is linear ; and therefore, if y_1, y_2, ..., y_n, be n different particular solutions of (ii),

$$y = A_1y_1 + A_2y_2 + \ldots\ldots + A_ny_n,$$

where A_1, A_2, ..., A_n, are arbitrary constants, is also a solution. If now the solutions y_1, y_2, ..., y_n, be independent of one another, so that no one of them can be expressed by means of a linear function of all, or of any of, the others, then the foregoing value of y is a solution involving n arbitrary constants; it is therefore the Complementary Function. In order that this may be the case, there must exist no equation of the form

$$\lambda_1 y_1 + \lambda_2 y_2 + \ldots\ldots + \lambda_n y_n = 0,$$

for any values whatever of the constants $\lambda_1, \lambda_2, \ldots\ldots, \lambda_n$, other than zero for each of them. If all the constants λ be not zero, we have the derived equations

$$\lambda_1 \frac{d^{n-1}y_1}{dx^{n-1}} + \lambda_2 \frac{d^{n-1}y_2}{dx^{n-1}} + \ldots\ldots + \lambda_n \frac{d^{n-1}y_n}{dx^{n-1}} = 0,$$

$$\lambda_1 \frac{d^{n-2}y_1}{dx^{n-2}} + \lambda_2 \frac{d^{n-2}y_2}{dx^{n-2}} + \ldots\ldots + \lambda_n \frac{d^{n-2}y_n}{dx^{n-2}} = 0,$$

$$\ldots\ldots\ldots\ldots\ldots\ldots\ldots\ldots\ldots\ldots\ldots\ldots\ldots\ldots$$

$$\lambda_1 \frac{dy_1}{dx} + \lambda_2 \frac{dy_2}{dx} + \ldots\ldots + \lambda_n \frac{dy_n}{dx} = 0;$$

and, since the λ's do not all vanish, the determinant obtained by eliminating the λ's must vanish, that is,

$$\Delta = \begin{vmatrix} \dfrac{d^{n-1}y_1}{dx^{n-1}}, & \dfrac{d^{n-1}y_2}{dx^{n-1}}, & \ldots\ldots, & \dfrac{d^{n-1}y_n}{dx^{n-1}} \\[2ex] \dfrac{d^{n-2}y_1}{dx^{n-2}}, & \dfrac{d^{n-2}y_2}{dx^{n-2}}, & \ldots\ldots, & \dfrac{d^{n-2}y_n}{dx^{n-2}} \\[2ex] \ldots\ldots\ldots\ldots\ldots\ldots\ldots\ldots\ldots \\[1ex] \dfrac{dy_1}{dx}, & \dfrac{dy_2}{dx}, & \ldots\ldots, & \dfrac{dy_n}{dx} \\[2ex] y_1, & y_2, & \ldots\ldots, & y_n \end{vmatrix} = 0.$$

Hence the condition that the y's should be independent or, in other words, that the foregoing value of y should be the Complementary Function, is that Δ should not vanish.

73. It is easily proved that, if Δ be zero, then some equation of the form

$$\lambda_1 y_1 + \lambda_2 y_2 + \ldots\ldots + \lambda_n y_n = 0$$

must exist. For otherwise let the value of the left-hand side be denoted by u; multiply the columns in Δ by $\lambda_1, \lambda_2, \ldots\ldots, \lambda_n$

respectively and add them together, substituting their sum for any one column—as the first. Then we have

$$\begin{vmatrix} \dfrac{d^{n-1}u}{dx^{n-1}}, & \dfrac{d^{n-1}y_2}{dx^{n-1}}, & \cdots\cdots, & \dfrac{d^{n-1}y_n}{dx^{n-1}} \\[2ex] \dfrac{d^{n-2}u}{dx^{n-2}}, & \dfrac{d^{n-2}y_2}{dx^{n-2}}, & \cdots\cdots, & \dfrac{d^{n-2}y_n}{dx^{n-2}} \\[2ex] \cdots\cdots\cdots\cdots\cdots\cdots\cdots\cdots \\[1ex] u, & y_2, & \cdots\cdots, & y_n \end{vmatrix} = 0,$$

an equation of order $n-1$ which determines u. Now this is satisfied by $u = y_1, y_2, \ldots, y_n$, that is, it has n particular solutions which are supposed independent. But the number of independent particular solutions which an equation can have is equal to its order, a property which is violated by the preceding result. The foregoing equation in u must therefore be an identity so that u is zero; and therefore, on the supposition that Δ is zero, there is a relation between the n quantities y.

74. The value of Δ, when different from zero, can be found as follows. Let the values $y = y_1, y_2, \ldots, y_n$ be substituted in (ii) and from the n resulting equations let the coefficients $X_2, X_3, \ldots, X_n$ be eliminated; then we have

$$X_0 \begin{vmatrix} \dfrac{d^n y_1}{dx^n}, & \dfrac{d^n y_2}{dx^n}, & \cdots\cdots, & \dfrac{d^n y_n}{dx^n} \\[2ex] \dfrac{d^{n-2}y_1}{dx^{n-2}}, & \dfrac{d^{n-2}y_2}{dx^{n-2}}, & \cdots\cdots, & \dfrac{d^{n-2}y_n}{dx^{n-2}} \\[2ex] \dfrac{d^{n-3}y_1}{dx^{n-3}}, & \dfrac{d^{n-3}y_2}{dx^{n-3}}, & \cdots\cdots, & \dfrac{d^{n-3}y_n}{dx^{n-3}} \\[2ex] \cdots\cdots\cdots\cdots\cdots\cdots\cdots\cdots \\[1ex] y_1, & y_2, & \cdots\cdots, & y_n \end{vmatrix} + X_1\Delta = 0.$$

The determinant which is multiplied by X_0 is $\dfrac{d\Delta}{dx}$, and therefore this equation is

$$X_0 \frac{d\Delta}{dx} + X_1\Delta = 0,$$

which when integrated gives

$$\Delta = Ce^{-\int X_1 X_0^{-1} dx}.$$

Since Δ and $\int X_1 X_0^{-1} dx$ are determinate functions of x, the constant C must be determined by some other method; comparison of particular terms is often effective. The value of C will evidently change with a change in the set of fundamental solutions $y_1, y_2, \ldots, y_n$; but C is never zero.

Ex. Let y_1 be a particular solution of the equation

$$X_0 \frac{d^m y}{dx^m} + X_1 \frac{d^{m-1}y}{dx^{m-1}} + \ldots\ldots + X_{m-1}\frac{dy}{dx} + X_m y = 0 ;$$

when we write $y_1 \int z\,dx$ for y, the equation determining z is (§ 76, *post*) of order $m-1$. Let z_1 be a particular solution of this, so that $y_1 \int z_1 dx$ is a second particular solution of the y-equation; and let $z_1 \int u\,dx$ be substituted for z. Thus the equation in u is of order $m-2$. Let u_1 be a particular solution of this equation; then $y_1 \int z_1 dx \int u_1 dx$ is a third particular solution of the original equation. Proceeding in this way by $m-1$ successive substitutions, we shall arrive at an equation of the form

$$\frac{dw}{dx} = tw,$$

of which a solution can be found ; and there will be, in all, m particular solutions y.

Prove that these particular solutions y are independent of one another ; and shew that for this set of particular solutions

$$\Delta = (-1)^{\frac{1}{2}m(m-1)} y_1{}^m z_1{}^{m-1} u_1{}^{m-2}\ldots\ldots\ldots w_1.$$

(Fuchs.)

75. The Particular Integral may now be deduced by means of the method of the variation of parameters; this is the most symmetrical method, but another will be indicated in the next section. In the equation

$$y = A_1 y_1 + A_2 y_2 + \ldots\ldots\ldots + A_n y_n,$$

let the A's be supposed functions of x instead of constants; then the value of $\frac{dy}{dx}$ is given by

$$\frac{dy}{dx} = A_1 \frac{dy_1}{dx} + A_2 \frac{dy_2}{dx} + \ldots\ldots\ldots + A_n \frac{dy_n}{dx}$$
$$+ y_1 \frac{dA_1}{dx} + y_2 \frac{dA_2}{dx} + \ldots\ldots\ldots + y_n \frac{dA_n}{dx}.$$

Now as we have n functions A, while the only condition as yet attached to them is that they are such as to make the preceding value of y satisfy the differential equation (i), we may make them satisfy $n-1$ other conditions assigned at pleasure, provided these are not inconsistent. Let us assume as one of these conditions

$$y_1 \frac{dA_1}{dx} + y_2 \frac{dA_2}{dx} + \ldots\ldots\ldots + y_n \frac{dA_n}{dx} = 0\,;$$

we then have

$$\frac{dy}{dx} = A_1 \frac{dy_1}{dx} + A_2 \frac{dy_2}{dx} + \ldots\ldots\ldots + A_n \frac{dy_n}{dx}.$$

Differentiating this again, we have

$$\frac{d^2y}{dx^2} = A_1 \frac{d^2y_1}{dx^2} + A_2 \frac{d^2y_2}{dx^2} + \ldots\ldots\ldots + A_n \frac{d^2y_n}{dx^2},$$

provided we assign as another condition

$$\frac{dy_1}{dx}\frac{dA_1}{dx} + \frac{dy_2}{dx}\frac{dA_2}{dx} + \ldots\ldots\ldots + \frac{dy_n}{dx}\frac{dA_n}{dx} = 0.$$

Proceeding in this way, and assuming that the A's are such as to satisfy

$$\frac{d^2y_1}{dx^2}\frac{dA_1}{dx} + \frac{d^2y_2}{dx^2}\frac{dA_2}{dx} + \ldots\ldots\ldots + \frac{d^2y_n}{dx^2}\frac{dA_n}{dx} = 0,$$

$$\frac{d^3y_1}{dx^3}\frac{dA_1}{dx} + \frac{d^3y_2}{dx^3}\frac{dA_2}{dx} + \ldots\ldots\ldots + \frac{d^3y_n}{dx^3}\frac{dA_n}{dx} = 0,$$

$$\ldots\ldots\ldots\ldots\ldots\ldots\ldots\ldots\ldots\ldots\ldots\ldots\ldots\ldots\ldots\ldots\ldots$$

$$\frac{d^{n-2}y_1}{dx^{n-2}}\frac{dA_1}{dx} + \frac{d^{n-2}y_2}{dx^{n-2}}\frac{dA_2}{dx} + \ldots\ldots\ldots + \frac{d^{n-2}y_n}{dx^{n-2}}\frac{dA_n}{dx} = 0,$$

(which, with the previous two, make up $n-1$ assigned conditions, not inconsistent), we have

$$\frac{d^3y}{dx^3} = A_1 \frac{d^3y_1}{dx^3} + A_2 \frac{d^3y_2}{dx^3} + \ldots\ldots\ldots + A_n \frac{d^3y_n}{dx^3},$$

$$\ldots\ldots\ldots\ldots\ldots\ldots\ldots\ldots\ldots\ldots\ldots\ldots\ldots\ldots\ldots\ldots\ldots$$

$$\frac{d^{n-1}y}{dx^{n-1}} = A_1 \frac{d^{n-1}y_1}{dx^{n-1}} + A_2 \frac{d^{n-1}y_2}{dx^{n-1}} + \ldots\ldots\ldots + A_n \frac{d^{n-1}y_n}{dx^{n-1}}.$$

The last of these, when differentiated, gives

$$\frac{d^n y}{dx^n} = A_1 \frac{d^n y_1}{dx^n} + A_2 \frac{d^n y_2}{dx^n} + \ldots\ldots\ldots + A_n \frac{d^n y_n}{dx^n}$$

$$+ \frac{d^{n-1}y_1}{dx^{n-1}}\frac{dA_1}{dx} + \frac{d^{n-1}y_2}{dx^{n-1}}\frac{dA_2}{dx} + \ldots\ldots\ldots + \frac{d^{n-1}y_n}{dx^{n-1}}\frac{dA_n}{dx}\,;$$

but, as all the conditions which were assignable have been used, the second part of the right-hand side does not vanish. We multiply the differential coefficients of y thus expressed by the algebraical coefficients, which are attached to them in the equation (i) of § 71, and add the results; since y is a solution of (i), and $y_1, y_2, \ldots, y_n$, are solutions of (ii) of § 71, we have

$$V = X_0 \left(\frac{d^{n-1}y_1}{dx^{n-1}} \frac{dA_1}{dx} + \frac{d^{n-1}y_2}{dx^{n-1}} \frac{dA_2}{dx} + \ldots + \frac{d^{n-1}y_n}{dx^{n-1}} \frac{dA_n}{dx} \right).$$

Let Δ_r be the minor of $\dfrac{d^{n-1}y_r}{dx^{n-1}}$ in Δ, for the values $r = 1, 2, \ldots, n$; then the n equations giving the values of $\dfrac{dA_1}{dx}, \dfrac{dA_2}{dx}, \ldots, \dfrac{dA_n}{dx}$, have as their solution the equation

$$X_0 \Delta \frac{dA_r}{dx} = V \Delta_r,$$

for all the values of r. Hence

$$\frac{dA_r}{dx} = \frac{V\Delta_r}{X_0\Delta},$$

and therefore

$$A_r = C_r + \int \frac{V\Delta_r}{X_0\Delta} \, dx,$$

where C_r is an arbitrary constant. The value of y is therefore

$$y = \sum_{r=1}^{r=n} y_r \left\{ C_r + \int \frac{V\Delta_r}{X_0\Delta} \, dx \right\},$$

the Particular Integral being

$$y_1 \int \frac{V\Delta_1}{X_0\Delta} \, dx + y_2 \int \frac{V\Delta_2}{X_0\Delta} \, dx + \ldots + y_n \int \frac{V\Delta_n}{X_0\Delta} \, dx.$$

Ex. 1. Shew that, if $f_1(x)$, $f_2(x)$, $f_3(x)$, be three particular solutions of the equation

$$\frac{d^3y}{dx^3} + \phi(x)\frac{d^2y}{dx^2} + Q\frac{dy}{dx} + Sy = 0,$$

in which Q and S are functions of x only, then the complete integral of

$$\frac{d^3y}{dx^3} + \phi(x)\frac{d^2y}{dx^2} + Q\frac{dy}{dx} + Sy = \psi(x)$$

is given by

$$y = C_1 f_1(x) + C_2 f_2(x) + C_3 f_3(x) + \int^x \psi(\xi) e^{\int_a^\xi \phi(z)dz} \begin{vmatrix} \dfrac{df_1(\xi)}{d\xi}, & \dfrac{df_2(\xi)}{d\xi}, & \dfrac{df_3(\xi)}{d\xi} \\ f_1(\xi), & f_2(\xi), & f_3(\xi) \\ f_1(x), & f_2(x), & f_3(x) \end{vmatrix} d\xi,$$

where C_1, C_2, C_3, are arbitrary constants and a is a determinate constant.

Ex. 2. Solve the equations

    (i) $\quad x^2 \dfrac{d^2y}{dx^2} - 2x(1+x)\dfrac{dy}{dx} + 2(1+x)y = x;$

    (ii) $\quad (x^2+2)\dfrac{d^3y}{dx^3} - 2x\dfrac{d^2y}{dx^2} + (x^2+2)\dfrac{dy}{dx} - 2xy = x^2+2.$

76. When we know one or several particular solutions of the equation (ii) of § 71, the order of the equation can be depressed by a number equal to the number of particular solutions known. Thus suppose we know that y_1 is a particular solution of the equation; when we change the variable from y to $y_1 u$, the equation becomes

$$X_0 y_1 \frac{d^n u}{dx^n} + X_1' \frac{d^{n-1}u}{dx^{n-1}} + \ldots\ldots + X'_{n-1}\frac{du}{dx}$$

$$+ u\left(X_0 \frac{d^n y_1}{dx^n} + X_1 \frac{d^{n-1}y_1}{dx^{n-1}} + \ldots\ldots + X_{n-1}\frac{dy_1}{dx} + X_n y_1\right) = 0$$

or, what is the same thing,

$$X_0 y_1 \frac{d^n u}{dx^n} + X_1' \frac{d^{n-1}u}{dx^{n-1}} + \ldots\ldots + X'_{n-1}\frac{du}{dx} = 0,$$

in which X_1', X_2', $\ldots\ldots$, X'_{n-1}, are functions of X_0, X_1, $\ldots\ldots$, X_{n-1}, and differential coefficients of y_1. If now for $\dfrac{du}{dx}$ we substitute v, the resulting equation is of order $n-1$; the original equation has therefore had its order depressed by unity.

If y_2 be another particular solution of (ii), then y_2/y_1 is a value of u, and therefore $\dfrac{d}{dx}\left(\dfrac{y_2}{y_1}\right)$ is a solution of the equation in v; this equation can therefore have its order depressed by unity and the order of the new equation will be less by two than that of (ii). It therefore is possible by proceeding in this way to diminish the order of an equation by m, when m particular solutions are known. Each depressed equation remains linear.

77. When $n-1$ particular solutions of an equation of the nth order are known, the equation can be depressed so as to be a linear equation of the first order, and as the latter can be solved, it follows that we can *obtain the primitive of an equation of the nth order when $n-1$ particular solutions are known.* The following method of obtaining the primitive avoids the process of successive depressions of the differential equation.

Let the $n-1$ particular solutions of the equation (ii) be represented by $y_1, y_2, \ldots\ldots, y_{n-1}$; and let $C_1, C_2, \ldots\ldots, C_{n-1}$, be $n-1$ functions of x such that

$$y = C_1 y_1 + C_2 y_2 + \ldots\ldots + C_{n-1} y_{n-1}$$

is a solution of (ii); as this is the only relation between the $n-1$ functions, we may assign at pleasure $n-2$ other relations, provided they are not inconsistent. Let these be

$$y_1 \frac{dC_1}{dx} + y_2 \frac{dC_2}{dx} + \ldots\ldots + y_{n-1} \frac{dC_{n-1}}{dx} = 0,$$

$$\frac{dy_1}{dx} \frac{dC_1}{dx} + \frac{dy_2}{dx} \frac{dC_2}{dx} + \ldots\ldots + \frac{dy_{n-1}}{dx} \frac{dC_{n-1}}{dx} = 0,$$

$$\ldots\ldots\ldots\ldots\ldots\ldots\ldots\ldots\ldots\ldots\ldots\ldots$$

$$\frac{d^{n-3}y_1}{dx^{n-3}} \frac{dC_1}{dx} + \frac{d^{n-3}y_2}{dx^{n-3}} \frac{dC_2}{dx} + \ldots\ldots + \frac{d^{n-3}y_{n-1}}{dx^{n-3}} \frac{dC_{n-1}}{dx} = 0;$$

then the values of the successive differential coefficients of y are given by

$$\frac{dy}{dx} = C_1 \frac{dy_1}{dx} + C_2 \frac{dy_2}{dx} + \ldots\ldots + C_{n-1} \frac{dy_{n-1}}{dx},$$

$$\frac{d^2y}{dx^2} = C_1 \frac{d^2y_1}{dx^2} + C_2 \frac{d^2y_2}{dx^2} + \ldots\ldots + C_{n-1} \frac{d^2y_{n-1}}{dx^2},$$

$$\ldots\ldots\ldots\ldots\ldots\ldots\ldots\ldots\ldots\ldots\ldots\ldots$$

$$\frac{d^{n-2}y}{dx^{n-2}} = C_1 \frac{d^{n-2}y_1}{dx^{n-2}} + C_2 \frac{d^{n-2}y_2}{dx^{n-2}} + \ldots\ldots + C_{n-1} \frac{d^{n-2}y_{n-1}}{dx^{n-2}},$$

$$\frac{d^{n-1}y}{dx^{n-1}} = C_1 \frac{d^{n-1}y_1}{dx^{n-1}} + C_2 \frac{d^{n-1}y_2}{dx^{n-1}} + \ldots\ldots + C_{n-1} \frac{d^{n-1}y_{n-1}}{dx^{n-1}}$$
$$+ \sum_{r=1}^{r=n-1} \frac{dC_r}{dx} \frac{d^{n-2}y_r}{dx^{n-2}},$$

$$\frac{d^n y}{dx^n} = C_1 \frac{d^n y_1}{dx^n} + C_2 \frac{d^n y_2}{dx^n} + \ldots\ldots + C_{n-1} \frac{d^n y_{n-1}}{dx^n}$$
$$+ 2 \sum_{r=1}^{r=n-1} \frac{dC_r}{dx} \frac{d^{n-1}y_r}{dx^{n-1}} + \sum_{r=1}^{r=n-1} \frac{d^2 C_r}{dx^2} \frac{d^{n-2}y_r}{dx^{n-2}}.$$

The substitution of these values in the equation (ii) gives

$$X_0 \left\{ \sum_{r=1}^{r=n-1} \left(\frac{d^2C_r}{dx^2} \frac{d^{n-2}y_r}{dx^{n-2}} + 2 \frac{dC_r}{dx} \frac{d^{n-1}y_r}{dx^{n-1}} \right) \right\} + X_1 \sum_{r=1}^{r=n-1} \frac{dC_r}{dx} \frac{d^{n-2}y_r}{dx^{n-2}} = 0,$$

since $y_1, y_2, \ldots\ldots, y_{n-1}$, are particular solutions.

Let Δ denote the determinant

$$\begin{vmatrix} \dfrac{d^{n-2}y_1}{dx^{n-2}}, & \dfrac{d^{n-2}y_2}{dx^{n-2}}, & \cdots\cdots, & \dfrac{d^{n-2}y_{n-1}}{dx^{n-2}} \\[2mm] \dfrac{d^{n-3}y_1}{dx^{n-3}}, & \dfrac{d^{n-3}y_2}{dx^{n-3}}, & \cdots\cdots, & \dfrac{d^{n-3}y_{n-1}}{dx^{n-3}} \\[2mm] \multicolumn{4}{c}{\cdots\cdots\cdots\cdots\cdots\cdots\cdots\cdots\cdots} \\[1mm] y_1, & y_2, & \cdots\cdots, & y_{n-1} \end{vmatrix};$$

and let the minor of $\dfrac{d^{n-2}y_r}{dx^{n-2}}$ in Δ be denoted by Δ_r, for the values $r = 1\ 2, \ldots\ldots, n-1$. Then we have

$$\frac{\dfrac{dC_1}{dx}}{\Delta_1} = \frac{\dfrac{dC_2}{dx}}{\Delta_2} = \ldots\ldots = \frac{\dfrac{dC_{n-1}}{dx}}{\Delta_{n-1}} = z \text{ say,}$$

and therefore, for these values of r,

$$\frac{dC_r}{dx} = z\Delta_r.$$

Hence

$$\sum_{r=1}^{r=n-1} \frac{dC_r}{dx} \frac{d^{n-1}y_r}{dx^{n-1}} = z \sum_{r=1}^{r=n-1} \Delta_r \frac{d^{n-1}y_r}{dx^{n-1}} = z \frac{d\Delta}{dx},$$

and

$$\sum_{r=1}^{r=n-1} \frac{dC_r}{dx} \frac{d^{n-2}y_r}{dx^{n-2}} = z \sum_{r=1}^{r=n-1} \Delta_r \frac{d^{n-2}y_r}{dx^{n-2}} = z\Delta.$$

Also

$$\frac{d^2C_r}{dx^2} = \frac{dz}{dx} \Delta_r + z \frac{d\Delta_r}{dx},$$

and

$$\sum_{r=1}^{r=n-1} \frac{d\Delta_r}{dx} \frac{d^{n-2}y_r}{dx^{n-2}} = 0;$$

so that

$$\sum_{r=1}^{r=n-1} \frac{d^2C_r}{dx^2} \frac{d^{n-2}y_r}{dx^{n-2}} = \frac{dz}{dx} \sum_{r=1}^{r=n-1} \Delta_r \frac{d^{n-2}y_r}{dx^{n-2}} = \Delta \frac{dz}{dx};$$

the transformed equation therefore is

$$X_0 \Delta \frac{dz}{dx} + 2X_0 \frac{d\Delta}{dx} z + X_1 \Delta z = 0.$$

Dividing by $X_0 \Delta$, we have

$$\frac{dz}{dx} + \left(\frac{2}{\Delta} \frac{d\Delta}{dx} + \frac{X_1}{X_0} \right) z = 0,$$

the integral of which is

$$z = A \Delta^{-2} e^{-\int \frac{X_1}{X_0} dx}.$$

The corresponding value of C_r is derivable from

$$\frac{dC_r}{dx} = z \Delta_r = A \frac{\Delta_r}{\Delta^2} e^{-\int \frac{X_1}{X_0} dx},$$

and therefore

$$C_r = A_r + A \int \frac{\Delta_r}{\Delta^2} e^{-\int \frac{X_1}{X_0} dx} \, dx,$$

for the values $r = 1, 2, \ldots\ldots, n - 1$. Hence we have n arbitrary constants, viz. $A, A_1, A_2, \ldots\ldots, A_{n-1}$; and the primitive of (ii) is thus

$$y = \sum_{r=1}^{r=n-1} A_r y_r + A \sum_{r=1}^{r=n-1} y_r \int \frac{\Delta_r}{\Delta^2} e^{-\int \frac{X_1}{X_0} dx} \, dx.$$

Ex. Solve completely

$$\frac{d^3 y}{dx^3} = P \left(x^2 \frac{d^2 y}{dx^2} - 2x \frac{dy}{dx} + 2y \right) + Q,$$

where P and Q are functions of x.

GEOMETRICAL APPLICATION: TRAJECTORIES.

78. It has already been noticed that a differential equation is the appropriate analytical expression of any property of a curve which is connected with its direction and its curvature; and it therefore follows that the investigation of many geometrical questions ultimately depends upon the solution of a differential equation. In the higher parts of mathematics, differential equations are of almost universal occurrence; but in other subjects it is less possible than it is in geometry to give examples, as there is no necessarily general method of arriving at the differential equation, while its deduction in geometrical problems is obtained almost immediately by the use of the formulæ of the differential calculus. There will be no attempt to give here any complete classification of applications to geometry; only a single general problem will be discussed, that of Trajectories.

A *Trajectory* is defined to be a line which, at its points of intersection with the members of a family of curves expressed by one equation, cuts them according to some given law.

79. As the most general form possible, let

$$f(x, y, a) = 0$$

denote a family of curves of which a is the parameter; through *any* point on one curve a trajectory will pass, and there will thus be a second system of curves representing these trajectories. Let ξ and η be the current coordinates of this second system; and suppose the analytical expression of the law which holds at each point of intersection to be

$$F\left\{x, y, \frac{dy}{dx}, \frac{d^2y}{dx^2}, \ldots\ldots\ldots, \xi, \eta, \frac{d\eta}{d\xi}, \frac{d^2\eta}{d\xi^2}, \ldots\ldots\ldots\right\} = 0.$$

In this equation, ξ and η at a point of intersection are respectively the same as x and y, being the coordinates of that point; but $\frac{d\eta}{d\xi}, \ldots\ldots\ldots$ are not the same as $\frac{dy}{dx}, \ldots\ldots\ldots$, for they indicate the direction and the curvature of the two intersecting curves.

We proceed as follows.

From the equation

$$f(x, y, a) = 0$$

we obtain the values of all the differential coefficients of y, which occur in the relation $F = 0$, as functions of x, y, and a; and in each of these expressions we substitute the value of a as a function of x and y derived from the equation of the curve. This will be equivalent to eliminating a between $f = 0$ and the equation giving each differential coefficient. Let these values of the differential coefficients of y be substituted in $F = 0$; it then becomes an equation which involves x, y, ξ, η, and differential coefficients of η with respect to ξ. But we have seen that x and y are the same as ξ and η, since both sets are the coordinates of the same point; therefore $F = 0$ becomes a differential equation in η and ξ only.

80. The most frequent example of trajectories is that in which a system of curves is to be obtained cutting a given system at a constant angle. If this angle be a right angle, the trajectory is called *orthogonal*; if other than a right angle, the trajectory is called *oblique*.

In the case of orthogonal trajectories, the tangents at a common point are to be perpendicular, and therefore

$$1 + \frac{dy}{dx}\frac{d\eta}{d\xi} = 0,$$

which is the form of $F = 0$ for this case. For the given system of curves, we have

$$f(x, y, a) = 0,$$
$$\frac{\partial f}{\partial x} + \frac{\partial f}{\partial y}\frac{dy}{dx} = 0,$$

from which we eliminate a and obtain a relation between x, y and $\frac{dy}{dx}$, which is really the differential equation of this system of curves; let this relation be

$$\psi\left(x, y, \frac{dy}{dx}\right) = 0.$$

Now, for the trajectory, we have

$$\xi = x, \quad y = \eta,$$
and
$$\frac{dy}{dx} = -\frac{1}{\dfrac{d\eta}{d\xi}};$$

therefore the differential equation of the trajectory is

$$\psi\left(\xi, \eta, -\frac{1}{\dfrac{d\eta}{d\xi}}\right) = 0.$$

The elimination of the parameter is immediate when the equation of the given family of curves occurs in the form

$$\phi(x, y) = a.$$

For we then have

$$\frac{\partial \phi}{\partial x} + \frac{\partial \phi}{\partial y} \frac{dy}{dx} = 0,$$

which at once gives $\frac{dy}{dx}$ independent of a, and is the form of $\psi = 0$ for this case.

81. When the equation of the curve is given in polar coordinates, the same method may be applied. For we then have

$$\chi(r, \theta, c) = 0$$

as the equation of the family of curves. If ϕ be the angle between the radius vector and the part of the tangent to the curve drawn from the point back towards the line from which θ is measured, we have

$$\tan \phi = r \frac{d\theta}{dr};$$

while, if Φ be the same quantity for the trajectory, and R and Θ be the polar coordinates of a point on it,

$$\tan \Phi = R \frac{d\Theta}{dR}.$$

Since the tangents are at right angles,

$$\Phi \sim \phi = \tfrac{1}{2}\pi,$$

and therefore

$$r \frac{d\theta}{dr} R \frac{d\Theta}{dR} + 1 = 0,$$

where R and r, Θ and θ (but not their derivatives), are the same.

Now

$$\frac{\partial \chi}{\partial r} + \frac{\partial \chi}{\partial \theta} \frac{d\theta}{dr} = 0;$$

eliminating c between this equation and the equation of the curve, we find a relation of the form

$$\psi\left(r, \theta, \frac{d\theta}{dr}\right) = 0.$$

For the trajectory

$$R = r, \ \Theta = \theta, \text{ and } \frac{d\theta}{dr} = -\frac{1}{R^2 \dfrac{d\Theta}{dR}} = -\frac{1}{R^2} \frac{dR}{d\Theta};$$

the differential equation of the trajectory is therefore

$$\psi\left(R, \ \Theta, \ -\frac{1}{R^2} \frac{dR}{d\Theta}\right) = 0.$$

This, when integrated, gives the equation of the system of curves possessing the required property.

Ex. 1. Find the orthogonal trajectory of the series of straight lines

$$y = mx.$$

We have

$$\frac{dy}{dx} = m,$$

and therefore the differential equation of these lines is

$$x \frac{dy}{dx} = y.$$

Hence, by our rule, the differential equation of the system of orthogonal trajectories is

$$\xi = -\eta \frac{d\eta}{d\xi},$$

which on integration gives

$$\xi^2 + \eta^2 = c^2,$$

a series of concentric circles having for common centre the common point of the lines.

Ex. 2. Find the orthogonal trajectory of

$$r^n = a^n \sin n\theta.$$

Taking logarithms and differentiating, we have

$$\frac{n}{r} \frac{dr}{d\theta} = n \frac{\cos n\theta}{\sin n\theta},$$

which is the differential equation of the family of curves. For the trajectory, we have

$$\frac{1}{r} \frac{dr}{d\theta} = -R \frac{d\Theta}{dR};$$

and therefore the differential equation of the trajectory is

$$R \frac{d\Theta}{dR} + \frac{\cos n\Theta}{\sin n\Theta} = 0.$$

The variables may be separated; we have

$$n \frac{dR}{R} = -n \frac{\sin n\Theta}{\cos n\Theta} d\Theta,$$

so that

$$R^n = A^n \cos n\Theta,$$

the family required.

Ex. 3. Prove that, whatever be the value of n, a non-parametric constant, the orthogonal trajectory of the curves included in

$$y = cx^n$$

is a family of conics.

Ex. 4. Shew that the orthogonal trajectory of a system of confocal ellipses is a system of hyperbolas confocal with the ellipses.

Ex. 5. Obtain the orthogonal trajectory of the system of curves

(i) $r^n \sin n\theta = a^n$;

(ii) $r^2 = a^2 \log (c \tan \theta)$, c being arbitrary.

Ex. 6. Shew that, if $f(x+iy)$ be denoted by $u+iv$, where u and v are real, and i denotes $\sqrt{-1}$, then the families of curves $u=$const., $v=$const., are the orthogonal trajectories of each other; and the families $u \cos a + v \sin a =$const.. for different values of a, are oblique trajectories of each other.

In particular, shew that, if u and v, so obtained, be homogeneous of order n, then the value of u is

$$nu = x \frac{\partial v}{\partial y} - y \frac{\partial v}{\partial x}.$$

How may the value of u be found when n is zero?

Ex. 7. Find a system of curves cutting, at a constant angle other than right, a system of concentric circles.

82. If one of the variables be given as an explicit function of the other and the parameter, the equation will be of the form

$$y = \phi(x, a);$$

instead of eliminating a we may proceed as follows. Let the equation of the orthogonal trajectory be

$$\eta = \phi(\xi, a),$$

where in the last a is to be considered an unknown function of ξ to be determined so that the curve may be the orthogonal trajectory. We now have

$$\frac{dy}{dx} = \frac{\partial\phi}{\partial x},$$

$$\frac{d\eta}{d\xi} = \frac{\partial\phi}{\partial\xi} + \frac{\partial\phi}{\partial a}\frac{da}{d\xi},$$

and therefore

$$\frac{\partial\phi}{\partial x}\left(\frac{\partial\phi}{\partial\xi} + \frac{\partial\phi}{\partial a}\frac{da}{d\xi}\right) + 1 = 0.$$

Now, as no further differentiations are to take place, we may write $\frac{\partial\phi}{\partial\xi}$ in place of $\frac{\partial\phi}{\partial x}$, since x is equal to ξ; hence we have

$$1 + \left(\frac{\partial\phi}{\partial\xi}\right)^2 + \frac{\partial\phi}{\partial a}\frac{\partial\phi}{\partial\xi}\frac{da}{d\xi} = 0.$$

This is an equation between two variables a and ξ; when integrated, it determines the value of a which, being substituted in

$$\eta = \phi(\xi, a),$$

gives the orthogonal trajectory.

Ex. Obtain the orthogonal trajectory of the ellipses represented by

$$y = a(1 - x^2)^{\frac{1}{2}}.$$

Here

$$\frac{\partial\phi}{\partial\xi} = -a\xi(1 - \xi^2)^{-\frac{1}{2}},$$

$$\frac{\partial\phi}{\partial a} = (1 - \xi^2)^{\frac{1}{2}};$$

and the equation determining a is

$$1 + a^2\frac{\xi^2}{1 - \xi^2} - \xi a\frac{da}{d\xi} = 0,$$

which gives

$$\frac{da^2}{d\xi} - \frac{2a^2\xi}{1 - \xi^2} = \frac{2}{\xi}.$$

This on integration leads to the equation

$$a^2(1 - \xi^2) = A - \xi^2 + \log\xi^2;$$

therefore the orthogonal trajectory required is

$$\eta^2 = A - \xi^2 + \log\xi^2.$$

MISCELLANEOUS EXAMPLES.

1. Solve the equations:

(i) $\dfrac{d^2y}{dx^2} + \dfrac{2}{x}\dfrac{dy}{dx} + \left(n^2 - \dfrac{2}{x^2}\right)y = 0$;

(ii) $x^2\dfrac{d^2y}{dx^2} - 2x\dfrac{dy}{dx} + 2y\,(1+x^2) = 0$;

(iii) $2y\dfrac{d^2y}{dx^2} - 3\left(\dfrac{dy}{dx}\right)^2 - 4y^2 = 0$;

(iv) $y^3 + \left\{y^2 + \left(\dfrac{dy}{dx}\right)^2\right\}\dfrac{d^2y}{dx^2} = 0$;

(v) $x\dfrac{d^2y}{dx^2} + \{2 - x\phi\,(x)\}\dfrac{dy}{dx} = y\phi\,(x)$;

(vi) $\left\{y - x\dfrac{dy}{dx}\right\}\dfrac{d^2y}{dx^2} = 4\left(\dfrac{dy}{dx}\right)^2$;

(vii) $\dfrac{d^2y}{dx^2} + 2n\cot nx\,\dfrac{dy}{dx} + (m^2 - n^2)y = 0$;

(viii) $x\,(x+y)\dfrac{d^2y}{dx^2} + (x-y)\dfrac{dy}{dx} + x\left(\dfrac{dy}{dx}\right)^2 - y = 0$;

(ix) $\left(\dfrac{dy}{dx}\right)^2 - y\dfrac{d^2y}{dx^2} = n\left\{\left(\dfrac{dy}{dx}\right)^2 + a^2\left(\dfrac{d^2y}{dx^2}\right)^2\right\}^{\frac{1}{2}}$;

(x) $\sin^2 x\,\dfrac{d^2y}{dx^2} = 2y$;

(xi) $\dfrac{d^2y}{dx} + \left(\dfrac{dy}{dx}\right)^n f(x) + \dfrac{dy}{dx}\,\phi\,(x) = 0.$

2. Assuming that the primitive of
$$\dfrac{d^2y}{dx^2} + \left(1 - \dfrac{2}{x^2}\right)y = 0$$

is of the form $y = u + \dfrac{v}{x}$, prove that it is given by

$$u = A\sin(x+a), \quad v = A\cos(x+a).$$

Obtain the primitive of

$$\dfrac{d^2y}{dx^2} + \left(1 - \dfrac{2}{x^2}\right)y = x^2.$$

3. By the method of variation of parameters, deduce the primitive of

$$\frac{d^2y}{dx^2} - 2\left(n - \frac{1}{x}\right)\frac{dy}{dx} + \left(n^2 - \frac{2n}{x}\right)y = 0.$$

4. Prove that the equation

$$(a_2 + b_2 x)\frac{d^2y}{dx^2} + (a_1 + b_1 x)\frac{dy}{dx} + (a_0 + b_0 x)y = 0$$

has a particular solution of the form $e^{\lambda x}$, provided

$$(a_0 b_1 - a_1 b_0)(a_1 b_2 - a_2 b_1) = (a_0 b_2 - a_2 b_0)^2;$$

and hence solve the equation, assuming this condition satisfied.

(Schlömilch.)

5. Integrate

$$\sin^2 x \frac{d^2u}{dx^2} + \sin x \cos x \frac{du}{dx} = u.$$

If $u = 0$ when $x = 0$, and $u = 1$ when $x = \dfrac{\pi}{2}$, then $u = \sqrt{2} - 1$ when $x = \dfrac{\pi}{4}$.

Also solve the differential equation

$$\frac{d^2y}{dx^2} + y = 2y\left(1 - \frac{n+1}{2}\frac{y^{2n}}{a^{2n}}\right),$$

determining the arbitrary constants by the conditions that $y = a$ and $\dfrac{dy}{dx} = 0$ when $x = 0$.

6. The equations

$$\frac{d^2y}{dx^2} + P\frac{dy}{dx} + Qy = 0,$$

$$\frac{d^2y}{dx^2} + P'\frac{dy}{dx} + Q'y = 0,$$

have a solution in common; find the primitive of each, and the necessary relation between P, P', Q, Q' supposed to be functions of x.

7. Prove that the equation

$$\frac{1}{v}\frac{d^2v}{dx^2} = \left(\mathfrak{a}, \mathfrak{b}, \mathfrak{c}, \tfrac{1}{2}(\mathfrak{a} - \mathfrak{b} - \mathfrak{c}), \tfrac{1}{2}(\mathfrak{b} - \mathfrak{c} - \mathfrak{a}), \tfrac{1}{2}(\mathfrak{c} - \mathfrak{a} - \mathfrak{b})\right)\left(\frac{1}{x-a}, \frac{1}{x-b}, \frac{1}{x-c}\right)^2$$

can be integrated by the method of § 68, provided the relation

$$(\mathfrak{a} + \tfrac{1}{4})^{\frac{1}{2}} + (\mathfrak{b} + \tfrac{1}{4})^{\frac{1}{2}} + (\mathfrak{c} + \tfrac{1}{4})^{\frac{1}{2}} = \tfrac{1}{2}$$

be satisfied for some one set of signs given to the radicals.

Find the solution when this condition is satisfied.

8. Solve the equation

$$(x+a)^2(x+b)^2\frac{d^2y}{dx^2} = k^2y,$$

where a, b, k are constants, by assuming

$$y = (x+a)^m (x+b)^n ;$$

and obtain the general solution.

Solve similarly the equation

$$(a+x)(b+x)\frac{d^2y}{dx^2} + \tfrac{1}{2}(a+2b+3x)\frac{dy}{dx} + \tfrac{1}{4}\frac{a-b}{a+x}y = 0 ;$$

also Ex. 1 in § 68.

9. Prove that, if $\phi(x)$ be a particular solution of the equation

$$\frac{d^2z}{dx^2} = ax^{n-2}z,$$

then $x\phi\left(\dfrac{1}{x}\right)$ is a particular solution of the equation

$$\frac{d^2z}{dx^2} = ax^{-n-2}z.$$

Hence solve the equation

$$x^4\frac{d^2z}{dx^2} = Az.$$

10. Prove that, if $z = \phi(x)$ be a solution of

$$\frac{d^2z}{dx^2} = z\psi(x),$$

then $\zeta = (cx+d)\phi\left(\dfrac{ax+b}{cx+d}\right)$ is a solution of

$$(cx+d)^4\frac{d^2\zeta}{dx^2} = \zeta\psi\left(\frac{ax+b}{cx+d}\right),$$

the constants a, b, c, d, being connected by the relation

$$ad - bc = 1.$$

Hence solve the first equation in question 8.

11. Shew how to solve the equation

$$\frac{d^ny}{dx^n} + \frac{A_1}{a+bx}\frac{d^{n-1}y}{dx^{n-1}} + \frac{A_2}{(a+bx)^2}\frac{d^{n-2}y}{dx^{n-2}} + \ldots + \frac{A_ny}{(a+bx)^n} = X,$$

where X is a function of x only, and A_1, ..., A_n, are constants.

12. Integrate the equation

$$\frac{d^2y}{dx^2} + (2X+a)\frac{dy}{dx} + \left(\frac{dX}{dx} + X^2 + aX + b\right)y = 0,$$

X being any function of x.

13. Shew that, if a particular solution of the equation

$$\frac{dy}{dx} + X_1 y^2 + X_2 = 0$$

be known, X_1 and X_2 being functions of x, the primitive can be obtained. Hence solve the equation

$$\frac{dy}{dx} + y^2 \sin x = 2 \frac{\sin x}{\cos^2 x}.$$

14. The primitive of

$$\frac{d^2 y}{dx^2} + p \frac{dy}{dx} + qy = 0$$

being

$$y = Ay_1 + By_2,$$

shew that the differential equation which has for its primitive

$$z = A' y_1{}^m + B' y_2{}^m$$

is

$$F(x) \frac{d^2 z}{dx^2} + \left\{ pF(x) - (m-1) \frac{dF}{dx} \right\} \frac{dz}{dx}$$

$$+ m \left\{ \tfrac{1}{2}(m-1) \frac{d^2 F}{dx^2} + \tfrac{1}{2}(m-1) p \frac{dF}{dx} + mq F(x) \right\} z = 0,$$

where

$$F(x) = y_1 y_2.$$

(Hermite.)

15. Given an equation

$$\frac{d^2 y}{dx^2} + Iy = 0,$$

shew that the equation, whose integral is given by

$$u = y^2,$$

is

$$\frac{d^3 u}{dx^3} + 4I \frac{du}{dx} + 2 \frac{dI}{dx} u = 0,$$

and that the equation, whose integral is given by

$$v = y^3,$$

is

$$\frac{d^4 v}{dx^4} + 10I \frac{d^2 v}{dx^2} + 10 \frac{dI}{dx} \frac{dv}{dx} + \left(3 \frac{d^2 I}{dx^2} + 9I^2 \right) v = 0.$$

Account for the respective orders of these equations; and state their primitives when the primitive of the first equation is known.

16. Prove that, if y_1 and y_2 be two particular solutions of the equation

$$\frac{d^2 y}{dx^2} + P \frac{dy}{dx} + Qy = 0,$$

the roots of $y_1 = 0$ and $y_2 = 0$ separate each other so long as both of these solutions remain continuous.

(Sturm.)

17. Solve the differential equations:

(i) $\sin^2\theta\dfrac{d^2y}{d\theta^2}+\sin\theta\cos\theta\dfrac{dy}{d\theta}-y=\theta-\sin\theta$;

(ii) $\dfrac{d^2y}{dx^2}+\dfrac{1}{x^2\log x}y=e^x\left(\dfrac{2}{x}+\log x\right)$;

(iii) $(1+ax^2)\dfrac{d^2y}{dx^2}+ax\dfrac{dy}{dx}=n^2y$;

(iv) $x^2(x^2+a)\dfrac{d^2y}{dx^2}+x(2x^2+a)\dfrac{dy}{dx}=n^2y$;

(v) $\dfrac{d^2y}{dx^2}-2\left(n-\dfrac{a}{x}\right)\dfrac{dy}{dx}+\left(n^2-\dfrac{2na}{x}\right)y=e^{nx}$;

(vi) $(a^2-x^2)\dfrac{d^2y}{dx^2}-8x\dfrac{dy}{dx}-12y=0$;

(vii) $(3-x)\dfrac{d^2y}{dx^2}-(9-4x)\dfrac{dy}{dx}+(6-3x)y=0$.

18. Solve the equation

$$P\dfrac{d^2y}{dx^2}+Q\dfrac{dy}{dx}-Ry=0,$$

where P, Q, and R, are functions of x which satisfy the relation

$$R\left(\dfrac{dQ}{dx}-R\right)=Q\dfrac{dR}{dx}.$$

When this relation is not satisfied, can the equation be solved by the introduction of a factor μ so chosen that the new coefficients satisfy the relation?

19. Solve the equation

$$\dfrac{d^2y}{dx^2}=a-\dfrac{by}{(2cx-x^2)^2}.$$

(Stokes.)

20. Find the form of ϕ such that, if $x=\phi(z)$ be substituted in the equation

$$x^4\dfrac{d^2y}{dx^2}+2x^3\dfrac{dy}{dx}+n^2y=0,$$

it will become

$$\dfrac{d^2y}{dz^2}+n^2y=0;$$

and thence solve the former equation.

21. Prove that the equation $\dfrac{d^2y}{dx^2}+P\dfrac{dy}{dx}+Qy=0$ can be transformed into

$$\dfrac{d^2y}{dz^2}+F(z)\dfrac{dy}{dz}+y\Phi(z)=0,$$

when the relation between z and x is given by

$$\int dz e^{-\int F(z)\,dz} = \int dx e^{-\int P\,dx},$$

and $\Phi(z)$ is given by

$$Q e^{2\int P\,dx} = \Phi(z)\, e^{2\int F(z)\,dz}.$$

Hence reduce the equation $\dfrac{d^2y}{dx^2} - \dfrac{1}{x}\dfrac{dy}{dx} = x^2 y\,(n^2 - e^{x^2})$ to the form

$$\frac{d^2y}{dz^2} + \frac{1}{z}\frac{dy}{dz} + \left(1 - \frac{n^2}{z^2}\right)y = 0.$$

22. Solve the equation

$$\frac{d^2v}{dz^2} + 2\left(\frac{1}{z} + \frac{B}{z^2}\right)\frac{dv}{dz} + \frac{A}{z^4}\,v = 0,$$

where A and B are constants.

Verify that the equation

$$\frac{d^2y}{dx^2} + 2\left(\frac{B'}{x^3} + \frac{3}{2x}\right)\frac{dy}{dx} + \frac{\mu y}{x^6} = 0$$

is transformable into the foregoing equation by the substitution

$$x = \sin(\arc\tan z^{\frac{1}{2}}),$$

provided $\qquad B'^2 = 4B^2 - 4A + \mu\,;$

and find the relation between y and v. Hence solve the second equation.

23. By transforming the dependent variable from y to e^z, solve the equation

$$P\frac{d^2y}{dx^2} - \frac{dP}{dx}\frac{dy}{dx} = a^2 P^3 y.$$

Hence solve the equation

$$\frac{d^2v}{dx^2} + 2\frac{dv}{dx}\,X = \tfrac{1}{2}v\left\{\tfrac{1}{2}\left(\frac{1}{P}\frac{dP}{dx}\right)^2 - \frac{d}{dx}\left(\frac{1}{P}\frac{dP}{dx}\right) + 2\,(a^2 P^2 - X^2) - 2\frac{dX}{dx}\right\}.$$

(Sparre.)

24. Prove that the primitive of the equation

$$\frac{d^2\sigma}{dx^2} - \frac{5}{4\sigma}\left(\frac{d\sigma}{dx}\right)^2 + \tfrac{2}{3}\sigma^2 = 0,$$

where σ is the Schwarzian derivative of y with regard to x, is

$$y\,(A' + B'x + C'x^2) = A + Bx + Cx^2\,;$$

and shew that this is also the primitive of

$$\begin{vmatrix} y_3, & 3y_2, & 3y_1 \\ y_4, & 4y_3, & 6y_2 \\ y_5, & 5y_4, & 10y_3 \end{vmatrix} = 0.$$

where $y_1, y_2, \ldots$ are the first, second, $\ldots\ldots$ differential coefficients of y.

25. Prove that the primitive of the equation

$$\{s,\ x\} = \theta\,(\alpha-\beta)^2\,(x-\alpha)^{-2}\,(x-\beta)^{-2}$$

is

$$\frac{as+b}{cs+d} = \left(\frac{x-\alpha}{x-\beta}\right)^n,$$

where $n^2 = 1 - 2\theta$. Discuss the case in which θ, supposed constant, is equal to $\frac{1}{2}$.

26. The arc of a plane curve measured from a fixed point A up to a point P, whose rectangular coordinates are x and y, is denoted by s; obtain the general Cartesian equations of the curves for which the following equations respectively hold:

    (i) $\ s = (x^2 + y^2)^{\frac{1}{2}}$;
    (ii) $\ s = c \arctan\left(\dfrac{y}{x}\right)$;

    (iii) $\ \left(\dfrac{dy}{ds}\right)^2 = a\,\dfrac{dx}{ds}\,\dfrac{d^2x}{ds^2}$;
    (iv) $\ s = a\,\dfrac{dy}{dx}$;

    (v) $\ \dfrac{ds}{dy} + 3y\,\dfrac{d^2s}{dy^2} = 0$;
    (vi) $\ s = (x^2 + 2cx)^{\frac{1}{2}}$;
    (vii) $\ s = (y^2 + mx^2)^{\frac{1}{2}}$.

27. Find the general differential equation of all parabolas touching the axes and having their chord of contact of constant length. Solve the equation obtained.

Obtain also the differential equation of all parabolas touching the axes.

28. Shew that the differential equation of a general conic is

$$\frac{d^3}{dx^3}\left\{\left(\frac{d^2y}{dx^2}\right)^{-\frac{2}{3}}\right\} = 0,$$

and of a general parabola is

$$\frac{d^2}{dx^2}\left\{\left(\frac{d^2y}{dx^2}\right)^{-\frac{2}{3}}\right\} = 0.$$

(Monge and Halphen.)

29. Find (i) the curve in which the radius of curvature is proportional to the arc measured from a fixed point; (ii) the curve in which the product of the perpendiculars from two fixed points on the tangent is constant; (iii) the curve which has an evolute similar to itself.

30. Find a differential equation of the first order of the curve, whose radius of curvature is equal to n times the normal; and shew that it is always integrable when n is an integer. In particular, shew that when $n = 2$ the curve is a cycloid, when $n = 1$ a circle, when $n = -1$ a catenary, and when $n = -2$ a parabola.

31. Shew that the system of curves, cutting a system of confocal ellipses at a constant angle α other than right, is given by

$$x = c\cos\phi\cosh n\,(\lambda+\phi), \quad y = c\sin\phi\sinh n\,(\lambda+\phi),$$

where $2c$ is the distance between the foci and n is $\tan\alpha$.

(Mainardi and Mukhopadhyay.)

32. Obtain the orthogonal trajectories of the curves

(i) $x^2 + y^2 = cx$;   (ii) $x^2 + y^2 + c^2 = 1 + 2cxy$;

(iii) $x^3 + y^3 = 3axy$;   (iv) $rr' = c^2$;

in the last r and r' are the distances from two fixed points.

33. The curve, for which the ordinate and the abscissa of the centre of gravity of the area included between the ordinates $x = a$ and $x = x$ are in the same ratio as the bounding ordinate y and the abscissa x, is given by the equation

$$\frac{a^3}{x^3} - \frac{b^3}{y^3} = 1.$$

34. The curve whose polar equation is $r^m \cos m\theta = a^m$ rolls on a fixed straight line. Assuming that straight line to be the axis of x, shew that the locus of the curve described by the pole in the rolling curve will have for its equation

$$dx = \left\{ \left(\frac{y}{a} \right)^{\frac{2m}{1-m}} - 1 \right\}^{-\frac{1}{2}} dy.$$

In particular, shew that, when $2m = 1$, the described curve is a catenary; when $m = 2$, the described curve is an elastica.

(Frenet.)

CHAPTER V

83. It may happen that a differential equation, the solution of which is required, comes under none of the preceding classes which are all of some particular form, and therefore that the methods applicable to these fail. Recourse can then be had to approximation, in order to obtain the value of the dependent variable. The form of approximation which is most frequently adopted is that derived from converging series; by retaining a large number of terms the error can be made small, and the series may be considered to represent the value of the variable. That this method is *à priori* justifiable may be seen as follows.

The given equation is a relation between the successive differential coefficients of y and may be considered as giving the one of highest order in terms of those of lower orders; thus if it were of the second order, it would give $\frac{d^2y}{dx^2}$ in terms of $\frac{dy}{dx}$ and y. When differentiated once, it would give $\frac{d^3y}{dx^3}$ in terms of $\frac{d^2y}{dx^2}$, $\frac{dy}{dx}$ and y, that is, in terms of $\frac{dy}{dx}$ and y, since $\frac{d^2y}{dx^2}$ is expressible in terms of these two, and so for each of the differential coefficients of higher order, which can thus be expressed in terms of $\frac{dy}{dx}$ and y. But the differential equation will not give any relation between

$\dfrac{dy}{dx}$ and y, which are thus independent of one another. Suppose now that a value a be assigned to x and that for this value of x we make $y = A$ and $\dfrac{dy}{dx} = B$, which constants are, in general, arbitrary; then the equations derived by successive differentiation furnish the values for $x = a$ of the differential coefficients of y of successive orders. Let these be denoted by $C, D, E, \ldots.$ Now if the value of y be $\phi(x)$, which we assume is a function expansible by Taylor's theorem in a converging series of ascending powers of $x - a$, we have

$$\phi(x) = \phi\{a + (x - a)\}$$
$$= \phi(a) + (x - a)\frac{d\phi(a)}{da} + \frac{(x-a)^2}{2!}\frac{d^2\phi(a)}{da^2} + \frac{(x-a)^3}{3!}\frac{d^3\phi(a)}{da^3} + \ldots,$$

where $\dfrac{d^r\phi(a)}{da^r}$ stands for the value of $\dfrac{d^r\phi(x)}{dx^r}$, when a is written for x after differentiation. Inserting now for the various coefficients their values, we obtain

$$y = \phi(x) = A + B(x - a) + C\frac{(x-a)^2}{2!} + D\frac{(x-a)^3}{3!} + \ldots\ldots$$

This series, if it converges, is a solution of the given equation.

It should be remarked that, for some particular value of x, the differential equation may determine not the coefficient of highest order but one of lower order. Thus the equation

$$\frac{d^2y}{dx^2} + \frac{2n}{x}\frac{dy}{dx} - m^2y = 0$$

would for values of x other than zero determine $\dfrac{d^2y}{dx^2}$; but for $x = 0$ would give $\dfrac{dy}{dx} = 0$, if we consider infinite values of any coefficient excluded.

The foregoing method and another, which is in practice substituted for it and which will be explained in the next article, are almost impracticable in the case of equations which neither are linear nor can be transformed so as to become linear; for such equations, the determination of more than the first few terms of the expansion often entails great labour.

Ex. 1. Let us apply the foregoing method to the equation

$$\frac{d^2y}{dx^2}+xy=0.$$

When differentiated n times, the equation gives

$$\frac{d^{n+2}y}{dx^{n+2}}+x\frac{d^ny}{dx^n}+n\frac{d^{n-1}y}{dx^{n-1}}=0 ;$$

and therefore, when $x=0$,

$$\frac{d^{n+2}y}{dx_0^{n+2}}=-n\frac{d^{n-1}y}{dx_0^{n-1}}.$$

Now the given equation leaves y arbitrary, say $=A$, and $\frac{dy}{dx}$ arbitrary, say $=B$,

when $x=0$; but $\frac{d^2y}{dx_0^2}=0$. Hence we have

$$\frac{d^{3p+2}y}{dx_0^{3p+2}}=-3p\frac{d^{3p-1}y}{dx_0^{3p-1}}$$

$$=3p\,(3p-3)\,\frac{d^{3p-4}y}{dx_0^{3p-4}}$$

$$=(-1)^p\,3p\,(3p-3)...6\,.\,3\,\frac{d^2y}{dx_0^2}$$

$$=0 ;$$

similarly

$$\frac{d^{3p+1}y}{dx_0^{3p+1}}=(-1)^p\,(3p-1)\,(3p-4)....\,5\,.\,2\,\frac{dy}{dx_0}$$

$$=(-1)^p\,(3p-1)\,(3p-4)....\,5\,.\,2\,.\,B ;$$

and

$$\frac{d^{3p}y}{dx_0^{3p}}=(-1)^p\,(3p-2)\,(3p-5)....\,4\,.\,1\,.\,y_0$$

$$=(-1)^p\,(3p-2)\,(3p-5)....\,4\,.\,1\,.\,A.$$

The expansion of y is, by Maclaurin's theorem,

$$y=y_0+x\frac{dy}{dx_0}+\frac{x^2}{2!}\frac{d^2y}{dx_0^2}+\frac{x^3}{3!}\frac{d^3y}{dx_0^3}+\frac{x^4}{4!}\frac{d^4y}{dx_0^4}+...$$

$$=A\left[1-\frac{1}{3!}x^3+\frac{1\,.\,4}{6!}x^6-\frac{1\,.\,4\,.\,7}{9!}x^9+...\right]$$

$$+Bx\left[1-\frac{2}{4!}x^3+\frac{2\,.\,5}{7!}x^6-\frac{2\,.\,5\,.\,8}{10!}x^9+...\right].$$

This expression is the sum of two converging series. It contains two arbitrary constants, and is thus the primitive of the equation.

Ex. 2. Solve

$$\text{(i)}\quad x\frac{d^2y}{dx^2}+2\frac{dy}{dx}+c^3x^2y=0 ;$$

$$\text{(ii)}\quad \frac{d^2y}{dx^2}+ax^2y=0.$$

Ex. 3. Obtain an integral of the equation

$$x \frac{d^2y}{dx^2} + \frac{dy}{dx} + my = 0$$

in the form

$$y = A \left[1 - \frac{mx}{1^2} + \frac{m^2x^2}{1^2 \cdot 2^2} - \frac{m^3x^3}{1^2 \cdot 2^2 \cdot 3^2} + \frac{m^4x^4}{1^2 \cdot 2^2 \cdot 3^2 \cdot 4^2} - \cdots \right].$$

84. The preceding investigation shews that, by means of the differential equation, and by the expansion of a function in terms of the independent variable as given in Taylor's or in Maclaurin's theorem, an expression in the form of a series can be obtained for the dependent variable; but, instead of working through what is sometimes a troublesome process, it is convenient to accept the principle that a series can be obtained, and so to assume for y some series arranged according to powers of x with indeterminate coefficients and indices. This series is then substituted for the dependent variable in the differential equation; as it is to be a solution of that equation, it must make the equation an identity. A comparison of the indices of the independent variable will shew the law of their progression; and a comparison of the coefficients of the different terms involving the same powers of the variable will give the required relations between the coefficients in the expression assumed. The latter will then, for such values of the independent variable as leave the series converging, be a solution.

85. As the method just indicated is really equivalent to the earlier one, it is not better suited to the solution of non-linear equations; but much labour is saved by it when the differential equation to be solved is linear. One of the most important forms to which it is specially applicable is that which may be written

$$\left\{ \phi \left(x \frac{d}{dx} \right) + \frac{1}{x} \psi \left(x \frac{d}{dx} \right) \right\} y = 0,$$

where ϕ and ψ are polynomial functions of their argument. To solve this equation, assume

$$y = A_1 x^{m_1} + A_2 x^{m_2} + A_3 x^{m_3} + \ldots,$$

where $m_1, m_2, m_3, \ldots$ are exponents in ascending order of magnitude; since

$$\phi \left(x \frac{d}{dx} \right) x^n = \phi(n) x^n,$$

the equation, when the value of y is substituted in it, gives

$$A_1 \phi(m_1) x^{m_1} + A_2 \phi(m_2) x^{m_2} + \ldots$$
$$+ A_1 \psi(m_1) x^{m_1-1} + A_2 \psi(m_2) x^{m_2-1} + \ldots = 0.$$

In this equation, $m_1 - 1$ is the lowest exponent and it occurs in only a single term; as the left-hand side is to vanish identically, this term must disappear, and therefore

$$A_1 \psi(m_1) = 0;$$

or, since A_1 is a coefficient of a term actually occurring and so is not zero, we must have

$$\psi(m_1) = 0.$$

A comparison of the indices of the remaining terms shews that

$$m_1 = m_2 - 1, \text{ and therefore } m_2 = m_1 + 1,$$
$$m_2 = m_3 - 1, \quad \text{,,} \quad \text{,,} \quad m_3 = m_1 + 2,$$

and so on; while a comparison of the coefficients of terms involving the same indices gives

$$A_1 \phi(m_1) + A_2 \psi(m_2) = 0,$$
$$A_2 \phi(m_2) + A_3 \psi(m_3) = 0,$$

and so on. Take now any value of m_1 as given by the equation $\psi(m_1) = 0$, say $m_1 = a$; as A_1 is quite arbitrary, denote it by A. The remaining coefficients are given by

$$A_2 = -\frac{\phi(a)}{\psi(a+1)} A,$$
$$A_3 = -\frac{\phi(a+1)}{\psi(a+2)} A_2 = +\frac{\phi(a)\phi(a+1)}{\psi(a+1)\psi(a+2)} A,$$

and so for the higher coefficients; the corresponding value of y is thus

$$A x^a \left[1 - \frac{\phi(a)}{\psi(a+1)} x + \frac{\phi(a)\phi(a+1)}{\psi(a+1)\psi(a+2)} x^2 \right.$$
$$\left. - \frac{\phi(a)\phi(a+1)\phi(a+2)}{\psi(a+1)\psi(a+2)\psi(a+3)} x^3 + \ldots \right].$$

The expressions connected with the other roots may be similarly obtained; and as the equation is linear, the sum of all these values of y is a solution,

Of this general form the most important example is that equation which has for a solution the series known as the hypergeometric series; it is discussed in full detail in the next chapter.

Ex. 1. Prove that the primitive of the equation

$$\frac{d^2y}{dx^2}+\frac{2n}{x}\frac{dy}{dx}+my=0$$

is given by

$$y=A\left[1-\frac{mx^2}{2\,(2n+1)}+\frac{m^2x^4}{2\,.\,4\,(2n+1)\,(2n+3)}-\cdots\right]$$
$$+Bx^{1-2n}\left[1-\frac{mx^2}{2\,(3-2n)}+\frac{m^2x^4}{2\,.\,4\,(3-2n)\,(5-2n)}-\cdots\right].$$

Ex. 2. In the case when $2n=1$, the separate parts involving the arbitrary constants in the preceding example become the same, each being

$$1-\frac{mx^2}{2^2}+\frac{m^2x^4}{2^2\,.\,4^2}-\cdots.$$

If this be denoted by v, and $y-uv=w$, where u and w are to be determined, we have on substituting, since v is a solution of the original equation,

$$\frac{d^2w}{dx^2}+\frac{1}{x}\frac{dw}{dx}+mw+v\left(\frac{d^2u}{dx^2}+\frac{1}{x}\frac{du}{dx}\right)+2\frac{du}{dx}\frac{dv}{dx}=0.$$

As there are two quantities u and w, we may assign any one condition we please; let this be

$$\frac{d^2u}{dx^2}+\frac{1}{x}\frac{du}{dx}=0.$$

The value of u hence derived is $A+B\log x$, and thus

$$\frac{d^2w}{dx^2}+\frac{1}{x}\frac{dw}{dx}+mw+\frac{2B}{x}\frac{dv}{dx}=0,$$

or

$$\frac{d^2w}{dx^2}+\frac{1}{x}\frac{dw}{dx}+mw=2Bm\left\{\frac{1}{2}-\frac{mx^2}{2^2\,.\,4}+\frac{m^2x^4}{2^2\,.\,4^2\,.\,6}-\frac{m^3x^6}{2^2\,.\,4^2\,.\,6^2\,.\,8}+\cdots\right\}.$$

The value of y is now

$$v\,(A+B\log x)+w,$$

and therefore contains two arbitrary constants, the total number necessary for the primitive; hence we require only a particular integral of the equation in w. To obtain this, write

$$w=B'+B_1x+B_2x^2+B_3x^3+B_4x^4+\ldots;$$

then

$$\frac{d^2w}{dx^2}+\frac{1}{x}\frac{dw}{dx}=\frac{B_1}{x}+2^2B_2+3^2B_3x+\ldots+n^2B_nx^{n-2}+\ldots.$$

Substituting and equating coefficients of different powers of x, we have from the

coefficient of x^{-1}.........$B_1 = 0$;

.... x^0 ....$mB' + 2^2 B_2 = Bm$;

.... x^1 ....$3^2 B_3 + mB_1 = 0$;

.... x^{2n-1} ......$(2n+1)^2 B_{2n+1} + mB_{2n-1} = 0$;

.... x^2 ....$4^2 B_4 + mB_2 = -\dfrac{Bm^2}{2 \cdot 4}$;

.... x^{2n-2} ......$(2n)^2 B_{2n} + mB_{2n-2} = (-1)^{n-1} \dfrac{Bm^n}{2 \cdot 4^2 \cdot 6^2 \ldots (2n-2)^2 \cdot 2n}$.

These equations give
$$B_1 = 0 = B_3 = \ldots = B_{2n-1} = \ldots,$$

so that no terms involving odd powers of x occur in w. For the coefficients of even powers, we have

$$B_2 = B \frac{m}{4} - B' \frac{m}{4} ;$$

$$B_4 = -B \frac{m^2}{2 \cdot 4^3} - B_2 \frac{m}{4^2}$$

$$= -B \frac{m^2}{2^2 \cdot 4^2} (\tfrac{1}{2} + 1) + B' \frac{m^2}{2^2 \cdot 4^2} ;$$

$$B_6 = +B \frac{m^3}{2 \cdot 4^2 \cdot 6^3} - B_4 \frac{m}{6^2}$$

$$= B \frac{m^3}{2^2 \cdot 4^2 \cdot 6^2} (\tfrac{1}{3} + \tfrac{1}{2} + 1) - B' \frac{m^3}{2^2 \cdot 4^2 \cdot 6^2} ;$$

and, generally,

$$B_{2n} = (-1)^{n-1} B \frac{m^n}{2^2 \cdot 4^2 \cdot 6^2 \ldots (2n)^2} \left(\frac{1}{n} + \frac{1}{n-1} + \ldots + \tfrac{1}{2} + 1 \right) + (-1)^n \frac{B' \cdot m^n}{2^2 \cdot 4^2 \cdot 6^2 \ldots (2n)^2}.$$

Hence the value of y is

$$(A + B \log x) \left\{ 1 - \frac{mx^2}{2^2} + \frac{m^2 x^4}{2^2 \cdot 4^2} - \frac{m^3 x^6}{2^2 \cdot 4^2 \cdot 6^2} + \ldots \right\}$$

$$+ B' \left\{ 1 - \frac{mx^2}{2^2} + \frac{m^2 x^4}{2^2 \cdot 4^2} - \frac{m^3 x^6}{2^2 \cdot 4^2 \cdot 6^2} + \ldots \right\}$$

$$+ B \sum_{n=1}^{n=\infty} \left\{ \frac{(-1)^{n-1} m^n x^{2n}}{2^2 \cdot 4^2 \cdot 6^2 \ldots (2n)^2} \left(\frac{1}{n} + \frac{1}{n-1} + \ldots + \tfrac{1}{3} + \tfrac{1}{2} + 1 \right) \right\}.$$

As B' is undetermined, there are apparently three arbitrary constants. But it will be seen that the expression multiplied by B' is the same as that multiplied by A ; and therefore these two constants coalesce into one new arbitrary constant A', which takes the place of $A + B'$.

Ex. 3. Obtain the primitive of the equation

$$x\frac{d^2y}{dx^2}+\frac{dy}{dx}+y=0$$

in the form

$$y=2B\left(x-\frac{3x^2}{2^3}+\frac{11x^3}{2^3.3^3}-\frac{50x^4}{2^3.3^3.4^3}+\dots\right)$$
$$+\left(1-\frac{x}{1^2}+\frac{x^2}{1^2.2^2}-\frac{x^3}{1^2.2^2.3^2}+\frac{x^4}{1^2.2^2.3^2.4^2}-\dots\right)(A+B\log x).$$

(Fourier.)

Ex. 4. Integrate in series, and express in a finite form, the primitives of the following equations :—

(i) $(1-x^2)\dfrac{d^2y}{dx^2}-x\dfrac{dy}{dx}+a^2y=0$;

(ii) $(x-x^3)\dfrac{d^2y}{dx^2}+(1-3x^2)\dfrac{dy}{dx}-xy=0.$

86. There are two special cases which occur in the integration of some differential equations. They owe their origin to the same cause, but they require to be dealt with separately.

As an example of one of them, let us recur to the series obtained as a solution of the equation

$$\left\{\phi\left(x\frac{d}{dx}\right)+\frac{1}{x}\psi\left(x\frac{d}{dx}\right)\right\}y=0,$$

which was

$$Ax^a\left[1-\frac{\phi(a)}{\psi(a+1)}x+\frac{\phi(a)\phi(a+1)}{\psi(a+1)\psi(a+2)}x^2-\dots\right],$$

the constant a being some root of the equation

$$\psi(m)=0.$$

This equation will usually have more than one root; let some other root be denoted by b. Then, in the case when b is greater than a by some integer k, the solution in the form above adopted ceases to be available; for in the denominator of the coefficient of x^k within the bracket there occurs the factor $\psi(a+k)$ or $\psi(b)$ which is zero, so that, unless there be a zero factor in the numerator, the coefficient apparently becomes infinite.

When such a zero factor does not occur in the numerator, we must recur to the fundamental equations from which the series was derived. These are

$$A_1\phi(a) + A_2\psi(a+1) = 0,$$
$$A_2\phi(a+1) + A_3\psi(a+2) = 0,$$
$$\dots\dots\dots\dots\dots\dots\dots\dots\dots\dots\dots$$
$$A_k\phi(a+k-1) + A_{k+1}\psi(a+k) = 0.$$

Now since $\psi(a+k)$ vanishes and A_{k+1} is not infinite, being a coefficient in a series supposed converging, it follows that either A_k or $\phi(a+k-1)$ is zero. Rejecting the latter on account of the hypothesis that no zero factor occurs in the numerator, we have $A_k = 0$, and thence from the preceding equations we find that all the coefficients $A_1, A_2, \dots, A_{k-1}$, are zero. Hence the part of the series which precedes the term x^k inside the bracket is, on account of its coefficients, evanescent, and the series actually must begin with the term Cx^{a+k}, that is, with Cx^b; and this will be the series derived from the root b of the equation $\psi(m) = 0$. One of the particular solutions has thus disappeared; but to obtain one in its place, we may proceed as in Ex. 2 in § 85. Denoting by v the one which remains and has absorbed the other, we may write

$$y = uv + w,$$

and, after substitution, assign some one relation which shall serve to determine u and w and render the differential equation easier to solve; this relation will usually be determined by the special form of the equation.

Ex. 1. Consider the differential equation

$$x^2\frac{d^2y}{dx^2} - (x^2+4x)\frac{dy}{dx} + 4y = 0.$$

Substituting $\qquad y = A_0x^m + A_1x^{m+1} + A_2x^{m+2} + \dots$

which is easily seen to be the necessary form, we find as the equation determining m

$$m(m-1) - 4m + 4 = 0,$$

that is, $\qquad (m-1)(m-4) = 0.$

Hence $a = 1$ and $b = 4$, so that the roots differ by an integer. It will be found that, on taking the root $m = 1$, the equation is of the form discussed and that the terms up to, but exclusive of, x^4 disappear; while the series derived from the root $m = 4$ is Ax^4e^x.

Complete the solution.

Ex. 2. Solve

$$x^2\frac{d^2y}{dx^2} + x(1+x)\frac{dy}{dx} + (3x-1)y = 0.$$

87. We now proceed to consider the other special case. Hitherto it has been assumed that no vanishing factor occurred in the numerator; and the result of the necessary alternative was indicated. But a vanishing factor may occur in the numerator of some of the coefficients of the terms within the bracket, either in that term in which there is a vanishing factor in the denominator, or in an earlier term. In the latter event, all the terms which do not have a vanishing factor in the denominators of the respective coefficients disappear; and if such a factor never occurs in a later term, the series will end at the term next before the first which contains that vanishing factor in the numerator, and the solution will thus be expressed in a finite form. But some vanishing factor may appear in the denominator of a later term, and the coefficient of this term will then take the indeterminate form 0/0, while the intervening terms will disappear; and all the terms after this will contain this indeterminate coefficient. The series will then be of the form

$$A x^a + B x^{a+1} + \ldots + F x^{a+f} + \frac{0}{0}(K x^{a+k} + L x^{a+k+1} + \ldots),$$

where $k-1$ is not less than f. This may be written

$$A\left(x^a + \frac{B}{A}x^{a+1} + \ldots + \frac{F}{A}x^{a+f}\right) + M\left(x^{a+k} + \frac{L}{K}x^{a+k+1} + \ldots\right),$$

where A is arbitrary, and $B/A, \ldots, F/A$, are determinate; M, being equal to $K \times 0/0$, is arbitrary (on account of the indeterminateness of $0/0$) and $L/K, \ldots$ are determinate. This series is a solution of the corresponding differential equation; it therefore will be a solution when a particular value is substituted for the arbitrary constant; hence

$$A\left(x^a + \frac{B}{A}x^{a+1} + \ldots + \frac{F}{A}x^{a+f}\right),$$

obtained by writing $M=0$, is a solution. In such a case there is therefore a *solution of the equation expressible in a finite form.*

Ex. 1. Consider as an example

$$x^2\frac{d^2y}{dx^2} + (x+x^2)\frac{dy}{dx} + (x-9)y = 0.$$

When we write $\qquad\qquad y = Ax^m + Bx^{m+1} + \ldots,$

the equation to determine m is

$$m^2 - 9 = 0,$$

and therefore $m = -3$ or $+3$.

For the root -3, it is not difficult to obtain the series

$$Ax^{-3}\left[1 - \frac{2}{5}x + \frac{2 \cdot 1}{5 \cdot 8}x^2 + \text{terms in } x^3, x^4, x^5 \text{ which vanish}\right]$$

$$+ Ax^{-3}\left[\frac{-2. -1.0.1.2.3}{-5. -8. -9. -8. -5.0}x^6 - \frac{-2. -1.0.1.2.3.4}{-5. -8. -9. -8. -5.0.7}x^7 + \cdots\right].$$

Write M instead of

$$\frac{-2. -1.0.1.2.3}{-5. -8. -9. -8. -5.0}A \ ;$$

then the series is

$$Ax^{-3}\left[1 - \frac{2}{5}x + \frac{2 \cdot 1}{5 \cdot 8}x^2\right]$$

$$+ Mx^3\left[1 - \frac{4}{7}x + \frac{4 \cdot 5}{(4^2 - 3^2)(5^2 - 3^2)}x^2 - \frac{4 \cdot 5 \cdot 6}{(4^2 - 3^2)(5^2 - 3^2)(6^2 - 3^2)}x^3 + \cdots\right],$$

thus verifying the theorem that one solution of the equation is expressible in a finite form.

Ex. 2. Verify the general theorem in the case of the equation

$$x^2\frac{d^2y}{dx^2} + x(1 + 2x)\frac{dy}{dx} = 4y.$$

Ex. 3. Solve the equation

$$\frac{d^2y}{dx^2} + (q - 2m)\frac{dy}{dx} + \left(m^2 - qm - \frac{2}{x^2}\right)y = 0.$$

88. Further illustrations of these special cases will occur later, and they need not therefore now be considered in greater detail; various other matters arise which will be discussed in connection with special equations. Thus it has not been stated that a series must always proceed in ascending or in descending powers of the independent variable; but the comparison of the terms in the differential equation after the expression for the dependent variable has been therein substituted will indicate the nature of the series. In the case when one of the solutions becomes evanescent one method has been pointed out, which will be useful for supplying the deficiency thus caused; another will be indicated below. In fact, the difficulties that arise are usually connected with special equations and not with the general equation; and therefore some special equations will be considered. Of equations of a particular form, there are four which reckon as the most important among those included in the class soluble by means of series; they are :—

First, the differential equation of the hypergeometric series which will be discussed separately in the next chapter;

Second, Legendre's equation;

Third, Bessel's equation;

Fourth, Riccati's equation.

The last three of these will now be discussed in order. It must of course be understood that what is carried out here is merely the complete solution of the differential equations. There is no attempt at an exhaustive investigation of the properties of the respective functions determined by the dependent variables.

LEGENDRE'S EQUATION.

89. This differential equation is

$$(1 - x^2)\frac{d^2y}{dx^2} - 2x\frac{dy}{dx} + n(n+1)y = 0,$$

or, what is the same equation,

$$\frac{d}{dx}\left\{(1 - x^2)\frac{dy}{dx}\right\} + n(n+1)y = 0,$$

in which the quantity n is a constant. The equation is one which frequently occurs in investigations connected with questions in most of the branches of applied mathematics; in these cases, n is usually, but not always, a positive integer. The equation is one of the second order, and has therefore two independent particular solutions; and every other particular solution can be expressed in terms of these two. It will be found that the form of these fundamental particular solutions is different in the two cases when n is, and when n is not, a positive integer.

We proceed to obtain these solutions. In accordance with the general method of integration by series, we write

$$y = A_1 x^{m_1} + A_2 x^{m_2} + A_3 x^{m_3} + \ldots,$$

and substitute this value of y. Then we have

$$n(n+1)(A_1 x^{m_1} + A_2 x^{m_2} + A_3 x^{m_3} + \ldots)$$

$$= \frac{d}{dx}\{(x^2-1)(m_1 A_1 x^{m_1-1} + m_2 A_2 x^{m_2-1} + m_3 A_3 x^{m_3-1} + \ldots)\}$$

$$= m_1(m_1+1) A_1 x^{m_1} - m_1(m_1-1) A_1 x^{m_1-2}$$

$$+ m_2(m_2+1) A_2 x^{m_2} - m_2(m_2-1) A_2 x^{m_2-2} + \ldots\ldots;$$

and this must be an identity. An inspection of the equation shews that, so far as powers of x are concerned, we have

$$m_2 = m_1 - 2,$$

$$m_3 = m_2 - 2,$$

$$\ldots\ldots\ldots\ldots\ldots$$

or the series must be one in *descending* powers of x; we therefore now assume that $m_1, m_2, m_3, \ldots$ are arranged in descending order of magnitude, their common difference being 2. A comparison of coefficients of the same powers of x gives, for those of x^{m_1},

$$\{m_1(m_1+1) - n(n+1)\} A_1 = 0,$$

or $$(m_1 - n)(m_1 + n + 1) A_1 = 0.$$

Now A_1 is not zero, being the coefficient of the highest term in y; hence either

$$m_1 = n,$$

or $$m_1 = -(n+1).$$

The relation between the coefficients of consecutive terms arises from equating the coefficients of $x^{m_1 - 2r + 2}$ on the two sides; it is, for values of r greater than unity,

$$n(n+1) A_r = (m_1 - 2r + 2)(m_1 - 2r + 3) A_r$$

$$- (m_1 - 2r + 4)(m_1 - 2r + 3) A_{r-1},$$

and this gives

$$(n - m_1 + 2r - 2)(n + m_1 - 2r + 3) A_r$$

$$= -(m_1 - 2r + 4)(m_1 - 2r + 3) A_{r-1}.$$

90. Consider first the solution corresponding to

$$m_1 = n.$$

The highest term is then $A_1 x^n$; and the relation between the successive A's is

$$(2r-2)(2n-2r+3)A_r = -(n-2r+4)(n-2r+3)A_{r-1},$$

so that

$$A_r = -\frac{(n-2r+4)(n-2r+3)}{2(r-1)(2n-2r+3)} A_{r-1}$$

$$= (-1)^{r-1} \frac{n(n-1)(n-2)\ldots(n-2r+4)(n-2r+3)}{2^{r-1}.1.2.3\ldots(r-1)(2n-1)(2n-3)\ldots(2n-2r+3)} A_1,$$

and therefore the series becomes

$$A_1 \left\{ x^n - \frac{n(n-1)}{2(2n-1)} x^{n-2} + \frac{n(n-1)(n-2)(n-3)}{2.4(2n-1)(2n-3)} x^{n-4} - \ldots \right\}.$$

Let the series within the bracket be denoted by y_1, which is therefore a particular solution. When n is a positive integer, the series is finite. When n is even, the last term is

$$(-1)^{\frac{1}{2}n} \frac{n(n-1)(n-2)\ldots\ldots\ldots 2.1}{2.4\ldots(n-2)n(2n-1)(2n-3)\ldots(n+1)},$$

or, what is the same thing,

$$(-1)^{\frac{1}{2}n} \frac{n!\,n!\,n!}{\frac{1}{2}n!\,\frac{1}{2}n!\,2n!};$$

while, when n is uneven, the last term is

$$(-1)^{\frac{1}{2}(n-1)} \frac{n(n-1)(n-2)\ldots\ldots\ldots 3.2}{2.4\ldots(n-3)(n-1)(2n-1)(2n-3)\ldots(n+4)(n+2)} x,$$

or, what is the same thing,

$$(-1)^{\frac{1}{2}(n-1)} \frac{n!\,n!\,(n-1)!}{\frac{1}{2}(n-1)!\,\frac{1}{2}(n-1)!\,(2n-1)!} x.$$

The numbers of terms in the two cases are respectively $\frac{1}{2}n+1$ and $\frac{1}{2}(n+1)$.

When n is an integer, $2n$ is an even integer, and therefore a zero factor can never enter into the denominator in this case; thus the series considered will never come under the class considered in § 87 which yields two solutions.

The series y_1, multiplied by

$$\frac{2n\,!}{2^n\,.\,n\,!\,n\,!},$$

n being a positive integer, is usually denoted by P_n, and sometimes is called a Zonal Harmonic. This function is an extremely important one in physical applications.

Ex. 1. Verify that

$$2^n\,.\,n\,!\,P_n = \frac{d^n}{dx^n}\{(x^2-1)^n\}.$$

Ex. 2. Shew that P_n is the coefficient of z^n in the expansion in ascending powers of z of $(1-2xz+z^2)^{-\frac{1}{2}}$.

Hence shew that $v=(1-2xz+z^2)^{-\frac{1}{2}}$ is a solution of the equation

$$z\frac{\partial^2(zv)}{\partial z^2}+\frac{\partial}{\partial x}\left\{(1-x^2)\frac{\partial v}{\partial x}\right\}=0.$$

Ex. 3. Prove that the roots of the equation $y_1=0$ are all real and numerically less than unity.

Ex. 4. Prove that the sum of the coefficients in P_n with their proper signs is unity.

Ex. 5. Obtain the equations

$$\text{(i)} \quad nP_n=(2n-1)\,xP_{n-1}-(n-1)\,P_{n-2};$$

$$\text{(ii)} \quad (x^2-1)\frac{dP_n}{dx}=nxP_n-nP_{n-1};$$

$$\text{(iii)} \quad \frac{dP_n}{dx}-\frac{dP_{n-2}}{dx}=(2n-1)\,P_{n-1}.$$

Ex. 6. Prove that

$$P_n=\frac{1}{\pi}\int_0^\pi\{x\pm(x^2-1)^{\frac{1}{2}}\cos\theta\}^n d\theta,$$

$$P_n=\frac{1}{\pi}\int_0^\pi\{x\pm(x^2-1)^{\frac{1}{2}}\cos\phi\}^{-n-1}d\phi.$$

Ex. 7. Shew that

$$\int_{-1}^1 P_m(x)\,P_n(x)\,dx=0,$$

when m and n are different positive integers; and that

$$\int_{-1}^1 P_n{}^2(x)\,dx=\frac{2}{2n+1}.$$

In the case when n is not a positive integer, the series y_1 proceeds to infinity; and for convergence, it is necessary that x should be greater than unity. But in particular when $2n$ is equal to some positive odd integer, say $2r-1$, then the coefficient of x^{n-2r} has a zero factor in the denominator, and no zero factor occurs in the numerator either of that term or of any subsequent term; hence (by § 86) the terms whose indices are higher than $n-2r$ do not exist in this solution of the differential equation, which will therefore begin with x^{n-2r} multiplied by some new arbitrary constant. But since $2n = 2r-1$, therefore $n-2r = -(n+1)$, or the solution degenerates into an infinite series of descending powers of x beginning with $x^{-(n+1)}$. To the consideration of this solution we therefore proceed.

91. We take now the second solution of the equation determining the value of m_1; this is $-(n+1)$, so that the term with highest index may be taken to be $A_1 x^{-(n+1)}$. The relation between the successive coefficients is

$$(2n+2r-1)(2r-2)A_r = (n+2r-3)(n+2r-2)A_{r-1}$$

for values of r greater than unity; and therefore

$$A_r = \frac{(n+1)(n+2)\ldots\ldots\ldots(n+2r-2)}{2^{r-1}.1.2.3\ldots(r-1)(2n+3)(2n+5)\ldots(2n+2r-1)}A_1,$$

so that the series is

$$A_1 \left\{ x^{-(n+1)} + \frac{(n+1)(n+2)}{2(2n+3)}x^{-(n+3)} \right.$$
$$\left. + \frac{(n+1)(n+2)(n+3)(n+4)}{2.4(2n+3)(2n+5)}x^{-(n+5)} + \ldots \right\}.$$

Let the series within the bracket be denoted by y_2, which is a particular solution; the series y_2 multiplied by

$$\frac{2^n.n!n!}{(2n+1)!},$$

n being a positive integer, is usually denoted by Q_n. For convergence, it is necessary that x should be greater than unity. This function Q_n, like the function P_n, is of great importance in physical investigations.

When n is a positive integer, the series proceeds to infinity.

When n is a negative integer, y_2 is a finite series; if $n = -2p$, the series begins with x^{2p-1} and proceeds for p terms; if $n = -(2p+1)$, the series begins with x^{2p} and proceeds for $p+1$ terms.

When $2n$ is equal to an odd negative integer other than -1, say $-(2r+1)$, then the coefficient of $x^{-(n+2r+1)}$ has a zero factor in the denominator, and no zero factor occurs in the numerator of any term in the series; hence, as before, the preceding terms do not exist, and the series begins with $x^{-(n+2r+1)}$ multiplied by some new arbitrary constant. But since $2n = -(2r+1)$, therefore $-(n+2r+1) = n$, or the solution y_2 becomes an infinite series of descending powers of x beginning with x^n, i.e. y_2 degenerates into y_1.

92. We thus have the following results.

I. When n *is a positive integer*, there are two independent solutions of the differential equation; (1) y_1, a finite series, (2) y_2, an infinite series; and the primitive is

$$y = Ay_1 + By_2.$$

II. When n *is a negative integer*, there are two independent solutions; (1) y_1, an infinite series, (2) y_2, a finite series; and the primitive is

$$y = Ay_1 + By_2.$$

III. When n *is not integral and* $2n$ *is not equal to some odd positive or negative integer*, there are two independent solutions; (1) y_1, an infinite series, (2) y_2, an infinite series; and the primitive is

$$y = Ay_1 + By_2.$$

IV. When $2n$ *is equal to an odd positive integer*, there has been obtained only one solution of the differential equation, for y_1 degenerates into y_2, this solution being an infinite series; the primitive is thus not expressible in terms of y_1 and y_2 alone.

V. When $2n$ *is equal to an odd negative integer other than* -1, there has been obtained only one solution of the differential equation, for y_2 degenerates into y_1, this solution being an infinite series; the primitive again is not expressible in terms of y_1 and y_2 alone.

VI. When $2n$ *is equal to* -1, there has been obtained only one solution of the differential equation, for y_1 and y_2 are the same infinite series beginning with $x^{-\frac{1}{2}}$; the primitive again is not expressible in terms of y_1 and y_2 alone.

It therefore remains to obtain the primitive in the last three cases.

93 (i). Consider first the case when $2n$ is equal to an odd positive integer; then

$$y_2 = x^{-n-1} + \frac{(n+1)(n+2)}{2(2n+3)} x^{-n-3} + \frac{(n+1)(n+2)(n+3)(n+4)}{2.4(2n+3)(2n+5)} x^{-n-5} + \dots$$

is a definite solution, and we have to find a second and different particular solution. In the first instance, assume

$$2n = 2p + 1 + \theta,$$

where θ is an infinitesimal quantity which will ultimately be made zero. Then, so long as θ is not zero, the quantity

$$y_1 = x^n - \frac{n(n-1)}{2(2n-1)} x^{n-2} + \dots$$

is also a definite solution; and it ceases to be so by the vanishing of θ, since θ enters as a factor into the denominator of the coefficient of x^{n-2p-2} and all lower powers. Now we have

$$\begin{aligned}
Ay_1 = \quad & A \left\{ x^n - \frac{n(n-1)}{2(2n-1)} x^{n-2} + \dots \right. \\
& + (-1)^p \frac{n(n-1)\dots(n-2p+1)}{2.4\dots 2p (2n-1)(2n-3)\dots(2n-2p+1)} x^{n-2p} \\
& + (-1)^{p+1} \frac{n(n-1)\dots(n-2p-1)}{2.4\dots(2p+2)(2n-1)(2n-3)\dots(2n-2p-1)} x^{n-2p-2} \\
& \left. + (-1)^{p+2} \frac{n(n-1)\dots(n-2p-3)}{2.4\dots(2p+4)(2n-1)(2n-3)\dots(2n-2p-3)} x^{n-2p-4} + \dots \right\} \\
= \quad & A \left\{ x^n - \frac{n(n-1)}{2(2n-1)} x^{n-2} + \dots \right. \\
& \left. + (-1)^p \frac{n(n-1)\dots(n-2p+1)}{2.4\dots 2p (2n-1)(2n-3)\dots(2n-2p+1)} x^{n-2p} \right\} \\
& + \frac{C x^{n-2p-2}}{2n-2p-1} \left\{ 1 - \frac{(n-2p-2)(n-2p-3)}{(2p+4)(2n-2p-3)} x^{-2} + \dots \right\},
\end{aligned}$$

where

$$C = (-1)^{p+1} A \frac{n(n-1)\dots(n-2p-1)}{2.4\dots(2p+2)(2n-1)(2n-3)\dots(2n-2p+1)},$$

and so is determinate and finite. But

$$n-2p-2=-(n+1)+\theta,$$

and therefore

$$x^{n-2p-2}=x^{-(n+1)}.x^{\theta}=x^{-n-1}(1+\theta\log x),$$

approximately. Also the coefficient of x^{-2r} within the second bracket is

$$(-1)^r\frac{(n-2p-2)(n-2p-3)...(n-2p-2r-1)}{(2p+4)(2p+6)...(2p+2r+2)(2n-2p-3)(2n-2p-5)...(2n-2p-2r-1)},$$

i.e., is

$$(-1)^r\frac{(n+1-\theta)(n+2-\theta)...(n+2r-\theta)}{(2n+3-\theta)(2n+5-\theta)...(2n+2r+1-\theta)(\theta-2)(\theta-4)...(\theta-2r)},$$

i.e., is

$$\frac{(n+1-\theta)(n+2-\theta)...(n+2r-\theta)}{(2n+3-\theta)(2n+5-\theta)...(2n+2r+1-\theta)(2-\theta)(4-\theta)...(2r-\theta)},$$

i.e., is

$$\frac{(n+1)(n+2)...(n+2r)}{(2n+3)(2n+5)...(2n+2r+1)2.4.6...2r}(1+C_r\theta),$$

where

$$C_r=\sum_{s=1}^{s=r}\frac{1}{2s}+\sum_{s=1}^{s=r}\frac{1}{2n+2s+1}-\sum_{s=1}^{s=2r}\frac{1}{n+s}.$$

Hence

$$Ay_1=A\left\{x^n-\frac{n(n-1)}{2(2n-1)}x^{n-2}+...\right.$$

$$\left.+(-1)^p\frac{n(n-1)...(n-2p+1)}{2.4...2n(2n-1)(2n-3)...(2n-2p+1)}x^{n-2p}\right\}$$

$$+Cx^{-(n+1)}\frac{1+\theta\log x}{\theta}$$

$$\times\left[1+\sum_{r=1}^{r=\infty}\left\{\frac{(n+1)(n+2)...(n+2r)}{2.4...2r(2n+3)(2n+5)...(2n+2r+1)}(1+C_r\theta)x^{-2r}\right\}\right].$$

When the second part of the right-hand side is expanded the aggregate of terms which involve $\frac{1}{\theta}$ is $\frac{C}{\theta}y_2$; the aggregate of terms which involve $\log x$ is

$$Cy_2\log x;$$

and there remains the aggregate of terms independent of θ (and also as it appears of $\log x$), as well as a further aggregate of terms multiplied by positive powers of θ, most of which have been omitted and all of which disappear when θ is made to vanish. From the first part of the right-hand side there is an aggregate of terms independent of θ, as well as an aggregate of terms which disappear when θ is made zero. Hence the primitive of the equation is

$$y=By_2+Ay_1$$

$$=\left(B+\frac{C}{\theta}\right)y_2+C(y_2\log x+T_n+R_n)$$

$$=Dy_2+C(y_2\log x+T_n+R_n),$$

on changing the arbitrary constants. Here T_n stands for

$$\frac{A}{C}\left\{x^n - \frac{n(n-1)}{2(2n-1)}x^{n-2}+\dots \right.$$
$$\left. +(-1)^p \frac{n(n-1)\dots(n-2p+1)}{2.4\dots 2p(2n-1)(2n-3)\dots(2n-2p+1)}x^{n-2p}\right\},$$

and R_n stands for

$$x^{-n-1}\sum_{r=1}^{r=\infty}\left\{\frac{(n+1)(n+2)\dots(n+2r)}{2.4\dots 2r(2n+3)(2n+5)\dots(2n+2r+1)}C_r x^{-2r}\right\},$$

the value of C_r being

$$\sum_{s=1}^{s=r}\left(\frac{1}{2s}+\frac{1}{2n+2s+1}-\frac{1}{n+2s-1}-\frac{1}{n+2s}\right).$$

The value of the coefficient A/C, which occurs in T_n, is

$$\left\{\frac{4.8.12\dots(4n-2)}{1.3.5\dots 2n}\right\}^2(8n+4),$$

so that we may write T_n in the form

$$\left\{\frac{4.8.12\dots(4n-2)}{1.3.5\dots 2n}\right\}^2(8n+4)\left[x^n - \frac{n(n-1)}{2(2n-1)}x^{n-2}+\dots\right.$$
$$\left. -n(n-1)\left\{\frac{\frac{1}{2}.\frac{3}{2}\dots(n-2)}{2.4\dots(2n-1)}\right\}^2 x^{-n+1}\right].$$

The second particular solution of the equation is thus

$$y_2\log x + T_n + R_n;^*$$

and it will be noticed that that part of it, which is expansible in descending powers of x, begins with a term involving x^n and contains no term involving x^{-n-1}.

But in the special case when $2n$ is equal to unity, so that p is zero in the preceding investigation, then the form of T_n, now $T_{\frac{1}{2}}$ say, is limited to the first term; and we have

$$C=-A\frac{n(n-1)}{2}=\tfrac{1}{8}A,$$

so that

$$T_{\frac{1}{2}}=8x^{\frac{1}{2}}.$$

The remaining parts are unchanged in form.

93 (ii). Consider now the case when $2n$ is equal to an odd negative integer other than -1; the solution y_1 is definite, but

$$y_2=x^{-n-1}+\frac{(n+1)(n+2)}{2(2n+3)}x^{-n-3}+\dots$$

is not a definite solution.

---

* The solution thus given corresponds to that for Bessel's equation, Ex. 1, § 105, due to Hankel.

Before assuming n to be half an odd integer, write

$$-n = m+1,$$

so that $2m$ is a positive odd integer when the assumption as to the special value of n is made. Then

$$y_1 = x^{-m-1} + \frac{(m+1)(m+2)}{2(2m+3)} x^{-m-3} + \ldots$$
$$= Y_2,$$

and

$$y_2 = x^m - \frac{m(m-1)}{2(2m-1)} x^{m-2} + \ldots$$
$$= Y_1,$$

where Y_1 and Y_2 are the special solutions of

$$\frac{d}{dx}\left\{(1-x^2)\frac{dy}{dx}\right\} + m(m+1)\,y = 0,$$

m being positive. When $2m$ is an odd positive integer we know, from the preceding investigation, that the primitive of this is

$$y = BY_2 + A\,(Y_2 \log x + T_m + R_m),$$

where

$$T_m = \left\{\frac{4.8.12\ldots(4m-2)}{1.3.5\ldots 2m}\right\}^2 (8m+4)\left[x^m - \frac{m(m-1)}{2(2m-1)} x^{m-2} + \ldots \right.$$
$$\left. - m(m-1)\left\{\frac{\frac{1}{2}.\frac{3}{2}.\ldots(m-2)}{2.4\ldots(2m-1)}\right\}^2 x^{-m+1}\right],$$

and

$$R_m = x^{-m-1} \sum_{r=1}^{r=\infty} \left\{\frac{(m+1)(m+2)\ldots(m+2r)}{2.4\ldots 2r\,(2m+3)(2m+5)\ldots(2m+2r+1)} A_r x^{-2r}\right\},$$

the value of A_r being

$$\sum_{s=1}^{s=r} \left(\frac{1}{2s} + \frac{1}{2m+2s+1} - \frac{1}{m+2s-1} - \frac{1}{m+2s}\right)$$

Hence the primitive of

$$\frac{d}{dx}\left\{(1-x^2)\frac{dy}{dx}\right\} + n(n+1)\,y = 0,$$

in the case when $2n$ is an odd negative integer other than -1, is

$$y = By_1 + A\,(y_1 \log x + V_n + U_n),$$

where

$$y_1 = x^n - \frac{n(n-1)}{2(2n-1)} x^{n-2} + \frac{n(n-1)(n-2)(n-3)}{2.4(2n-1)(2n-3)} x^{n-4} - \ldots,$$

$$V_n = \left\{\frac{4.8.12\ldots(-4n-6)}{1.3.5\ldots(-2n-2)}\right\}^2 (-8n-4)\left[x^{-n-1} + \frac{(n+1)(n+2)}{2(2n+3)} x^{-n-3} + \ldots \right.$$
$$\left. \ldots - (n+1)(n+2)\left\{\frac{\frac{1}{2}.\frac{3}{2}.\ldots(-n-3)}{2.4\ldots(-2n-3)}\right\}^2 x^{n+2}\right],$$

and

$$U_n = x^n \sum_{r=1}^{r=\infty} \left\{ \frac{n(n-1)\dots(n-2r+1)}{2\,.\,4\dots2r\,(2n-1)\,(2n-3)\dots(2n-2r+1)} (-1)^r E_r x^{-2r} \right\},$$

where in U_n the value of E_r is

$$\sum_{s=1}^{s=r} \left(\frac{1}{2s} + \frac{1}{2s-2n-1} - \frac{1}{2s-n-2} - \frac{1}{2s-n-1} \right).$$

The second particular solution of the equation in this case is thus

$$y_1 \log x + V_n + U_n\,;$$

and it will be noticed that the part of it, which is expansible in descending powers of x, begins with a term involving x^{-n-1} and contains no term involving x^n.

93 (iii). Lastly, for the special case in which $2n$ is equal to -1, we proceed in a manner similar to that adopted in § 93 (i); and we find that the primitive of the equation is

$$A y_1 + B(y_1 \log x - w_{-\frac{1}{2}}),$$

where y_1 is the series

$$x^{-\frac{1}{2}} + \frac{\frac{1}{2}\,.\,\frac{3}{2}}{2\,.\,2} x^{-\frac{5}{2}} + \frac{\frac{1}{2}\,.\,\frac{3}{2}\,.\,\frac{5}{2}\,.\,\frac{7}{2}}{2\,.\,4\,.\,2\,.\,4} x^{-\frac{9}{2}} + \dots,$$

and

$$w_{-\frac{1}{2}} = x^{-\frac{1}{2}} \sum_{r=1}^{r=\infty} \frac{\frac{1}{2}\,.\,\frac{3}{2}\,.\,\dots\,.\,\frac{4r-1}{2}}{r!\,r!\,2^{2r}} D_r x^{-2r}.$$

and

$$D_r = 2 \sum_{s=1}^{s=r} \left(\frac{1}{2r+2s-1} + \frac{1}{2s-1} - \frac{1}{2s} \right).$$

94. Since in all these cases $2n$ is an odd integer, the equation can be written

$$\frac{d}{dx} \left\{ (1-x^2) \frac{dy}{dx} \right\} + (p^2 - \tfrac{1}{4}) y = 0,$$

where p is an integer.

The case of p positive is that considered in § 93 (i); the case of p negative is that considered in § 93 (ii); and the case of p zero is that considered in § 93 (iii). Properties of the functions defined by the differential equation in the present form have been discussed by Mr W. M. Hicks in his memoir on "Toroidal Functions," *Phil. Trans. Roy. Soc.* (1881), pp. 609—652.

Ex. 1. Assuming the result of Ex. 1 in § 64, shew how the solution of

$$\frac{d}{dx} \left\{ (1-x^2) \frac{dy}{dx} \right\} = \tfrac{1}{4} y$$

can be derived from that of

$$(1-k^2) \frac{d^2 v}{dk^2} + \frac{1-3k^2}{k} \frac{dv}{dk} = v,$$

which is the differential equation for the quarter-period in elliptic functions.

Ex. 2. Prove that the Particular Integral of the equation

$$(1-x^2)\frac{d^2w}{dx^2}+n(n+1)w=\frac{dP_n}{dx}$$

is λP_{n-1}, where λ is a constant; and that the Particular Integral of the equation

$$(1-x^2)\frac{d^2w}{dx^2}+n(n+1)w=\frac{dQ_n}{dx}$$

is $\lambda'Q_{n+1}$, where λ' is a constant.

95. In the general case of the differential equation, as represented by I., II., III. of § 92, it is possible to express the second particular solution in terms of that already obtained and of similar functions. Let v denote the particular solution already obtained, so that for instance v would be P_n in I.; and let

$$y=uv-w,$$

where u and w are as yet indeterminate. When this is substituted in the differential equation, we have

$$-\left[\frac{d}{dx}\left\{(1-x^2)\frac{dw}{dx}\right\}+n(n+1)w\right]+v\left\{(1-x^2)\frac{d^2u}{dx^2}-2x\frac{du}{dx}\right\}$$

$$+2\frac{du}{dx}(1-x^2)\frac{dv}{dx}+u\left[\frac{d}{dx}\left\{(1-x^2)\frac{dv}{dx}\right\}+n(n+1)v\right]=0.$$

Since v is a solution, the last term disappears; and, as the only condition imposed on u and w is that y must satisfy the equation, we may arbitrarily assign another. Assigning it so that the coefficient of v may vanish, we have

$$(1-x^2)\frac{d^2u}{dx^2}-2x\frac{du}{dx}=0,$$

and therefore

$$(x^2-1)\frac{du}{dx}=\text{constant.}$$

As we are seeking a particular solution, it is convenient to have it as simple as possible; and therefore, giving a special value to the constant, we may write

$$(x^2-1)\frac{du}{dx}=-1,$$

so that a value of u is given by

$$u=\tfrac{1}{2}\log\left(\frac{x+1}{x-1}\right).$$

The equation to determine w now becomes

$$\frac{d}{dx}\left\{(1-x^2)\frac{dw}{dx}\right\}+n(n+1)w=2\frac{dv}{dx}.$$

When the Particular Integral, say w_1, of this is obtained, the second solution of the original equation is

$$y = \tfrac{1}{2} v \log \left(\frac{x+1}{x-1} \right) - w_1.$$

The value of w_1 as a series of descending powers of x is easily obtained. Thus in the case when n is a positive integer, we take

$$v = x^n - \frac{n(n-1)}{2(2n-1)} x^{n-2} + \frac{n(n-1)(n-2)(n-3)}{2.4(2n-1)(2n-3)} x^{n-4} - \ldots,$$

and at once have the equation, which determines w_1, in the form

$$\frac{d}{dx} \left\{ (1-x^2) \frac{dw_1}{dx} \right\} + n(n+1) w_1 = 2n \left[x^{n-1} - \frac{(n-1)(n-2)}{2(2n-1)} x^{n-3} + \ldots \right].$$

Let

$$w_1 = C_1 x^{n-1} + C_2 x^{n-3} + C_3 x^{n-5} + \ldots;$$

then, substituting and equating the coefficients of the highest term, we have

$$C_1 \{ n(n+1) - n(n-1) \} = 2n,$$

or

$$C_1 = 1;$$

and equating the coefficients of the terms involving x^{n-2r+1}, we have

$$C_r \{ n(n+1) - (n-2r+1)(n-2r+2) \} + (n-2r+3)(n-2r+2) C_{r-1}$$

$$= (-1)^{r-1} 2 \frac{n(n-1)(n-2)\ldots(n-2r+2)}{2.4\ldots(2r-2)(2n-1)\ldots(2n-2r+3)}.$$

The general value of C_r, deducible from this, is complicated; the values of the earlier coefficients are

$$C_2 = - \frac{(n-1)(n-2)(3n-1)}{3(2n-1)(2n-2)},$$

$$C_3 = \frac{(n-1)(n-2)(n-3)(n-4)(30n^2 - 50n + 12)}{3.4.5(2n-1)(2n-2)(2n-3)(2n-4)},$$

and so on; but there is no advantage in writing down more of the coefficients, as the expression for w_1 will soon be put into a different form.

Relation between the particular solutions.

96. We have now obtained the primitive of Legendre's equation in all cases when n is a real constant, by deducing two solutions which are linearly independent (§ 72) of one another. But we know (§ 65) that when one solution of a differential equation of the second order has been found, the primitive can be expressed in terms of it and, if necessary, of other functions, and therefore any other solution is so expressible; we proceed to obtain this relation for the cases—viz. I., II., III. above—in which it has not been obtained. The first form in which it may be given is derived by means of § 65. We may define P_n and Q_n by the generalised equations

$$P_n = \frac{\Pi(2n)}{2^n \Pi(n)\Pi(n)} \left\{ x^n - \frac{n(n-1)}{2(2n-1)} x^{n-2} + \ldots \right\},$$

and

$$Q_n = \frac{2^n \Pi(n)\Pi(n)}{\Pi(2n+1)} \left\{ x^{-(n+1)} + \frac{(n+1)(n+2)}{2(2n+3)} x^{-(n+3)} + \ldots \right\},$$

whether n be integral or not; $\Pi(n)$ is Gauss's Π function and is $\Gamma(n+1)$, and (see next chapter, § 126) is $n!$ when n is a positive integer; and P_n and Q_n are still solutions of the Legendre's equation, since they are respectively constant multiples of y_1 and y_2. We therefore have

$$(1-x^2)\frac{d^2 P_n}{dx^2} - 2x\frac{dP_n}{dx} + n(n+1)P_n = 0,$$

$$(1-x^2)\frac{d^2 Q_n}{dx^2} - 2x\frac{dQ_n}{dx} + n(n+1)Q_n = 0;$$

multiplying the former by Q_n and subtracting the latter multiplied by P_n, we have

$$(x^2-1)\left(Q_n\frac{d^2 P_n}{dx^2} - P_n\frac{d^2 Q_n}{dx^2}\right) + 2x\left(Q_n\frac{dP_n}{dx} - P_n\frac{dQ_n}{dx}\right) = 0,$$

or

$$(x^2-1)\left(Q_n\frac{dP_n}{dx} - P_n\frac{dQ_n}{dx}\right) = A,$$

where A is a constant, which is definite and not arbitrary since Q_n and P_n are definite functions. To find A, we consider the terms containing the highest powers of x; these are

in Q_n

$$\frac{2^n \Pi(n)\Pi(n)}{\Pi(2n+1)} x^{-(n+1)},$$

and in P_n

$$\frac{\Pi(2n)}{2^n \Pi(n)\Pi(n)} x^n;$$

hence

$$A = \frac{\Pi(2n)}{\Pi(2n+1)}\{n+(n+1)\} = 1,$$

since $\Pi(2n+1) = (2n+1)\Pi(2n)$; and therefore

$$Q_n\frac{dP_n}{dx} - P_n\frac{dQ_n}{dx} = \frac{1}{x^2-1}.$$

This gives

$$\frac{d}{dx}\left(\frac{P_n}{Q_n}\right) = \frac{1}{(x^2-1)Q_n{}^2},$$

or, its equivalent

$$\frac{d}{dx}\left(\frac{Q_n}{P_n}\right) = \frac{1}{(1-x^2)P_n{}^2},$$

and therefore

$$\frac{Q_n}{P_n} = -\int_{\infty}^{x} \frac{dx}{(x^2-1)\,P_n{}^2} = \int_{x}^{\infty} \frac{dx}{(x^2-1)\,P_n{}^2},$$

no constant being needed, as may be seen by comparing the coefficients of the highest powers of x in the expansion of the two sides in descending powers of x.

97. The result may be written in a different form, when n is an integer ; but it is first necessary to prove two relations between the functions given by Legendre's equation for different values of n.

From the expressions given in the preceding article, we find that the coefficient of x^{n+1-2r} in $P_{n+1} - P_{n-1}$ is

$$(-1)^r \frac{\Pi(2n-2)}{2^{n-1}\Pi(n-1)\Pi(n-1)} \frac{(n-1)(n-2)\ldots(n-2r+2)}{2\,.\,4\ldots2r\,(2n+1)(2n-1)\ldots(2n-2r+1)}$$

$$\times \left\{ \frac{(2n+2)(2n+1)\,2n\,(2n-1)}{4\,(n+1)\,n\,(n+1)\,n}(n+1)\,n\,(2n-2r+1)+2r\,(2n+1)(2n-1) \right\};$$

the last factor is easily simplified into

$$(2n+1)^2\,(2n-1),$$

and therefore the coefficient is

$$(-1)^r \frac{\Pi(2n)}{2^n\Pi(n)\Pi(n)} \frac{n(n-1)(n-2)\ldots(n-2r+2)}{2\,.\,4\ldots2r\,(2n-1)(2n-3)\ldots(2n-2r+3)(2n-2r+1)}(2n+1).$$

Hence the coefficient of x^{n-2r} in

$$\frac{dP_{n+1}}{dx} - \frac{dP_{n-1}}{dx}$$

is $(-1)^r (2n+1) \dfrac{\Pi(2n)}{2^n\Pi(n)\Pi(n)} \dfrac{n(n-1)\ldots(n-2r+2)(n-2r+1)}{2\,.\,4\ldots2r\,(2n-1)(2n-3)\ldots(2n-2r+1)},$

that is, is the coefficient of the same power in $(2n+1)\,P_n$. These two expressions are thus equal term by term ; and therefore

$$\frac{dP_{n+1}}{dx} - \frac{dP_{n-1}}{dx} = (2n+1)\,P_n,$$

or

$$\frac{dP_n}{dx} - \frac{dP_{n-2}}{dx} = (2n-1)\,P_{n-1}.$$

In the present case n is a positive integer, so that this leads to a finite series for $\dfrac{dP_n}{dx}$, viz. :

$$\frac{dP_n}{dx} = (2n-1)\,P_{n-1} + (2n-5)\,P_{n-3} + (2n-9)\,P_{n-5} + \ldots;$$

the last term of the series $3P_1$ or P_0 (i.e. 1), according as n is even or odd.

98. Now by § 95 we see that

$$\tfrac{1}{2}P_n \log\left(\frac{x+1}{x-1}\right) - w$$

is a solution of the differential equation, if w be determined as the Particular Integral of

$$\frac{d}{dx}\left\{(1-x^2)\frac{dw}{dx}\right\} + n(n+1)w = 2\frac{dP_n}{dx} = 2\{(2n-1)P_{n-1} + (2n-5)P_{n-3} + \ldots\},$$

by the formula just obtained. To obtain this Particular Integral, we write

$$w = a_1 P_{n-1} + a_3 P_{n-3} + \ldots + a_{2r-1}P_{n-2r+1} + \ldots,$$

and substitute; since

$$\frac{d}{dx}\left\{(1-x^2)\frac{dP_m}{dx}\right\} = -m(m+1)P_m,$$

the left-hand side has, as the coefficient of $a_{2r-1}P_{n-2r+1}$,

$$n(n+1) - (n-2r+1)(n-2r+2)$$
$$= 2(2r-1)(n-r+1);$$

and therefore

$$a_{2r-1}(2r-1)(n-r+1) = 2n-4r+3.$$

The value of w is therefore now definite; and the corresponding solution of Legendre's equation is

$$\tfrac{1}{2}P_n \log\left(\frac{x+1}{x-1}\right) - \left\{\frac{2n-1}{1 \cdot n}P_{n-1} + \frac{2n-5}{3(n-1)}P_{n-3} + \frac{2n-9}{5(n-2)}P_{n-5} + \ldots\ldots\right\},$$

the last term being

$$\frac{3}{(n-1)(\tfrac{1}{2}n+1)}P_1,$$

when n is even, and

$$\frac{1}{\tfrac{1}{2}n(n+1)}P_0, \text{ i.e., } \frac{1}{\tfrac{1}{2}n(n+1)},$$

when n is odd.

99. We have now to compare this solution with Q_n. Let it be supposed expanded in a series of descending powers of x; it must then be of the form

$$AP_n + BQ_n,$$

where A and B are constants. Now in the series the term involving x^n does not occur, since

$$\tfrac{1}{2}\log\left(\frac{x+1}{x-1}\right) = \frac{1}{x} + \frac{1}{3x^3} + \frac{1}{5x^5} + \ldots\ldots$$

and therefore A must be zero; hence the coefficients of the powers between x^n and $x^{-(n+1)}$ exclusive of the latter disappear; this is easily verified for the first few. The above solution is therefore a constant multiple of Q_n, and thus

$$BQ_n = \tfrac{1}{2} P_n \log \left(\frac{x+1}{x-1} \right) - \left\{ \frac{2n-1}{1 \cdot n} P_{n-1} + \frac{2n-5}{3(n-1)} P_{n-3} + \frac{2n-9}{5(n-2)} P_{n-5} + \ldots \ldots \right\}$$

$$= \tfrac{1}{2} P_n \log \left(\frac{x+1}{x-1} \right) - Z_n,$$

where Z_n stands for the series which, when n is integral, is a function of degree $n-1$. Hence

$$B \frac{Q_n}{P_n} = \tfrac{1}{2} \log \frac{x+1}{x-1} - \frac{Z_n}{P_n},$$

and therefore

$$B \frac{d}{dx} \left(\frac{Q_n}{P_n} \right) = - \frac{1}{x^2-1} - \frac{U}{P_n^2},$$

where U is an integral function of x of degree not higher than $2n-2$. When we substitute on the left-hand side from § 96, it becomes

$$\frac{B}{(x^2-1) P_n^2} = \frac{1}{x^2-1} + \frac{U}{P_n^2},$$

or

$$B = P_n^2 + (x^2-1) U,$$

where the right-hand side is a finite integral function of x. This is true for all values of x; writing $x=1$, we have $B=$ value of P_n^2 when x is unity. Now in Ex. 2 of § 90, P_n was indicated as the coefficient of z^n in the expansion of $(1 - 2xz + z^2)^{-\frac{1}{2}}$ in ascending powers of z; and therefore the value of P_n when $x=1$ is the coefficient of z^n in the expansion of $(1 - 2z + z^2)^{-\frac{1}{2}}$, i.e., of $(1-z)^{-1}$. This coefficient is unity, so that P_n is unity when $x=1$; thus $B=1$, and the equation becomes

$$Q_n = \tfrac{1}{2} P_n \log \left(\frac{x+1}{x-1} \right) - Z_n.$$

Ex. 1. Discuss the significance of this expression for Q_n when n is not an integer.

Ex. 2. The following properties, analogous to those of P_n, hold for Q_n:

(i) $\dfrac{d^{n+1} Q_n}{dx^{n+1}} = - \dfrac{(-2)^n \, \Pi(n)}{(x^2-1)^{n+1}}$;

(ii) $\dfrac{dQ_{n+1}}{dx} - \dfrac{dQ_{n-1}}{dx} = (2n+1) Q_n$;

(iii) $n \dfrac{dQ_{n+1}}{dx} + (n+1) \dfrac{dQ_{n-1}}{dx} = (2n+1) x \dfrac{dQ_n}{dx}$.

Ex. 3. Obtain the properties of the integrals Q, corresponding to those of the integrals P given in the examples in § 90.

Ex. 4. Prove that, if $y>1>x>-1$,

$$(y-x)^{-1} = \sum_{n=0}^{n=\infty} (2n+1) P_n(x) Q_n(y).$$

The further development of the properties of the functions which are the particular solutions of Legendre's equations does not depend merely upon the differential equation; the student will find most ample investigation of their analytical properties and their applications to mathematical physics in the excellent treatise by Heine—*Handbuch der Kugelfunctionen.* The treatises by Todhunter, *The Functions of Laplace, Lamé and Bessel,* and by Ferrers, *Spherical Harmonics,* will prove useful.

BESSEL'S EQUATION

100. This differential equation is

$$x^2 \frac{d^2y}{dx^2} + x \frac{dy}{dx} + (x^2 - n^2) y = 0,$$

or, what is the same thing,

$$x \frac{d}{dx} \left(x \frac{dy}{dx} \right) + (x^2 - n^2) y = 0,$$

in which n is a constant; it will be assumed that n is real. The equation, like Legendre's, occurs in investigations in applied mathematics, and n is usually an integer in such applications; but, as in the case of the preceding differential equation, this limitation will not be imposed on the value of n.

To solve the equation, we write

$$y = A_1 x^{m_1} + A_2 x^{m_2} + A_3 x^{m_3} + \cdots\cdots\cdots$$

and substitute; we then have

$$(m_1^2 - n^2) A_1 x^{m_1} + (m_2^2 - n^2) A_2 x^{m_2} + (m_3^2 - n^2) A_3 x^{m_3} + \cdots\cdots$$
$$+ A_1 x^{m_1+2} + A_2 x^{m_2+2} + \cdots\cdots\cdots = 0,$$

which must be identically satisfied. Hence, from a comparison of the indices, we have

$$m_2 = m_1 + 2,$$
$$m_3 = m_2 + 2,$$

or the series is one in *ascending* powers of x, the common difference of the indices of the powers being 2; and thus $m_r = m_1 + 2(r-1)$. Taking the term in x with the lowest index, we have

$$m_1{}^2 = n^2,$$

since A_1 is not zero; and therefore

$$m_1 = +n, \quad \text{or} \quad m_1 = -n.$$

The coefficient of x^{m_1+2r} on the left-hand side must be zero; and therefore

$$\{(m_1 + 2r)^2 - n^2\} A_{r+1} + A_r = 0,$$

or, since

$$m_1{}^2 = n^2,$$

$$A_{r+1} = -\frac{A_r}{2^2 r} \frac{1}{m_1 + r}.$$

101. Consider, first, the solution corresponding to

$$m_1 = +n.$$

The coefficients A are then given by

$$A_{r+1} = -\frac{A_r}{2^2 r (n+r)},$$

so that

$$A_r = (-1)^{r-1} \frac{A_1}{(r-1)!\, 2^{2(r-1)} (n+1)(n+2)\ldots(n+r-1)},$$

for values of r greater than unity; and the series, which is a solution of the differential equation, becomes

$$A_1 x^n \left[1 - \frac{x^2}{2^2 (n+1)} + \frac{x^4}{2!\, 2^4 (n+1)(n+2)} - \frac{x^6}{3!\, 2^6 (n+1)(n+2)(n+3)} + \cdots \right],$$

where A_1 is an arbitrary constant.

When to A_1 is assigned the particular value $\dfrac{1}{2^n \Pi(n)}$, where $\Pi(n)$ is Gauss's function Π and is the same as $\Gamma(n+1)$, then the expression is denoted by J_n, so that

$$J_n = \frac{x^n}{2^n \Pi(n)} \left[1 - \frac{x^2}{2^2 (n+1)} + \frac{x^4}{2!\, 2^4 (n+1)(n+2)} - \cdots \right]$$

$$= \sum_{r=0}^{r=\infty} \frac{(-1)^r}{\Pi(n+r)\, \Pi(r)} \left(\frac{x}{2} \right)^{n+2r},$$

which is usually called the Bessel's function of order n.

When n is positive, whether an integer or not, the series proceeds to infinity and, for finite values of the variable, obviously converges. Thus AJ_n, where A is an arbitrary constant, is one solution of the differential equation.

Before considering the form of J_n, where n is a negative integer, it is convenient to obtain the solution corresponding to the case

$$m_1 = -n.$$

The work is the same as before with the change of sign of n, and the solution is

$$B_1 x^{-n} \left[1 - \frac{x^2}{2^2(-n+1)} + \frac{x^4}{2!\,2^4(-n+1)(-n+2)} \right.$$
$$\left. - \frac{x^6}{3!\,2^6(-n+1)(-n+2)(-n+3)} + \cdots \right],$$

where B_1 is an arbitrary constant. To B_1 assign the value $\dfrac{1}{2^{-n}\,\Pi(-n)}$; then the resulting expression is exactly the same function of $-n$ as J_n is of $+n$, and it may therefore be denoted by J_{-n}, so that

$$J_{-n} = \frac{x^{-n}}{2^{-n}\,\Pi(-n)} \left[1 - \frac{x^2}{2^2(-n+1)} + \frac{x^4}{2!\,2^4(-n+1)(-n+2)} - \cdots \right]$$
$$= \sum_{r=0}^{r=\infty} \frac{(-1)^r}{\Pi(-n+r)\,\Pi(r)} \left(\frac{x}{2}\right)^{-n+2r}$$

If now n be negative, whether an integer or not, or be positive but not an integer, this series proceeds to infinity and, for finite values of the variable, converges. In this case, BJ_{-n} is another solution of the differential equation.

If then n be not an integer, whether it be a positive or a negative quantity, J_n and J_{-n} are two independent and determinate particular solutions of the differential equation; and the primitive is

$$y = AJ_n + BJ_{-n}.$$

102. If n be an integer other than zero, two cases arise. First, if n be a negative integer and equal to $-p$, a zero factor occurs in the coefficient of all terms after x^{2p} inclusive within the bracket; and therefore (by § 86) the terms which precede this disappear, and J_n becomes

$$\sum_{r=p}^{r=\infty} \frac{(-1)^r}{\Pi(n+r)\,\Pi(r)} \left(\frac{x}{2}\right)^{n+2r},$$

or, what is the same thing,

$$\sum_{s=0}^{s=\infty} \frac{(-1)^{s+p}}{\Pi(s)\,\Pi(s+p)} \left(\frac{x}{2}\right)^{p+2s},$$

since $n+p=0$. Now this last expression is $(-1)^p J_p$, that is, is $(-1)^{-n} J_{-n}$; so that, in the case when n is a negative integer, one of the particular solutions, J_n, degenerates into a constant multiple of the other, J_{-n}.

Similarly it may be proved, or it may be at once deduced from the foregoing, that when n *is a positive integer* one of the particular solutions, J_{-n}, degenerates into a constant multiple of the other, J_n.

When n *is zero*, the two solutions coincide. Hence in every case when n is an integer, whether positive, zero, or negative, we may write

$$J_n = (-1)^n J_{-n}.$$

But that this equation may be valid, it must be remembered that it refers to the respective *limiting* forms of the particular solution of the differential equation when the superfluous terms of the latter for the special value of n have been removed from the expression in the general case; and the relation merely *gives this limiting form*. It however shews that, when n is an integer, it is sufficient to take the positive square root of n^2 and to consider, as the corresponding particular solution, the function associated with that square root.

It thus remains to find a second particular solution in two cases, in order to have the primitive; and these two cases are:—

First, when n is zero :

Second, when n is an integer which (from the above explanation) may be considered positive.

103. To obtain these particular solutions, it is convenient to have some fundamental properties proved.

It may be at once verified that

$$\text{(i)} \quad \frac{dJ_0}{dx} = -J_1 ;$$

$$\text{(ii)} \quad \frac{d}{dx}(x^n J_n) = x^n J_{n-1} ;$$

$$\text{(iii)} \quad \frac{d}{dx}(x^{-n} J_n) = -x^{-n} J_{n+1} .$$

From the last two, we have

$$x^n \frac{dJ_n}{dx} + nx^{n-1} J_n = x^n J_{n-1},$$

$$x^{-n} \frac{dJ_n}{dx} - nx^{-n-1} J_n = -x^{-n} J_{n+1}.$$

Dividing the first of these throughout by x^{n-1} and the second by x^{-n-1}, and subtracting the latter from the former, we have.

$$2n J_n = x (J_{n-1} + J_{n+1}),$$

or

$$J_{n-1} + J_{n+1} = \frac{2}{x} n J_n.$$

Similarly

$$-J_{n+1} - J_{n+3} = -\frac{2}{x}(n+2) J_{n+2},$$

$$J_{n+3} + J_{n+5} = \frac{2}{x}(n+4) J_{n+4},$$

$$\dots\dots\dots\dots\dots\dots\dots\dots\dots\dots\dots$$

Now it is evident, from the general value of J_n, that $J_\infty = 0$; hence the preceding equations give

$$J_{n-1} = \frac{2}{x} \{n J_n - (n+2) J_{n+2} + (n+4) J_{n+4} - \dots \text{ad inf.}\} ;$$

this series converges.

Ex. 1. Prove that

$$\frac{dJ_n}{dx} = \frac{2}{x} \{\tfrac{1}{2} n J_n - (n+2) J_{n+2} + (n+4) J_{n+4} - \dots \text{ad inf.}\}.$$

Ex. 2. Prove that, when n is a positive integer, $J_n(x)$ is the coefficient of z^n in the expansion of

$$e^{\frac{1}{2}x\left(z - \frac{1}{z}\right)}$$

in ascending and descending powers of z.

Hence deduce the relations (i), (ii), (iii), given at the beginning of § 103.

Ex. 3. Shew that, when n is a positive integer,

$$J_n(x) = \frac{1}{\pi} \int_0^\pi \cos(n\theta - x\sin\theta)\, d\theta.$$

Prove also that

$$J_0(x) = \frac{1}{\pi} \int_0^\pi \cos(x\cos\theta)\, d\theta\,;$$

and hence (or otherwise) verify that

$$\int_0^\infty e^{-ax} J_0(bx)\, dx = (a^2 + b^2)^{-\frac{1}{2}}.$$

104. To obtain the desired particular solution in the case when n *is zero*, we substitute

$$y = u J_0 + w$$

in the differential equation

$$\frac{d^2y}{dx^2} + \frac{1}{x}\frac{dy}{dx} + y = 0\,;$$

the result is

$$\frac{d^2w}{dx^2} + \frac{1}{x}\frac{dw}{dx} + w = -J_0\left(\frac{d^2u}{dx^2} + \frac{1}{x}\frac{du}{dx}\right) - 2\frac{du}{dx}\frac{dJ_0}{dx}.$$

To make the coefficient of J_0 vanish, we take

$$\frac{d^2u}{dx^2} + \frac{1}{x}\frac{du}{dx} = 0,$$

which is satisfied by

$$u = \log x.$$

The equation determining w is now

$$\frac{d^2w}{dx^2} + \frac{1}{x}\frac{dw}{dx} + w = -\frac{2}{x}\frac{dJ_0}{dx}$$

$$= \frac{2}{x}J_1$$

$$= \frac{4}{x^2}\{2J_2 - 4J_4 + 6J_6 - 8J_8 + \ldots\}.$$

Now from the equation

$$\frac{d^2J_n}{dx^2} + \frac{1}{x}\frac{dJ_n}{dx} + J_n = \frac{n^2}{x^2}J_n,$$

it follows that

$$y = \lambda J_n$$

is the Particular Integral of

$$\frac{d^2y}{dx^2} + \frac{1}{x}\frac{dy}{dx} + y = \frac{\lambda n^2}{x^2}J_n.$$

The general term in the right-hand side of the equation determining w is

$$\frac{4}{x^2}(-1)^{\frac{1}{2}n-1}nJ_n\,;$$

we have therefore for this term

$$\lambda = \frac{4}{n}(-1)^{\frac{1}{2}n-1}.$$

Hence

$$w = 2\left\{J_2 - \tfrac{1}{2}J_4 + \tfrac{1}{3}J_6 - \tfrac{1}{4}J_8 + \tfrac{1}{5}J_{10} - \ldots\ldots\right\};$$

and therefore a solution of the original equation is

$$J_0\log x + 2\left\{J_2 - \tfrac{1}{2}J_4 + \tfrac{1}{3}J_6 - \tfrac{1}{4}J_8 + \tfrac{1}{5}J_{10} - \ldots\ldots\right\}.$$

Let this be denoted by Y_0; then the primitive of the equation

$$\frac{d^2y}{dx^2} + \frac{1}{x}\frac{dy}{dx} + y = 0$$

is

$$y = AJ_0 + BY_0,$$

where A and B are arbitrary constants.

105. To obtain the second particular solution in the case where n *is an integer other than zero*, we write

$$y = J_n\log x - w,$$

so that

$$\frac{d^2w}{dx^2} + \frac{1}{x}\frac{dw}{dx} + \left(1 - \frac{n^2}{x^2}\right)w = \frac{2}{x}\frac{dJ_n}{dx}$$

$$= \frac{2n}{x^2}J_n - \frac{4}{x^2}\left\{(n+2)J_{n+2} - (n+4)J_{n+4} + (n+6)J_{n+6} - \ldots\right\}.$$

Now

$$\frac{d^2(\lambda J_m)}{dx^2} + \frac{1}{x}\frac{d(\lambda J_m)}{dx} + \left(1 - \frac{n^2}{x^2}\right)\lambda J_m = \frac{m^2 - n^2}{x^2}\lambda J_m,$$

λ being a constant; and therefore a value of w satisfying

$$\frac{d^2w}{dx^2} + \frac{1}{x}\frac{dw}{dx} + \left(1 - \frac{n^2}{x^2}\right)w = \frac{(-1)^r}{x^2}4(n+2r)J_{n+2r}$$

is

$$w = (-1)^r\frac{n+2r}{r(n+r)}J_{n+2r}.$$

Let w_1 be a quantity satisfying

$$\frac{d^2w_1}{dx^2} + \frac{1}{x}\frac{dw_1}{dx} + \left(1 - \frac{n^2}{x^2}\right)w_1 = \frac{2n}{x^2}J_n\,;$$

then a suitable value of w is

$$w = w_1 + \sum_{r=1}^{r=\infty}(-1)^r\frac{n+2r}{r(n+r)}J_{n+2r}.$$

The right-hand side of the equation giving w_1 must be transformed. By the general relation between three successive Bessel's functions, we have

$$\frac{2}{x} J_1 - J_0 = J_2 ;$$

hence

$$2 \left(\frac{2}{x}\right)^2 J_1 - 2 \left(\frac{2}{x}\right) J_0 - J_1 = 2 \left(\frac{2}{x}\right) J_2 - J_1 = J_3 ;$$

hence also

$$2 \cdot 3 \left(\frac{2}{x}\right)^3 J_1 - 2 \cdot 3 \left(\frac{2}{x}\right)^2 J_0 - 3 \left(\frac{2}{x}\right) J_1 - J_2 = 2 \frac{3}{x} J_3 - J_2 = J_4 ;$$

also

$$2 \cdot 3 \cdot 4 \left(\frac{2}{x}\right)^4 J_1 - 2 \cdot 3 \cdot 4 \left(\frac{2}{x}\right)^3 J_0 - 3 \cdot 4 \left(\frac{2}{x}\right)^2 J_1 - 4 \left(\frac{2}{x}\right) J_2 - J_3 = 2 \frac{4}{x} J_4 - J_3 = J_5,$$

and so on; and the general relation is

$$\Pi (n-1) \left(\frac{2}{x}\right)^{n-1} J_1 - \Pi(n-1) \left(\frac{2}{x}\right)^{n-2} J_0 - \frac{\Pi(n-1)}{\Pi(2)} \left(\frac{2}{x}\right)^{n-3} J_1 - \frac{\Pi(n-1)}{\Pi(3)} \left(\frac{2}{x}\right)^{n-4} J_2$$

$$- \ldots - \frac{\Pi(n-1) \, 2}{\Pi(n-2) \, x} J_{n-3} - J_{n-2} = J_n,$$

or, what is the same relation,

$$\frac{2n}{x^2} J_n = \tfrac{1}{2} \Pi(n) \left\{ \left(\frac{2}{x}\right)^{n+1} J_1 - \sum_{p=0}^{p=n-2} \left(\frac{2}{x}\right)^{n-p} \frac{J_p}{\Pi(p+1)} \right\}.$$

Also, by actual substitution, we have

$$\left\{ \frac{d^2}{dx^2} + \frac{1}{x} \frac{d}{dx} + 1 - \frac{n^2}{x^2} \right\} \frac{J_p}{x^m} = \frac{1}{x^m} \left[\frac{d^2 J_p}{dx^2} + \frac{1}{x} \frac{d J_p}{dx} + \left(1 + \frac{m^2 - n^2}{x^2}\right) J_p - \frac{2m}{x} \frac{d J_p}{dx} \right]$$

$$= \frac{1}{x^m} \left[-\frac{2m}{x} \frac{d J_p}{dx} + \frac{p^2 + m^2 - n^2}{x^2} J_p \right],$$

so that, on writing $m = n - p$,

$$\left\{ \frac{d^2}{dx^2} + \frac{1}{x} \frac{d}{dx} + 1 - \frac{n^2}{x^2} \right\} \lambda_p \frac{J_p}{x^m} = -\frac{2(n-p)}{x^{n-p}} \lambda_p \left(\frac{1}{x} \frac{d J_p}{dx} + \frac{p}{x^2} J_p \right),$$

λ_p being a constant. If p be not zero, the right-hand side is

$$-\frac{2(n-p)}{x^{n-p+1}} \lambda_p J_{p-1} ;$$

while if p be zero, the right-hand side is

$$+\frac{2n}{x^{n+1}} \lambda_0 J_1.$$

If now we substitute in the equation for w_1 the value

$$w_1 = \sum_{p=0}^{p=n-1} \lambda_p \frac{J_p}{x^{n-p}},$$

a comparison of the two sides of the equation gives

$$-2(n-p)\lambda_p = -\tfrac{1}{2}\Pi(n)\frac{2^{n-p+1}}{\Pi(p)}$$

if p be not zero, and gives

$$2n\lambda_0 = \tfrac{1}{2}\Pi(n)\,2^{n+1}$$

if p be zero; and therefore, whatever p may be,

$$\lambda_p = \frac{2^{n-p}}{n-p}\frac{1}{\Pi(p)}\frac{\Pi(n)}{2}.$$

Hence the value of w_1 is

$$w_1 = \tfrac{1}{2}\Pi(n)\sum_{p=0}^{p=n-1}\frac{1}{n-p}\left(\frac{2}{x}\right)^{n-p}\frac{J_p}{\Pi(p)};$$

and therefore *the second particular solution of Bessel's equation, in the case when n is a positive integer other than zero, is*

$$y = J_n\log x - \sum_{r=1}^{r=\infty}(-1)^r\frac{n+2r}{r(n+r)}J_{n+2r}$$

$$-\tfrac{1}{2}\Pi(n)\sum_{p=0}^{p=n-1}\frac{1}{n-p}\left(\frac{2}{x}\right)^{n-p}\frac{J_p}{\Pi(p)}.$$

Let the right-hand side be denoted by Y_n; then the primitive is given by

$$y = AJ_n + BY_n.$$

Note. In a Supplementary Note (I, pp. 243 sqq.) at the end of Chapter VI., an account is given of the method devised by Frobenius for the integration of linear equations in series. The method is applied to construct the primitive of the Bessel's equation of order zero (Ex. 3, p. 252) and the primitive of the Bessel's equation of order n (Ex. 4, p. 253). The process there given will be found more general, simpler, and more direct, than the preceding analysis.

Ex. 1. Another method of obtaining a second particular solution is employed by Hankel as follows. Any linear function of the particular solutions is also a particular solution; hence, in the general case, such a solution is given by

$$2\pi e^{n\pi i}\frac{J_n\cos n\pi - J_{-n}}{\sin 2n\pi},$$

which is then perfectly determinate; while, in the particular case when n is an integer, it takes the form $0/0$ since $(-1)^n J_n = J_{-n}$. Prove that, when evaluated, this assumes the form

$$-\left(\frac{2}{x}\right)^n\sum_{p=0}^{p=n-1}\frac{\Pi(n-p-1)}{\Pi(p)}\left(\frac{x}{2}\right)^{2p}$$

$$\left(\frac{x}{2}\right)^n\sum_{p=0}^{p=\infty}\frac{(-1)^p}{\Pi(n+p)\Pi(p)}\left(\frac{x}{2}\right)^{2p}\left\{\log\left(\frac{x^2}{4}\right)-\Psi(n+p)-\Psi(p)\right\},$$

where $$\Psi(z) = \frac{d}{dz} \log \Pi(z);$$

and compare this with the solution already obtained.

(*Math. Ann.* I. p. 469.)

Ex. 2. The series for J_n is always a converging series; but, when z is large, the convergence is slow, and it is convenient to have a series proceeding in descending powers of z. Prove that

$$J_n = \left(\frac{2}{\pi z}\right)^{\frac{1}{2}} \left\{ 1 - \frac{(1^2 - 4n^2)(3^2 - 4n^2)}{2!(8z)^2} + \ldots \right\} \cos\left(z - \frac{\pi}{4} - n\frac{\pi}{2}\right)$$

$$+ \left(\frac{2}{\pi z}\right)^{\frac{1}{2}} \left\{ \frac{1^2 - 4n^2}{8z} - \frac{(1^2 - 4n^2)(3^2 - 4n^2)(5^2 - 4n^2)}{3!(8z)^3} + \ldots \right\} \sin\left(z - \frac{\pi}{4} - n\frac{\pi}{2}\right),$$

so that the series terminates, if $2n$ be equal to an odd integer.

(Lommel.)

106. The relation between the two linearly independent integrals J_n and J_{-n} may be found as in § 96. We have

$$\frac{d^2 J_n}{dx^2} + \frac{1}{x}\frac{dJ_n}{dx} + \left(1 - \frac{n^2}{x^2}\right) J_n = 0,$$

and

$$\frac{d^2 J_{-n}}{dx^2} + \frac{1}{x}\frac{dJ_{-n}}{dx} + \left(1 - \frac{n^2}{x^2}\right) J_{-n} = 0;$$

and therefore

$$\left(\frac{d^2 J_n}{dx^2} J_{-n} - \frac{d^2 J_{-n}}{dx^2} J_n\right) + \frac{1}{x}\left(\frac{dJ_n}{dx} J_{-n} - \frac{dJ_{-n}}{dx} J_n\right) = 0,$$

which gives

$$\frac{dJ_n}{dx} J_{-n} - \frac{dJ_{-n}}{dx} J_n = \frac{A}{x},$$

where A is a constant which, however, is not arbitrary since J_n and J_{-n} are definite functions. To obtain the value of A, it is sufficient to consider the lowest terms only in the left-hand side; when these are substituted, we find that

$$A = \frac{1}{2^n \Pi(n)} \frac{1}{2^{-n} \Pi(-n)} (n + n)$$

$$= \frac{2n}{\Pi(n) \Pi(-n)}$$

$$= \frac{2}{\Pi(n-1) \Pi(-n)}$$

$$= \frac{2 \sin n\pi}{\pi}.$$

Therefore

$$\frac{dJ_n}{dx} J_{-n} - \frac{dJ_{-n}}{dx} J_n = \frac{2}{\pi x} \sin n\pi,$$

or, what is the same thing,

$$\frac{d}{dx} \left(\frac{J_{-n}}{J_n}\right) = -\frac{2 \sin n\pi}{\pi x J_n^2}.$$

Ex. Obtain the corresponding equation when n is an integer.

Relation between the Equations of Legendre and Bessel.

107. It is possible to derive Bessel's equation from that of Legendre. For, differentiating the equation

$$(1 - x^2) \frac{d^2y}{dx^2} - 2x \frac{dy}{dx} + n(n+1) y = 0$$

m times, and writing

$$z = \frac{d^m y}{dx^m},$$

we have

$$(1 - x^2) \frac{d^2z}{dx^2} - (2m + 2) x \frac{dz}{dx} + \{n(n+1) - m(m+1)\} z = 0.$$

Let the dependent variable be changed to ξ, where

$$\xi = (1 - x^2)^{\frac{1}{2}m} z;$$

the equation now becomes

$$(1 - x^2) \frac{d^2\xi}{dx^2} - 2x \frac{d\xi}{dx} + \left\{n(n+1) - \frac{m^2}{1 - x^2}\right\} \xi = 0.$$

Let the independent variable be changed from x to ϕ, where

$$\phi^2 = n^2 (1 - x^2);$$

then, after slight reductions, the equation becomes

$$\left(1 - \frac{\phi^2}{n^2}\right) \frac{d^2\xi}{d\phi^2} + \left(1 - \frac{2\phi^2}{n^2}\right) \frac{1}{\phi} \frac{d\xi}{d\phi} + \left(1 + \frac{1}{n} - \frac{m^2}{\phi^2}\right) \xi = 0.$$

When we make n infinite, we have

$$\frac{d^2\xi}{d\phi^2} + \frac{1}{\phi}\frac{d\xi}{d\phi} + \left(1 - \frac{m^2}{\phi^2}\right)\xi = 0,$$

which is Bessel's differential equation.

When all these operations are combined, we have, as the result, that the limit of

$$(-\phi)^m \left\{\frac{(n^2-\phi^2)^{\frac{1}{2}}}{\phi}\frac{d}{d\phi}\right\}^m P_n\left\{\left(1 - \frac{\phi^2}{n^2}\right)^{\frac{1}{2}}\right\},$$

when n is infinite, is Bessel's function of order m, ϕ being the independent variable.

It would appear from the foregoing process that ϕ is infinite; this however is avoided by making x approach indefinitely closely to the value unity. The geometrical analogue of this relation between ϕ and x is that whereby any very small portion of a spherical (or other) surface in the neighbourhood of a point is studied, by assuming it ultimately to coincide with the tangent plane of the surface at that point and to be magnified in that plane.

Ex. Verify that the above expression becomes, in the limit, a multiple of J_m.

In this connection the student may consult Heine, *Theorie der Kugelfunctionen*, 2nd edition, vol. I. p. 182 ; Lord Rayleigh, *Proc. Lond. Math. Soc.* vol. IX. p. 61, *Scientific Papers*, vol. I. pp. 338—341.

The primitive of Bessel's differential equation has been obtained for every case ; the further development of the properties of the functions, which occur in that primitive, cannot be given here. The student will find the functions fully treated by Lommel in his *Studien über die Bessel'sche Functionen* and in several papers by the same writer in the *Mathematische Annalen*, vols. II. III. IV. IX. XIV. XVI. ; in particular, the paper in vol. XIV. deals with differential equations which are integrable by Bessel's functions. Reference should also be made to Neumann's *Theorie der Bessel'schen Functionen* and to Heine's *Theorie der Kugelfunctionen*, 2nd edition, where (vol. I. p. 189) a list of memoirs referring to the functions is given. The most modern and comprehensive book on the subject is G. N. Watson's *Treatise on Bessel Functions*.

For a general property of all linear differential equations similar to those which have just been discussed and which give rise to functions depending upon a constant parameter, the student may consult, in addition to the foregoing, Sturm, *Liouville*, vol. I. ; and Routh, *Proc. Lond. Math. Soc.* vol. X.

RICCATI'S EQUATION.

108. Riccati's differential equation is

$$\frac{dy}{dx} + by^2 = cx^m;$$

but it is convenient to consider first the more general form

$$x\frac{dy}{dx} - ay + by^2 = cx^n.$$

If in the latter the independent variable be changed from x to z, where $z = x^a$, and the dependent variable be changed from y to u, where $y = uz$, the equation becomes

$$\frac{du}{dz} + \frac{b}{a}u^2 = \frac{c}{a}z^{\frac{n}{a}-2},$$

which is Riccati's form.

Consider now the more general form.

Firstly, *it can be integrated in finite terms when* $n = 2a$.

For assuming $y = ux^a$, we find on substitution

$$x^{a+1}\frac{du}{dx} + bx^{2a}u^2 = cx^n,$$

so that

$$x^{1-a}\frac{du}{dx} + bu^2 = cx^{n-2a}.$$

In the case when $n = 2a$, this becomes

$$x^{1-a}\frac{du}{dx} = c - bu^2;$$

the variables are separable, and u is expressible in terms of exponential, or circular, functions according as b and c have, or have not, like signs.

Secondly, *it can be integrated in finite terms when* $(n \pm 2a)/2n$ *is a positive integer.*

Let the dependent variable be changed from y to y_1, where $A + \dfrac{x^n}{y_1} = y$, and A is a constant the value of which has yet to be determined. When substitution takes place and the terms are rearranged, the equation becomes

$$-aA + bA^2 + (n - a + 2bA)\frac{x^n}{y_1} + b\frac{x^{2n}}{y_1^2} - \frac{x^{n+1}}{y_1^2}\frac{dy_1}{dx} = cx^n.$$

We choose A so that the constant term vanishes; thus $A = 0$ or $A = a/b$.

Taking the value a/b for A and substituting in this new form, we have, after a slight change,

$$x\frac{dy_1}{dx} - (a + n)\,y_1 + cy_1^2 = bx^n.$$

Now this equation is of the same *form* as that with which we began; and the changes, that have taken place, are in the coefficients—the original a has changed to $a + n$, and b and c have changed places. In this last equation we write

$$y_1 = \frac{a + n}{c} + \frac{x^n}{y_2};$$

the foregoing analysis then shews that the equation in y_2 will be

$$x\frac{dy_2}{dx} - (a + 2n)\,y_2 + by_2^2 = cx^n.$$

And the result of i successive transformations will be to reduce the given equation either to

$$x\frac{dy_i}{dx} - (a + in)\,y_i + cy_i^2 = bx^n,$$

or to

$$x\frac{dy_i}{dx} - (a + in)\,y_i + by_i^2 = cx^n,$$

according as the integer i is odd or even.

Now, by the case first considered, this equation is integrable in finite terms, if

$$n = 2\,(a + in),$$

that is, if

$$\frac{n - 2a}{2n}$$

is a positive integer.

Taking next the value zero for A, we can easily transform the equation into

$$x\frac{dy_1}{dx} - (n - a)\,y_1 + cy_1^2 = bx^n,$$

an equation which differs from the former in y_1 only so far as regards the sign of a. Adopting now for this the preceding series of transformations, we write

$$y_1 = \frac{n - a}{c} + \frac{x^n}{y_2},$$

and the equation in y_2 is

$$x \frac{dy_2}{dx} - (2n - a) y_2 + b y_2^2 = cx^n.$$

Hence after $i - 1$ transformations of this series (and therefore after i transformations in all) the given equation is reduced either to

$$x \frac{dy_i}{dx} - (in - a) y_i + c y_i^2 = bx^n,$$

or to $\qquad x \frac{dy_i}{dx} - (in - a) y_i + b y_i^2 = cx^n.$

In either case, the equation is integrable in finite terms, if

$$n = 2 (in - a),$$

that is, if $\qquad \dfrac{n + 2a}{2i}$

is a positive integer.

Combining then these two results, we infer that *the equation*

$$x \frac{dy}{dx} - ay + by^2 = cx^n$$

is integrable in finite terms when $(n \pm 2a)/2n$ *is a positive integer.*

In each case, the integral is given in the form of a finite continued fraction, the last denominator of which involves either exponential or circular functions.

109. We can now obtain conditions that Riccati's equation shall be integrable in finite terms. From § 108 it follows that

$$\frac{du}{dx} + bu^2 = cx^m$$

is transformed by the substitution $u = y/x$ into

$$x \frac{dy}{dx} - y + by^2 = cx^n,$$

where $m = n - 2$. Now the latter equation is so integrable when

$$n \pm 2 = 2ni,$$

where i is a positive integer; and therefore Riccati's equation is integrable in finite terms if

$$m + 2 \pm 2 = 2i (m + 2).$$

Taking the negative sign, we have

$$m = -\frac{4i}{2i-1};$$

while the positive sign gives

$$m = \frac{-4(i-1)}{2i-1},$$

or (what is the same thing in the case of the latter)

$$m = \frac{-4i}{2i+1},$$

by merely changing the integer i.

Hence *Riccati's equation is integrable in finite terms, if*

$$m = \frac{-4i}{2i \pm 1},$$

i being zero or a positive integer.

Ex. Prove that the equation

$$\frac{du}{dx} + bx^k u^2 = cx^m$$

is integrable in finite terms, if

$$\frac{m+1}{k+1} = \frac{-2i+1}{2i+1} \text{ or } \frac{-2i-1}{2i-1},$$

i being an integer; and obtain the limitations upon the value of i in the respective cases.

110. Both Riccati's equation, and the more general form discussed in § 108, are instances of the equation

$$\frac{dy}{dx} = P + Qy + Ry^2,$$

where P, Q, R, are functions of x.

Writing

$$y = -\frac{1}{R}\frac{1}{u}\frac{du}{dx},$$

we find that the equation for u is

$$\frac{d^2u}{dx^2} - \left(Q + \frac{d\log R}{dx}\right)\frac{du}{dx} + PRu = 0.$$

The complete primitive of this equation is of the form

$$u = Au_1 + Bu_2;$$

and the corresponding value of y is

$$y = -\frac{1}{R} \frac{1}{cu_1 + u_2} \left(c \frac{du_1}{dx} + \frac{du_2}{dx} \right)$$

which, as it contains one arbitrary constant, is the primitive of the original equation. In other words, the primitive of

$$\frac{dy}{dx} = P + Qy + Ry^2$$

is of the form

$$y = \frac{cp + q}{cr + s},$$

where p, q, r, s, are appropriate functions of x, and c is an arbitrary constant.

Moreover, by §§ 76, 77, it follows that, if one particular solution of the equation for u is known, the most general solution can be obtained by means of quadratures. A particular value of u leads to a particular value of y; and therefore it may be expected that, if a particular solution of the equation

$$\frac{dy}{dx} = P + Qy + Ry^2$$

is known, the most general solution can be obtained by quadratures. To establish this inference, let y_1 denote the particular solution that is supposed known; and write

$$y = y_1 + \frac{1}{v}.$$

Then we have

$$P + Qy_1 + Ry_1^2 - \frac{1}{v^2} \frac{dv}{dx}$$

$$= \frac{dy_1}{dx} - \frac{1}{v^2} \frac{dv}{dx}$$

$$= \frac{dy}{dx}$$

$$= P + Q\left(y_1 + \frac{1}{v}\right) + R\left(y_1 + \frac{1}{v}\right)^2;$$

and therefore

$$\frac{dv}{dx} + (Q + 2Ry_1)\,v = -R,$$

so that

$$v e^{\int (Q+2Ry_1)\,dx} = A - \int R e^{\int (Q+2Ry_1)\,dx}\,dx.$$

When the two quadratures required to give v as an explicit function have been effected, and when this value of v is substituted in the expression for y, we obtain the most general value of y which satisfies the equation. (It may be added that the process just indicated is frequently of practical use in the solution of differential equations of this type; its effectiveness depends upon the knowledge of a particular solution.)

It has been assumed that one solution is known. If another solution is known, we can avoid one of the quadratures in the expression for v. In fact, a second particular solution implies that a particular value of v is known. Denoting this by v_1, and writing $v = v_1 w$, so that

$$y = y_1 + \frac{1}{v_1 w},$$

we have

$$w\frac{dv_1}{dx} + v_1\frac{dw}{dx} + (Q + 2Ry_1)\,v_1 w = -R.$$

But as v_1 is a particular value of v, we have

$$\frac{dv_1}{dx} + (Q + 2Ry_1)\,v_1 = -R,$$

so that

$$v_1\frac{dw}{dx} = R\,(w-1),$$

and therefore

$$w - 1 = Ce^{\int \frac{1}{v_1} R\,dx},$$

which requires only one quadrature. When the quadrature has been effected, and the value of w is substituted, we again have the most general value of y which satisfies the equation.

The last result depends upon a knowledge of two particular solutions. If it should happen that three particular solutions are known, then we can obtain the primitive without requiring to perform any quadratures. For we have seen that the primitive is of the form

$$y = \frac{cp + q}{cr + s},$$

where p, q, r, s, are appropriate functions of x, and c is an arbitrary constant. Let the supposed three integrals be denoted by y_1, y_2, y_3; and let the corresponding values of c be c_1, c_2, c_3; then

$$\frac{(y - y_1)(y_2 - y_3)}{(y - y_2)(y_3 - y_1)} = \frac{(c - c_1)(c_2 - c_3)}{(c - c_2)(c_3 - c_1)}$$
$$= B,$$

where B is an arbitrary constant because c is arbitrary. This relation expresses y in terms of known quantities, and it contains an arbitrary constant. Manifestly it is the primitive, which accordingly can be obtained without any quadratures.

Ex. 1. Solve the equation

$$\frac{dy}{dx} = P(1 - xy) - y^2,$$

where P is any function of x. Manifestly, a particular solution is given by

$$y = \frac{1}{x};$$

accordingly, we write

$$y = \frac{1}{x} + \frac{1}{v},$$

and we find the equation in v to be

$$\frac{dv}{dx} - \left(xP + \frac{2}{x}\right)v = 1.$$

Hence

$$\frac{v}{x^2} e^{-\int xP dx} = A - \int^x \frac{dx}{x^3} e^{-\int xP dx};$$

and therefore the primitive of the equation is

$$\frac{x}{xy - 1} = A x^2 e^{\int xP dx} - x^2 e^{\int xP dx} \int^x \frac{dx}{x^2} e^{-\int xP dx}.$$

Ex. 2. Solve the equations

(i) $\quad x(1 - x^3)\dfrac{dy}{dx} = x^2 + y - 2xy^2$;

(ii) $\quad \dfrac{dy}{dx} = 2x - (x^2 + 1)y + y^2$;

(iii) $\quad \dfrac{dy}{dx} = 2 + \tfrac{1}{2}\left(x - \dfrac{1}{x}\right)y - \tfrac{1}{2}y^2$;

(iv) $\quad \dfrac{dy}{dx} = \cos x - y \sin x + y^2$.

RELATION BETWEEN THE EQUATIONS OF BESSEL AND RICCATI.

111. The equations of § 108 in the form in which they have been discussed are of the first order, but are not linear; there are some important transformations which render them linear of the second order.

In Riccati's equation, let the dependent variable be changed from u to v, where

$$bu = \frac{1}{v}\frac{dv}{dx},$$

so that, if u is expressible in finite terms, v often will be so also; the equation then becomes

$$\frac{d^2v}{dx^2} - bcvx^m = 0,$$

which might be taken as a standard form, equivalent to Riccati's equation.

If b and c have the same sign (in which case exponential functions occur in u), this equation may be written

$$\frac{d^2v}{dx^2} - a^2x^mv = 0.$$

If their signs be unlike (in which case circular functions occur in u), the equation is

$$\frac{d^2v}{dx^2} + a^2x^mv = 0.$$

Both of these are integrable in a finite form for the same value of m that renders Riccati's equation thus integrable.

Change the independent variable from x to z, where

$$qz = x^q,$$

and

$$q = \tfrac{1}{2}m + 1 = \frac{1}{n}\text{ say};$$

the equation then becomes

$$\frac{d^2v}{dz^2} - \frac{n-1}{z}\frac{dv}{dz} - bcv = 0.$$

This equation therefore is integrable in a finite form, if

$$\frac{1}{n} = \tfrac{1}{2}m + 1 = 1 - \frac{2i}{2i \pm 1} = \frac{\pm 1}{2i \pm 1},$$

whence it follows that n must be equal to an odd integer; and so, if the equation be written

$$\frac{d^2v}{dz^2} - \frac{2p}{z}\frac{dv}{dz} - bcv = 0,$$

the condition of integrability in a finite form is that p should be an integer.

This is reducible to its normal form by the substitution

$$vz^{-p} = w;$$

and the equation for w is

$$\frac{d^2w}{dz^2} - bcw = \frac{p(p+1)}{z^2}\, w,$$

which is integrable in a finite form if p be an integer.

Lastly, let $w = z^{\frac{1}{2}}t$ be substituted; the equation for t is

$$\frac{d^2t}{dz^2} + \frac{1}{z}\frac{dt}{dz} - bct - (p + \tfrac{1}{2})^2 \frac{t}{z^2} = 0,$$

the primitive of which is

$$t = A J_{p+\frac{1}{2}} \{z(-bc)^{\frac{1}{2}}\} + B J_{-(p+\frac{1}{2})} \{z(-bc)^{\frac{1}{2}}\}.$$

If $p + \tfrac{1}{2}$ be an integer, this ceases to be the primitive; we then have for the primitive

$$t = A J_{p+\frac{1}{2}} \{z(-bc)^{\frac{1}{2}}\} + B Y_{p+\frac{1}{2}} \{z(-bc)^{\frac{1}{2}}\}.$$

Hence the *solution of Riccati's equation can be expressed in terms of Bessel's functions; and, in particular, the primitive of*

$$\frac{d^2v}{dx^2} + \lambda v x^m = 0$$

is given by

$$v = x^{\frac{1}{2}} \left[A J_{\frac{1}{m+2}} (z\lambda^{\frac{1}{2}}) + B J_{-\frac{1}{m+2}} (z\lambda^{\frac{1}{2}}) \right],$$

or
$$v = x^{\frac{1}{2}} \left[A J_{\frac{1}{m+2}} (z\lambda^{\frac{1}{2}}) + B Y_{\frac{1}{m+2}} (z\lambda^{\frac{1}{2}}) \right],$$

according as $m + 2$ is not, or is, the reciprocal of an integer.

This is immediately derivable from a combination of the preceding transformations.

The only case of failure is that in which $m + 2$ is zero, that is, when m is -2; the equation is then

$$x^2 \frac{d^2v}{dx^2} + \lambda v = 0,$$

which can be solved by the method of § 47.

For further information upon this equation a memoir by J. W. L. Glaisher in the *Phil. Trans.*, 1881, pp. 759—828, should be consulted, where full references to authorities are given; and the connection between Riccati's equation and Bessel's will be found fully discussed in the book and papers of Lommel to which reference has already (p. 187) been made.

Some examples of the solution expressed by series occur in the Miscellaneous Examples.

Symbolical Solutions.

112. In cases, when the solution of a differential equation in series consists of a function in a finite form, or when it consists of a terminating series together with some function or functions in a finite form, it is sometimes possible to obtain a solution of a symbolical nature which will, when the operations therein indicated are performed, prove equivalent to the solution otherwise obtained.

As an example, consider the differential equation

$$\frac{d^2y}{dx^2} - n^2y = \frac{m(m+1)}{x^2}\, y,$$

the solution of which has been proved to be expressible in a finite form when m is an integer. When the dependent variable is transformed from y to u by means of the relation

$$y = ux^{m+1},$$

the equation becomes

$$\frac{d^2u}{dx^2} + 2(m+1)\frac{1}{x}\frac{du}{dx} - n^2u = 0.$$

Consider now the differential equation

$$\frac{d^2v}{dx^2} - n^2v = 0,$$

the general integral of which is

$$v = Ae^{nx} + Be^{-nx},$$

and change the independent variable from x to z, where z stands for $\frac{1}{2}x^2$; the equation becomes

$$2z\frac{d^2v}{dz^2} + \frac{dv}{dz} - n^2v = 0.$$

Let this be differentiated $m+1$ times with regard to z, and let t denote $\dfrac{d^{m+1}v}{dz^{m+1}}$; then we have

$$2z\frac{d^2t}{dz^2} + (2m+3)\frac{dt}{dz} - n^2t = 0.$$

Let now the independent variable be rechanged from z to x; the equation becomes

$$\frac{d^2t}{dx^2} + \frac{2(m+1)}{x}\frac{dt}{dx} - n^2t = 0.$$

Hence we have

$$u = t$$

$$= \frac{d^{m+1}v}{dz^{m+1}}$$

$$= \left(\frac{1}{x}\frac{d}{dx}\right)^{m+1}(Ae^{nx} + Be^{-nx});$$

the primitive of the original equation in y therefore is

$$y = x^{m+1}\left(\frac{1}{x}\frac{d}{dx}\right)^{m+1}(Ae^{nx} + Be^{-nx}).$$

A slightly different form may be given to this, for

$$\frac{1}{x}\frac{d}{dx}(Ae^{nx} + Be^{-nx}) = \frac{nAe^{nx} - nBe^{-nx}}{x}$$

$$= \frac{A'e^{nx} + B'e^{-nx}}{x},$$

on changing the arbitrary constants; and the primitive may be written in the form

$$y = x^{m+1} \left(\frac{1}{x} \frac{d}{dx} \right)^m \left(\frac{A' e^{nx} + B' e^{-nx}}{x} \right).$$

Since the differential equation remains unaltered, when for m is substituted $-(m+1)$, the primitive may be expressed in the additional forms

$$y = x^{-m} \left(\frac{1}{x} \frac{d}{dx} \right)^{-m} (A e^{nx} + B e^{-nx})$$

and

$$y = x^{-m} \left(\frac{1}{x} \frac{d}{dx} \right)^{-m-1} \left(\frac{A' e^{nx} + B' e^{-nx}}{x} \right).$$

Ex. 1. From the foregoing, it can be at once deduced that the primitive of

$$\frac{d^2 y}{dx^2} + n^2 y = \frac{6}{x^2} y$$

(an equation arising in investigations connected with the Figure of the Earth) is expressible in the form

$$y = C \left\{ \left(1 - \frac{3}{n^2 x^2} \right) \sin (nx + a) + \frac{3}{nx} \cos (nx + a) \right\}.$$

Ex. 2. Prove that the primitive of the differential equation

$$\frac{d^2 v}{dz^2} - n^2 z^{2q-2} v = 0$$

can, in the case when q is the reciprocal of an odd integer $2i + 1$, be exhibited in the forms

$$v = z \left(z^{-2q+1} \frac{d}{dz} \right)^{i+1} \left(A e^{\frac{n}{q} z^q} + B e^{-\frac{n}{q} z^q} \right),$$

$$v = \left(z^{-2q+1} \frac{d}{dz} \right)^{-i} \left(A e^{\frac{n}{q} z^q} + B e^{-\frac{n}{q} z^q} \right),$$

$$v = z \left(z^{-2q+1} \frac{d}{dz} \right)^{i} \left\{ z^{-q} \left(A e^{\frac{n}{q} z^q} + B e^{-\frac{n}{q} z^q} \right) \right\},$$

$$v = \left(z^{-2q+1} \frac{d}{dz} \right)^{-i-1} \left\{ z^{-q} \left(A e^{\frac{n}{q} z^q} + B e^{-\frac{n}{q} z^q} \right) \right\}.$$

(Glaisher.)

Ex. 3. Prove that the primitive of the equation

$$\frac{d^2 u}{dx^2} + a^2 u = \frac{p(p+1)}{x^2} u$$

is given by

$$u = Cx^{-p} \left(\frac{d}{dr}\right)^p \frac{\cos(r^{\frac{1}{2}} x + a)}{r^{\frac{1}{2}}},$$

where r is to be put equal to a^2 after the performance of the differentiations.

(Gaskin.)

In all these cases where the solution of the equation is thus given symbolically, it is not difficult to identify the solution in this form with that obtained in any other form, such as one in series by the earlier methods of this chapter, or as one by means of definite integrals to be indicated in Chapter VII. The student, who wishes for fuller information on the subject of these symbolical solutions and their connection with solutions in other forms, will find a full discussion in the memoir (Section VI.) by J. W. L. Glaisher already (p. 197) quoted.

MISCELLANEOUS EXAMPLES.

1. Integrate in series, and express in a finite form, the integrals of the equations

$$\text{(i)} \quad x^{\frac{4}{3}} \frac{d^2 y}{dx^2} - c^2 y = 0; \qquad \text{(ii)} \quad x^{\frac{8}{5}} \frac{d^2 y}{dx^2} - c^2 y = 0;$$

and integrate

$$\frac{d^2 y}{dx^2} + \frac{2}{x} \frac{dy}{dx} = \left(n^2 + \frac{2}{x^2}\right) y.$$

2. Solve the equations

(i) $\quad x \dfrac{d^2 y}{dx^2} - \dfrac{dy}{dx} + y = 0;$

(ii) $\quad x^3 \dfrac{d^3 y}{dx^3} + (x^3 + 3x^2) \dfrac{d^2 y}{dx^2} + (5x^2 - 30x) \dfrac{dy}{dx} + (4x + 30) y = 0;$

(iii) $\quad \dfrac{d^3 y}{dx^3} - x^2 \dfrac{d^2 y}{dx^2} - x \dfrac{dy}{dx} = aby - (a+b) x \dfrac{dy}{dx};$

(iv) $\quad (x^2 + qx^3) \dfrac{d^2 y}{dx^2} + \{(a+3) qx^2 + (b-c+1) x\} \dfrac{dy}{dx} + \{(a+1) qx - bc\} y = 0.$

3. Integrate in series the differential equation

$$x(1 - 4x) \frac{d^2 u}{dx^2} + \{(4p - 6) x - p + 1\} \frac{du}{dx} - p(p-1) u = 0;$$

and express the integral in the finite form

$$A \{1 - (1 - 4x)^{\frac{1}{2}}\}^p + B \{1 + (1 - 4x)^{\frac{1}{2}}\}^p.$$

(Glaisher.)

4. Verify that any root of the equation

$$y^3 + y + x = 0$$

satisfies

$$(\tfrac{1}{4}x^2 + \tfrac{1}{27})\frac{d^2y}{dx^2} + \tfrac{1}{4}x\frac{dy}{dx} - \tfrac{1}{36}y = 0.$$

(Spitzer.)

5. Transform the equation

$$n\frac{d^2y}{dx^2} + (m + x - n\alpha)\frac{dy}{dx} + \{p - \alpha(m + x)\}y = 0,$$

by assuming $y = e^{\alpha x}\zeta$ and $m + x + n\alpha = \xi(-n)^{\frac{1}{2}}$, into

$$\frac{d^2\zeta}{d\xi^2} = \xi\frac{d\zeta}{d\xi} + p\zeta;$$

and integrate the last equation in series.

6. Obtain the primitive of the equation

$$\frac{d^2y}{dx^2} + q\frac{dy}{dx} = \frac{2y}{x^2}$$

in the form

$$qxy = A(qx - 2) + B(qx + 2)e^{-qx}.$$

7. Obtain the primitive of the equation

$$\frac{d^3y}{dx^3} + q^3y = \frac{6}{x^2}\frac{dy}{dx}$$

in the form

$$y = Ae^{-qx}\left(1 + \frac{2}{qx}\right) + Be^{\frac{1}{2}qx}\left\{\left(1 - \frac{1}{qx}\right)\sin\left(\frac{3^{\frac{1}{2}}}{2}qx + a\right) + \frac{3^{\frac{1}{2}}}{qx}\cos\left(\frac{3^{\frac{1}{2}}}{2}qx + a\right)\right\}.$$

(Leslie Ellis.)

8. Prove that the coefficient of a^m in the expansion, in ascending powers of a, of

$$(1 - 2ax + a^2)^{-n}$$

is a solution of

$$\frac{d}{dx}\left\{(1 - x^2)^{n + \frac{1}{2}}\frac{dy}{dx}\right\} + m(m + 2n)(1 - x^2)^{n - \frac{1}{2}}y = 0.$$

9. Prove that, with the notation used for the solution of Legendre's equation, $\{P_n(\cos\theta)\}^2$ is a solution of the differential equation

$$\left(\frac{d}{d\theta}\sin\theta\right)^2\frac{dU}{d\theta} + 4n(n + 1)\sin\theta\left(\frac{d}{d\theta}\sin\theta\right)U = 0.$$

10. Prove that, with the notation of §§ 90, 91,

$$P_{n+1}Q_n - Q_{n+1}P_n = \frac{1}{n + 1}.$$

(Trinity Fellowship Examination, 1884.)

11. Prove that the primitive of the equation

$$(1-x^2)\frac{d^2y}{dx^2} - 2(m+1)x\frac{dy}{dx} + (n+m+1)(n-m)y = 0$$

is given by

$$y = A\frac{d^mP_n}{dx^m} + B\frac{d^mQ_n}{dx^m},$$

provided m be not greater than n.

What is the primitive when m is greater than n?

(Heine.)

12. Shew that the solution of the equation

$$\frac{d}{dx}\left\{(1-x^2)\frac{dy}{dx}\right\} + n(n+1)y = \frac{k^2}{1-x^2}y,$$

where k is an integer, may be expressed in the form

$$y = (1-x^2)^{\frac{1}{2}k}\frac{d^ky_n}{dx^k},$$

where y_n is the solution of Legendre's equation.

13. Obtain the primitive of the equation

$$(1-x^2)\frac{d^2y}{dx^2} + 2(m-1)x\frac{dy}{dx} + (n-m+1)(n+m)y = 0.$$

(Heine.)

14. Prove that the equation

$$x\frac{d^2y}{dx^2} + n\frac{dy}{dx} + \tfrac{1}{4}y = 0$$

has, in the case when n is an integer, for its primitive

$$y = x^{-\frac{1}{2}(n-1)}\{AJ_{n-1}(x^{\frac{1}{2}}) + BY_{n-1}(x^{\frac{1}{2}})\}.$$

(Lommel.)

15. Obtain the primitive of the equation

$$x^2\frac{d^2y}{dx^2} + nx\frac{dy}{dx} + (b+cx^{2m})y = 0$$

in the form

$$y = x^{-\frac{1}{2}(n-1)}\left[AJ_\mu\left(c^{\frac{1}{2}}\frac{x^m}{m}\right) + BJ_{-\mu}\left(c^{\frac{1}{2}}\frac{x^m}{m}\right)\right],$$

where $\mu^2m^2 = \tfrac{1}{4}(n-1)^2 - b.$

(Lommel.)

16. Verify that the primitive of

$$x^m\frac{d^{2m}y}{dx^{2m}} = y$$

is $y = x^{\frac{1}{2}m}\sum\limits_{p=0}^{p=m-1}[A_pJ_m\{2(-a_px)^{\frac{1}{2}}\} + B_pY_m\{2(-a_px)^{\frac{1}{2}}\}],$

where $a_0, a_1, \ldots, a_{m-1}$, are the roots of the equation $a^m = 1$; and that of

$$x^{m+\frac{1}{2}} \frac{d^{2m+1}y}{dx^{2m+1}} = y$$

is

$$y = x^{\frac{1}{2}m+\frac{1}{4}} \sum_{p=0}^{p=2m} C_p \{ J_{-m-\frac{1}{2}}(2a_p x^{\frac{1}{2}}) + i J_{m+\frac{1}{2}}(2a_p x^{\frac{1}{2}}) \},$$

where $a_0, a_1, \ldots, a_{2m}$, are the roots of $a^{2m+1} = -i$.

(Lommel.)

17. The primitive of the equation

$$\frac{d^2 y}{dx^2} + y e^{2x} = 0$$

is

$$y = A J_0(e^x) + B Y_0(e^x);$$

and that of

$$x^4 \frac{d^2 y}{dx^2} + e^{\frac{2}{x}} y = 0$$

is

$$y = x \{ A J_0(e^{\frac{1}{x}}) + B Y_0(e^{\frac{1}{x}}) \}.$$

(Lommel.)

(See, for connection between these two equations, Ex. 10, p. 142.)

18. Prove that, with the notation of §§ 101, 105,

$$J_n J_{1-n} + J_{n-1} J_{-n} = \frac{2}{\pi x} \sin n\pi,$$

n not being an integer, and that

$$Y_n J_{n+1} - Y_{n+1} J_n = \frac{1}{x}.$$

(Lommel.)

19. The differential equation

$$\frac{d^2 u}{dx^2} + 2Q \frac{du}{dx} + \left\{ Q^2 + \frac{dQ}{dx} + a - \frac{m(m+1)}{x^2} \right\} u = 0$$

is integrable in finite terms, whatever function of x is denoted by Q, provided m be an integer.

20. The equation

$$\frac{d^2 u}{dx^2} + \frac{r}{x} \frac{du}{dx} = \left(bx^m + \frac{c}{x^2} \right) u$$

is integrable in finite terms, if

$$m + 2 = \frac{2 \{ (1-r)^2 + 4c \}^{\frac{1}{2}}}{2i+1},$$

where i is a positive integer or zero.

(Malmsten.)

21. Prove that the coefficient of h^{p+1} in the expansion of $e^{a(x^2+xh)^{\frac{1}{2}}}$ satisfies the differential equation

$$\frac{d^2u}{dx^2} - a^2u = \frac{p(p+1)}{x^2}u.$$

(Glaisher.)

22. Shew that, if $y = X$ be a solution of the equation

$$\frac{d^m y}{dx^m} + k^m y = 0$$

(k being a constant), then a solution of

$$\frac{d^m y}{dx^m} + k^m y = \frac{pm}{x}\frac{d^{m-1}y}{dx^{m-1}}$$

is given by

$$y = x^{m(p+1)-1}\left(\frac{1}{x^{m-1}}\frac{d}{dx}\right)^p \frac{X}{x^{n-1}}.$$

Hence solve the equation

$$\frac{d^2y}{dx^2} - \frac{4}{x}\frac{dy}{dx} + k^2 y = 0.$$

(Leslie Ellis.)

23. The equation

$$(1-ax^2)\frac{d^2y}{dx^2} - bx\frac{dy}{dx} - cy = 0$$

is integrable in finite terms in the following cases:

(i)   when $\dfrac{b}{a}$ is an odd integer;

(ii) when $\left\{\left(1-\dfrac{b}{a}\right)^2 - 4\dfrac{c}{a}\right\}^{\frac{1}{2}}$ is an odd integer;

(iii) when $\dfrac{b}{a} \pm \left\{\left(1-\dfrac{b}{a}\right)^2 - 4\dfrac{c}{a}\right\}^{\frac{1}{2}}$ is an odd integer.

24. Prove that the equation

$$(a+bx^n)x^2\frac{d^2u}{dx^2} + (c+ex^n)x\frac{du}{dx} + (f+gx^n)u = X$$

admits of finite solution,

(i) when any one of the four quantities $\alpha - \beta$ is an even integer,

(ii) when any two of the quantities

$$\alpha_1 - \alpha_2, \quad \beta_1 - \beta_2, \quad \alpha_1 + \alpha_2 - \beta_1 - \beta_2,$$

are odd integers; where α_1, α_2, and β_1, β_2, are the roots of the respective quadratic equations

$$\tfrac{1}{4}bn(\alpha-2)(n\alpha-2n-2) + \tfrac{1}{2}en(\alpha-2) + g = 0,$$

and

$$\tfrac{1}{4}an\beta(n\beta-2) + \tfrac{1}{2}cn\beta + f = 0.$$

(Pfaff.)

25. Prove that the three expressions

$$x^{-p}\left\{1-\frac{1}{p-\frac{1}{2}}\frac{a^2x^2}{2^2}+\frac{1}{(p-\frac{1}{2})(p-\frac{3}{2})}\frac{a^4x^4}{2!\,2^4}-\frac{1}{(p-\frac{1}{2})(p-\frac{3}{2})(p-\frac{5}{2})}\frac{a^6x^6}{3!\,2^6}+\cdots\right\},$$

$$e^{ax}x^{-p}\left\{1-\frac{p}{p}\,ax+\frac{p\,(p-1)}{p\,(p-\frac{1}{2})}\frac{a^2x^2}{2!}-\frac{p\,(p-1)\,(p-2)}{p\,(p-\frac{1}{2})(p-1)}\frac{a^3x^3}{3!}+\cdots\right\},$$

$$e^{-ax}x^{-p}\left\{1+\frac{p}{p}\,ax+\frac{p\,(p-1)}{p\,(p-\frac{1}{2})}\frac{a^2x^2}{2!}+\frac{p\,(p-1)\,(p-2)}{p\,(p-\frac{1}{2})(p-1)}\frac{a^3x^3}{3!}+\cdots\right\},$$

are particular solutions of the equation

$$\frac{d^2u}{dx^2}-a^2u=\frac{p\,(p+1)}{x^2}\,u.$$

Shew that, when p is not an integer, these three expressions are equal to one another; and obtain, in this case, a second and independent particular solution.

26. Prove that the primitive of

$$\frac{d^2y}{dx^2}-a^2y=\frac{p\,(p+1)}{x^2}\,y$$

may be written in either of the forms

$$y=x^{-p-1}\left(x^3\frac{d}{dx}\right)^p\{x^{-2p+1}(Ae^{ax}+Be^{-ax})\},$$

$$y=x^{-p-3}\left(x^3\frac{d}{dx}\right)^{p+1}\{x^{-2p}(Ae^{ax}+Be^{-ax})\}.$$

(Boole.)

Prove that the primitive of the same equation may also be written in the form

$$y=x^p\left(\frac{d}{dx}\frac{1}{x}\right)^p(Ae^{ax}+Be^{-ax}).$$

(Donkin.)

27. The primitive of the equation

$$x\frac{d^2y}{dx^2}+\{m+n+(a+\beta)\,x\}\frac{dy}{dx}+(m\beta+na+a\beta x)\,y=0$$

can be expressed in the form

$$y=Ae^{-ax}\frac{d^{m-1}}{dx^{m-1}}\{x^{-n}e^{x(a-\beta)}\}+Be^{-\beta x}\frac{d^{n-1}}{dx^{n-1}}\{x^{-m}e^{x(\beta-a)}\}.$$

Obtain that of

$$\frac{d^2y}{dx^2}+my=x\frac{dy}{dx}$$

in the form

$$y=Ae^{\frac{1}{2}x^2}\frac{d^m}{dx^m}(e^{-\frac{1}{2}x^2})+Be^{\frac{1}{2}x^2}\frac{d^m}{dx^m}\{e^{-\frac{1}{2}x^2}\int e^{\frac{1}{2}x^2}\,dx\}.$$

(Spitzer.)

28. The orthogonal trajectory of the system of surfaces of revolution given by $P_n = cr^{n+1}$, where P_n is the solution of Legendre's equation and its argument x is the cosine of the vectorial angle of any point, is given by the equation

$$P_{n+1} - P_{n-1} = ar^n.$$

29. Prove that, if the equation

$$\frac{d^3y}{dx^3} + yf(x) = 0$$

be transformed by the relations $z(cx+d) = ax+b$ and $y = u(cx+d)^2$, so that u is the new dependent variable, the new equation is

$$\frac{d^3u}{dz^3} + uF(z) = 0,$$

where

$$F(z) = \left(\frac{dx}{dz}\right)^3 f(x).$$

Hence, or otherwise, solve the equation

$$\frac{d^3y}{dx^3} = \frac{y}{(Ax^2 + 2Bx + C)^3}.$$

30. Obtain the primitive of the equation

$$\frac{dy}{dx} + \frac{y^2}{x} = \frac{x}{x^2 - 1}$$

in the form

$$y = \frac{A(E-K) + BE'}{AE + B(E'-K')},$$

where $A \div B$ is an arbitrary constant, while K, K', E, E', are the complete elliptic integrals of the first and the second kind respectively with modulus x.

CHAPTER VI

HYPERGEOMETRIC SERIES

113. THE series

$$1 + \frac{\alpha\beta}{1 \cdot \gamma} x + \frac{\alpha(\alpha+1)\beta(\beta+1)}{1 \cdot 2 \cdot \gamma(\gamma+1)} x^2$$

$$+ \frac{\alpha(\alpha+1)(\alpha+2)\beta(\beta+1)(\beta+2)}{1 \cdot 2 \cdot 3 \cdot \gamma(\gamma+1)(\gamma+2)} x^3 + \ldots$$

is called the hypergeometric series and is usually denoted by $F(\alpha, \beta, \gamma, x)$; the four quantities α, β, γ, x, are called the elements, and of these x alone is variable. The elements α and β may be interchanged without affecting the value of F; if either of them be a negative integer, the series will consist of a finite number of terms; otherwise it will proceed to infinity. It will be assumed that γ is not a negative integer, so that infinite terms may be excluded.

The series converges* for all real values of x such that $-1 < x < 1$. It diverges if $x > 1$ and if $x < -1$. When $x = 1$, the series converges if $\gamma > \alpha + \beta$, and it diverges if $\gamma \leqslant \alpha + \beta$. When $x = -1$, the series converges if $\gamma + 1 > \alpha + \beta$, and it diverges if $\gamma + 1 \leqslant \alpha + \beta$. It is important to consider the actual range of convergence of the series; for the most part, we shall not consider the conditions of convergence for values of x at the limits of the range.

The series is one of very great generality, and it includes as particular examples very many of the series which occur in analysis. The following examples admit of easy verification :

I. $\qquad (1+x)^n = F(-n, \beta, \beta, -x)$.

II. $\quad (1+x)^n + (1-x)^n = 2F(-\tfrac{1}{2}n, -\tfrac{1}{2}n + \tfrac{1}{2}, \tfrac{1}{2}, x^2)$.

---

* Bromwich, *An introduction to the theory of infinite series*, pp. 33, 35.

III. $\log (1 + x) = xF(1, 1, 2, -x)$.

IV. $\log \dfrac{1 + x}{1 - x} = 2xF(\frac{1}{2}, 1, \frac{3}{2}, x^2)$.

V. $e^x = F\left(1, \beta, 1, \dfrac{x}{\beta}\right)$, when $\beta = \infty$.

VI. $\cosh x = F\left(\alpha, \beta, \frac{1}{2}, \dfrac{x^2}{4\alpha\beta}\right)$, when $\alpha = \infty = \beta$.

VII. $\cos nx = F(\frac{1}{2}n, -\frac{1}{2}n, \frac{1}{2}, \sin^2 x)$.

Ex. 1. Prove that all the differential coefficients of the series diverge for the value $x = 1$ if the series itself diverges for that value; and that all the differential coefficients, from and after one of some order, diverge for the value $x = 1$ though the series converges for that value.

Ex. 2. Express as hypergeometric series

    (i) $\sin t$, the variable element in the series being t^2;

    (ii) $\sin nt$, the variable element in the series being $\sin^2 t$;

    (iii) $\cos nt$, the variable element in the series being $-\tan^2 t$.

Other examples are given by Gauss at the beginning of his earlier memoir referred to in § 134.

114. Let the coefficient of x^r be written A_r; then the relation connecting consecutive A's is

$$(1 + r)(\gamma + r) A_{r+1} = (\alpha + r)(\beta + r) A_r.$$

Consider the differential equation

$$\left\{(\vartheta + \alpha)(\vartheta + \beta) - \frac{1}{x}\vartheta(\vartheta + \gamma - 1)\right\} y = 0 \dots\dots\dots (i),$$

in which ϑ stands for the operator $x\dfrac{d}{dx}$. A solution of this equation can be obtained in a series: let the series be

$$y = B_0 x^\mu + B_1 x^{\mu+1} + B_2 x^{\mu+2} + \dots\dots$$

Substitute this value in the differential equation, which must be identically satisfied; each separate power of x must therefore disappear in virtue of the constant multiplying it being zero. Thus, for the lowest power, we have

$$-\mu(\mu + \gamma - 1) B_0 = 0;$$

and from the vanishing of the coefficients of the higher powers, the relation between the successive quantities B is given by

$$(\mu + r + 1)(\mu + r + \gamma) B_{r+1} - (\mu + r + \alpha)(\mu + r + \beta) B_r = 0.$$

We shall assume that the first coefficient B_0 is not zero, because the relation $B_0 = 0$ would make all the B's zero; and thus the former equation is satisfied by either

$$\mu = 0,$$

or

$$\mu = 1 - \gamma.$$

115. Take first the value $\mu = 0$; the relation connecting the quantities B becomes

$$(1 + r)(\gamma + r) B_{r+1} = (\alpha + r)(\beta + r) B_r.$$

Now when $B_0 = 1 = A_0$, the relation just proved, compared with that which connects the A's, shews that $B_r = A_r$; and therefore the series assumed for y becomes the hypergeometric series. Thus one solution of the differential equation (i) is $F(\alpha, \beta, \gamma, x)$.

Let the operating factors in (i) be expanded and terms of the same order collected; then the equation may be written

$$[(1 - x) \vartheta^2 + \{\gamma - 1 - x(\alpha + \beta)\} \vartheta - \alpha\beta x] y = 0.$$

But

$$\vartheta = x \frac{d}{dx},$$

$$\vartheta^2 = x^2 \frac{d^2}{dx^2} + x \frac{d}{dx};$$

when these values are inserted, the above equation, after rearrangement and division by $x^2 (1 - x)$, becomes

$$\frac{d^2 y}{dx^2} + \frac{\gamma - (\alpha + \beta + 1) x}{x(1 - x)} \frac{dy}{dx} - \frac{\alpha\beta}{x(1 - x)} y = 0 \ldots\ldots(1),$$

which is the differential equation satisfied by $F(\alpha, \beta, \gamma, x)$.

Take next the value $\mu = 1 - \gamma$; the relation connecting the quantities B becomes

$$(1 + r)(2 - \gamma + r) B_{r+1} = (\alpha + 1 - \gamma + r)(\beta + 1 - \gamma + r) B_r.$$

Let $B_0 = 1$; this equation shews that the quantities B are the successive coefficients in a hypergeometric series whose constant elements are respectively $\alpha + 1 - \gamma$, $\beta + 1 - \gamma$, $2 - \gamma$. The series assumed for y begins with $x^{1-\gamma}$; hence the value of y is

$$x^{1-\gamma} F(\alpha + 1 - \gamma, \beta + 1 - \gamma, 2 - \gamma, x),$$

so that this also is a solution of the differential equation (1).

We have thus two particular solutions of this differential equation; and therefore any other particular solution, which is finite for values of x such that $-1 < x < 1$, may be represented by

$$AF(\alpha, \beta, \gamma, x) + Bx^{1-\gamma} F(\alpha + 1 - \gamma, \beta + 1 - \gamma, 2 - \gamma, x),$$

where A and B are constants, the values of which may be determined by comparing powers of x. If in this expression A and B denote arbitrary constants, it furnishes the primitive of (1).

116. To reduce (1) to its normal form, we must compare it with the general linear equation of the second order. We then have

$$P = \frac{\gamma - (\alpha + \beta + 1) x}{x (1 - x)} = \frac{\gamma}{x} + \frac{\gamma - \alpha - \beta - 1}{1 - x},$$

$$Q = \frac{-\alpha\beta}{x (1 - x)};$$

and therefore the invariant I, being

$$\tfrac{1}{4}\left[4Q - 2\frac{dP}{dx} - P^2 \right],$$

becomes, after some reductions,

$$\tfrac{1}{4}\frac{1 - \lambda^2}{x^2} + \tfrac{1}{4}\frac{1 - \nu^2}{(x-1)^2} + \tfrac{1}{4}\frac{\lambda^2 - \mu^2 + \nu^2 - 1}{x(x-1)},$$

where

$$\lambda^2 = (1 - \gamma)^2; \quad \mu^2 = (\alpha - \beta)^2; \quad \nu^2 = (\gamma - \alpha - \beta)^2.$$

Let this invariant be denoted either by I or $\psi(x)$; the latter form will be convenient when the independent variable comes to be changed.

Thus equation (1), by the substitution

$$v = ye^{\tfrac{1}{2}\int P dx}$$

$$= yx^{\tfrac{1}{2}\gamma} (1 - x)^{\tfrac{1}{2}(\alpha + \beta + 1 - \gamma)},$$

becomes

$$\frac{d^2 v}{dx^2} + v\psi(x) = 0 \ \dots\dots\dots\dots\dots\dots(2),$$

in which $\psi(x)$ denotes the foregoing function of x.

SET OF 24 PARTICULAR SOLUTIONS.

117. We now proceed to find some further particular solutions of this differential equation. It follows from the investigation of § 64 that the conditions, which must be satisfied in order that the equations

$$\frac{d^2v}{dx^2} + v\psi(x) = 0$$

and

$$\frac{d^2z}{dt^2} + z\psi_1(t) = 0 \quad \ldots\ldots\ldots\ldots\ldots(3)$$

should be transformable into one another, are firstly,

$$v = z\left(\frac{dt}{dx}\right)^{-\frac{1}{2}} = zu,$$

and secondly,

$$\tfrac{1}{2}\{t,\,x\} + \left(\frac{dt}{dx}\right)^2 \psi_1(t) - \psi(x) = 0 \quad \ldots\ldots\ldots(4).$$

Hence, if we consider $\psi_1(t)$ as a given function of t, the latter equation will give the value of t in terms of x; and when this value is found, the former will furnish the relation between v and z.

Now assume that the function $\psi_1(t)$ is such as to make equation (3) the normal form of the equation satisfied by a hypergeometric series with constant elements α', β', γ'; and suppose that we can obtain from (4) a value of t in terms of x. Then, since the value of u will be at once derivable from that of t, we have a solution of (2) in the form

$$u t^{\frac{1}{2}\gamma'}(1-t)^{\frac{1}{2}(\alpha'+\beta'+1-\gamma')} F(\alpha',\,\beta',\,\gamma',\,t);$$

and this is distinct from the value of v in § 116.

118. The primitive of (4) will give t as a function of x, α, β, γ, α', β', γ'; let us select those forms of this function, which make t dependent on x alone and independent of the two sets of constant elements. We may, to obtain these, write

$$\{t,\,x\} = 0,$$

$$\psi_1(t)\left(\frac{dt}{dx}\right)^2 = \psi(x).$$

The former of these, on multiplication by $t'^{-\frac{1}{2}}$, is directly integrable in the form

$$t''t'^{-\frac{3}{2}} = C;$$

proceeding with the integration, we have

$$t = A - \frac{4}{C(Cx + C')}$$

$$= \frac{ax + b}{cx + d},$$

on changing the constants. This is the *general* value of t which makes the function $\{t, x\}$ vanish. But the conditions require that

$$\psi_1(t) \left(\frac{dt}{dx}\right)^2 = \psi(x),$$

or $$\frac{(ad - bc)^2}{(cx + d)^4} \psi_1\left(\frac{ax + b}{cx + d}\right) = \psi(x);$$

and this will not be satisfied for arbitrary values of these constants, which must therefore be determined so as to be independent of the constant elements of the series. Now

$$\psi(x) = \tfrac{1}{4} \frac{Ax^2 + Bx + C}{x^2(1 - x)^2},$$

where $$A = 1 - \mu^2,$$

$$B = \lambda^2 + \mu^2 - \nu^2 - 1,$$

$$C = 1 - \lambda^2;$$

and we may write

$$\psi_1(t) = \tfrac{1}{4} \frac{A't^2 + B't + C'}{t^2(1 - t)^2}.$$

Hence the constants a, b, c, d, must be such as to satisfy

$$\frac{Ax^2 + Bx + C}{x^2(1 - x)^2}$$

$$= (ad - bc)^2 \frac{A'(ax + b)^2 + B'(ax + b)(cx + d) + C'(cx + d)^2}{(ax + b)^2(cx + d)^2\{(c - a)x + d - b\}^2}.$$

The quantities α, β, γ (and therefore A, B, C which are functions of them) are arbitrary, and thus the numerator and denominator of the left-hand fraction can have no common factor except a constant; and similarly for the right-hand side. Hence we may write

$$m\left(Ax^2 + Bx + C\right)$$

$$= (ad - bc)^2 \left[A'\left(ax + b\right)^2 + B'\left(ax + b\right)\left(cx + d\right) + C'\left(cx + d\right)^2\right],$$

$$mx^2\left(1 - x\right)^2 = (ax + b)^2\left(cx + d\right)^2 \left\{(c - a)\,x + d - b\right\}^2,$$

in which m is constant. The latter of these equations will determine the values of a, b, c, d which are admissible; the former will then serve to indicate the relations of α', β', γ' to α, β, γ, in order that the expression at the end of § 117 may be a solution of (1).

119. Comparing now the coefficients of the different powers of x on the two sides of the latter equation, we find that the following six sets of values for the constants make the equation identically satisfied:

$$\text{(i)} \quad c = 0 = b \quad\quad = a - d; \; m = a^6;$$

$$\text{(ii)} \quad c = 0 = d - b = a + b; \; m = a^6;$$

$$\text{(iii)} \quad a = 0 = d \quad\quad = c - b; \; m = b^6;$$

$$\text{(iv)} \quad a = 0 = d - b = c + d; \; m = b^6;$$

$$\text{(v)} \quad b = 0 = c - a = c + d; \; m = a^6;$$

$$\text{(vi)} \quad d = 0 = c - a = a + b; \; m = b^6.$$

These values, substituted successively in the expression for t in terms of x, give:

$$\text{(i)} \quad t = x; \quad\quad \text{(ii)} \quad t = 1 - x; \quad\quad \text{(iii)} \quad t = \frac{1}{x};$$

$$\text{(iv)} \quad t = \frac{1}{1 - x}; \quad\quad \text{(v)} \quad t = \frac{x}{x - 1}; \quad\quad \text{(vi)} \quad t = \frac{x - 1}{x},$$

respectively; and these form the complete system of values of t required.

120. We now transform the first of the two equations by means of each of these in turn, and obtain the necessary relations between α', β', γ' and α, β, γ.

Consider first the set of values (i). We have

$$Ax^2 + Bx + C = A'x^2 + B'x + C',$$

so that

$$A = A', \quad B = B', \quad C = C';$$

or, what is an equivalent set of equations,

$$\lambda^2 = \lambda'^2; \quad \mu^2 = \mu'^2; \quad \nu^2 = \nu'^2.$$

When expressed in terms of the constant elements, these relations are

$$(1 - \gamma')^2 = (1 - \gamma)^2,$$
$$(\alpha' - \beta')^2 = (\alpha - \beta)^2,$$
$$(\gamma' - \alpha' - \beta')^2 = (\gamma - \alpha - \beta)^2;$$

and (remembering that an interchange of the first and second constant elements makes no change in a hypergeometric series), we find that these are satisfied by

(1) $\alpha' = \alpha$ ....$\beta' = \beta$ ....$\gamma' = \gamma$;

(2) $\alpha' = \gamma - \alpha$ ....$\beta' = \gamma - \beta$ ....$\gamma' = \gamma$;

(3) $\alpha' = \alpha - \gamma + 1$........$\beta' = \beta - \gamma + 1$ ......$\gamma' = 2 - \gamma$;

(4) $\alpha' = 1 - \alpha$ ....$\beta' = 1 - \beta$ ....$\gamma' = 2 - \gamma$.

Since $t = x$, $\dfrac{dt}{dx}$ is unity and therefore u is unity for this value of t; and the particular solutions of the v-equation, which correspond to these four sets of values, are respectively

$$x^{\frac{1}{2}\gamma} (1 - x)^{\frac{1}{2}(\alpha + \beta + 1 - \gamma)} F(\alpha, \beta, \gamma, x),$$

$$x^{\frac{1}{2}\gamma} (1 - x)^{\frac{1}{2}(\gamma - \alpha - \beta + 1)} F(\gamma - \alpha, \gamma - \beta, \gamma, x),$$

$$x^{1 - \frac{1}{2}\gamma} (1 - x)^{\frac{1}{2}(\alpha + \beta + 1 - \gamma)} F(\alpha - \gamma + 1, \beta - \gamma + 1, 2 - \gamma, x),$$

$$x^{1 - \frac{1}{2}\gamma} (1 - x)^{\frac{1}{2}(\gamma - \alpha - \beta + 1)} F(1 - \alpha, 1 - \beta, 2 - \gamma, x).$$

Now these are solutions of equation (2). In order to obtain the corresponding solutions of equation (1), we must (§ 116) multiply each of them by

$$x^{-\frac{1}{2}\gamma} (1 - x)^{-\frac{1}{2}(\alpha + \beta + 1 - \gamma)};$$

and therefore four particular solutions of equation (1) are

(I) $\quad y = F(\alpha, \beta, \gamma, x)$;

(II) $\quad y = (1 - x)^{\gamma - \alpha - \beta} F(\gamma - \alpha, \gamma - \beta, \gamma, x)$;

(III) $\quad y = x^{1 - \gamma} F(\alpha - \gamma + 1, \beta - \gamma + 1, 2 - \gamma, x)$;

(IV) $\quad y = x^{1 - \gamma} (1 - x)^{\gamma - \alpha - \beta} F(1 - \alpha, 1 - \beta, 2 - \gamma, x)$.

Treating now the relation $t = 1 - x$ in the same way, we find other four particular solutions in the forms

(V) $y = F(\alpha, \beta, \alpha + \beta - \gamma + 1, 1 - x)$;

(VI) $y = x^{1-\gamma} F(\alpha - \gamma + 1, \beta - \gamma + 1, \alpha + \beta - \gamma + 1, 1 - x)$;

(VII) $y = (1 - x)^{\gamma - \alpha - \beta} F(\gamma - \alpha, \gamma - \beta, \gamma - \alpha - \beta + 1, 1 - x)$;

(VIII) $y = x^{1-\gamma} (1 - x)^{\gamma - \alpha - \beta} F(1 - \alpha, 1 - \beta, \gamma - \alpha - \beta + 1, 1 - x)$.

And from the relation $t = \dfrac{1}{x}$ we have as one particular solution

(IX) $y = x^{-\alpha} F\left(\alpha, \alpha - \gamma + 1, \alpha - \beta + 1, \dfrac{1}{x}\right)$.

121. All the particular solutions for the different values of t can be found in the above manner. Each value of t leads to four particular solutions, so that there are in all 24 of these. But this laborious method of obtaining the remainder need not now be adopted; it is possible to write down, from the nine foregoing, the following fifteen to complete the set:

(X) $y = x^{-\beta} F\left(\beta, \beta - \gamma + 1, \beta - \alpha + 1, \dfrac{1}{x}\right)$;

(XI) $y = x^{\alpha - \gamma} (1 - x)^{\gamma - \alpha - \beta} F\left(1 - \alpha, \gamma - \alpha, \beta - \alpha + 1, \dfrac{1}{x}\right)$;

(XII) $y = x^{\beta - \gamma} (1 - x)^{\gamma - \alpha - \beta} F\left(1 - \beta, \gamma - \beta, \alpha - \beta + 1, \dfrac{1}{x}\right)$;

(XIII) $y = (1 - x)^{-\alpha} F\left(\alpha, \gamma - \beta, \alpha - \beta + 1, \dfrac{1}{1 - x}\right)$;

(XIV) $y = (1 - x)^{-\beta} F\left(\beta, \gamma - \alpha, \beta - \alpha + 1, \dfrac{1}{1 - x}\right)$;

(XV) $y = x^{1-\gamma} (1 - x)^{\gamma - \alpha - 1} F\left(\alpha - \gamma + 1, 1 - \beta, \alpha - \beta + 1, \dfrac{1}{1 - x}\right)$;

(XVI) $y = x^{1-\gamma} (1 - x)^{\gamma - \beta - 1} F\left(\beta - \gamma + 1, 1 - \alpha, \beta - \alpha + 1, \dfrac{1}{1 - x}\right)$;

(XVII) $y = (1 - x)^{-\alpha} F\left(\alpha, \gamma - \beta, \gamma, \dfrac{x}{x - 1}\right)$;

(XVIII) $\quad y = (1-x)^{-\beta} F\left(\beta, \gamma - \alpha, \gamma, \dfrac{x}{x-1}\right);$

(XIX) $\quad y = x^{1-\gamma}(1-x)^{\gamma-\alpha-1} F\left(\alpha - \gamma + 1, 1 - \beta, 2 - \gamma, \dfrac{x}{x-1}\right);$

(XX) $\quad y = x^{1-\gamma}(1-x)^{\gamma-\beta-1} F\left(\beta - \gamma + 1, 1 - \alpha, 2 - \gamma, \dfrac{x}{x-1}\right);$

(XXI) $\quad y = x^{-\alpha} F\left(\alpha, \alpha - \gamma + 1, \alpha + \beta - \gamma + 1, \dfrac{x-1}{x}\right);$

(XXII) $\quad y = x^{-\beta} F\left(\beta, \beta - \gamma + 1, \alpha + \beta - \gamma + 1, \dfrac{x-1}{x}\right);$

(XXIII) $\quad y = x^{\alpha-\gamma}(1-x)^{\gamma-\alpha-\beta} F\left(1 - \alpha, \gamma - \alpha, \gamma - \alpha - \beta + 1, \dfrac{x-1}{x}\right);$

(XXIV) $\quad y = x^{\beta-\gamma}(1-x)^{\gamma-\alpha-\beta} F\left(1 - \beta, \gamma - \beta, \gamma - \alpha - \beta + 1, \dfrac{x-1}{x}\right).$

RELATIONS BETWEEN THE PARTICULAR SOLUTIONS.

122. Let these solutions be denoted by

$$y_1, y_2, \ldots\ldots\ldots, y_{23}, y_{24},$$

the suffixes and the numbers of the foregoing equations corresponding to one another. These quantities y are not independent:
for, by the ordinary property of a linear differential equation of
the second order (of which they all are solutions), there is between
any three of them y_λ, y_μ, y_ν, a relation of the form

$$y_\lambda = A y_\mu + B y_\nu;$$

and we must find these relations for the different combinations of
the solutions. But certain cases will arise in which either A or B
will be zero, and therefore the corresponding solutions will differ
from one another only by a constant factor; and these can be
recognised by the application of the following lemma:—

If there be two solutions of the differential equation (1) *developed
in the same ascending powers of* x, *and if both series converge, then
they differ from one another only by a constant factor.*

For the sake of simplicity suppose one of the solutions to be $F(\alpha, \beta, \gamma, x)$ and the other, when developed in ascending powers of x, to be given by

$$y = A + Bx + Cx^2 + \ldots\ldots\ldots$$

Substituting this value of y in the differential equation we should, by a process similar to that in § 114, find $y = AF(\alpha, \beta, \gamma, x)$, which proves the lemma.

123. Let us apply this lemma to obtain the particular solutions which are equal to y_1; this we shall suppose to be a converging series, so that $-1 < x < 1$. Then y_2 is also a converging series proceeding in the same ascending powers of x as y_1; the first term in each is unity; the constant factor of the lemma is therefore 1, and we have

$$y_1 = y_2.$$

The next one in the list which, expanded in ascending powers of x, begins with x^0 is y_5; when we select from

$$F(\alpha, \beta, \alpha + \beta - \gamma + 1, 1 - x)$$

the coefficient of x^n, we find it to be

$$(-1)^n \frac{\alpha(\alpha+1)\ldots\ldots(\alpha+n-1)\beta(\beta+1)\ldots\ldots(\beta+n-1)}{1 \cdot 2 \ldots\ldots n \cdot (\alpha+\beta-\gamma+1)(\alpha+\beta-\gamma+2)\ldots\ldots(\alpha+\beta-\gamma+n)}$$
$$\times F(\alpha+n, \beta+n, \alpha+\beta-\gamma+n+1, 1).$$

But in this coefficient F is converging (and so has a finite value) only if

$$\alpha + \beta - \gamma + n + 1 - (\alpha+n) - (\beta+n)$$

be positive (see § 113), that is, if $1 - \gamma - n$ be positive. Hence from and after some definite term the coefficients of the powers of x will be diverging series; and we cannot then consider the series $F(\alpha, \beta, \alpha + \beta - \gamma + 1, 1 - x)$ to be converging though expansible in ascending powers of x. Hence y_5 is not equal to y_1.

Dealing with $y_7, y_{13}, y_{14}, y_{17}, y_{18}$, in the same way, it will be found that the last two alone are converging series at the same time as $F(\alpha, \beta, \gamma, x)$; and hence we have

$$y_1 = y_2 = y_{17} = y_{18} \quad\ldots\ldots\ldots\ldots\ldots\ldots\ldots(i).$$

Again y_3 and y_1, y_4 and y_2, y_{19} and y_{17}, y_{20} and y_{18}, are derived from each other by exactly similar transformations of elements;

thus to pass from y_1 to y_3 the former is multiplied by $x^{1-\gamma}$, the new first and second elements being obtained by subtracting the old third from the old first and second and adding unity to each result, and the new third element by subtracting the old third element from 2. This process is seen to be the same for all; and therefore

$$y_3 = y_4 = y_{19} = y_{20} \quad\text{......................(ii)}.$$

Ex. Prove that

$$y_5 = y_6 = y_{21} = y_{22} \quad\text{..............................(iii)},$$

$$y_7 = y_8 = y_{23} = y_{24} \quad\text{.............................(iv)},$$

$$(-1)^{-\alpha} y_9 = (-1)^{\beta-\gamma} y_{12} = y_{13} = (-1)^{1-\gamma} y_{15} \quad\text{...............(v)},$$

$$(-1)^{-\beta} y_{10} = (-1)^{\alpha-\gamma} y_{11} = y_{14} = (-1)^{1-\gamma} y_{16} \quad\text{............(vi)}.$$

It thus appears that the 24 solutions can be divided into six classes; and the equal members of these classes we may denote respectively by Y_1, Y_2, Y_3, Y_4, Y_5, Y_6, corresponding to the above sets of quantities in order. It remains to find such relations as there may be between these owing to the fact that they are solutions of the differential equation.

124. Account, however, must be taken of the ranges in which the respective integrals have significance.

The range of convergence of the hypergeometric series, which occur in y_1, y_2, y_3, y_4, is $-1 < x < 1$. For those which occur in y_5, y_6, y_7, y_8, it is $-1 < 1-x < 1$, that is, the range is $0 < x < 2$. For those which occur in y_9, y_{10}, y_{11}, y_{12}, it is $-1 < \frac{1}{x} < 1$; that is, the range consists of the two parts, $-\infty < x < 1$ and $1 < x < +\infty$. For those which occur in y_{13}, y_{14}, y_{15}, y_{16}, it is $-1 < \frac{1}{1-x} < 1$; that is, the range consists of the two parts, $-\infty < x < 0$ and $2 < x < +\infty$. For those which occur in y_{17}, y_{18}, y_{19}, y_{20}, it is $-1 < \frac{x}{x-1} < 1$, that is, the range is $-\infty < x < \frac{1}{2}$. Finally, for those which occur in y_{21}, y_{22}, y_{23}, y_{24}, it is $-1 < \frac{x-1}{x} < 1$, that is, the range is $\frac{1}{2} < x < +\infty$.

Thus in Y_1, the quantities y_1 and y_2 are equal over the range $-1 < x < 1$; the quantities y_{17} and y_{18} are equal over the range

$-\infty < x < \frac{1}{2}$; all the four quantities are equal over the range $-1 < x < \frac{1}{2}$. Similarly, for the sets of four quantities in Y_2, Y_3, Y_4, Y_5, Y_6, for their respective ranges.

Now, when γ is not equal to unity*, the two solutions y_1 and y_3 are linearly independent of one another; and they coexist in the range $-1 < x < 1$. They therefore suffice for the expression of the primitive of the equation, and also for the expression of any special integral of the equation, existing in that range or in any portion of that range.

The primitive of the equation in the whole of that range is

$$y = Ay_1 + By_3,$$

where A and B are arbitrary constants.

A special integral of the equation is y_5, which exists in the range $0 < x < 2$ and therefore exists in the portion $0 < x < 1$ of the range of existence for y_1 and y_3; hence there must be a linear relation between y_1, y_3, and y_5. This may be taken in the form

$$y_1 = My_3 + Ny_5,$$

which is valid over the range $0 < x < 1$. We may also write it in the form

$$Y_1 = MY_2 + NY_3;$$

but for the determination of the constants M and N, it is convenient to take the relation in the earlier form

$$y_1 = My_3 + Ny_5.$$

In the same way, we shall have relations of the forms

$$Y_1 = M_1 Y_2 + N_1 Y_4,$$
$$Y_1 = M_2 Y_2 + N_2 Y_5,$$
$$Y_1 = M_3 Y_2 + N_3 Y_6,$$

selecting out of the typical quantities Y_1, Y_2, Y_3, Y_4, Y_5, Y_6, such of the respectively equal members as can coexist in any relation. Thus the relation

$$Y_1 = M_1 Y_2 + N_1 Y_4$$

could be taken in the form

$$y_2 = M_1 y_4 + N_1 y_7,$$

---

* The case when $\gamma = 1$ is dealt with later, p. 251, Ex. 1; the primitive of the equation changes its character.

valid over the range $0 < x < 1$. The relation
$$Y_1 = M_2 Y_2 + N_2 Y_5$$
could be taken in the form
$$y_1 = M_2 y_3 + N_2 y_{13},$$
valid over the range $-1 < x < 0$; but as the range of equality of $(-1)^{-a} y_9$ and y_{13} is $-\infty < x < -1$ and $2 < x < \infty$, the relation could not be taken in the form
$$y_1 = M_2 y_3 + N_2 y_9,$$
for there is no range in which the three integrals y_1, y_3, y_9 coexist. Similarly the relation
$$Y_1 = M_3 Y_2 + N_3 Y_6$$
could be taken in the form
$$y_1 = M_3 y_3 + N_3 y_{14},$$
valid over the range $-1 < x < 0$; but it could not be taken in the form
$$y_1 = M_3 y_3 + N_3 y_{10},$$
for there is no range in which the three integrals y_1, y_3, y_{10} coexist.

Similarly, it is possible to have a typical relation
$$Y_2 = P Y_3 + Q Y_4,$$
which could be taken in the form
$$y_3 = P y_5 + Q y_7,$$
valid over the range $0 < x < 1$. There are, in fact, twenty such typical relations; but in substituting a selected y from the four represented by the corresponding Y, it is always necessary to have the integrals in special form coexistent over some range of x.

For our purpose*, it is sufficient to consider the four typical relations
$$Y_1 = M Y_2 + N Y_3,$$
$$Y_1 = M_1 Y_2 + N_1 Y_4,$$
$$Y_1 = M_2 Y_2 + N_2 Y_5,$$
$$Y_1 = M_3 Y_2 + N_3 Y_6.$$

---

* A full discussion of the relations requires that x should be allowed to become a complex variable. Such a discussion will be found in Goursat's memoir, quoted in § 134.

We have to find the values of all the constants M and N; and we shall see that, when M and N are known for the first relation, the values of the other constants can be deduced by using the properties of the hypergeometric series.

Accordingly, to determine M and N in the first relation, which we shall take in the form

$$y_1 = My_3 + Ny_5,$$

the substitution of any two particular values of x will be sufficient. Let these be $x = 1$ and $x = 0$, and suppose $1 - \gamma$ a positive quantity so that $x^{1-\gamma}$ is zero when $x = 0$; we have for the two values

$$F(\alpha, \beta, \gamma, 1) = MF(\alpha - \gamma + 1, \beta - \gamma + 1, 2 - \gamma, 1) + N,$$
$$1 = NF(\alpha, \beta, \alpha + \beta - \gamma + 1, 1).$$

To evaluate M and N we must obtain the relations between the series for argument unity, to which we now proceed.

INTRODUCTION OF GAUSS'S Π FUNCTION.

125. The coefficient of x^m in

$$F(\alpha, \beta, \gamma, x) - F(\alpha, \beta, \gamma - 1, x)$$

is

$$\frac{\alpha(\alpha+1)\ldots\ldots(\alpha+m-1)\,\beta(\beta+1)\ldots\ldots(\beta+m-1)}{1.2\ldots\ldots m.\gamma(\gamma+1)\ldots\ldots(\gamma+m-1)}\left\{1 - \frac{\gamma+m-1}{\gamma-1}\right\}$$

$$= -\frac{\alpha\beta}{\gamma(\gamma-1)}.\frac{(\alpha+1)(\alpha+2)\ldots\ldots(\alpha+m-1)(\beta+1)\ldots(\beta+m-1)}{1.2.3\ldots\ldots(m-1).(\gamma+1)\ldots(\gamma+m-1)}$$

$$= \text{coefficient of } x^m \text{ in } -\frac{\alpha\beta x}{\gamma(\gamma-1)}F(\alpha+1, \beta+1, \gamma+1, x);$$

and the term on the left-hand side independent of x vanishes, so that

$$F(\alpha, \beta, \gamma, x) - F(\alpha, \beta, \gamma - 1, x)$$

$$= -\frac{\alpha\beta x}{\gamma(\gamma-1)}F(\alpha+1, \beta+1, \gamma+1, x)$$

$$= -\frac{x}{\gamma-1}\frac{d}{dx}F(\alpha, \beta, \gamma, x).$$

But from the differential equation satisfied by $F(\alpha, \beta, \gamma, x)$ we have

$$\frac{dF}{dx}\{\gamma - (\alpha + \beta + 1)\, x\} = \alpha\beta F - x\,(1 - x)\,\frac{d^2 F}{dx^2}.$$

Let the value of $F(\alpha, \beta, \gamma, x)$ when x is made unity be denoted by $F_1(\alpha, \beta, \gamma)$; the value of $\dfrac{d^2 F}{dx^2}$ when x is made unity is finite, and therefore

$$F_1(\alpha, \beta, \gamma) - F_1(\alpha, \beta, \gamma - 1) = -\frac{1}{\gamma - 1}\left[\frac{dF}{dx}\right]_{x=1}$$

$$= -\frac{\alpha\beta}{(\gamma - 1)(\gamma - \alpha - \beta - 1)}\,F_1(\alpha, \beta, \gamma),$$

so that $\quad F_1(\alpha, \beta, \gamma - 1) = \dfrac{(\gamma - 1)(\gamma - \alpha - \beta - 1) + \alpha\beta}{(\gamma - 1)(\gamma - \alpha - \beta - 1)}\,F_1(\alpha, \beta, \gamma)$

$$= \frac{(\gamma - 1 - \alpha)(\gamma - 1 - \beta)}{(\gamma - 1)(\gamma - \alpha - \beta - 1)}\,F_1(\alpha, \beta, \gamma),$$

or, changing γ into $\gamma + 1$, we have

$$F_1(\alpha, \beta, \gamma) = \frac{(\gamma - \alpha)(\gamma - \beta)}{\gamma\,(\gamma - \alpha - \beta)}\,F_1(\alpha, \beta, \gamma + 1).$$

Similarly

$$F_1(\alpha, \beta, \gamma + 1) = \frac{(\gamma + 1 - \alpha)(\gamma + 1 - \beta)}{(\gamma + 1)(\gamma + 1 - \alpha - \beta)}\,F_1(\alpha, \beta, \gamma + 2);$$

and therefore

$$F_1(\alpha, \beta, \gamma) = \frac{(\gamma - \alpha)(\gamma + 1 - \alpha)(\gamma - \beta)(\gamma + 1 - \beta)}{\gamma\,(\gamma + 1)(\gamma - \alpha - \beta)(\gamma + 1 - \alpha - \beta)}\,F_1(\alpha, \beta, \gamma + 2)$$

$$= \frac{(\gamma - \alpha)(\gamma + 1 - \alpha)\ldots(\gamma + k - 1 - \alpha)(\gamma - \beta)(\gamma + 1 - \beta)\ldots(\gamma + k - 1 - \beta)}{\gamma\,(\gamma + 1)\ldots(\gamma + k - 1)(\gamma - \alpha - \beta)(\gamma + 1 - \alpha - \beta)\ldots(\gamma + k - 1 - \alpha - \beta)}$$

$$\times F_1(\alpha, \beta, \gamma + k).$$

Let

$$\frac{1\,.\,2\,.\,3 \ldots\ldots k}{(z + 1)(z + 2)\ldots\ldots(z + k)}\, k^z \text{ be denoted by } \Pi\,(k, z);$$

then

$$F_1(\alpha, \beta, \gamma) = \frac{\Pi\,(k, \gamma - 1)\,\Pi\,(k, \gamma - \alpha - \beta - 1)}{\Pi\,(k, \gamma - \alpha - 1)\,\Pi\,(k, \gamma - \beta - 1)}\,F_1(\alpha, \beta, \gamma + k).$$

Since

$$1.2.3...k.(k+1)...(k+z) = 1.2.3...z.(z+1)(z+2)...(z+k),$$

we have

$$1.2.3...k.k^z\left(1+\frac{1}{k}\right)\left(1+\frac{2}{k}\right)...\left(1+\frac{z}{k}\right)$$
$$= 1.2.3...z.(z+1)(z+2)...(z+k);$$

and so

$$\Pi(k,z) = \frac{1.2.3......k}{(z+1)(z+2)...(z+k)}k^z$$
$$= \frac{1.2.3......z}{\left(1+\frac{1}{k}\right)\left(1+\frac{2}{k}\right)...\left(1+\frac{z}{k}\right)},$$

on the supposition that z is an integer. From this transformation and from the original definition alike, we have

$$\Pi(k,z+1) = \Pi(k,z)\frac{1+z}{1+\frac{1+z}{k}}.$$

These equations shew that for a given value of z the function $\Pi(k,z)$ tends towards a limiting value as k approaches infinity, and that this limiting value is finite. As then $\Pi(\infty,z)$ is a function of z alone, let it be denoted by $\Pi(z)$; the last equation shews that

$$\Pi(z+1) = (z+1)\Pi(z),$$

and the former shews that, if z be an integer,

$$\Pi(z) = z!,$$

while in any case we have

$$\Pi(z) = \Gamma(z+1),$$

where $\Gamma(z+1)$ is the Gamma Function of Euler.

In the equation giving F_1, let k become infinite; then every term of the series $F_1(\alpha, \beta, \gamma+\infty)$ is zero except the first, which is unity. If we substitute for $\Pi(\infty, \gamma-1)$ and the other functions their values $\Pi(\gamma-1)$, we have

$$F_1(\alpha, \beta, \gamma) = \frac{\Pi(\gamma-1)\Pi(\gamma-\alpha-\beta-1)}{\Pi(\gamma-\alpha-1)\Pi(\gamma-\beta-1)}.$$

Ex. 1. From the expansion of t in a series of ascending powers of $\sin t$, prove that

$$\Pi\left(\tfrac{1}{2}\right)=\tfrac{1}{2}\pi^{\frac{1}{2}}.$$

Ex. 2. Prove that

$$\Pi\left(-z\right)\Pi\left(z-1\right)=\pi\,\operatorname{cosec}\,z\pi.$$

Ex. 3. Obtain the relations

      (i) $F_1(\alpha,\,\beta,\,\gamma)\,F_1(-\alpha,\,\beta,\,\gamma-\alpha)=1$;

      (ii) $F_1(\alpha,\,\beta,\,\gamma)\,F_1(\alpha,\,-\beta,\,\gamma-\beta)=1$.

Ex. 4. Prove that

$$n^{nz+\frac{1}{2}}\,\Pi\left(z\right)\Pi\left(z-\frac{1}{n}\right)\Pi\left(z-\frac{2}{n}\right)\ldots\ldots\ldots\Pi\left(z-\frac{n-1}{n}\right)=(2\pi)^{\frac{1}{2}(n-1)}\,\Pi\left(nz\right).$$

$$\text{(Gauss.)}$$

Determination of Constants in the relations of § 124.

126. The equations of § 124 now become

$$N=\frac{1}{F_1\left(\alpha,\,\beta,\,\alpha+\beta-\gamma+1\right)}$$

$$=\frac{\Pi\left(\beta-\gamma\right)\Pi\left(\alpha-\gamma\right)}{\Pi\left(\alpha+\beta-\gamma\right)\Pi\left(-\gamma\right)};$$

and therefore

$$M\frac{\Pi\left(1-\gamma\right)\Pi\left(\gamma-\alpha-\beta-1\right)}{\Pi\left(-\alpha\right)\Pi\left(-\beta\right)}+\frac{\Pi\left(\beta-\gamma\right)\Pi\left(\alpha-\gamma\right)}{\Pi\left(\alpha+\beta-\gamma\right)\Pi\left(-\gamma\right)}$$

$$=\frac{\Pi\left(\gamma-1\right)\Pi\left(\gamma-\alpha-\beta-1\right)}{\Pi\left(\gamma-\alpha-1\right)\Pi\left(\gamma-\beta-1\right)},$$

from which, with the use of Example 2 (§ 125), it is not difficult to deduce that

$$M=\frac{\Pi\left(\gamma-1\right)\Pi\left(\alpha-\gamma\right)\Pi\left(\beta-\gamma\right)}{\Pi\left(1-\gamma\right)\Pi\left(\alpha-1\right)\Pi\left(\beta-1\right)}.$$

These then are the values of the constants in the equation

$$Y_1=MY_2+NY_3.$$

127. Next, the relation just obtained can be written in the form

$$y_1=M\left(\alpha,\,\beta,\,\gamma\right)y_3+N\left(\alpha,\,\beta,\,\gamma\right)y_5.$$

Change α, β, $\alpha + \beta - \gamma + 1$, into $\gamma' - \alpha'$, $\gamma' - \beta'$, $\gamma' - \alpha' - \beta' + 1$ respectively; that is, leave γ unaltered, and change α and β into $\gamma - \alpha'$ and $\gamma - \beta'$ respectively. Also, multiply throughout by $(1 - x)^{\gamma - \alpha' - \beta'}$, that is, by $(1 - x)^{\gamma - \alpha' - \beta'}$; the foregoing relation becomes

$$F(\gamma - \alpha', \gamma - \beta', \gamma, x)(1 - x)^{\gamma - \alpha' - \beta'}$$

$$= M(\alpha, \beta, \gamma) x^{1-\gamma} (1 - x)^{\gamma - \alpha' - \beta'} F(\alpha' - \gamma + 1, \beta' - \gamma + 1, 2 - \gamma) x$$

$$+ N(\alpha, \beta, \gamma)(1 - x)^{\gamma - \alpha' - \beta'} F(\gamma - \alpha', \gamma - \beta', \gamma - \alpha' - \beta' + 1, 1 - x),$$

which is of the form

$$y_2' = M(\alpha, \beta, \gamma) y_4' + N(\alpha, \beta, \gamma) y_7',$$

where y_2' denotes y_2 with α', β' as elements instead of α, β. Now

$$M(\alpha, \beta, \gamma) = \frac{\Pi(\gamma - 1)\Pi(\alpha - \gamma)\Pi(\beta - \gamma)}{\Pi(1 - \gamma)\Pi(\alpha - 1)\Pi(\beta - 1)}$$

$$= \frac{\Pi(\gamma - 1)\Pi(-\alpha')\Pi(-\beta')}{\Pi(1 - \gamma)\Pi(\gamma - \alpha' - 1)\Pi(\gamma - \beta' - 1)},$$

$$N(\alpha, \beta, \gamma) = \frac{\Pi(\beta - \gamma)\Pi(\alpha - \gamma)}{\Pi(\alpha + \beta - \gamma)\Pi(-\gamma)}$$

$$= \frac{\Pi(-\beta')\Pi(-\alpha')}{\Pi(\gamma' - \alpha' - \beta')\Pi(-\gamma')}.$$

Hence, putting

$$M_1 = \frac{\Pi(\gamma - 1)\Pi(-\alpha)\Pi(-\beta)}{\Pi(1 - \gamma)\Pi(\gamma - \alpha - 1)\Pi(\gamma - \beta - 1)},$$

$$N_1 = \frac{\Pi(-\beta)\Pi(-\alpha)}{\Pi(\gamma - \alpha - \beta)\Pi(-\gamma)},$$

we can take our deduced relation in the form

$$y_2 = M_1 y_4 + N_1 y_7,$$

or in the form

$$Y_1 = M_1 Y_2 + N_1 Y_4.$$

To find the constants in the relation

$$Y_1 = M_2 Y_2 + N_2 Y_5,$$

we note that y_{17} in Y_1, y_{19} in Y_2, and y_{13} in Y_5, coexist, the range of their coexistence being $-\infty < x < 0$; so the constants will be found from the relation

$$y_{17} = M_2 y_{19} + N_2 y_{13}.$$

Change the variable x into x', where

$$x = \frac{x'}{x' - 1};$$

then the relation becomes

$$(1 - x')^\alpha F(\alpha, \gamma - \beta, \gamma, x')$$
$$= M_2 \left(\frac{x'}{x' - 1}\right)^{1-\gamma} (1 - x')^{-\gamma + \alpha + 1} F(\alpha - \gamma + 1, 1 - \beta, 2 - \gamma, x')$$
$$+ N_2 (1 - x')^\alpha F(\alpha, \gamma - \beta, \alpha - \beta + 1, 1 - x'),$$

that is, on division by $(1 - x')^\alpha$,

$$F(\alpha, \gamma - \beta, \gamma, x') = M_2 (-x')^{1-\gamma} F(\alpha - \gamma + 1, 1 - \beta, 2 - \gamma, x')$$
$$+ N_2 F(\alpha, \gamma - \beta, \alpha - \beta + 1, 1 - x').$$

Let $\gamma - \beta = \beta'$; we now have

$$F(\alpha, \beta', \gamma, x') = - M_2 (-1)^{-\gamma} x'^\gamma F(\alpha - \gamma + 1, \beta' - \gamma + 1, 2 - \gamma, x')$$
$$+ N_2 F(\alpha, \beta', \alpha + \beta' - \gamma + 1, 1 - x').$$

Comparing this with the relation

$$y_1 = M_2 y_3 + N_2 y_5,$$

we have

$$M_2 = -(-1)^\gamma \frac{\Pi(\gamma - 1)\,\Pi(\alpha - \gamma)\,\Pi(\beta' - \gamma)}{\Pi(1 - \gamma)\,\Pi(\alpha - 1)\,\Pi(\beta' - 1)}$$
$$= -(-1)^\gamma \frac{\Pi(\gamma - 1)\,\Pi(\alpha - \gamma)\,\Pi(-\beta)}{\Pi(1 - \gamma)\,\Pi(\alpha - 1)\,\Pi(\gamma - \beta - 1)},$$

$$N_2 = \frac{\Pi(\beta' - \gamma)\,\Pi(\alpha - \gamma)}{\Pi(\alpha + \beta' - \gamma)\,\Pi(-\gamma)}$$
$$= \frac{\Pi(-\beta)\,\Pi(\alpha - \gamma)}{\Pi(\alpha - \beta)\,\Pi(-\gamma)};$$

and with these values we have

$$Y_1 = M_2 Y_2 + N_2 Y_5.$$

We proceed in the same way to the determination of the constants in the relation

$$Y_1 = M_3 Y_2 + N_3 Y_6,$$

taking it in the form

$$y_{18} = M_3 y_{20} + N_3 y_{14},$$

the special integrals coexisting in the range $-\infty < x < 0$. The analysis is exactly similar to the analysis in the last case; and we find

$$M_3 = -(-1)^\gamma \frac{\Pi(\gamma-1)\,\Pi(\beta-\gamma)\,\Pi(-\alpha)}{\Pi(1-\gamma)\,\Pi(\beta-1)\,\Pi(\gamma-\alpha-1)},$$

$$N_3 = \frac{\Pi(-\alpha)\,\Pi(\beta-\gamma)}{\Pi(\beta-\alpha)\,\Pi(-\gamma)}.$$

Ex. 1. Establish the following results, indicating the respective ranges of values of x over which the relations are valid:—

(i) $Y_1 = A_{34}Y_3 + B_{34}Y_4,$

where

$$A_{34} = \frac{\Pi(\gamma-1)\,\Pi(\gamma-\alpha-\beta-1)}{\Pi(\gamma-\alpha-1)\,\Pi(\gamma-\beta-1)}, \qquad B_{34} = \frac{\Pi(\gamma-1)\,\Pi(\alpha+\beta-\gamma-1)}{\Pi(\alpha-1)\,\Pi(\beta-1)};$$

(ii) $Y_1 = A_{35}Y_3 + B_{35}Y_5,$

where

$$A_{35} = (-1)^\beta \frac{\Pi(\gamma-1)\,\Pi(\alpha-\gamma)}{\Pi(\alpha+\beta-\gamma)\,\Pi(\gamma-\beta-1)}, \qquad B_{35} = (-1)^{\beta-\gamma} \frac{\Pi(\gamma-1)\,\Pi(\alpha-\gamma)}{\Pi(\beta-1)\,\Pi(\alpha-\beta)};$$

(iii) $Y_1 = A_{36}Y_3 + B_{36}Y_6,$

where

$$A_{36} = (-1)^\alpha \frac{\Pi(\gamma-1)\,\Pi(\beta-\gamma)}{\Pi(\alpha+\beta-\gamma)\,\Pi(\gamma-\alpha-1)}, \qquad B_{36} = (-1)^{\alpha-\gamma} \frac{\Pi(\gamma-1)\,\Pi(\beta-\gamma)}{\Pi(\alpha-1)\,\Pi(\beta-\alpha)};$$

(iv) $Y_1 = A_{45}Y_4 + B_{45}Y_5,$

where

$$A_{45} = (-1)^{\gamma-\alpha} \frac{\Pi(\gamma-1)\,\Pi(-\beta)}{\Pi(\alpha-1)\,\Pi(\gamma-\alpha-\beta)}, \qquad B_{45} = (-1)^\alpha \frac{\Pi(\gamma-1)\,\Pi(-\beta)}{\Pi(\alpha-\beta)\,\Pi(\gamma-\alpha-1)};$$

(v) $Y_1 = A_{46}Y_4 + B_{46}Y_6,$

where

$$A_{46} = (-1)^{\gamma-\beta} \frac{\Pi(\gamma-1)\,\Pi(-\alpha)}{\Pi(\beta-1)\,\Pi(\gamma-\alpha-\beta)}, \qquad B_{46} = (-1)^\beta \frac{\Pi(\gamma-1)\,\Pi(-\alpha)}{\Pi(\beta-\alpha)\,\Pi(\gamma-\beta-1)};$$

and

(vi) $Y_1 = A_{56}Y_5 + B_{56}Y_6,$

where

$$A_{56} = \frac{\Pi(\gamma-1)\,\Pi(\beta-\alpha-1)}{\Pi(\beta-1)\,\Pi(\gamma-\alpha-1)}, \qquad B_{56} = \frac{\Pi(\gamma-1)\,\Pi(\alpha-\beta-1)}{\Pi(\alpha-1)\,\Pi(\gamma-\beta-1)}.$$

(These six relations, together with the four in the text, are the full set of linear relations into which Y_1 enters.)

Ex. 2. Shew that
$$Y_2 = PY_3 + QY_4,$$
where
$$P = -\frac{\Pi(1-\gamma)\,\Pi(\gamma-\alpha-\beta-1)}{\Pi(-\alpha)\,\Pi(-\beta)}, \qquad Q = \frac{\Pi(1-\gamma)\,\Pi(\alpha+\beta-\gamma-1)}{\Pi(\alpha-\gamma)\,\Pi(\beta-\gamma)}.$$

Ex. 3. Prove that, by an appropriate selection of individual integrals from the groups $Y_1, \ldots, Y_6$, made by taking one from a group, it is possible to establish linear relations between any three of the quantities $Y_1, \ldots, Y_6$.

Obtain the nine linear relations, other than those given in the text or in the preceding two examples.

128. We now pass to a different set of equations which connect any two of the particular solutions and their differential coefficients.

It has been proved that, if Y_1 and Y_2 be two particular solutions of the equation
$$\frac{d^2y}{dx^2} + P\frac{dy}{dx} + Qy = 0,$$
then
$$Y_1\frac{dY_2}{dx} - Y_2\frac{dY_1}{dx} = Ce^{-\int P\,dx},$$
where C has a constant value which depends upon the pair of particular solutions selected. In the case when the equation is that satisfied by the hypergeometric series, we have
$$P = \frac{\gamma - (\alpha+\beta+1)x}{x(1-x)} = \frac{\gamma}{x} + \frac{\gamma-\alpha-\beta-1}{1-x},$$
and therefore
$$Y_1\frac{dY_2}{dx} - Y_2\frac{dY_1}{dx} = Cx^{-\gamma}(1-x)^{\gamma-\alpha-\beta-1}.$$

The value of C in any equation may be determined, either by a comparison of coefficients of the same power of x on the two sides, or by the substitution of a particular value of x.

Example 1. Let
$$Y_1 = y_3 = x^{1-\gamma}\,F(\alpha-\gamma+1,\ \beta-\gamma+1,\ 2-\gamma,\ x);$$
$$Y_2 = y_1 = F(\alpha,\ \beta,\ \gamma,\ x).$$

Let each side be expanded in ascending powers of x; the term involving the lowest power of x in
$$Y_1\frac{dY_2}{dx}$$

is $\dfrac{\alpha\beta}{\gamma}\, x^{1-\gamma}$; the term involving the lowest power of x in

$$- Y_2 \frac{dY_1}{dx}$$

is $-(1-\gamma)\, x^{-\gamma}$: hence, equating the coefficients of the lowest powers, we have

$$C = -(1-\gamma) = \gamma - 1,$$

and therefore

$$y_3 \frac{dy_1}{dx} - y_1 \frac{dy_3}{dx} = (\gamma - 1)\, x^{-\gamma}\, (1-x)^{\gamma - \alpha - \beta - 1}.$$

Example 2. Let

$$Y_1 = y_5 = F(\alpha,\, \beta,\, \alpha + \beta - \gamma + 1,\, 1 - x);$$
$$Y_2 = y_1 = F(\alpha,\, \beta,\, \gamma,\, x).$$

We have already proved that

$$y_1 = M y_3 + N y_5,$$

in which M and N are definite constants. This relation gives, on differentiation,

$$\frac{dy_1}{dx} = M \frac{dy_3}{dx} + N \frac{dy_5}{dx};$$

and therefore

$$M\left(y_3 \frac{dy_1}{dx} - y_1 \frac{dy_3}{dx}\right) = N\left(y_1 \frac{dy_5}{dx} - y_5 \frac{dy_1}{dx}\right),$$

or $\qquad y_5 \dfrac{dy_1}{dx} - y_1 \dfrac{dy_5}{dx} = -\dfrac{M}{N}(\gamma - 1)\, x^{-\gamma}\,(1-x)^{\gamma - \alpha - \beta - 1},$

from the result of the last example. Now, from the values of M and N, we have

$$\frac{M}{N} = \frac{\Pi(\gamma - 1)\, \Pi(-\gamma)\, \Pi(\alpha + \beta - \gamma)}{\Pi(1 - \gamma)\, \Pi(\alpha - 1)\, \Pi(\beta - 1)}.$$

But $\qquad \Pi(1-\gamma) = (1-\gamma)\,\Pi(-\gamma) = -(\gamma - 1)\,\Pi(-\gamma),$

and therefore

$$-\frac{M}{N}(\gamma - 1) = \frac{\Pi(\gamma - 1)\, \Pi(\alpha + \beta - \gamma)}{\Pi(\alpha - 1)\, \Pi(\beta - 1)};$$

the equation becomes

$$y_5 \frac{dy_1}{dx} - y_1 \frac{dy_5}{dx} = \frac{\Pi(\gamma - 1)\, \Pi(\alpha + \beta - \gamma)}{\Pi(\alpha - 1)\, \Pi(\beta - 1)}\, x^{-\gamma}\,(1-x)^{\gamma - \alpha - \beta - 1}.$$

Ex. Prove that

$$y_5 \frac{dy_3}{dx} - y_3 \frac{dy_5}{dx} = \frac{\Pi(\alpha+\beta-\gamma)\,\Pi(1-\gamma)}{\Pi(\alpha-\gamma)\,\Pi(\beta-\gamma)}\, x^{-\gamma}(1-x)^{\gamma-\alpha-\beta-1};$$

and that

$$(-1)^\beta \left(y_{10} \frac{dy_1}{dx} - y_1 \frac{dy_{10}}{dx} \right) = \frac{\Pi(\gamma-1)\,\Pi(\beta-\alpha)}{\Pi(\beta-1)\,\Pi(\gamma-\alpha-1)}\, x^{-\gamma}(1-x)^{\gamma-\alpha-\beta-1}.$$

129. In all the foregoing investigations, the quantities α, β, γ, have been supposed to be independent, and the series have consequently retained their most general form; but many important applications are made, by assigning either one or two relations between the three constant elements, or by giving numerical values to one or more of them. Such applications (as for instance to elliptic integrals) cannot be discussed here; but the student, who wishes for information on these matters, will find at the end of the chapter a list of the more important memoirs dealing with hypergeometric series.

SPECIAL CASES OF INTEGRATION IN A FINITE FORM.

130. We pass now to consider some special cases when the hypergeometric series can be expressed in a finite form.

It has been proved (§ 61) that the quotient s of any two particular solutions of the equation

$$\frac{d^2y}{dx^2} + Iy = 0$$

satisfies the equation

$$\tfrac{1}{2}\{s,\, x\} = I,$$

where I is a function of x only; and it has been further shewn that, from any particular value of s which satisfies this equation, the value of the two particular solutions of the former equation can be obtained. In the case of the hypergeometric series, the value of I is

$$\tfrac{1}{4}\left[\frac{1-\lambda^2}{x^2} + \frac{1-\nu^2}{(x-1)^2} + \frac{\lambda^2-\mu^2+\nu^2-1}{x(x-1)} \right]\quad\ldots\ldots\ldots(\mathrm{A}),$$

λ, μ, ν, being definite functions of the constants α, β and γ; and therefore the corresponding differential equation, which gives s, may be written

$$\{s,\, x\} = \tfrac{1}{2}\frac{1-\lambda^2}{x^2} + \tfrac{1}{2}\frac{1-\nu^2}{(x-1)^2} + \tfrac{1}{2}\frac{\lambda^2-\mu^2+\nu^2-1}{x(x-1)}.$$

If then a relation between s and x can be found which is expressible in finite terms, it follows from the formulæ of § 62 that the hypergeometric series will be expressible in finite terms. This cannot be expected to occur in the case when the parameters are general; from the few instances now to be given, it will be seen that the values of λ, μ, ν, are definite numerical constants.

There are in all fifteen separate cases, and no more; for the proof of this, reference should be made in the first place to the memoirs of Schwarz (see § 134) to whom the investigation, in a completely different form, is originally due.

It is convenient to recapitulate here the general formulæ of transformation of the function $\{s, x\}$ for the changes of the variables; the special examples given in Ex. 3, § 62, are particular cases of the general relations which are

$$\{s, x\} = \left(\frac{dS}{dx}\right)^2 \{s, S\} - \left(\frac{dX}{dx}\right)^2 \{x, X\} + \left(\frac{dX}{dx}\right)^2 \{S, X\} \dots\dots\dots\text{(i)},$$

and

$$\left\{\frac{ax+b}{cx+d}, x\right\} = 0 \dots\dots\dots\dots\dots\dots\dots\dots\dots\text{(ii)}.$$

As additional examples we may take

$$\{s, x\} = \frac{(\alpha\delta - \beta\gamma)^2}{(\gamma x + \delta)^4} \left\{s, \frac{\alpha x + \beta}{\gamma x + \delta}\right\} \dots\dots\dots\dots\dots\text{(iii)},$$

and

$$\left\{\frac{as+b}{cs+d}, \frac{\alpha x + \beta}{\gamma x + \delta}\right\} = \frac{(\gamma x + \delta)^4}{(\alpha\delta - \beta\gamma)^2} \{s, x\} \dots\dots\dots\dots\text{(iv)}.$$

Another formula, which will prove useful, is that which arises by supposing $s^n = x$; then we have

$$s' = \frac{1}{n} x^{\frac{1}{n}-1},$$

so that

$$\frac{s''}{s'} = \frac{\frac{1}{n}-1}{x};$$

therefore

$$\frac{s'''}{s'} - \left(\frac{s''}{s'}\right)^2 = \frac{1 - \frac{1}{n}}{x^2},$$

and

$$\frac{1}{2}\left(\frac{s''}{s'}\right)^2 = \frac{\frac{1}{2} - \frac{1}{n} + \frac{1}{2n^2}}{x^2},$$

so that

$$\{s, x\} = \frac{1 - \frac{1}{n^2}}{2x^2},$$

which may be written in either of the forms

$$\left.\begin{aligned}\{x^{\frac{1}{n}},\ x\} &= \frac{1-\dfrac{1}{n^2}}{2x^2}\\[2ex]\{s,\ s^n\} &= \frac{1-\dfrac{1}{n^2}}{2s^{2n}}\end{aligned}\right\} \quad\dots\dots\dots\dots\dots\dots\text{(v)}.$$

131. CASE I.

By writing $X = x$ in (i) in the formulæ just enumerated, we have

$$\{s,\ x\} = \{S,\ x\} + \left(\frac{dS}{dx}\right)^2 \{s,\ S\}.$$

By a series of proper substitutions we may pass from this equation to the corresponding equation for the hypergeometric series.

Firstly, let

$$S = \frac{\sigma - 1}{\sigma + 1} = \frac{s^n - 1}{s^n + 1};$$

then

$$\{s,\ x\} = \{S,\ x\} + \left(\frac{dS}{dx}\right)^2 \left\{s,\ \frac{\sigma - 1}{\sigma + 1}\right\};$$

while, by (iii),

$$\left\{s,\ \frac{\sigma - 1}{\sigma + 1}\right\} = \tfrac{1}{4}(\sigma + 1)^4 \{s,\ \sigma\}.$$

But $\sigma = s^n$; therefore

$$\{s,\ \sigma\} = \frac{1 - \dfrac{1}{n^2}}{2\sigma^2},$$

and thus

$$\left\{s,\ \frac{\sigma - 1}{\sigma + 1}\right\} = \frac{1 - \dfrac{1}{n^2}}{2}\frac{(\sigma + 1)^4}{4\sigma^2}.$$

Secondly, let

$$T = S^2 = 1 - x,$$

so that the relation between s and x is

$$\left(\frac{s^n - 1}{s^n + 1}\right)^2 = 1 - x;$$

then

$$\left(\frac{dS}{dx}\right)^2 = \frac{\frac{1}{4}}{1 - x}.$$

Again, using (i), we have
$$\{S, x\} = \{T, x\} + \left(\frac{dT}{dx}\right)^2 \{S, T\};$$
but
$$\frac{dT}{dx} = -1,$$
$$\{T, x\} = \{1 - x, x\} = 0,$$
$$\{S, T\} = \{S, S^2\} = \frac{1 - \frac{1}{4}}{2T^2};$$
so that we have
$$\{S, x\} = \tfrac{1}{2} \frac{1 - \frac{1}{4}}{(1-x)^2}.$$

Also, since
$$\left(\frac{\sigma - 1}{\sigma + 1}\right)^2 = 1 - x,$$
we have
$$x = \frac{4\sigma}{(\sigma + 1)^2},$$
and therefore
$$\left\{s, \frac{\sigma - 1}{\sigma + 1}\right\} = \left(1 - \frac{1}{n^2}\right)\frac{2}{x^2}.$$

When these substitutions are made in the original equation which gave $\{s, x\}$, it becomes
$$\{s, x\} = \tfrac{1}{2}\left[\frac{1 - \frac{1}{4}}{(1-x)^2} + \frac{1 - \frac{1}{n^2}}{x^2(1 - x)}\right]$$
$$= \tfrac{1}{2}\left[\frac{1 - \frac{1}{4}}{(1-x)^2} + \frac{1 - \frac{1}{n^2}}{x^2} + \frac{\frac{1}{n^2} - 1}{x(x - 1)}\right].$$

This is of the same form as the equation (A) in the general case, and is identical with it when we write
$$\lambda = \frac{1}{n}, \quad \nu = \tfrac{1}{2}, \quad \mu = \tfrac{1}{2};$$
and then the relation between s and x is
$$\left(\frac{s^n - 1}{s^n + 1}\right)^2 = 1 - x,$$
or
$$x = \frac{4s^n}{(s^n + 1)^2}.$$

Now $\lambda^2 = (1 - \gamma)^2$, $\mu^2 = (\alpha - \beta)^2$, $\nu^2 = (\gamma - \alpha - \beta)^2$; remembering that $\gamma - \alpha - \beta$ must be positive in order that the series may converge even when the variable is equal to unity, and assuming that α is greater than β (which is permissible), we may take

$$\alpha = \tfrac{1}{2} - \frac{1}{2n}, \quad \beta = -\frac{1}{2n}, \quad \gamma = 1 - \frac{1}{n}.$$

If it be desired to have β positive, we can change the sign of n; and then the elements of the hypergeometric series are

$$\alpha = \tfrac{1}{2} + \frac{1}{2n}, \quad \beta = \frac{1}{2n}, \quad \gamma = 1 + \frac{1}{n},$$

and the relation between s and x becomes

$$\frac{1 - s^n}{1 + s^n} = (1 - x)^{\frac{1}{2}}.$$

The latter gives

$$s^n = \frac{1 - (1 - x)^{\frac{1}{2}}}{1 + (1 - x)^{\frac{1}{2}}};$$

and therefore

$$s = \frac{x^{\frac{1}{n}}}{\{1 + (1 - x)^{\frac{1}{2}}\}^{\frac{2}{n}}},$$

while $\qquad s'^{-\frac{1}{2}} = n^{\frac{1}{2}} (1 - x)^{\frac{1}{4}} x^{\frac{1}{2} - \frac{1}{2n}} \{1 + (1 - x)^{\frac{1}{2}}\}^{\frac{1}{n}}.$

Now the two particular solutions, when the equation is in its normal form, are

$$C_1 s'^{-\frac{1}{2}} \text{ and } C_2 s'^{-\frac{1}{2}} s,$$

and the relation between the dependent variable v in this case and the dependent variable in the ordinary differential equation is (§ 116)

$$y = v x^{-\frac{1}{2}\gamma} (1 - x)^{-\frac{1}{2}(\alpha + \beta + 1 - \gamma)},$$

which becomes

$$y = v x^{-\left(\frac{1}{2} + \frac{1}{2n}\right)} (1 - x)^{-\frac{1}{4}}$$

in the special case.

Hence the primitive of the differential equation

$$x(1 - x)\frac{d^2 y}{dx^2} + \frac{dy}{dx}\left\{1 + \frac{1}{n} - x\left(\frac{3}{2} + \frac{1}{n}\right)\right\} - \frac{1}{4n}\left(1 + \frac{1}{n}\right) y = 0$$

is $\qquad y = C_1 x^{-\frac{1}{n}} \{1 + (1 - x)^{\frac{1}{2}}\}^{\frac{1}{n}} + C_2 \{1 + (1 - x)^{\frac{1}{2}}\}^{-\frac{1}{n}}.$

Moreover, on comparing these two particular solutions

$$\{1 + (1 - x)^{\frac{1}{2}}\}^{-\frac{1}{n}} \quad \text{and} \quad x^{-\frac{1}{n}} \{1 + (1 - x)^{\frac{1}{2}}\}^{\frac{1}{n}}$$

with the set of particular solutions, we find that they correspond respectively to I. and III. of § 120; in fact, the relations are

$$F\left\{\frac{1}{2} + \frac{1}{2n}, \ \frac{1}{2n}, \ 1 + \frac{1}{n}, \ x\right\} = 2^{\frac{1}{n}} \{1 + (1 - x)^{\frac{1}{2}}\}^{-\frac{1}{n}} \quad \dots\dots(\text{I.}),$$

and $\quad F\left\{\frac{1}{2} - \frac{1}{2n}, \ -\frac{1}{2n}, \ 1 - \frac{1}{n}, \ x\right\} = 2^{-\frac{1}{n}} \{1 + (1 - x)^{\frac{1}{2}}\}^{\frac{1}{n}} \dots\dots(\text{II.}),$

the common factor $x^{-\frac{1}{n}}$ having been removed from the latter. These two relations are of course equivalent to one another.

132. CASE II. From what has been proved in the last case it follows that, when we assign the particular value 2 to n, we have the relation

$$\xi = \frac{4\sigma^2}{(\sigma^2 + 1)^2}$$

as a solution of

$$\{\sigma, \xi\} = \frac{1}{2} \left[\frac{1 - \frac{1}{4}}{(1 - \xi)^2} + \frac{1 - \frac{1}{4}}{\xi^2} + \frac{\frac{1}{4} - 1}{\xi(\xi - 1)} \right]$$

$$= \frac{3}{8} \frac{\xi^2 - \xi + 1}{\xi^2 (1 - \xi)^2}.$$

Firstly, let

$$\xi(\xi_1 + 1) = 1 ;$$

then $\qquad \{\sigma, \xi_1\} = \left\{\sigma, \ \frac{1 - \xi}{\xi}\right\} = \xi^4 \{\sigma, \xi\}$

$$= \frac{3}{8} \frac{\xi^2 (\xi^2 - \xi + 1)}{(1 - \xi)^2}$$

$$= \frac{3}{8} \frac{\xi_1^2 + \xi_1 + 1}{\xi_1^2 (\xi_1 + 1)^2},$$

and $\qquad\qquad \xi_1 = \frac{(\sigma^2 - 1)^2}{4\sigma^2}.$

Secondly, let

$$\xi_1 = \xi_2^2 ;$$

then $\qquad \{\sigma, \xi_2\} = \left(\frac{d\xi_1}{d\xi_2}\right)^2 [\{\sigma, \xi_1\} - \{\xi_2, \xi_1\}],$

and
$$\{\xi_2,\ \xi_1\} = \{\xi_2,\ \xi_2{}^2\} = \frac{1-\frac{1}{4}}{2\xi_1{}^2} = \frac{3}{8\xi_1{}^2},$$

so that
$$\{\sigma,\ \xi_2\} = \tfrac{3}{2}\xi_2{}^2\left[\frac{\xi_1{}^2+\xi_1+1}{\xi_1{}^2(\xi_1+1)^2} - \frac{1}{\xi_1{}^2}\right]$$

$$= \tfrac{3}{2}\xi_1\frac{-\xi_1}{\xi_1{}^2(\xi_1+1)^2} = -\tfrac{3}{2}\frac{1}{(\xi_2{}^2+1)^2};$$

and the relation is
$$\xi_2 = \frac{\sigma^2-1}{2\sigma}.$$

Thirdly, by writing
$$\xi_2 = \sqrt{3}\,\xi_3,$$

we at once have
$$\{\sigma,\ \xi_3\} = 3\,\{\sigma,\ \xi_2\} = -\tfrac{9}{2}\frac{1}{(1+3\xi_3{}^2)^2},$$

where
$$\xi_3 = \frac{\sigma^2-1}{2\sigma\sqrt{3}}.$$

Fourthly, let $\sigma = s^2$; then
$$\{s,\ \xi_3\} = \left(\frac{d\sigma}{d\xi_3}\right)^2\{s,\ \sigma\} + \{\sigma,\ \xi_3\}.$$

Now
$$\{s,\ \sigma\} = \{s,\ s^2\} = \frac{3}{8\sigma^2};$$

and
$$2\sqrt{3} = \frac{\sigma^2+1}{\sigma^2}\frac{d\sigma}{d\xi_3},$$

so that
$$\frac{1}{\sigma^2}\left(\frac{d\sigma}{d\xi_3}\right)^2 = \frac{12\sigma^2}{(\sigma^2+1)^2} = \frac{3}{1+3\xi_3{}^2}.$$

Hence
$$\{s,\ \xi_3\} = \tfrac{27}{8}\frac{\xi_3{}^2-1}{(1+3\xi_3{}^2)^2};$$

and the relation is
$$\xi_3 = \frac{s^4-1}{2s^2\sqrt{3}}.$$

Fifthly, let
$$\xi_3 = \frac{\xi_4+1}{\xi_4-1};$$

then $\quad \{s, \xi_4\} = \dfrac{4}{(\xi_4 - 1)^4} \{s, \xi_3\}$

$$= \dfrac{4}{(\xi_4 - 1)^4} \cdot \dfrac{27}{8} \cdot \dfrac{4\xi_4}{(\xi_4 - 1)^2} \cdot \dfrac{(\xi_4 - 1)^4}{16(\xi_4{}^2 + \xi_4 + 1)^2}$$

$$= \dfrac{27}{8} \dfrac{\xi_4}{(\xi_4{}^3 - 1)^2},$$

and the relation is

$$\xi_4 = \dfrac{s^4 + 2s^2\sqrt{3} - 1}{s^4 - 2s^2\sqrt{3} - 1}.$$

Sixthly, let

$$\xi_5 = \xi_4{}^3;$$

then $\quad \{s, \xi_5\} = \left(\dfrac{d\xi_4}{d\xi_5}\right)^2 [\{s, \xi_4\} - \{\xi_5, \xi_4\}].$

Also $\qquad \dfrac{d\xi_4}{d\xi_5} = \dfrac{\frac{1}{3}}{\xi_5{}^{\frac{2}{3}}},$

and $\qquad \{\xi_5, \xi_4\} = -\dfrac{4}{\xi_4{}^2} = -\dfrac{4}{\xi_5{}^{\frac{2}{3}}}.$

Hence $\qquad \{s, \xi_5\} = \dfrac{\frac{3}{8}}{\xi_5(\xi_5 - 1)^2} + \dfrac{\frac{4}{9}}{\xi_5{}^2}$

$$= \dfrac{\frac{3}{8}}{(\xi_5 - 1)^2} + \dfrac{\frac{4}{9}}{\xi_5{}^2} + \dfrac{\frac{3}{8}}{\xi_5(1 - \xi_5)};$$

and the relation is

$$\xi_5 = \left[\dfrac{s^4 + 2s^2\sqrt{3} - 1}{s^4 - 2s^2\sqrt{3} - 1}\right]^3.$$

It therefore follows that a solution of

$$\{s, x\} = \tfrac{1}{2}\left[\dfrac{1 - \lambda^2}{x^2} + \dfrac{1 - \nu^2}{(1 - x)^2} + \dfrac{\lambda^2 - \mu^2 + \nu^2 - 1}{x(x - 1)}\right],$$

in the case when

$$\lambda = \tfrac{1}{3} = \mu, \quad \nu = \tfrac{1}{2},$$

is given by

$$x = \left(\dfrac{s^4 + 2s^2\sqrt{3} - 1}{s^4 - 2s^2\sqrt{3} - 1}\right)^3.$$

From this relation the value of s can be found (it is a somewhat complicated function of x) and thence s'; and this will lead (§ 62) to the solution of the equation

$$x(1 - x)\dfrac{d^2y}{dx^2} + \left(\dfrac{2}{3} - \dfrac{7}{6}x\right)\dfrac{dy}{dx} + \dfrac{1}{48}y = 0.$$

133. CASE III. From the two preceding cases a new one can be constructed.

For, in Case II., let

$$\frac{4z}{(z+1)^2} = x;$$

then

$$\{z, x\} = \frac{\frac{3}{8}}{(1-x)^2}$$

by Case I.; and so

$$\{s, z\} = \left(\frac{dx}{dz}\right)^2 \left[\{s, x\} - \{z, x\}\right]$$

$$= \left(\frac{dx}{dz}\right)^2 \left[\frac{\frac{4}{9}}{x^2} + \frac{\frac{3}{8}}{x(1-x)}\right]$$

$$= \frac{\frac{4}{9}}{z^2} + \frac{-\frac{5}{18}}{z(z+1)^2}.$$

Now change z into $-z$, so that

$$\frac{4z}{(z-1)^2} = -x = \left(\frac{s^4 + 2s^2\sqrt{3} - 1}{-s^4 + 2s^2\sqrt{3} + 1}\right)^3;$$

then

$$\{s, z\} = \{s, -z\} = \frac{\frac{4}{9}}{z^2} + \frac{\frac{5}{18}}{z(1-z)^2}$$

$$= \frac{\frac{4}{9}}{z^2} + \frac{\frac{5}{18}}{(1-z)^2} + \frac{\frac{5}{18}}{z(1-z)}.$$

A comparison with the general formula shews that the last relation between z and s is a solution, provided

$$\lambda = \tfrac{1}{3}, \quad \nu = \tfrac{2}{3}, \quad \mu = \tfrac{1}{3};$$

and therefore

$$\alpha = -\beta = \tfrac{1}{6}, \quad \gamma = \tfrac{2}{3}.$$

Hence by means of the preceding relation we can obtain the primitive of

$$z(1-z)\frac{d^2y}{dz^2} + (\tfrac{2}{3} - z)\frac{dy}{dz} + \tfrac{1}{36}y = 0$$

in a finite form.

Ex. 1. Shew that from Case II. can be derived in a finite form the solution of

$$x(1-x)\frac{d^2y}{dx^2} + (\tfrac{4}{3} - \tfrac{11}{6}x)\frac{dy}{dx} - \tfrac{7}{48}y = 0.$$

Ex. 2. Shew that from Case III. can be derived in a finite form the solution of

$$x(1-x)\frac{d^2y}{dx^2}+(\tfrac{4}{3}-\tfrac{5}{3}x)\frac{dy}{dx}-\tfrac{1}{12}y=0.$$

Further cases will be found in the Miscellaneous Examples at the end of the chapter.

It may easily be verified that, for all the examples given, we have on taking positive values of λ, μ, ν, the inequality

$$\lambda+\mu+\nu>1;$$

the case of $\lambda+\mu+\nu=1$ is integrable by the simpler method of § 68. See Ex. 7, p. 141.

134. An entirely different method of treatment of the matter contained in §§ 130—133 will be found in the author's *Theory of Differential Equations*, vol. IV. §§ 59—62. For further information on the subject of the hypergeometric series the following memoirs should be consulted:—

GAUSS, "Disquisitiones generales circa seriem infinitam

$$1+\frac{\alpha\beta}{1.\gamma}x+\frac{\alpha(\alpha+1)\beta(\beta+1)}{1.2.\gamma(\gamma+1)}x^2+\ldots\ldots,"$$

Ges. Werke, t. III. pp. 123—163;

"Determinatio seriei nostræ per æquationem differentialem secundi ordinis," *id.* pp. 207—230.

KUMMER, "Ueber die hypergeometrische Reihe," *Crelle*, t. XV. pp. 39—83 and 127—172.

SCHWARZ, "Ueber einige Abbildungsaufgaben," *Crelle*, t. LXX. pp. 105—120;

"Ueber diejenigen Fälle in welchen die *Gauss*ische hypergeometrische Reihe eine algebraische Function ihres vierten Elementes darstellt," *Crelle*, t. LXXV. pp. 292—335.

CAYLEY, "On the Schwarzian derivative and the Polyhedral Functions," *Camb. Phil. Trans.* vol. XIII.; *Coll. Math. Papers*, vol. XI. pp. 148—216;

in the last of which references will be found to further memoirs.

There is a memoir by GOURSAT which may be consulted with great advantage—"Sur l'équation différentielle qui admet pour intégrale la série hypergéométrique" (*Annales de l'école normale supérieure*, Sér. II. t. X.)—in which, by developing a method due originally to Jacobi, he obtains the results of Kummer and Schwarz.

The Gamma-function, and the function defined by the hypergeometric series, are discussed from the functional point of view by Whittaker and Watson, *Modern Analysis*, Chapters XII. and XIV.

MISCELLANEOUS EXAMPLES.

1. Prove that, if

$$(a^2+b^2-2ab\cos\phi)^{-n}=A_0+2A_1\cos\phi+2A_2\cos 2\phi+2A_3\cos 3\phi+\dots,$$

then A_r may be written in any of the forms

(i) $\dfrac{(n+r-1)!}{r!\,(n-1)!}\,a^{-2n-r}b^r\,F\left(n,\,n+r,\,r+1,\,\dfrac{b^2}{a^2}\right)$;

(ii) $\dfrac{(n+r-1)!}{r!\,(n-1)!}\,\dfrac{a^r b^r}{(a^2+b^2)^{n+r}}\,F\left(\tfrac{1}{2}n+\tfrac{1}{2}r,\,\tfrac{1}{2}n+\tfrac{1}{2}r+\tfrac{1}{2},\,r+1,\,\dfrac{4a^2b^2}{(a^2+b^2)^2}\right)$;

(iii) $\dfrac{(n+r-1)!}{r!\,(n-1)!}\,\dfrac{a^r b^r}{(a+b)^{2(n+r)}}\,F\left(n+r,\,r+\tfrac{1}{2},\,2r+1,\,\dfrac{4ab}{(a+b)^2}\right)$;

(iv) $\dfrac{(n+r-1)!}{r!\,(n-1)!}\,\dfrac{a^r b^r}{(a-b)^{2(n+r)}}\,F\left(n+r,\,r+\tfrac{1}{2},\,2r+1,\,-\dfrac{4ab}{(a-b)^2}\right)$.

(Gauss.)

2. Obtain a solution of the equation

$$(A+Bx+Cx^2)\frac{d^2y}{dx^2}+(D+Ex)\frac{dy}{dx}+Fy=0,$$

as a hypergeometric series ; $A,\,B,\,C,\,D,\,E,\,F$, are supposed to be constants.

(Gauss.)

3. A function is said to be contiguous to $F(a,\,\beta,\,\gamma,\,x)$ when it is derived from it by changing one and only one of the constant elements by unity. Let $F(a+1,\,\beta,\,\gamma,\,x)$ be denoted by F_{a+} ; $F(a-1,\,\beta,\,\gamma,\,x)$ by F_{a-} ; and $F(a,\,\beta,\,\gamma,\,x)$ by F. Then prove the following relations :—

(i) $0=(\beta-a)\,F+aF_{a+}-\beta F_{\beta+}$;

(ii) $0=(\gamma-a-1)\,F+aF_{a+}-(\gamma-1)\,F_{\gamma-}$;

(iii) $0=\{\gamma-2a-(\beta-a)\,x\}\,F+a(1-x)\,F_{a+}-(\gamma-a)\,F_{a-}$;

(iv) $0=\gamma\{a-(\gamma-\beta)\,x\}\,F-a\gamma(1-x)\,F_{a+}+(\gamma-a)\,(\gamma-\beta)\,xF_{\gamma+}$;

(v) $0=(\gamma-a-\beta)\,F+a(1-x)\,F_{a+}-(\gamma-\beta)\,F_{\beta-}$.

(Gauss.)

4. Prove that

$$(1-x)\,F(a,\,\beta,\,\gamma,\,x)\,F(1-a,\,1-\beta,\,1-\gamma,\,x)-1$$
$$=\frac{(\gamma-a)\,(\gamma-\beta)}{\gamma(1-\gamma)}\,xF(a,\,\beta,\,\gamma+1,\,x)\,F(1-a,\,1-\beta,\,2-\gamma,\,x).$$

(Gauss.)

5. By changing the independent variable in the differential equation, verify the following relations:—

(i) $(1+y)^{2a}F(2a, 2a+1-\gamma, \gamma, y)=F\left(a, a+\tfrac{1}{2}, \gamma, \dfrac{4y}{(1+y)^2}\right).$

(Gauss.)

(ii) $(1+y)^{2a}F(a, a+\tfrac{1}{2}-\beta, \beta+\tfrac{1}{2}, y^2)=F\left(a, \beta, 2\beta, \dfrac{4y}{(1+y)^2}\right).$

(Gauss.)

(iii) $F(a, \beta, a+\beta+\tfrac{1}{2}, \sin^2\theta)=F\left(2a, 2\beta, a+\beta+\tfrac{1}{2}, \sin^2\dfrac{\theta}{2}\right).$

(Kummer.)

Prove also that, by changing the variable from x to $-8x\{1+(1-x)^{\frac{1}{2}}\}^{-3}$,

$$F\left(\frac{a}{2}, \frac{a+1}{6}, \frac{2a+2}{3}, \sin^2 2\theta\right)=\cos^{-2a}\theta F\left(\frac{a}{2}, \frac{a+1}{6}, \frac{2a+2}{3}, \frac{-4\sin^2\theta}{\cos^4\theta}\right).$$

(Kummer.)

6. Shew that the functions P_n and Q_n, which are the independent solutions of Legendre's equation, may be expressed by hypergeometric series in the forms

$$P_n(x)=\frac{\Pi(2n)}{2^{2n}\Pi(n)\Pi(n)}\xi^{-n}F(\tfrac{1}{2}, -n, \tfrac{1}{2}-n, \xi^2),$$

$$Q_n(x)=\frac{2^{2n+1}\Pi(n)\Pi(n)}{\Pi(2n+1)}\xi^{n+1}F(\tfrac{1}{2}, n+1, n+\tfrac{3}{2}, \xi^2),$$

the variable x of Legendre's equation being connected with ξ by the relation

$$2x=\xi+\xi^{-1}.$$

(Heine.)

7. Shew that, if the independent variable in Legendre's equation be restricted to be less than unity, the primitive may be represented by

$$AF(-\tfrac{1}{2}n, \tfrac{1}{2}n+\tfrac{1}{2}, \tfrac{1}{2}, x^2)+BxF(-\tfrac{1}{2}n+\tfrac{1}{2}, \tfrac{1}{2}n+1, \tfrac{3}{2}, x^2),$$

where the series, if infinite, converge.

(Heine.)

8. Denoting the series

$$1+\frac{a\beta\gamma}{\theta\epsilon}x+\frac{a.a+1.\beta.\beta+1.\gamma.\gamma+1}{1.2.\theta.\theta+1.\epsilon.\epsilon+1}x^2+\dots$$

by $F\left\{\begin{pmatrix}a, \beta, \gamma \\ \theta, \epsilon\end{pmatrix}, x\right\}$, prove that F satisfies the differential equation

$$(1-x)x^2\frac{d^3F}{dx^3}+\{\theta+\epsilon+1-(a+\beta+\gamma+3)x\}x\frac{d^2F}{dx^2}$$

$$+\{\theta\epsilon-x(a\beta+\beta\gamma+\gamma a+a+\beta+\gamma+1)\}\frac{dF}{dx}-a\beta\gamma F=0.$$

9. Obtain integrals of the equation

$$\frac{d^2y}{dx^2} + \frac{1-a-a'-(1+\beta+\beta')\,x}{x\,(1-x)}\frac{dy}{dx} + \frac{aa'-(aa'+\beta\beta'-\gamma\gamma')\,x+\beta\beta'\,x^2}{x^2\,(1-x)^2}\,y = 0,$$

(i) in powers of x, (ii) in powers of $1-x$, (iii) in powers of x^{-1}; and indicate relations between these integrals, it being given that

$$a+a'+\beta+\beta'+\gamma+\gamma'=1.$$

10. Shew that the differential equation of the hypergeometric series possesses two integrals, whose product is a polynomial in x, in the following cases :

   (i) $a = -\tfrac{1}{2}n$,

   (ii) $\beta = -\tfrac{1}{2}n$,

   (iii) $a+\beta = -n$, and $\gamma = \tfrac{1}{2}$, or $-\tfrac{1}{2}$, or $-\tfrac{3}{2}$, ..., or $-n+\tfrac{1}{2}$,

where n is an even integer in each case.

Are these all the cases in which the indicated property belongs to the equation ?

11. The equation

$$x\,(1-x)\frac{d^2y}{dx^2} + (\tfrac{3}{2}-2x)\frac{dy}{dx} - \tfrac{1}{4}y = 0$$

has a particular solution of the form x^n; determine n and obtain the primitive.

Hence express $\sin^{-1}x$ as a hypergeometric series.

(Goursat.)

12. Obtain in a finite form the primitive of

$$x\,(1-x)\frac{d^2y}{dx^2} + \tfrac{1}{3}\,(1-2x)\frac{dy}{dx} + \tfrac{20}{9}y = 0;$$

also of

$$x\,(1-x)\frac{d^2y}{dx^2} + \tfrac{1}{2}\frac{dy}{dx} + 2y = 0.$$

(Goursat.)

13. Prove that the relation

$$\frac{x}{x-1} = \frac{(s^8+14s^4+1)^3}{108s^4\,(s^4-1)^4}$$

satisfies the equation

$$\{s,\,x\} = \frac{\tfrac{4}{9}}{x^2} + \frac{\tfrac{15}{32}}{(1-x)^2} + \frac{\tfrac{155}{288}}{x\,(1-x)}.$$

Hence obtain in a finite form the primitives of the equations :

   (i) $x\,(1-x)\frac{d^2y}{dx^2} + (\tfrac{2}{3}-1\tfrac{7}{12}x)\frac{dy}{dx} + \tfrac{11}{576}y = 0;$

   (ii) $x\,(1-x)\frac{d^2y}{dx^2} + (\tfrac{4}{3}-2\tfrac{5}{12}x)\frac{dy}{dx} - \tfrac{133}{576}y = 0.$

14. Prove that the relation

$$-\frac{(z-1)^2}{4z} = \frac{(s^8+14s^4+1)^3}{108s^4(s^4-1)^4}$$

satisfies the equation

$$\{s,\, z\} = \frac{\frac{15}{32}}{z^2} + \frac{\frac{5}{18}}{(1-z)^2} + \frac{\frac{5}{18}}{z(1-z)}.$$

Hence obtain in a finite form the primitives of the equations:

(i) $x(1-x)\dfrac{d^2y}{dx^2} + (\tfrac{3}{4}-1\tfrac{3}{12}x)\dfrac{dy}{dx} + \tfrac{1}{72}y = 0$;

(ii) $x(1-x)\dfrac{d^2y}{dx^2} + (\tfrac{5}{4}-1\tfrac{9}{12}x)\dfrac{dy}{dx} - \tfrac{5}{72}y = 0$.

15. Prove that

$$F(a,\, a+\tfrac{1}{2},\, 2a+1,\, x) = 2^{2a}\{1+(1-x)^{\frac{1}{2}}\}^{-2a};$$

and that

$$F(2,\, \tfrac{5}{2},\, 5,\, x) = \{F(1,\, \tfrac{3}{2},\, 3,\, x)\}^2$$
$$= \{F(\tfrac{1}{2},\, 1,\, 2,\, x)\}^4.$$

SUPPLEMENTARY NOTES.

I.

INTEGRATION OF LINEAR EQUATIONS IN SERIES BY THE METHOD OF FROBENIUS.

The two methods, given in §§ 83 and 84, are equivalent to one another: and either of them is effectively a process for constructing the coefficients in a Taylor series. The inverse of the process is given in §§ 114 and 115, where a knowledge of the coefficients is used to determine the differential equation. The only integrals, that are thus obtained directly, are those which occur in the form of series of powers of the variable. When other integrals, which are not expressible solely in series of powers, e.g. integrals which involve logarithms, are required, supplementary substitutions have to be made and further investigations, sometimes of an elaborate character, prove necessary.

There is another method, quite distinct in character and significance, by which a single set of properly devised calculations can be used so as to give all the integrals of the kind indicated that are possessed by the differential equation. It is due to Frobenius*; for the full establishment of the validity of the various steps, various propositions in the general theory of functions are necessary. Consequently, only an outline will be given here; if a fuller account is desired, reference must be made either to the original memoir by Frobenius or to other places where a detailed exposition is given†. Further, the explanations will be given only in connection with linear equations of the second order; they apply to linear equations of any order. And a further limitation, to be indicated immediately, will be imposed on the form of the equation.

Let the equation be taken initially in the form

$$\frac{d^2y}{dx^2} + P\frac{dy}{dx} + Qy = 0.$$

If an integral or integrals exist, proceeding in powers of $x - a$, where both P and Q are finite when $x = a$, the method possesses no special advantage over the methods already given. If, however, either P or Q or both P and Q be infinite when $x = a$, then the method can be applied effectively to determine integrals. We shall assume that P, if infinite, has $x = a$ for an infinity of the first degree; we shall also assume that Q, if infinite, has $x = a$ for an infinity of degree not higher than the second‡. We write

$$x - a = t,$$
$$P = \frac{1}{t}p, \qquad Q = \frac{1}{t^2}q,$$

* *Crelle*, t. LXXVI. (1873), pp. 214—224.

† Such an exposition is given in the author's *Theory of Differential Equations*, vol. IV. (Cambridge, University Press, 1902), pp. 78 *et seq.*

‡ It will be observed that these assumptions are justified for Legendre's equation when $x = 1$ or $x = -1$; for Bessel's equation when $x = 0$; and for the equation of the hypergeometric series when $x = 0$ or $x = 1$. If we are dealing with large values of x, the natural expansion is in ascending powers of x^{-1}; and very slight calculation will shew that the assumption made in the text is justified for Legendre's equation and for the equation of the hypergeometric series (but not for Bessel's equation) when $x = \infty$.

The corresponding assumption, when the linear equation is of order m and has the form

$$\frac{d^m y}{dx^m} + \sum_{r=1}^{m} P_r \frac{d^{m-r}y}{dx^{m-r}} = 0,$$

is that P_r, if infinite when $x = a$, has $x = a$ for an infinity of degree not higher than r, for all the values of r.

where p and q are finite (including possibly a zero value) when $t = 0$, and are supposed expansible in converging or terminating series of positive integer powers of t. The equation thus is of the form

$$Dy = t^2 \frac{d^2y}{dt^2} + tp\, \frac{dy}{dt} + qy = 0.$$

If an integral exists which is of the type indicated, by being expansible in ascending powers of t, it may be taken in the form

$$y = t^\rho \sum_{n=0} c_n t^n = \sum_{n=0} c_n t^{\rho+n} = g\,(t,\, \rho),$$

where the summation is for positive integer values of n. Should this be an integral, it must satisfy the equation identically. Now

$$Dt^m = \{m\,(m-1) + mp + q\}\, t^m$$
$$= t^m f\,(t,\, m),$$

where $f(t,\, m)$ is of the second degree in m and, owing to the character of p and q, can be expanded in positive integer powers of t. When this expansion is effected, we have

$$f\,(t,\, m) = f_0\,(m) + tf_1\,(m) + t^2f_2\,(m) + \ldots\ldots,$$

where $f_0(m)$ is of the second degree in m, and the other co-efficients $f_1(m)$, $f_2(m)$, ... are of degree in m not higher than the first. Then

$$Dg\,(t,\, \rho) = \sum_{n=0} c_n Dt^{\rho+n}$$
$$= \sum_{n=0} c_n t^{\rho+n} f(t,\, \rho+n)$$
$$= \sum_{n=0} c_n t^{\rho+n} \{f_0\,(\rho+n) + tf_1\,(\rho+n) + t^2f_2\,(\rho+n) + \ldots\}$$
$$= \sum_{n=0} t^{\rho+n} \{c_n f_0\,(\rho+n) + c_{n-1} f_1\,(\rho+n-1) + \ldots + c_0 f_n\,(\rho)\},$$

on gathering together the terms with the same exponent. If the postulated expression is to be an integral of the equation, the right-hand side must vanish identically and therefore the co-efficients of the various powers of t must vanish; hence

$$0 = c_0 f_0\,(\rho),$$
$$0 = c_0 f_1\,(\rho) + c_1 f_0\,(\rho+1),$$
$$0 = c_0 f_2\,(\rho) + c_1 f_1\,(\rho+1) + c_2 f_0\,(\rho+2),$$

and so on. As c_0 is not zero, because it is the coefficient of the first term that occurs in y, the first of these relations shews that the values of ρ to be considered are given by

$$f_0(\rho) = 0,$$

which is a quadratic in ρ. The remaining relations give

$$c_1 = -c_0 \frac{f_1(\rho)}{f_0(\rho+1)},$$

$$c_2 = c_0 \frac{h_2(\rho)}{f_0(\rho+1) f_0(\rho+2)},$$

$$c_3 = c_0 \frac{h_3(\rho)}{f_0(\rho+1) f_0(\rho+2) f_0(\rho+3)},$$

and so on, where $f_1(\rho)$, $h_2(\rho)$, $h_3(\rho)$, ... are polynomials in ρ.

If the two roots of the quadratic $f_0(\rho) = 0$ are equal, then no denominator in the expressions for the successive coefficients c_n vanishes; these coefficients are finite, and the expression $g(x, \rho)$ is formally adequate for an integral*.

If the two roots of the quadratic are unequal, and their difference is not a whole number, then no denominator in the expressions for the successive coefficients c_n vanishes; these coefficients are finite and, for each of the values of ρ, the expression $g(x, \rho)$ is formally adequate for an integral.

If the two roots of the quadratic are unequal, and their difference be a whole number s, which will be assumed positive, the roots are of the form

$$\rho, \rho + s.$$

We then take

$$c_0 = f_0(\rho+1) f_0(\rho+2) \dots f_0(\rho+s) c,$$

and thus secure that no one of the coefficients c_n is infinite; moreover c_0 is undetermined, and therefore an arbitrary constant, so that c is an arbitrary constant. The expression $g(x, \rho)$, when c_0 is replaced by its modified value, is formally adequate for an integral.

---

* In order to render the proof of the method complete, it would be necessary to establish the convergence of the series $g(x, \rho)$ and, later, the convergence of allied series. For these, and corresponding omissions, we refer to the sources already quoted; the actual convergence of the series in particular examples will be manifest on inspection.

Thus far, the process coincides with the ordinary process; it is at this stage that modifications are introduced which give the distinctive character to the method of Frobenius. Instead of dealing with quantities ρ which are roots of a quadratic equation, we introduce a parametric quantity α; the quantities c_0 and c are regarded as arbitrary functions of α. Quantities c_1, c_2, ... are determined by the equations

$$0 = c_1 f_0(\alpha+1) + c_0 f_1(\alpha),$$
$$0 = c_2 f_0(\alpha+2) + c_1 f_1(\alpha+1) + c_0 f_2(\alpha),$$
$$\dots\dots\dots\dots$$

which in form are the same as the earlier set, if the first equation of that set is omitted and ρ is replaced by α. Moreover, we have

$$c_n(\alpha) = \frac{c_0(\alpha) h_n(\alpha)}{f_0(\alpha+1) f_0(\alpha+2) \dots f_0(\alpha+n)};$$

so that, if we choose α as different from a root of the quadratic $f_0(\rho) = 0$, the quantities $c_n(\alpha)$ are finite. We thus have an expression

$$z = g(t, \alpha) = \sum_{n=0} c_n t^{\alpha+n};$$

also

$$\begin{aligned}
D(z) &= \sum_{n=0} c_n D t^{\alpha+n} \\
&= \sum_{n=0} c_n t^{\alpha+n} f(t, \alpha+n) \\
&= \sum_{n=0} c_n t^{\alpha+n} \{ f_0(\alpha+n) + t f_1(\alpha+n) + t^2 f_2(\alpha+n) + \dots \} \\
&= \sum_{n=0} t^{\alpha+n} \{ c_n f_0(\alpha+n) + c_{n-1} f_1(\alpha+n-1) + \dots + c_0 f_n(\alpha) \}.
\end{aligned}$$

On account of the relations among the quantities c, the coefficient of $t^{\alpha+n}$ vanishes when $n = 1, 2, \dots$; and so

$$Dz = c_0 f_0(\alpha) t^{\alpha},$$

an equation satisfied by

$$z = g(t, \alpha),$$

where α is an arbitrary parametric quantity*.

We take the three possibilities in order.

* The series for z is one of the allied series indicated on p. 246, foot-note; its convergence is assumed in this exposition.

I. Let $f_0(\rho) = 0$ have a repeated root $\rho = \sigma$; then, as the coefficient of m^2 in $f_0(m)$ is unity, we have

$$f_0(\alpha) = (\alpha - \sigma)^2;$$

so that

$$Dz = c_0(\alpha - \sigma)^2 t^{\alpha}.$$

It is clear that the values of

$$z, \quad \frac{\partial z}{\partial \alpha},$$

when α is made equal to σ, give

$$Dz = 0, \quad D\frac{\partial z}{\partial \alpha} = 0;$$

in other words, two integrals of the equation

$$Dy = 0$$

are given by

$$y = [z]_{\alpha=\sigma}, \quad y = \left[\frac{\partial z}{\partial \alpha}\right]_{\alpha=\sigma}.$$

II. Let $f_0(\rho) = 0$ have two unequal roots $\rho = \sigma$ and $\rho = \tau$, where $\tau - \sigma$ is not a whole number: then

$$f_0(\alpha) = (\alpha - \tau)(\alpha - \sigma),$$

so that

$$Dz = c_0(\alpha - \tau)(\alpha - \sigma)t^{\alpha}.$$

It is clear that the value of z when α is made equal to τ, and its value when α is made equal to σ, give

$$Dz = 0;$$

in other words, two integrals of the equation

$$Dy = 0$$

are given by

$$y = [z]_{\alpha=\tau}, \quad y = [z]_{\alpha=\sigma}.$$

III. Let $f_0(\rho) = 0$ have two roots σ and $\sigma - s$, where s is a positive whole number greater than 0; then

$$f_0(\alpha) = (\alpha - \sigma)(\alpha - \sigma + s).$$

We now take

$$c_0 = f_0(\alpha + 1)f_0(\alpha + 2)\dots f_0(\alpha + s)c;$$

this makes the expression for z finite when the special values σ and $\sigma - s$ are assigned to α. Also

$$c_0 f_0(\alpha) = (\alpha - \sigma)(\alpha - \sigma + s)^2 A,$$

where A is a quantity that remains finite when α has either of the values σ and $\sigma - s$. Hence

$$Dz = (\alpha - \sigma)(\alpha - \sigma + s)^2 A t^a.$$

It is clear that the value of z when α is made equal to σ, the value of z when α is made equal to $\sigma - s$, and the value of $\dfrac{\partial z}{\partial \alpha}$ when α is made equal to $\sigma - s$, all give

$$Dz = 0 ;$$

in other words, there are apparently three integrals of the equation

$$Dy = 0$$

given by

$$y = [z]_{\alpha=\sigma},$$

$$y = [z]_{\alpha=\sigma-s}, \quad y = \left[\frac{\partial z}{\partial \alpha}\right]_{\alpha=\sigma-s}.$$

But consider

$$y = [z]_{\alpha=\sigma-s}.$$

We have

$$z = t^a \sum_{n=0} c_n t^n$$

$$= t^a \sum_{n=0}^{s-1} c_n t^n + t^a \sum_{n=s} c_n t^n$$

$$= t^a \sum_{n=0}^{s-1} c_n t^n + t^{a+s} \sum_{m=0} c_{s+m} t^m.$$

s regards the first sum, the coefficients c_n, for all the values

$$n = 0, 1, \ldots, s-1,$$

contain a factor $\alpha - \sigma + s$, because

$$c_n = \frac{c_0(\alpha) h_n(\alpha)}{f_0(\alpha+1)\ldots f_0(\alpha+n)}$$

$$= f_0(\alpha+n+1)\ldots f_0(\alpha+s) h_n(\alpha) c,$$

which, for $n = 0, \ldots, s-1$, contains $f_0(\alpha+s)$, and therefore also $\alpha + s - \sigma$, as a factor. Hence, when α is made equal to $\sigma - s$, the first sum vanishes. As regards the second sum, we write it in the full form

$$t^{a+s} c_s(\alpha) + t^{a+s+1} c_{s+1}(\alpha) + \ldots,$$

which, when α is made equal to $\sigma - s$, becomes

$$t^{\sigma} c_s (\sigma - s) + t^{\sigma+1} c_{s+1} (\sigma - s) + \dots,$$

a series that begins with t^{σ} and proceeds in ascending powers of t. But

$$y = [z]_{\alpha=\sigma}$$

is a series that begins with t^{σ} and proceeds in ascending powers of t; hence the second sum in $[z]_{\alpha=\sigma-s}$ is either another integral beginning with t^{σ}, or it is a constant multiple of $y = [z]_{\alpha=\sigma}$.

If it could be another integral, a combination

$$[z]_{\alpha=\sigma} - \lambda [z]_{\alpha=\sigma-s}$$

could be taken so that, by appropriate choice of the constant λ, the term in t^{σ} would vanish and the expression would begin with $t^{\sigma+1}$. It is a linear combination of two supposed integrals and therefore is an integral of the equation: so that, on the hypothesis assumed, we have an integral of the equation that begins with $t^{\sigma+1}$ and involves powers of t only. There is no such integral, for $\sigma + 1$ is not a root of $f_0(\rho) = 0$. Hence

$$y = [z]_{\alpha=\sigma-s}$$

is not an independent integral; it is a constant multiple (which may be a zero multiple) of

$$y = [z]_{\alpha=\sigma}.$$

Consequently, in the present case, two integrals of the equation $Dy = 0$ are given by

$$y = [z]_{\alpha=\sigma},$$

$$y = \left[\frac{\partial z}{\partial \alpha} \right]_{\alpha=\sigma-s}.$$

Summing up the results, we may state them in the following rule:—

The primitive of the differential equation

$$t^2 \frac{d^2 y}{dt^2} + pt \frac{dy}{dt} + qy = 0,$$

where p and q are uniform functions of t that remain finite when $t = 0$, can be obtained by constructing an expression

$$z = \Sigma c_n t^{\alpha+n}$$

such that all the terms in

$$t^2 \frac{d^2 z}{dt^2} + pt \frac{dz}{dt} + qz$$

vanish except only the term in t^a. If

$$t^2 \frac{d^2 z}{dt^2} + pt \frac{dz}{dt} + qz = c_0 t^a f_0(\alpha),$$

$f_0(\alpha)$ *being a quadratic function of α, there are three cases:*

I. *If the roots of $f_0(\rho) = 0$ be equal, and have σ as their common value, two linearly independent integrals are given by*

$$y = [z]_{a=\sigma}, \quad y = \left[\frac{\partial z}{\partial \alpha}\right]_{a=\sigma}.$$

II. *If the roots of $f_0(\rho) = 0$ be unequal and do not differ by an integer, and if they have σ and τ as their values, two linearly independent integrals are given by*

$$y = [z]_{a=\sigma}, \quad y = [z]_{a=\tau}.$$

III. *If the roots of $f_0(\rho) = 0$ be σ and $\sigma - s$, where s is a positive integer, the proper modification in c_0 must be made: and then two linearly independent integrals are given by*

$$y = [z]_{a=\sigma}, \quad y = \left[\frac{\partial z}{\partial \alpha}\right]_{a=\sigma-s}.$$

As the equation

$$f_0(\rho) = 0$$

determines the index of the initial term in the various expressions for y, it is often called the *indicial equation*; and the quantity $f_0(\rho)$ is often called the *indicial function*.

The general theory will now be illustrated by some particular examples.

Ex. 1. Consider the hypergeometric equation when $\gamma = 1$; it is

$$x(1-x) \frac{d^2 y}{dx^2} + \{1 - (a + \beta + 1)x\} \frac{dy}{dx} - a\beta y = 0.$$

Let D denote the operator

$$x(1-x) \frac{d^2}{dx^2} + \{1 - (a + \beta + 1)x\} \frac{d}{dx} - a\beta.$$

Taking

$$y = x^\rho \sum_{n=0}^{\infty} c_n x^n,$$

and the relations

$$(n+1+\rho)^2 c_{n+1} = (n + a + \rho)(n + \beta + \rho) c_n,$$

we have

$$y = c_0 x^\rho \left\{ 1 + \frac{(a+\rho)(\beta+\rho)}{(1+\rho)^2} x + \frac{(a+\rho)(a+\rho+1)(\beta+\rho)(\beta+\rho+1)^2}{(1+\rho)^2(2+\rho)^2} x^2 + \ldots \right\}$$

$$= c_0 x^\rho g(a, \beta, \rho, x),$$

say. We also have

$$Dy = x^{\rho-1} c_0 \rho^2.$$

Clearly

$$[y]_{\rho=0}, \quad \left[\frac{\partial y}{\partial \rho} \right]_{\rho=0},$$

are solutions.

In the present case, we can take c_0 unity. The integral $[y]_{\rho=0}$ gives

$$F(a, \beta, 1, x)$$

as one solution. The integral $\left[\dfrac{\partial y}{\partial \rho} \right]_{\rho=0}$ gives

$$F(a, \beta, 1, x) \log x$$
$$+ \frac{a\beta}{1^2} \left(\frac{1}{a} + \frac{1}{\beta} - \frac{2}{1} \right) x + \frac{a(a+1)\beta(\beta+1)}{1^2 \cdot 2^2} \left(\frac{1}{a} + \frac{1}{a+1} + \frac{1}{\beta} + \frac{1}{\beta+1} - \frac{2}{1} - \frac{2}{2} \right) x^2 + \ldots$$

as the other solution.

Ex. 2. Obtain the primitive of the hypergeometric equation, when γ is a negative integer.

Ex. 3. Consider Bessel's equation for functions of order zero, viz.

$$Dy = xy'' + y' + xy = 0.$$

It is not in the exact form of the general theory: to compare it with that form, we note that

$$p = x, \quad q = x^2:$$

an irrelevant factor x can be removed. We substitute

$$z = c_0 x^a + c_1 x^{a+1} + c_2 x^{a+2} + \ldots,$$

and choose relations among the constants c so that all terms other than those involving the lowest power of x vanish; and we find

$$x Dz = c_0 a^2 x^a,$$

provided

$$(a+1)^2 c_1 = 0,$$

$$(a+n+1)^2 c_{n+1} + c_{n-1} = 0,$$

the latter holding for $n = 1, 2, 3, \ldots$. When these relations are solved, and the values of the coefficients c are substituted, we have

$$z = c_0 x^a \left\{ 1 - \frac{x^2}{(a+2)^2} + \frac{x^4}{(a+2)^2(a+4)^2} - \ldots \right\}.$$

In the present instance, we have $f_0(a) = a^2$, so that the equation is included in the first case (p. 251): the two linearly independent integrals are

$$y_1 = [z]_{a=0}, \quad y_2 = \left[\frac{\partial z}{\partial a}\right]_{a=0}.$$

The first of these is

$$c_0 \left\{1 - \frac{x^2}{2^2} + \frac{x^4}{2^2 \cdot 4^2} - \dots\right\},$$

or, taking $c_0 = 1$, we have

$$y_1 = J_0(x).$$

For the second, we note that

$$\frac{\partial}{\partial a} \left\{\frac{c_0 x^{2p+a}}{(a+2)^2 (a+4)^2 \dots (a+2p)^2}\right\}$$

$$= \frac{\partial c_0}{\partial a} \frac{x^{2p+a}}{(a+2)^2 \dots (a+2p)^2} + c_0 \frac{x^{2p+a}}{(a+2)^2 \dots (a+2p)^2} \log x$$

$$- \frac{c_0 x^{2p+a}}{(a+2)^2 \dots (a+2p)^2} \left(\frac{2}{a+2} + \frac{2}{a+4} + \dots + \frac{2}{a+2p}\right);$$

and therefore, making $c_0 = 1$, we obtain the second integral in the form

$$y_2 = \left(1 - \frac{x^2}{2^2} + \frac{x^4}{2^2 \cdot 4^2} - \dots\right) \log x$$

$$+ \frac{x^2}{2^2} - \frac{x^4}{2^2 \cdot 4^2}(1 + \tfrac{1}{2}) + \frac{x^6}{2^2 \cdot 4^2 \cdot 6^2}(1 + \tfrac{1}{2} + \tfrac{1}{3}) - \dots$$

$$= J_0 \log x + \sum_{p=1}^{\infty} \frac{(-1)^{p-1}}{\{\Pi(p)\}^2} \left(\frac{x}{2}\right)^{2p} \psi(p),$$

where

$$\psi(p) = 1 + \tfrac{1}{2} + \dots + \frac{1}{p} = \left[\frac{\Pi'(z)}{\Pi(z)}\right]_{z=p}.$$

The expression for y_2 is usually denoted by $K_0(x)$, so that

$$y_2 = K_0(x);$$

the primitive of the equation is

$$y = A J_0(x) + B K_0(x).$$

Ex. 4. Consider Bessel's equation for functions of order n, viz.

$$Dy = x^2 y'' + xy' + (x^2 - n^2) y = 0.$$

Constructing an expression

$$z = c_0 x^a + c_1 x^{a+1} + c_2 x^{a+2} + \dots,$$

we have

$$Dz = c_0(a^2 - n^2) x^a,$$

provided

$$\{(a+1)^2 - n^2\} c_1 = 0,$$

$$\{(a+m)^2 - n^2\} c_m + c_{m-2} = 0,$$

the latter holding for $m = 1, 2, 3, \ldots$. When these relations are solved, and the values of the coefficients c are substituted, we have

$$z = c_0 x^a \left[1 - \frac{x^2}{(a+2)^2 - n^2} + \frac{x^4}{\{(a+2)^2 - n^2\}\{(a+4)^2 - n^2\}} - \cdots \right].$$

In the present instance, we have

$$f_0(a) = a^2 - n^2 = (a - n)(a + n),$$

so that the equation is included* in the second case (p. 251) when n is not a whole number, and it is included in the third case (p. 251) when n is a whole number other than zero.

First, suppose that n is not a whole number. Then, when $a = n$, take

$$c_0 = \frac{1}{2^n \Pi(n)};$$

and when $a = -n$, take

$$c_0 = \frac{1}{2^{-n} \Pi(-n)}.$$

The two integrals are

$$y_1 = [z]_{a=n}$$

$$= \sum_{r=0}^{\infty} \frac{(-1)^r}{\Pi(n+r)\Pi(r)} \left(\frac{x}{2}\right)^{n+2r} = J_n,$$

and

$$y_2 = [z]_{a=-n}$$

$$= \sum_{r=0}^{\infty} \frac{(-1)^r}{\Pi(-n+r)\Pi(r)} \left(\frac{x}{2}\right)^{-n+2r} = J_{-n};$$

and the primitive is

$$y = A J_n + B J_{-n}.$$

Secondly, suppose that n is a whole number which will be taken to be the positive square root of n^2, the quantity that occurs in the differential equation.

One integral is given by

$$y_1 = [z]_{a=n};$$

taking c_0 to be $\dfrac{1}{2^n \Pi(n)}$, we have

$$y_1 = \sum_{r=0}^{\infty} \frac{(-1)^r}{\Pi(n+r)\Pi(r)} \left(\frac{x}{2}\right)^{n+2r} = J_n.$$

---

* Strict comparison with the general case shews that the critical value arises when $2n$ is a whole number. But the vanishing denominator factor arises through a quantity of the form $(a+2p)^2 - n^2$, where p is a whole number: so that, whether a be $+n$ or $-n$, the factor can vanish only when n (and not merely $2n$) is a whole number.

The second integral arises when $a = -n$. One of the coefficients becomes formally infinite through the occurrence of a denominator factor $(a+2n)^2 - n^2$; accordingly, we write

$$c_0 = C\{(a+2n)^2 - n^2\},$$

$$(-1)^n C = E \prod_{r=1}^{n-1} \{(a+2r)^2 - n^2\};$$

and then

$$z = C\{(a+2n)^2 - n^2\} x^a \left[1 - \frac{x^2}{(a+2)^2 - n^2} + \ldots + (-1)^{n-1} \frac{x^{2n-2}}{\prod\limits_{r=1}^{n-1} \{(a+2r)^2 - n^2\}} \right]$$

$$+ E x^{a+2n} \left[1 - \frac{x^2}{(a+2n+2)^2 - n^2} + \frac{x^4}{\{(a+2n+2)^2 - n^2\} \{(a+2n+4)^2 - n^2\}} - \ldots \right]$$

$$= z_1 + z_2,$$

say, where z_1 denotes the first line and z_2 the second line in the expression for z. And now

$$Dz = C (a^2 - n^2) \{(a+2n)^2 - n^2\} x^a$$

$$= C (a-n)(a+n)^2 (a+3n) x^a.$$

Two integrals are given by $a = -n$; they are

$$[z]_{a=-n}, \quad \left[\frac{\partial z}{\partial a}\right]_{a=-n},$$

and the former, by the general theory, is to be a constant multiple of $[z]_{a=n}$. Now

$$[z_1]_{a=-n} = 0,$$

on account of the factor $a+n$ in each term of z_1; and

$$[z_2]_{a=-n} = \frac{E}{2^n \Pi(n)} J_n,$$

so that

$$[z]_{a=-n} = [z_1]_{a=-n} + [z_2]_{a=-n} = \frac{E}{2^n \Pi(n)} J_n;$$

and thus it provides no new integral.

For the other, we have

$$\left[\frac{\partial z}{\partial a}\right]_{a=-n} = \left[\frac{\partial z_1}{\partial a}\right]_{a=-n} + \left[\frac{\partial z_2}{\partial a}\right]_{a=-n}$$

Now

$$\left[\frac{\partial z_1}{\partial a}\right]_{a=-n} = \frac{1}{x^n} \frac{2Cn}{\Pi(n-1)} \sum_{m=0}^{n-1} \left(\frac{x}{2}\right)^{2m} \frac{\Pi(n-1-m)}{\Pi(m)} = \zeta_1,$$

say; and

$$\left[\frac{\partial z_2}{\partial a}\right]_{a=-n} = E x^n \log x \left[1 - \frac{x^2}{2^2 (n+1)} + \frac{x^4}{2^4 (n+1)(n+2) . 1 . 2} - \ldots \right]$$

$$+ \tfrac{1}{2} E x^n \sum_{r=0}^{\infty} \frac{(-1)^{r-1} \Pi(n)}{\Pi(r) \Pi(n+r)} \{\psi(r) + \psi(n+r) - \psi(n)\} \left(\frac{x}{2}\right)^{2r}$$

$$= \zeta_2,$$

say: so that the integral is $\zeta_1 + \zeta_2$.

In ζ_2, the part represented by

$$\tfrac{1}{2}Ex^n \sum_{r=0}^{\infty} \frac{(-1)^r \Pi(n)\,\psi(n)}{\Pi(r)\,\Pi(n+r)} \left(\frac{x}{2}\right)^{2r}$$

is a constant multiple of J_n; it therefore can be omitted, for J_n has been retained and any linear combination of integrals with constant coefficients is an integral. Rejecting this part, and choosing

$$C = -\frac{1}{n}\,2^{n-1}\Pi(n-1),$$

so that

$$E = \frac{1}{2^{n-1}\Pi(n)},$$

we have the second integral in the form

$$y_2 = -\left(\frac{2}{x}\right)^n \sum_{m=0}^{n-1} \frac{\Pi(n-1-m)}{\Pi(m)} \left(\frac{x}{2}\right)^{2m}$$
$$+ \left(\frac{x}{2}\right)^n \sum_{r=0}^{\infty} \frac{(-1)^r}{\Pi(r)\,\Pi(n+r)} \{2\log x - \psi(r) - \psi(n+r)\} \left(\frac{x^{2r}}{2}\right),$$

which differs only by a constant multiple of J_n from the expression given in § 105, Ex. 1.

The primitive is

$$AJ_n + By_2.$$

Ex. 5. Integrate the equation

$$Dy = x(2-x^2)\,y'' - (x^2+4x+2)\{(1-x)\,y' + y\} = 0$$

by means of integrals proceeding in powers of x. (The equation can be integrated by quadratures: it is here used to illustrate the method of Frobenius.)

We take

$$z = \sum_{n=0}^{\infty} c_n x^{a+n},$$

and easily find that

$$x\,Dz = 2a(a-2)\,c_0 x^a,$$

provided

$$0 = c_1(a^2-1) - c_0(a+1),$$
$$0 = 2c_2 a(a+2) - 2c_1(a+2) - c_0(a-2)^2,$$

and

$$2(n+a+1)\{(n+a-1)\,c_{n+1} - c_n\} = (n+a-3)\{(n+a-3)\,c_{n-1} - c_{n-2}\},$$

the last holding for $n = 2, 3, \ldots$.

The equation $f_0(\rho) = 0$ here is

$$\rho(\rho-2) = 0,$$

so that the two values of a are $2, 0$. When a is made equal to 0, the coefficient of c_2 vanishes as the equations stand; so that we write $c_0 = Ca$, and then

$$c_1 = \frac{Ca}{a-1},$$

$$c_2 = \tfrac{1}{2} \frac{Ca(a^2 - 5a + 10)}{a^2 + a - 2},$$

$$c_3 = \frac{c_2}{a+1},$$

$$(a+2)c_4 = c_3 + a^2 A,$$

where A is finite, and so on. Thus

$$z = Cax^a \left(1 + \frac{x}{a-1}\right) + c_2 x^{a+2} \left\{1 + \frac{x}{a+1} + \frac{x^2}{(a+1)(a+2)} + \ldots\right\} + a^2 R(x, a),$$

where $R(x, a)$ is a regular function of x and a, which remains finite if $a = 0$ or $a = 2$. For this value of z, we have

$$xDz = 2a^2(a-2) Cx^a.$$

First, let $a = 2$; then an integral is given by $[z]_{a=2}$, which is

$$2Cx^2\left(1 + x + \frac{x^2}{2!} + \frac{x^3}{3!} + \ldots\right);$$

or, making $2C = 1$, an integral is given by

$$y_1 = x^2 e^x.$$

Next, let $a = 0$; then integrals are given by

$$[z]_{a=0}, \quad \left[\frac{\partial z}{\partial a}\right]_{a=0}.$$

Taking account of the value of c_2, the former of these is evanescent, and the latter is

$$C(1 - x) - \tfrac{5}{2} Cx^2 e^x;$$

or, adding $\tfrac{5}{2}Cy_1$ and making $C = 1$, we have an integral

$$y_2 = 1 - x.$$

The primitive of the equation is

$$y = Ax^2 e^x + B(1 - x).$$

Ex. 6. Integrate the equations:

  (i)　$x^2(1+x)y'' - (1+2x)(xy' - y) = 0$;

  (ii)　$x^3(1+x)y''' - (2+4x)x^2y'' + (4+10x)xy' - (4+12x)y = 0$;

  (iii)　$x^3(1+x^2)y''' - (2+4x^2)x^2y'' + (4+10x^2)xy' - (4+12x^2)y = 0$;

  (iv)　$2(2-x)x^2y'' - (4-x)xy' + (3-x)y = 0$;

  (v)　$(1-x)x^2y'' + (5x-4)xy' + (6-9x)y = 0$;

  (vi)　$xy'' + (4x^2+1)y' + 4x(x^2+1)y = 0$;

obtaining integrals in each case in series of powers of x.

Ex. 7. Discuss the integrals of the equation

$$x(1-x)y'' + \{1 - (a+b+1)x\}y' - aby = 0,$$

which proceed in powers of $1-x$; indicating their form in particular when $a+b=1$.

Shew that the above equation includes Legendre's equation

$$(1-z^2)y'' - 2zy' + p(p+1)y = 0,$$

the relation between the independent variables being

$$z = 1 - 2x;$$

and hence obtain an expression for the primitive of Legendre's equation.

Ex. 8. Integrate the equations:

    (i) $x^2y'' + 4(x+a)y = 0;$

    (ii) $xy'' + (1+ax^2)y' + bxy = 0;$

where a and b are constants.

II.

EQUATIONS HAVING ALL THEIR INTEGRALS REGULAR.

This note is restricted, for the sake of simplicity, to linear equations of the second order. For the discussion of the corresponding questions affecting linear equations of any order, reference may be made to my *Theory of Differential Equations*, vol. IV., Chapters III., IV., VI., VII.

We have had numerous examples of linear equations of the second order in which the integrals belonged to one or other of two types. One type was represented by a power-series, say $R(x)$. The other type was represented by an expression involving power-series and a logarithm. Thus for Legendre's equation, in varying circumstances, we have had solutions given by

    (i) P_n and Q_n,

    (ii) Q_n and $Q_n \log x + R_n$,

    (iii) P_n and $P_n \log x + S_n$,

where P_n, Q_n, R_n, S_n, are power-series. Similarly for Bessel's equation, in ascending powers of x; and for the hypergeometric equation. Such integrals are called *regular*.

Some equations have all their integrals regular, whatever be the variable of expansion in the power-series. Thus all the integrals of the hypergeometric equation are regular, whether they proceed in powers of x, or $1 - x$, or $1/x$, or $1/(1 - x)$, or $x/(x-1)$, or $(x-1)/x$. All the integrals of the Legendre equation are regular. The integrals of the Bessel equation are regular when expanded in ascending powers of x; but (Ex. 2, § 105) they are not regular when expanded in descending powers of x, for the power-series in $1/x$ are multiplied by e^{xi} or e^{-xi}.

It follows that we may expect to obtain a special class of equations, characterised by the property that all their integrals (in whatever expansions they are taken) are of the type called regular. We proceed to obtain their general form, when we deal only with equations of the second order, represented by

$$\frac{d^2 y}{dx^2} + P \frac{dy}{dx} + Qy = 0,$$

where we shall assume that P and Q are uniform functions of x.

In the first place, if the equation has an integral

$$y = u \log (x - a) + v,$$

where u and v are series of powers in $x - a$, then

$$y = u$$

is also an integral. Let the former be substituted in the equation; then

$$\left(\frac{d^2 u}{dx^2} + P \frac{du}{dx} + Qu\right) \log (x - a)$$

$$+ P \left(\frac{dv}{dx} + \frac{u}{x - a}\right) + Qv + \frac{2}{x - a} \frac{du}{dx} - \frac{1}{(x - a)^2} u = 0,$$

which must be satisfied identically when the power-series for u and v are substituted. The only way, in which the terms involving $\log (x - a)$ can disappear, is by having

$$\frac{d^2 u}{dx^2} + P \frac{du}{dx} + Qu = 0$$

identically—which shews that $y = u$ is an integral of the equation.

Accordingly, we have two cases to consider, viz. (i) when there are two linearly independent integrals in the form of power-series, say

$$y_1 = u, \quad y_2 = w,$$

where u and w are series in ascending powers of x; (ii) when there are two linearly independent integrals

$$y_1 = u, \quad y_2 = u \log x + v,$$

where u and v are series in ascending powers of x.

In case (i), let the first term in u be $a_0 x^m$ and the first term in w be $c_0 x^n$. Without loss of generality, we can assume that m and n are unequal; for if they were equal, we should substitute

$$c_0 u - a_0 w$$

for either u or w, and the requirement would be satisfied.

In case (ii), let the first term in u be $a_0 x^m$ and the first term in v be $b_0 x^l$. Again, without loss of generality, we can assume that m and l are unequal; if they were equal, we should substitute

$$a_0 y_2 - b_0 y_1$$

for y_2, and then the requirement would be satisfied.

For both cases, we have

$$\frac{d^2 y_1}{dx^2} + P \frac{dy_1}{dx} + Q y_1 = 0,$$

$$\frac{d^2 y_2}{dx^2} + P \frac{dy_2}{dx} + Q y_2 = 0;$$

and therefore

$$\left(y_2 \frac{dy_1}{dx} - y_1 \frac{dy_2}{dx} \right) P = -\left(y_2 \frac{d^2 y_1}{dx^2} - y_1 \frac{d^2 y_2}{dx^2} \right),$$

$$\left(y_2 \frac{dy_1}{dx} - y_1 \frac{dy_2}{dx} \right) Q = -\left(\frac{dy_1}{dx} \frac{d^2 y_2}{dx^2} - \frac{dy_2}{dx} \frac{d^2 y_1}{dx^2} \right).$$

CASE I. Taking the expansions in ascending powers of x, we have

$$y_2 \frac{dy_1}{dx} - y_1 \frac{dy_2}{dx} = a_0 c_0 (m - n) x^{m+n-1} + \ldots,$$

$$y_2 \frac{d^2 y_1}{dx^2} - y_1 \frac{d^2 y_2}{dx^2} = a_0 c_0 (m - n)(m + n - 1) x^{m+n-2} + \ldots,$$

$$\frac{dy_1}{dx} \frac{d^2 y_2}{dx^2} - \frac{dy_2}{dx} \frac{d^2 y_1}{dx^2} = -a_0 c_0 mn (m - n) x^{m+n-3} + \ldots.$$

Now $m - n$ is not zero; hence

$$xP = -\frac{m + n - 1 + \text{powers of } x}{1 + \text{powers of } x},$$

that is, when P is expanded in powers of x, it may contain a term in $\dfrac{1}{x}$ as its lowest term; otherwise it will contain no negative power. Again,

$$x^2 Q = \frac{mn + \text{powers of } x}{1 + \text{powers of } x},$$

that is, when Q is expanded in powers of x, it may contain a term in $\dfrac{1}{x^2}$ as its lowest term, and also a term in $\dfrac{1}{x}$. Clearly P contains no term with a negative index, if $m + n = 1$; and Q does not contain the term in $\dfrac{1}{x^2}$, if m or n is zero.

CASE II. We have

$$\frac{d^2 u}{dx^2} + P \frac{du}{dx} + Qu = 0,$$

$$\frac{d^2 v}{dx^2} + \frac{2}{x} \frac{du}{dx} - \frac{u}{x^2} + P\left(\frac{dv}{dx} + \frac{u}{x}\right) + Qv = 0.$$

There are two sub-cases, according as $m < l$ or $m > l$.

We proceed as above. When $m < l$, the lowest term in the expansion of P is

$$-\frac{2m - 1}{x},$$

and the lowest term in the expansion of Q is

$$\frac{m^2}{x^2}.$$

When $m > l$, the lowest term in the expansion of P is

$$-\frac{l + m - 1}{x},$$

and the lowest term in the expansion of Q is

$$\frac{lm}{x^2}.$$

The forms of P and Q, as regards the index of the lowest power of x which they contain, are the same as before; the quantity xP contains no power of x with a negative index, and the quantity $x^2 Q$ contains no power of x with a negative index.

Hence for all the cases it follows that, *when the integrals of the equation*

$$\frac{d^2y}{dx^2} + P\frac{dy}{dx} + Qy = 0$$

are expanded in ascending powers of x, *and are of the type called regular for that expansion,* xP *and* x^2Q *must be devoid of terms with negative indices when they are expanded in ascending powers of* x. But xP does not necessarily contain a term in x^0, and x^2Q does not necessarily contain terms in x^0 and x^1.

Next, consider expansions in powers of $x - a$, where a is any finite constant; thus $a = 1$ for the hypergeometric equation. Writing $x - a = t$, we have the differential equation in the form

$$\frac{d^2y}{dt^2} + P'\frac{dy}{dt} + Q'y = 0,$$

where now P' and Q' are uniform functions of t. In order that the integrals, when expressed in powers of t and of a possible $\log t$, may be regular, tP' and t^2Q' must be devoid of terms with negative indices when they are expanded in ascending powers of t. Consequently, when P and Q are expanded in ascending powers of $x - a$, it is necessary that $(x - a)P$ and $(x - a)^2Q$ should contain no terms with negative indices, if the integrals of the equation are to be regular so far as concerns expansion in powers of $x - a$.

Further, consider expansions valid for very large values of x. We change the independent variable from x to z, where $xz = 1$; and we consider expansions in ascending powers of z. The equation becomes

$$\frac{d^2y}{dz^2} + P''\frac{dy}{dz} + Q''y = 0,$$

where

$$P'' = -\frac{1}{z^2}P + \frac{2}{z}, \quad Q'' = \frac{1}{z^4}Q.$$

In order that the integrals of the equation, when expressed in ascending powers of z and of a possible $\log z$, may be regular, zP'' and z^2Q'' must be devoid of terms with negative indices when they are expanded in ascending powers of z. Consequently, when P and Q are expanded in ascending powers of z, P must begin with a term involving at least the first power of z, and Q must begin with a term involving at least the second power of z. Therefore, also, when P and Q are expanded in descending powers of x, P must begin with a term involving at least the first power of $1/x$, and Q must begin with a term involving at least the second power of $1/x$.

We have considered the forms of P and Q for all the expansions that are to occur: the results may be gathered together so as to provide the complete explicit forms of P and Q. Let $a_1, a_2, \ldots, a_n$, be all the different finite values of x (including zero) that need to be considered; and write

$$T = (x - a_1)(x - a_2) \ldots (x - a_n).$$

Then, for expansions in ascending powers of $x - a_r$, the product TP contains no term in $x - a_r$ with a negative index; and this holds for each of the quantities $a_1, \ldots, a_n$. Hence TP will contain only terms with positive powers for each expansion; so denoting this quantity by M, we have

$$TP = M.$$

Similarly, the product T^2Q will contain only terms with positive powers for each expansion; so denoting this quantity by N, we have

$$T^2Q = N.$$

When the integrals are to be regular for large values of x, we imagine P and Q expanded in descending powers of x; and then P must begin at least with a term in $\dfrac{1}{x}$, while Q must begin at least with a term in $\dfrac{1}{x^2}$. Now

$$P = \frac{M}{T}, \quad Q = \frac{N}{T^2},$$

and T is of degree n in x. Now P is not to have any positive power of x and not to have any constant term; hence M is a polynomial in x of degree not greater than $n - 1$. Similarly, Q is not to have any positive power of x, no constant term, and no term in $\dfrac{1}{x}$; hence N is a polynomial in x of degree not greater than $2n - 2$.

We thus have the result:—

The most general form of linear equation of the second order, which has its integrals everywhere of the type called regular, is

$$\frac{d^2y}{dx^2} + \frac{M}{T}\frac{dy}{dx} + \frac{N}{T^2}y = 0,$$

where T is a polynomial in x of any degree n having no equal roots, while M and N are polynomials in x of degrees not greater than $n-1$ and $2n-2$ respectively.

These equations were first discussed by Fuchs* by a different method, and for any order; so they are often said to be of *Fuchsian type.*

It follows, from the theorem which has just been proved and from the analysis used in the proof, that, when an equation

$$\frac{d^2y}{dx^2} + P\frac{dy}{dx} + Qy = 0$$

is given, where P and Q are rational functions of x, we can at once settle by inspection what are the expansions that give regular integrals and what are the expansions that do not give regular integrals.

The regular integrals would be constructed by the method of Frobenius.

Ex. 1. Legendre's equation

$$(1-x^2)\frac{d^2y}{dx^2} - 2x\frac{dy}{dx} + p(p+1)y = 0$$

is of Fuchsian type. For $T = 1 - x^2$, having unequal roots; also $n=2$. Now $M = -2x$, so that its degree is not greater than $n-1$; and

$$N = p(p+1)T = p(p+1)(1-x^2),$$

so that its degree is not greater than $2n-2$. All the conditions are satisfied: the integrals are everywhere regular.

Similarly all the conditions are satisfied for the hypergeometric equation

$$x(1-x)\frac{d^2y}{dx^2} + \{\gamma - (a+\beta+1)x\}\frac{dy}{dx} - a\beta y = 0;$$

its integrals are everywhere regular.

Ex. 2. Consider Bessel's equation

$$\frac{d^2y}{dx^2} + \frac{1}{x}\frac{dy}{dx} + \frac{(x^2-m^2)}{x^2}y = 0.$$

Here $T = x$, with a single root, and $n = 1$.

* *Crelle*, t. LXVI. (1866), pp. 139—154.

The quantity M is unity. As regards expansions in ascending powers of x, the condition for M is satisfied; and for expansions in descending powers of x, the condition for M is satisfied.

The quantity N is $x^2 - m^2$. For expansions in ascending powers of x, we take $N = -m^2 + x^2$, and the condition is satisfied. For expansions in descending powers of x, the degree of the polynomial N should be not greater than $2(1-1)$, while it actually is equal to 2, so that the condition is not satisfied.

The integrals of Bessel's equation are regular for expansions in ascending powers of x; they are not regular for expansions in descending powers of x.

Ex. 3. Shew that the equation

$$\frac{d^2 y}{dx^2} + \frac{2x^2 + ax + \beta}{T}\frac{dy}{dx} + \frac{a'x^2 + \beta'x + \gamma'}{T^2}\, y = 0,$$

where

$$T = (x - a)(x - b)(x - c),$$

no two of the three constants a, b, c, being equal to one another, and a, β, a', β', γ', denoting constants, has all its integrals regular. Obtain these integrals, in expansions in $x - a$, $x - b$, $x - c$, $1/x$, respectively.

Ex. 4. Discuss the preceding equation for the cases

$$\text{(i)} \quad a = b \neq c;$$
$$\text{(ii)} \quad a = b = c;$$

obtaining in each case such integrals as are regular.

III.

EQUATIONS HAVING SOME (BUT NOT ALL) INTEGRALS REGULAR.

We have seen that an equation (such as Bessel's, for example) may satisfy the conditions that all the integrals may be regular for one expansion and not all be regular for another.

The question then arises; if not all the integrals are regular for one expansion, are there any that are regular for that expansion?

For simplicity, we shall consider an expansion in ascending powers of x; and we shall, as before, represent the equation by

$$Dy = y'' + Py' + Qy = 0.$$

When all the integrals are regular, the coefficients P and Q have the form

$$P = \frac{p_0}{x} + p_1 + p_2 x + \dots,$$

$$Q = \frac{q_0}{x^2} + \frac{q_1}{x} + q_2 + q_3 x + \dots;$$

and the index ρ for a regular integral

$$y = c_0 x^\rho + c_1 x^{\rho+1} + \ldots,$$

obtained by the Frobenius method with the relation

$$Dy = c_0 x^{\rho-2} \{\rho(\rho-1) + p_0\rho + q_0\},$$

is a root (p. 251) of the indicial equation

$$\rho(\rho-1) + p_0\rho + q_0 = 0,$$

which is a quadratic. The method of proceeding has already been explained.

Now suppose that not all the integrals are regular; thus either P, or Q, or both P and Q, must cease to have the foregoing form. We still contemplate the possibility of an integral of the form

$$y = c_0 x^\rho + c_1 x^{\rho+1} + \ldots.$$

First, let P have the same form as before, and let

$$Q = \frac{q_0}{x^3} + \frac{q_1}{x^2} + \frac{q_2}{x} + q_3 + q_4 x + \ldots;$$

then, when we use the Frobenius method for the construction of the integral, we have

$$Dy = c_0 x^{\rho-3} q_0.$$

The indicial equation is

$$q_0 = 0:$$

that is, there is no index ρ, and therefore no regular integral.

Next, let Q have the earlier form, and let

$$P = \frac{p_0}{x^2} + \frac{p_1}{x} + p_2 + p_3 x + \ldots;$$

when we use the Frobenius method for the construction of the integral, we have

$$Dy = c_0 x^{\rho-3} p_0 \rho,$$

so that the indicial equation is

$$p_0 \rho = 0.$$

Thus there is one index, viz. $\rho = 0$. The equation *may* have *one* regular integral; so we should construct the expression by the Frobenius method and, in any case where the series obtained is a converging series, we then do obtain one regular integral.

Next, let neither P nor Q have the earlier form; let them be

$$P = \frac{p_0}{x^2} + \frac{p_1}{x} + p_2 + p_3 x + \dots,$$

$$Q = \frac{q_0}{x^3} + \frac{q_1}{x^2} + \frac{q_2}{x} + q_3 + q_4 x + \dots.$$

When we use the Frobenius method for the construction of the integral, we have

$$Dy = c_0 x^{\rho-3} (p_0 \rho + q_0),$$

so that the indicial equation is

$$p_0 \rho + q_0 = 0.$$

There is one index, viz. $\rho = -q_0/p_0$. The equation may, or may not, have one regular integral; we proceed as in the last case.

So for other forms of P and Q. What is done in every case is to take the initial stage in the Frobenius method, so as to obtain the indicial equation.

If the indicial equation has one root, the differential equation may have one regular integral. We continue the Frobenius method; according as it does or does not give a significant expression in the form of a converging series, we infer the existence or the non-existence of one integral which is regular for the expansion.

If the indicial equation has no root, the differential equation has no integral which is regular for the expansion.

Ex. 1. Has the equation

$$y'' + \frac{1}{x^2}(2+x)y' - \frac{4}{x^2}y = 0$$

any regular integral proceeding in ascending powers of x ?

When we construct the indicial equation for a possible integral $y = c_0 x^\rho + \dots$, we find it in the form

$$2c_0 \rho = 0;$$

that is, there is one possible index, given by $\rho = 0$. So we substitute

$$y = c_0 + c_1 x + c_2 x^2 + \dots$$

as the expression of a regular integral, if it exists; and we proceed to make it satisfy the equation formally. The law among the coefficients is

$$(n^2 - 4) c_n + 2 (n+1) c_{n+1} = 0;$$

so that

$$c_1 = 2c_0, \quad c_2 = \tfrac{3}{4}c_1, \quad c_3 = 0, \quad c_4 = 0, \dots.$$

There is one regular integral, given by

$$y = c_0 (1 + 2x + \tfrac{3}{2}x^2).$$

Ex. 2. Has the equation

$$y'' + \frac{1}{x^2} y' - \frac{1}{x^2}(3+2x)\, y = 0$$

any regular integral proceeding in ascending powers of x ?

When we construct the indicial function for a possible integral $y = c_0 x^\rho + \dots$ we find it in the form

$$c_0\,(\rho - 3) = 0\,;$$

that is, there is one possible index, given by $\rho = 3$. So we substitute

$$y = c_0 x^3 + c_1 x^4 + c_2 x^5 + \dots$$

as the expression of a regular integral, if it exists; and, as before, we proceed to make it satisfy the equation formally. The law among the coefficients is

$$c_{n+1} = -(n+4)\, c_n,$$

so that

$$c_1 = -4c_0, \quad c_2 = -5c_1 = 20c_0, \quad \dots.$$

The series diverges, and thus is not a significant expression. The differential equation has no regular integral, proceeding in ascending powers of x.

Ex. 3. Discuss the following equations, in regard to their possession of regular integrals, proceeding in ascending powers of x :—

(i) $y'' + \dfrac{1}{x^2}(a+x)\, y' - \dfrac{b}{x^2}\, y = 0,$

where a and b are constants ;

(ii) $y'' + \dfrac{1}{x^2}(4a-x)\, y' - \dfrac{1}{x^3}\left(6a + \tfrac{5}{4}x\right) y = 0,$

where a is a constant ;

(iii) $y'' + \dfrac{1}{x^4}\, y' - \dfrac{1}{x^5}(5+2x^3)\, y = 0 ;$

(iv) $x^2 y'' - (1 - 2x + 2x^2)\, y' + (1-x)^2\, y = 0 ;$

(v) $x^2 y'' + (3x-1)\, y' + y = 0.$

In every case when the equation possesses a regular integral (which we denote by y_1), we can obtain an expression for the primitive of the equation. Proceeding as in § 58, we obtain this primitive in the form

$$y = A y_1 + B y_1 \int \frac{dx}{y_1{}^2} e^{-\int P\,dx},$$

where A and B are arbitrary constants.

IV.

EQUATIONS HAVING NO REGULAR INTEGRALS.

The question next arises as to whether an equation, which possesses no integrals of the type called regular, still possesses integrals of a recognised type or types in connection with which expansions can be used. As in the preceding discussion, we shall limit ourselves to expansions in ascending powers of x.

If a linear equation of the first order

$$\frac{dy}{dx} + Ry = 0$$

possesses a regular integral

$$y = c_0 \left(x^\rho + c_1 x^{\rho+1} + c_2 x^{\rho+2} + \dots \right),$$

then

$$R = -\frac{\rho}{x} - \frac{c_1 + 2c_2 x + \dots}{1 + c_1 x + c_2 x^2 + \dots}.$$

Hence, when expanded in ascending powers of x, R is of the form

$$R = -\frac{\rho}{x} + a_1 + a_2 x + a_3 x^2 + \dots,$$

that is, R may contain a term with index -1, but it cannot contain a term with an index lower than -1.

If then, for a given linear equation as above, the expansion of R in ascending powers of x is of the form

$$R = \frac{sk_s}{x^{s+1}} + \frac{(s-1)\,k_{s-1}}{x^s} + \dots + \frac{k_1}{x^2} - \frac{\rho}{x} + a_1 + a_2 x + a_3 x^2 + \dots\dots,$$

the integral of the equation will not be regular, in the previous sense of the term. Take

$$\Omega = \frac{k_1}{x} + \frac{k_2}{x^2} + \dots + \frac{k_s}{x^s},$$

$$\psi(x) = e^{-a_1 x - \frac{1}{2}a_2 x^2 - \frac{1}{3}a_3 x^3 - \dots},$$

so that Ω is a polynomial in $\dfrac{1}{x}$, and $\psi(x)$ can be expanded in a converging series of powers of x. We have

$$y = A e^{\Omega} x^{\rho} \psi(x),$$

where A is the arbitrary constant.

The expression for y in this integral consists of two factors. One of them is $x^{\rho} \psi(x)$, an expression of the type which has been called regular; it can be expanded in ascending powers of x. The other of them is e^{Ω}, which cannot be expanded* in ascending powers of x. Accordingly, we have an integral, quite definite in type, and quite distinct in type from regular integrals. It is customary to call such an integral *normal*.

Hence, for equations of the second order which have no regular integrals for the expansions in ascending powers of x, we proceed to enquire whether an equation

$$y'' + Py' + Qy = 0$$

has a normal integral. If it does possess such an integral, say in a form

$$y = e^{\Omega} u,$$

where Ω is a polynomial in $1/x$, the equation for u must possess a regular integral. Now, on substitution, we find that the equation satisfied by u is

$$u'' + P_1 u' + Q_1 u = 0,$$

where

$$P_1 = P + 2\Omega', \quad Q_1 = Q + P\Omega' + \Omega'' + \Omega'^2.$$

There are two ways of proceeding at this stage; but, in each of the ways, the effectiveness depends upon the forms of P and Q.

Firstly, suppose that Ω' can be chosen so that all the terms, in the expansion of P_1, in ascending powers of x and having indices in those powers less than -1, disappear. Let P_0 denote the remaining terms in P_1, so that P_0 is of the form

$$\frac{b_0}{x} + b_1 + b_2 x + b_3 x^2 + \ldots.$$

---

* The value $x = 0$ gives what, in the theory of functions, is called an *essential singularity* of the function e^{Ω}.

Let Q_0 denote the resulting value of Q_1 when this value of Ω' is inserted; and let the expansion of Q_0, in ascending powers of x, be

$$\frac{\alpha_s}{x^s} + \frac{\alpha_{s-1}}{x^{s-1}} + \dots + \frac{\alpha_1}{x} + \alpha_0 + \beta_1 x + \beta_2 x^2 + \dots.$$

If $s > 2$, the indicial equation possesses no root; the differential equation in u, for this determination of Ω, possesses no regular integral. But if $s \leqslant 2$, the indicial equation possesses two roots; the differential equation in u then possesses two regular integrals, which can be obtained by the customary Frobenius method. In the latter event, the original equation in y possesses two normal integrals.

Secondly, if the preceding way is not effective, suppose that the coefficients, and the initial index of the highest negative power of $1/x$ in Ω', can be chosen so as to make the lowest power of x and (if possible) succeeding powers in Q_1 acquire vanishing coefficients. For this value of Ω', let P_2 and Q_2 denote the resulting values of P_1 and Q_1; then the equation for u is

$$\frac{d^2u}{dx^2} + P_2 \frac{du}{dx} + Q_2 u = 0.$$

If this differential equation for u possesses a regular integral or regular integrals, then the original equation for y possesses a normal integral or normal integrals.

Thus, in any particular case, we have two definite alternative methods of trying to discover whether the given differential equation does or does not possess a normal integral.

Ex. 1. Is there any normal integral of the equation

$$y'' + \frac{1}{x^2}(2+x) y' + \frac{1}{x^4}(1-x-x^3) y = 0,$$

which is easily seen to possess no regular integral?

Here we have

$$P = \frac{2}{x^2} + \frac{1}{x}, \quad Q = \frac{1}{x^4} - \frac{1}{x^3} = \frac{1}{x}.$$

Accordingly, adopting the first method of proceeding which has been indicated, we choose

$$2\Omega' + \frac{2}{x^2} = 0,$$

that is,

$$\Omega' = -\frac{1}{x^2}, \quad \Omega = \frac{1}{x}.$$

Then

$$P_1 = P + 2\Omega' = \frac{1}{x},$$

$$Q_1 = Q + P\Omega' + \Omega'' + \Omega'^2 = -\frac{1}{x};$$

and therefore the equation for u is

$$u'' + \frac{1}{x} u' - \frac{1}{x} u = 0.$$

This equation has two regular integrals u_1 and u_2, where

$$u_1 = 1 + \frac{x}{(1!)^2} + \frac{x^2}{(2!)^2} + \frac{x^3}{(3!)^2} + \cdots,$$

$$u_2 = u_1 \log x - 2 \left\{ \frac{x}{(1!)^2} \frac{1}{1} + \frac{x^2}{(2!)^2} \left(\frac{1}{1} + \frac{1}{2} \right) + \frac{x^3}{(3!)^2} \left(\frac{1}{1} + \frac{1}{2} + \frac{1}{3} \right) + \cdots \right\};$$

and therefore the original equation has two normal integrals

$$u_1 e^{\frac{1}{x}}, \quad u_2 e^{\frac{1}{x}}.$$

Ex. 2. Has the equation

$$y'' - \frac{1}{x^4} y = 0$$

any normal integral?

Manifestly, as $P=0$, the first method of procedure cannot be adopted: so we adopt the second. The expression for Q_1 is now

$$-\frac{1}{x^4} + \Omega'^2 + \Omega'';$$

so we take

$$\Omega' = \frac{a}{x^2},$$

and assume $a^2 = 1$, so that $a = \pm 1$. We now have

$$P_2 = \frac{2a}{x}, \quad Q_2 = -\frac{2a}{x^3};$$

the equation for u is

$$u'' + \frac{2a}{x^3} (xu' - u) = 0.$$

This is satisfied by $u = x$ for both values of a; so the original equation in y has two normal integrals $xe^{\frac{1}{x}}$, $xe^{-\frac{1}{x}}$, and its primitive is

$$y = x (A e^{\frac{1}{x}} + B e^{-\frac{1}{x}}).$$

Ex. 3. Has the equation

$$y'' + \frac{1}{x} y' + \frac{1}{x^4} \{1 - (p - \tfrac{1}{2})^2 x^2\} y = 0,$$

where p is a positive or negative integer, any normal integrals? (It is easily seen to be Bessel's equation, with $1/x$ for variable in place of x; and its indicial equation shews that it has no regular integrals with expansions in ascending powers of x.)

Again, the first method of procedure cannot be adopted, owing to the form of P; for it would only give $\Omega'=0$. So we adopt the second method. The expression for Q_1 is

$$\frac{1}{x^4} - \frac{1}{x^2}(p-\tfrac{1}{2})^2 + \frac{1}{x}\Omega' + \Omega'' + \Omega'^2.$$

We take

$$\Omega' = \frac{a}{x^2},$$

where $a^2 = -1$. Accordingly, we choose $a = -i$; and then

$$P_2 = \frac{1}{x} - \frac{2i}{x^2}, \quad Q_2 = \frac{i}{x^3} - \frac{1}{x^2}(p-\tfrac{1}{2})^2,$$

so that the equation for u is

$$u'' - \frac{1}{x^2}(2i-x)u' + \frac{1}{x^3}\{i - (p-\tfrac{1}{2})^2 x\}u = 0.$$

The indicial equation for the u-differential equation is

$$-2i\rho + i = 0,$$

that is, a possible value for ρ is $\tfrac{1}{2}$. We therefore use the Frobenius method to see if there is a regular integral

$$u = x^{\frac{1}{2}}(c_0 + c_1 x + c_2 x^2 + \ldots);$$

and the application of the method shews that this expression will satisfy the differential equation, if

$$2sc_s i = \{(s-\tfrac{1}{2})^2 - (p-\tfrac{1}{2})^2\}c_{s-1},$$

for all the values of s. We thus have

$$c_s = -i\frac{(s-p)(s+p-1)}{8s}c_{s-1},$$

so that, whether p be positive or negative, the series for u becomes a polynomial. Let

$$U = 1 - \frac{(1-p)p \cdot (2-p)(p+1)}{1 \cdot 2}\left(\frac{x}{8}\right)^2 + \ldots,$$

$$V = (1-p)p\frac{x}{8} - \frac{(1-p)p \cdot (2-p)(p+1) \cdot (3-p)(p+2)}{1 \cdot 2 \cdot 3}\left(\frac{x}{8}\right)^3 + \ldots;$$

then the regular integral u becomes

$$c_0(U - iV)x^{\frac{1}{2}},$$

and the corresponding normal integral y is (on dropping the constant c_0)

$$e^{\frac{i}{x}}(U - iV)x^{\frac{1}{2}}.$$

A second normal integral is obtained by changing the sign of i; it is

$$e^{-\frac{i}{x}}(U+iV)x^{\frac{1}{2}}.$$

Thus the original equation has two normal integrals; and its primitive can be expressed in the form

$$Ax^{\frac{1}{2}}\left\{U\cos\left(\frac{1}{x}+a\right)+V\sin\left(\frac{1}{x}+a\right)\right\},$$

where A and a are arbitrary constants.

This result may be compared with the expression of J_n for large values of x given in Ex. 2, § 105.

Ex. 4. Discuss the following equations, as regards possession of normal integrals :—

    (i) $x^3y''+2xy'-y=0$;

    (ii) $x^4y''+2x^3y'-(a^2+2x^2)y=0$;

    (iii) $4x^4y''-(4+12x+3x^2)y=0$.

Ex. 5. Shew that the equation

$$y''+\frac{a}{x}y'+\frac{b}{x^4}y=0$$

possesses two normal integrals, if the constant a is an even integer.

Ex. 6. Shew that the equation

$$y''=\frac{1}{x^6}(1+ax^2)y$$

possesses a normal integral if a is equal to 3, or 1, or -1, or -3.

In the preceding methods for the determination of normal integrals, when they exist, there is the assumption that the quantity Ω is a polynomial in $1/x$. But in the second of the processes, where Ω is determined in connection with the quantity Q_1, where

$$Q_1=Q+P\Omega'+\Omega''+\Omega'^2,$$

in such a way as to make the lowest terms in the expansion of Q_1 disappear, it is the quantity Ω'^2, and not the quantity Ω'', which supplies the governing terms in the expression for Ω. For, if Ω begins with a term in x^{-n}, where $n>0$, then Ω' begins with a term in x^{-n-1} and Ω'' with a term in x^{-n-2}, so that the lowest term in Ω'^2 has a lower index than the lowest term in Ω''. Accordingly, we have to balance the orders of the lowest terms in Ω'^2, $P\Omega'$, Q.

Now it may happen that, in this comparison of lowest terms, the indices of the powers of $1/x$ which occur in Ω' become fractional, if the coefficients of the lowest power or powers in Q_1 are to disappear. In that event, it is customary to call the corresponding integral (if it exists) of the y-equation a *subnormal integral*. The general process of determining subnormal integrals, when they exist, is exactly similar to that adopted for the determination of normal integrals*. One or two examples will suffice to illustrate the process.

Ex. 7. The equation

$$y'' = \frac{1}{x^3}\left(a^2 + \tfrac{5}{16}x\right)y$$

has no regular integral and no normal integral, for expansions in ascending powers of x. Has it a subnormal integral for such expansions ?

The first method of proceeding is not effective; so we adopt the second method, by which Ω is to be chosen so that the lowest powers of x in

$$-\frac{a^2}{x^3} - \frac{5}{16x^2} + \Omega'' + \Omega'^2$$

shall vanish, Ω being kept infinite when $x = 0$. Manifestly we can take

$$\Omega' = \frac{a}{x^{\frac{3}{2}}},$$

with either sign for a; and then the differential equation for u, where

$$y = ue^{-2ax^{-\frac{1}{2}}},$$

is

$$u'' + \frac{2x}{x^{\frac{3}{2}}}u' - \left(\frac{3}{2}\frac{a}{x^{\frac{5}{2}}} + \frac{5}{16}\frac{1}{x^2}\right)u = 0.$$

Change the independent variable from x to z, where $x = z^2$; then the equation for u is

$$\frac{d^2u}{dz^2} + \left(\frac{4a}{z^2} - \frac{1}{z}\right)\frac{du}{dz} - \left(\frac{6a}{z^3} + \frac{5}{4}\frac{1}{z^2}\right)u = 0.$$

The indicial equation for expansions in ascending powers of z, for this differential equation, is

$$4a\rho - 6a = 0,$$

so that there is one possible index, viz. $\rho = \frac{3}{2}$. Accordingly, we take

$$u = z^{\frac{3}{2}}(c_0 + c_1 z + c_2 z^2 + \ldots),$$

* For a fuller discussion of the whole matter, reference may be made to my *Theory of Differential Equations*, vol. IV., ch. VII.

substitute it in the differential equation for u, and determine the coefficients so that the equation may be satisfied formally. We find

$$c_1 = \frac{1}{2a}\, c_0, \quad c_2 = 0, \quad c_3 = 0, \quad \dots;$$

and so the value of u, a normal integral of the u-equation, is

$$u = c_0 z^{\frac{3}{2}} \left(1 + \frac{1}{2a}\, z \right),$$

where c_0 is an arbitrary constant.

Moreover, a is either of the square roots of a^2. Hence the primitive of the original y-equation is

$$y = A e^{-2ax^{-\frac{1}{2}}} \left(x^{\frac{3}{4}} + \frac{1}{2a}\, x^{\frac{5}{4}} \right) + B e^{2ax^{-\frac{1}{2}}} \left(x^{\frac{3}{4}} - \frac{1}{2a}\, x^{\frac{5}{4}} \right),$$

where A and B are arbitrary constants.

Ex. 8. Shew that the equation

$$y'' = \frac{1}{x^3}(a^2 + bx)\, y$$

has two subnormal integrals, when the constant b is equal to $\frac{1}{16}(2n-1)(2n+3)$, where n is any positive integer.

Ex. 9. Shew that the equation

$$y'' + \frac{a}{2x}\, y' + \frac{b}{x^3}\, y = 0,$$

where a is an odd integer and b is any constant, has two linearly independent subnormal integrals.

Ex. 10. Does the equation

$$x^2 y'' + ay' + by = 0$$

possess any integral, which is regular or normal or subnormal for expansions in ascending powers of x?

CHAPTER VII

SOLUTION BY DEFINITE INTEGRALS

135. THE principal methods, which lead to expressions for the dependent variable in terms of the independent variable by means of what are ordinarily called *known* functions, have now been given. There is however another method which certainly leads to a solution of some differential equations though the full evaluation by the operations indicated may not be carried out. This method consists in expressing as a definite integral the value of the dependent variable; its chief application in ordinary differential equations arises in the case of a certain general class of linear equations which can otherwise be solved in series, though not in so concise a form. The method is however of primary importance in the solution of those linear partial differential equations of order higher than the first which arise in investigations in mathematical physics; in fact, in some questions these solutions by means of definite integrals constitute the only solutions hitherto obtained. At this stage, however, we are concerned with the application to ordinary differential equations.

136. The method applies with peculiar advantage to linear equations, into the coefficients of which x enters only in the first degree, and in which there is no term independent of y or of differential coefficients of y; such an equation, in its most general form, is

$$(a_0+b_0x)\frac{d^ny}{dx^n}+(a_1+b_1x)\frac{d^{n-1}y}{dx^{n-1}}+\ldots+(a_{n-1}+b_{n-1}x)\frac{dy}{dx}+(a_n+b_nx)y=0,$$

where the a's and b's are constants. This may be written

$$x\phi\left(\frac{d}{dx}\right)y + \psi\left(\frac{d}{dx}\right)y = 0,$$

where ϕ and ψ are rational integral functions of the order n in general, though the order of either may diminish through the vanishing of some of the coefficients. To solve this equation, we assume

$$y = \int e^{xt}\,T\,dt,$$

where T is a function of t but not of x; the form of this function and the limits of integration (supposed independent of x) are to be determined by substituting this proposed value of y in the differential equation. Since

$$\frac{dy}{dx} = \int t e^{xt}\,T\,dt,$$

$$\frac{d^2y}{dx^2} = \int t^2 e^{xt}\,T\,dt,$$

$$\dotsi\dotsi\dotsi\dotsi$$

the result of the substitution may be expressed in the form

$$\int x e^{xt}\phi(t)\,T\,dt + \int e^{xt}\psi(t)\,T\,dt = 0,$$

which must be identically satisfied. The former of the terms, on being integrated by parts, is superseded by

$$[e^{xt}\phi(t)\,T] - \int e^{xt}\frac{d}{dt}\{\phi(t)\,T\}\,dt;$$

and therefore the identity becomes

$$[e^{xt}\phi(t)\,T] - \int e^{xt}\left[\frac{d}{dt}\{\phi(t)\,T\} - \psi(t)\,T\right]dt = 0,$$

the first term being taken between the limits of the integral, which as yet are unknown. Now this will be satisfied, if we make

$$\frac{d}{dt}\{\phi(t)\,T\} - \psi(t)\,T = 0$$

for all values of t included within the range of integration, and if

$$[e^{xt}\phi(t)\,T] = 0$$

at the limits. The former of these equations determines T as a function of t; the latter will determine the limits of this assumed integral.

137. To derive the value of T, we write the first equation in the form

$$\frac{d}{dt}\{\phi(t)\,T\} - \frac{\psi(t)}{\phi(t)}\,\phi(t)\,T = 0,$$

and therefore

$$\phi(t)\,T = A e^{\int \frac{\psi(t)}{\phi(t)}dt},$$

where A is an arbitrary constant. Hence *the value of y is*

$$y = A \int \frac{e^{xt+\int \frac{\psi(t)}{\phi(t)}dt}}{\phi(t)}\,dt,$$

taken between limits of integration defined by the equation

$$A\left[e^{xt+\int \frac{\psi(t)}{\phi(t)}dt}\right] = 0,$$

these limits being independent of x.

138. We have now to determine the limits. Consider the equation

$$A e^{xt+\int \frac{\psi(t)}{\phi(t)}dt} = \mu_1,$$

where μ_1 is a constant. Let β_1 be a value of t independent of x and satisfying the equation; let $\mu_2, \ldots, \mu_r$, be other constants and $\beta_2, \ldots, \beta_r$, be corresponding values of t, all independent of x.

If the value

$$y = A_1 \int_{\beta_1}^{\beta_2} e^{xt}T\,dt + A_2 \int_{\beta_1}^{\beta_3} e^{xt}T\,dt + \ldots$$

be substituted in the equation, and if for each of these definite integrals (T being assumed to have the value before obtained) a single integration by parts be effected, as in the preceding analysis, then that the equation may be satisfied we must have

$$A_1\left[e^{xt+\int \frac{\psi(t)}{\phi(t)}dt}\right]_{\beta_1}^{\beta_2} + A_2\left[e^{xt+\int \frac{\psi(t)}{\phi(t)}dt}\right]_{\beta_1}^{\beta_3} + \ldots = 0;$$

and when this is identically satisfied the foregoing value of y is a solution of the equation. This last identity will indicate such necessary relations as may subsist among the arbitrary constants A, and so will fix the number of independent constants. When this number is the same as the order of the differential equation, the foregoing value of y is the primitive; but if it be less, the number of particular solutions necessary to make up the primitive must be otherwise determined. Examples will be given here-after.

139. This is the most general method of obtaining the limits; it includes as a particular set the limits obtained by taking those roots of the equation

$$e^{xt+\int \frac{\psi(t)}{\phi(t)}dt} = 0$$

which are independent of x; they obviously make

$$\left[e^{xt+\int \frac{\psi(t)}{\phi(t)}dt}\right] = 0,$$

and they are usually the simplest obtainable. When this equation indicates only two limits distinct from one another, these will give the only definite integral immediately derivable in such an example. If, however, more than two, say $r + 1$, limits be indicated, then r particular solutions may be constructed; in fact, denoting these limits by α, β_1, β_2, ..., β_r, we derive from them as the corresponding part of the primitive

$$y = \sum_{s=1}^{s=r} \left\{ A_s \int_\alpha^{\beta_s} e^{xt} T dt \right\}.$$

Ex. 1. We apply the foregoing to obtain the primitive of the equation

$$\frac{d^n y}{dx^n} - xy = 0.$$

Here we have, with the above notation,

$$\phi(t) = -1,$$
$$\psi(t) = t^n;$$

and therefore

$$-T = A_0 e^{-\int t^n dt},$$

or, changing the sign of the arbitrary constant, this is

$$T = A_0 e^{-\frac{t^{n+1}}{n+1}},$$

while, in accordance with the general rule, the equation determining the limits is

$$e^{xt-\frac{t^{n+1}}{n+1}} = -\frac{1}{A_0}\mu.$$

Now this is satisfied by $t = \infty$ when μ is zero, and by $t = 0$ when $\mu = -A_0$; hence we may take 0 and ∞ as the limits of the definite integral, which thus becomes

$$A_0 \int_0^\infty e^{-\frac{t^{n+1}}{n+1}+xt} dt$$

It must be noticed that, just as in the general case one of the definite integrals alone was not a solution of the differential equation, this is not a solution of the equation, because the terms outside the integral are

$$= A_0 \left[-e^{xt - \frac{t^{n+1}}{n+1}} \right]_0^\infty$$

$$= A_0,$$

instead of zero. This value of y is therefore the Particular Integral of the equation

$$\frac{d^n y}{dx^n} = xy + A_0.$$

Now the quantity T does not change, if for t we write ωt, where ω is a root of the equation

$$\xi^{n+1} = 1.$$

Moreover, the limits of the definite integral are unaltered since in the equation determining those limits the term xt in the exponent has changed into $x\omega t$ which, so far as this equation is concerned, is the same as changing x into $x\omega$, a change which has no effect on the limits since they are independent of x. Hence we have another definite integral in the form

$$A_1 \int_0^\infty e^{-\frac{t^{n+1}}{n+1} + \omega xt} d(\omega t),$$

or, when the ω is moved outside the sign of integration, it is

$$\omega A_1 \int_0^\infty e^{-\frac{t^{n+1}}{n+1} + \omega xt} dt.$$

Forming now these definite integrals for all the $(n+1)^{\text{th}}$ roots of unity and adding them together, we find as the expression for y, which has to be substituted,

$$y = A_0 \int_0^\infty e^{-\frac{t^{n+1}}{n+1} + xt} dt + \omega A_1 \int_0^\infty e^{-\frac{t^{n+1}}{n+1} + \omega xt} dt + \ldots + \omega^n A_n \int_0^\infty e^{-\frac{t^{n+1}}{n+1} + \omega^n xt} dt.$$

When this value is substituted, as in the general investigation, the terms which are under the integral sign vanish identically; and that part of the expression taken between the limits, which is furnished by the integral involving A_r, is A_r. Hence the resulting equation, when this value of y is substituted in the differential equation, is

$$A_0 + A_1 + \ldots + A_n = 0.$$

If then this single condition be satisfied among the $n+1$ arbitrary constants, the above expression for y is the primitive of the differential equation

$$\frac{d^n y}{dx^n} = xy.$$

Ex. 2. Prove that the above expression for y is the primitive of the equation

$$\frac{d^n y}{dx^n} = xy + a,$$

provided the constants A satisfy the condition

$$A_0 + A_1 + A_2 + \ldots + A_n = a.$$

Ex. 3. Prove that the primitive of the equation

$$x \frac{d^2 y}{dx^2} - (a + \beta)(1 + x)\frac{dy}{dx} + xa\beta y = 0$$

is, for positive values of x, given by

$$y = A \int_\alpha^\beta e^{ux}(u - a)^{-\left\{\frac{a(a+\beta)}{a-\beta}+1\right\}}(u - \beta)^{-\left\{\frac{\beta(\beta+a)}{\beta-a}+1\right\}} du$$

$$+ B \int_{-\infty}^a e^{ux}(u - a)^{-\left\{\frac{a(a+\beta)}{a-\beta}+1\right\}}(u - \beta)^{-\left\{\frac{\beta(\beta+a)}{\beta-a}+1\right\}} du.$$

Obtain the corresponding primitive for negative values of x.

(Petzval.)

Ex. 4. Solve

$$x \frac{d^2 y}{dx^2} + a \frac{dy}{dx} - q^2 xy = 0,$$

where a and q are constants. Here

$$\phi(t) = t^2 - q^2, \quad \psi(t) = at,$$

so that

$$\int \frac{\psi(t)}{\phi(t)} dt = \tfrac{1}{2} a \log(t^2 - q^2).$$

Hence one solution of the equation is

$$y = C \int e^{xt}(t^2 - q^2)^{\frac{1}{2}a - 1} dt,$$

taken between the limits given by

$$[e^{xt}(t^2 - q^2)^{\frac{1}{2}a}] = 0.$$

To obtain the limits, write

$$e^{xt}(t^2 - q^2)^{\frac{1}{2}a} = 0,$$

and suppose a positive; then two roots of the equation are given by

$$t = +q \text{ and } t = -q.$$

If now x be restricted to positive values, a third root is given by

$$t = -\infty,$$

while, when x is negative, a third root is given by

$$t = +\infty.$$

As in either case we have three values given by the limits equation, we can construct two distinct particular solutions and so have the primitive. Thus when x is positive, the primitive is

$$y = A \int_{-q}^{q} (t^2 - q^2)^{\frac{1}{2}a - 1} e^{tx} dt + B \int_{-\infty}^{-q} (t^2 - q^2)^{\frac{1}{2}a - 1} e^{tx} dt;$$

while, when x is negative, the primitive is

$$y = A \int_{-q}^{q} (t^2 - q^2)^{\frac{1}{2}a - 1} e^{tx} dt + B \int_{q}^{\infty} (t^2 - q^2)^{\frac{1}{2}a - 1} e^{tx} dt.$$

Ex. 5. Verify that, when a lies between zero and 2, the primitive of the equation is

$$y = C_1 \int_0^{\pi} e^{qx \cos \theta} \sin^{a-1} \theta \, d\theta + C_2 x^{1-a} \int_0^{\pi} e^{qx \cos \theta} \sin^{1-a} \theta \, d\theta,$$

unless a be unity, in which case the primitive may be written

$$y = \int_0^{\pi} e^{qx \cos \theta} \{A + B \log (x \sin^2 \theta)\} \, d\theta.$$

(Boole.)

Ex. 6. Obtain by means of definite integrals the primitive of Bessel's equation.

140. The foregoing general linear differential equation is one with variable coefficients which are of the first degree in the independent variable; and the definite-integral solution was obtained by means of a linear differential equation of the first order determining the unknown function T. It is not, however, the only type of differential equation to which the assumed form of integral is applicable; it is, in fact, a particular case of a more general process, indicated by the following proposition.

The solution, by means of definite integrals, of the general linear differential equation of the nth order, whose coefficients are not constant but are polynomial functions of the independent variable of degree not higher than m, can be made to depend upon the solution of a linear differential equation of order not higher than m, the coefficients of which are variable.

This proposition we proceed to prove. Let the differential equation be denoted by

$$X_n \frac{d^n y}{dx^n} + X_{n-1} \frac{d^{n-1} y}{dx^{n-1}} + X_{n-2} \frac{d^{n-2} y}{dx^{n-2}} + \ldots \ldots + X_1 \frac{dy}{dx} + X_0 y = 0,$$

where X_r (for all values of the suffix r) is a function of x only, of degree not higher than m, given by

$$X_r = a_r + b_r x + c_r x^2 + \ldots\ldots + k_r x^{m-1} + l_r x^m,$$

while for some values of r some of the coefficients of the highest powers of x may vanish. Taking a particular solution in the same form as before, we write

$$y = \int e^{tx} T dt$$

with the limits as yet undetermined, and T an unknown function of t. Now this value of y gives

$$\frac{d^s y}{dx^s} = \int e^{tx} t^s T dt;$$

and therefore the equation, when this expression for y is substituted in it, becomes

$$\int e^{tx} T [t^n X_n + t^{n-1} X_{n-1} + \ldots\ldots + t X_1 + X_0] dt = 0,$$

which must be identically satisfied. Rearranging the expression

$$t^n X_n + t^{n-1} X_{n-1} + \ldots\ldots + t X_1 + X_0,$$

so that it may proceed in powers of x, and writing

$$a_n t^n + a_{n-1} t^{n-1} + \ldots\ldots + a_1 t + a_0 = U_0,$$
$$b_n t^n + b_{n-1} t^{n-1} + \ldots\ldots + b_1 t + b_0 = U_1,$$
$$\ldots\ldots\ldots\ldots\ldots\ldots\ldots\ldots\ldots\ldots\ldots\ldots\ldots\ldots\ldots\ldots\ldots$$
$$k_n t^n + k_{n-1} t^{n-1} + \ldots\ldots + k_1 t + k_0 = U_{m-1},$$
$$l_n t^n + l_{n-1} t^{n-1} + \ldots\ldots + l_1 t + l_0 = U_m,$$

we transform the above equation into

$$\int e^{tx} T [U_0 + U_1 x + U_2 x^2 + \ldots\ldots + U_{m-1} x^{m-1} + U_m x^m] dt = 0.$$

Now the left-hand side is the sum of $m+1$ integrals of the form

$$\int e^{tx} T U_r x^r dt;$$

and each of these can be integrated by parts until the variable x ceases to occur except in the exponential. Thus we have

$$\int e^{tx} T U_r x^r \, dt = \left[e^{tx} \left\{ x^{r-1} T U_r - x^{r-2} \frac{d}{dt} (T U_r) + x^{r-3} \frac{d^2}{dt^2} (T U_r) - \ldots \right.\right.$$
$$\left.\left. + (-1)^{r-1} \frac{d^{r-1}}{dt^{r-1}} (T U_r) \right\} \right] + (-1)^r \int e^{tx} \frac{d^r}{dt^r} (T U_r) \, dt,$$

the part without the sign of integration being taken between the limits of the integral, as yet undetermined. Denoting the expression

$$x^{r-1} T U_r - x^{r-2} \frac{d}{dt} (T U_r) + \ldots + (-1)^{r-1} \frac{d^{r-1}}{dt^{r-1}} (T U_r)$$

by V_r for all values of r except zero (in which case no integration by parts is necessary), and applying the foregoing formula to each of the definite integrals on the left-hand side of the equation, we change the equation into

$$\left[e^{tx} \sum_{r=1}^{r=m} V_r \right]$$
$$+ \int e^{tx} \left\{ T U_0 - \frac{d}{dt} (T U_1) + \frac{d^2}{dt^2} (T U_2) - \ldots\ldots + (-1)^m \frac{d^m}{dt^m} (T U_m) \right\} dt = 0.$$

This will be identically satisfied if the unknown function T be chosen so as to satisfy the equation

$$T U_0 - \frac{d}{dt} (T U_1) + \frac{d^2}{dt^2} (T U_2) - \ldots\ldots + (-1)^m \frac{d^m}{dt^m} (T U_m) = 0$$

for all values of t between the limits of integration. These limits* must be determined by

$$\left[e^{tx} \sum_{r=1}^{r=m} V_r \right] = 0.$$

Now this equation determining T is linear with variable coefficients, and it is of the order m, but it may degenerate to one of lower order; when it is solved, a definite-integral solution of the original equation is derivable.

Hence the proposition follows as enunciated.

* The more precise determination of the limits of the definite integral and of its range is a matter of great difficulty in the general case. The most important contribution of this subject is due to Poincaré: an account of his investigations, leading up to what are known as asymptotic expansions, will be found in the author's *Theory of Differential Equations*, vol. IV., §§ 101—105.

Since the equation which determines T is of order m, it will have m independent particular solutions; these may be denoted by T_1, T_2, ..., T_m. Corresponding to these, there will be m particular solutions of the original equation obtained by substituting for T in

$$\int e^{tx} T \, dt$$

these m values in turn.

141. In the case when $m = 2$, the equation which determines T becomes

$$TU_0 - \frac{d}{dt}(TU_1) + \frac{d^2}{dt^2}(TU_2) = 0,$$

or, what is the same thing,

$$U_2 \frac{d^2 T}{dt^2} + \left(2 \frac{dU_2}{dt} - U_1\right)\frac{dT}{dt} + \left(\frac{d^2 U_2}{dt^2} - \frac{dU_1}{dt} + U_0\right) T = 0.$$

The following are some of the special cases in which this equation can be integrated very simply.

(1) Let the coefficients a, b, c, be such that the equation

$$\frac{d^2 U_2}{dt^2} - \frac{dU_1}{dt} + U_0 = 0$$

is satisfied for all values of t; then the value of T is easily proved to be

$$A \int \frac{dt}{U_2^2} e^{\int \frac{U_1}{U_2} dt}.$$

(2) On multiplying the equation throughout by U_2, we can rewrite it in the form

$$\frac{d}{dt}\left(U_2^2 \frac{dT}{dt}\right) - U_1 U_2 \frac{dT}{dt} - U_2\left(\frac{dU_1}{dt} - \frac{d^2 U_2}{dt^2} - U_0\right) T = 0,$$

the left-hand side of which is a perfect differential if

$$\frac{d}{dt}(U_1 U_2) = U_2\left(\frac{dU_1}{dt} - \frac{d^2 U_2}{dt^2} - U_0\right),$$

that is, if

$$U_2 \frac{d^2 U_2}{dt^2} + U_1 \frac{dU_2}{dt} + U_0 U_2 = 0.$$

If the values of a, b, c, be such as to make this an identity, then the value of T is given by

$$U_2{}^2 \frac{dT}{dt} - U_1 U_2 T = A,$$

which leads to the result

$$T e^{-\int \frac{U_1}{U_2} dt} = A \int \frac{dt}{U_2{}^2} e^{-\int \frac{U_1}{U_2} dt} + B.$$

(3) When the equation in T is reduced to its normal form by the substitution

$$T U_2 e^{-\frac{1}{2} \int \frac{U_1}{U_2} dt} = S,$$

the new equation is

$$\frac{d^2 S}{dt^2} + \mathfrak{T} S = 0,$$

where

$$\mathfrak{T} = \frac{U_0}{U_2} - \frac{1}{4} \left(\frac{U_1}{U_2}\right)^2 - \frac{1}{2} \frac{d}{dt} \left(\frac{U_1}{U_2}\right).$$

A solution of the equation is at once obtainable when $\mathfrak{T}$ vanishes, i.e. when

$$U_0 U_2 = \frac{1}{4} U_1{}^2 + \frac{1}{2} U_2{}^2 \frac{d}{dt} \left(\frac{U_1}{U_2}\right).$$

Further, immediately integrable cases are furnished when $\mathfrak{T}$ is a constant, or is of the form $\lambda (e + ft)^{-2}$, or of the form $\lambda (e + ft)^{-4}$.

In any case, whatever be the relations among the constants in the functions U, the solution of the equation determining T is of the form

$$T = C_1 T_1 + C_2 T_2.$$

The equation giving the limits of the definite integral is

$$\left[e^{tx} \left\{ x U_2 T - \frac{d}{dt} (U_2 T) + U_1 T \right\} \right] = 0,$$

which is satisfied by the values of t, if any, common to

$$T = 0, \quad \frac{dT}{dt} = 0.$$

Ex. Integrate, by means of a definite integral, the equation

$$x \frac{d^2y}{dx^2} - (1 + x^2) \frac{dy}{dx} + \mu(1 - \mu x - x^2) y = 0,$$

where μ is a constant.

142. Another set of equations to which the method of solution by definite integrals can be applied is the set derived from

$$\frac{d^2y}{dx^2} - \lambda x^n y = 0$$

for different values of n. To solve this we assume

$$y = \int e^{-pt} P \, dp,$$

where t denotes an unknown function of x alone and P an unknown function of p alone, both of which functions, as well as the limits of the integral, have to be determined. Differentiating the value of y twice and substituting in the equation, we find

$$\int e^{-pt} \left\{ p^2 \left(\frac{dt}{dx} \right)^2 - \lambda x^n \right\} P \, dp - \int e^{-pt} pP \frac{d^2t}{dx^2} dp = 0.$$

Choose the unknown function t so that

$$\left(\frac{dt}{dx} \right)^2 = \lambda x^n \, ;$$

and suppose that λ is positive and equal to c^2, so that the differential equation is

$$\frac{d^2y}{dx^2} = c^2 x^n y.$$

Then the equation which determines t is

$$\frac{dt}{dx} = cx^{\frac{1}{2}n},$$

and therefore

$$t = \frac{c}{\frac{1}{2}n + 1} x^{\frac{1}{2}n+1} = \frac{c}{m} x^m,$$

if m denote $\frac{1}{2}n + 1$. Hence we have

$$\frac{1}{t} \frac{dt}{dx} = \frac{m}{x}, \quad \frac{1}{t} \frac{d^2t}{dx^2} = \frac{m(m-1)}{x^2}.$$

Let the equation involving the integrals be multiplied throughout by x^2/mt; it becomes, after a very slight reduction,

$$m \int e^{-pt} (p^2 - 1) Pt \, dp - (m - 1) \int e^{-pt} Pp \, dp = 0.$$

Integrating the first term by parts, we have

$$-[me^{-pt}(p^2-1)P]+m\int e^{-pt}\frac{d}{dp}\{(p^2-1)P\}\,dp-(m-1)\int e^{-pt}Pp\,dp=0.$$

Now this relation will be identically satisfied if we make

$$m\frac{d}{dp}\{(p^2-1)P\}=(m-1)\,Pp,$$

for all values of p included between the limits of integration defined by

$$[e^{-pt}(p^2-1)P]=0.$$

The former equation serves to determine P as a function of p; it is of the first order and linear, and its solution is

$$P=A\,(p^2-1)^{-\frac{m+1}{2m}},$$

A being an arbitrary constant; and the equation which gives the limits is

$$[e^{-pt}(p^2-1)^{\frac{m-1}{2m}}]=0.$$

The latter equation is satisfied by $p=\infty$, and by $p=\pm 1$ provided the exponent of p^2-1 is positive; this requires that m should either be positive and greater than unity, or be negative, and therefore that n should be greater than zero or less than -2. Assuming that this condition is satisfied, we are in a position to construct two definite integrals; they are

$$\int_{-1}^{1}e^{-pt}(p^2-1)^{-\frac{m+1}{2m}}\,dp,$$

and

$$\int_{1}^{\infty}e^{-pt}(p^2-1)^{-\frac{m+1}{2m}}\,dp.$$

The former of these is equal to

$$\int_{0}^{1}e^{-pt}(p^2-1)^{-\frac{m+1}{2m}}\,dp+\int_{-1}^{0}e^{-pt}(p^2-1)^{-\frac{m+1}{2m}}\,dp,$$

$$=\int_{0}^{1}e^{-pt}(p^2-1)^{-\frac{m+1}{2m}}\,dp+\int_{0}^{1}e^{pt}(p^2-1)^{-\frac{m+1}{2m}}\,dp,$$

$$=\int_{0}^{1}(e^{pt}+e^{-pt})(p^2-1)^{-\frac{m+1}{2m}}\,dp.$$

Hence the primitive may be represented by

$$A' \int_0^1 (e^{pt} + e^{-pt})(p^2 - 1)^{-\frac{m+1}{2m}} dp + B \int_1^\infty e^{-pt}(p^2 - 1)^{-\frac{m+1}{2m}} dp;$$

substituting for t, we have

$$y = A \int_0^1 (1 - p^2)^{-\frac{n+4}{2n+4}} \cosh\left(\frac{2cp}{n+2} x^{\frac{1}{2}n+1}\right) dp$$

$$+ B \int_1^\infty e^{-\frac{2cp}{n+2} x^{\frac{1}{2}n+1}} (p^2 - 1)^{-\frac{n+4}{2n+4}} dp,$$

as the primitive of the equation

$$\frac{d^2y}{dx^2} = c^2 x^n y$$

for values of n such that $n > 0$ or $n < -2$.

Ex. Prove that the primitive of the same equation may be given in the form

$$y = A'x \int_0^1 (1 - p^2)^{-\frac{n}{2n+4}} \cosh\left(\frac{2cp}{n+2} x^{\frac{1}{2}n+1}\right) dp$$

$$+ B'x \int_1^\infty e^{-\frac{2cp}{n+2} x^{\frac{1}{2}n+1}} (p^2 - 1)^{-\frac{n}{2n+4}} dp,$$

provided $n > -2$ or $n < -4$.

(Lobatto.)

APPLICATION TO THE HYPERGEOMETRIC SERIES.

143. In order to obtain a definite integral which shall satisfy the differential equation of the hypergeometric series, we assume

$$y = \int (1 - vx)^m V dv,$$

where V is an unknown function of v only and m is a constant; the form of V, the value of m, and the limits of the integral, have to be determined. From this value of y we at once have

$$\frac{dy}{dx} = -m \int v V (1 - vx)^{m-1} dv,$$

$$\frac{d^2y}{dx^2} = m(m-1) \int v^2 V (1 - vx)^{m-2} dv;$$

so that, when these values are substituted in the equation

$$x(1-x)\frac{d^2y}{dx^2} + \{\gamma - (\alpha+\beta+1)x\}\frac{dy}{dx} - \alpha\beta y = 0,$$

it becomes

$$\int V(1-vx)^{m-2}[m(m-1)v^2x(1-x) - mv(1-vx)\{\gamma - (\alpha+\beta+1)x\}$$
$$- \alpha\beta(1-vx)^2]\,dv = 0.$$

The coefficient of x^2v^2 within the brackets is of the second degree in m, which is as yet an undetermined constant; let m be so chosen that this coefficient vanishes, so that m is given by

$$-m(m-1) - m(\alpha+\beta+1) - \alpha\beta = 0,$$
or $$m^2 + m(\alpha+\beta) + \alpha\beta = 0,$$

whence m may be taken equal to either $-\alpha$ or $-\beta$. As the differential equation is unaltered when α and β are interchanged, either of these roots may be taken; we shall take

$$m = -\alpha,$$

and then, substituting this value, we find that the equation

$$\int V(1-vx)^{-\alpha-2}[\alpha(\alpha+1)v^2x + \alpha v\{\gamma - x(\alpha+\beta+v\gamma+1)\}$$
$$- \alpha\beta(1-2vx)]\,dv = 0$$

must be identically satisfied. Rearranging the expression within the brackets under the sign of integration and dividing out by the factor α, we transform the equation into

$$\int V(1-vx)^{-\alpha-2}(\alpha+1)v(v-1)x\,dv$$
$$+ \int V(1-vx)^{-\alpha-2}(v\gamma - \beta)(1-vx)\,dv = 0.$$

Integrating the first term by parts, we have

$$-Vv(1-v)(1-vx)^{-\alpha-1} + \int(1-vx)^{-\alpha-1}\frac{d}{dv}\{v(1-v)V\}\,dv,$$

and therefore the equation becomes

$$-[Vv(1-v)(1-vx)^{-\alpha-1}]$$
$$+ \int(1-vx)^{-\alpha-1}\left[\frac{d}{dv}\{v(1-v)V\} - (\beta-v\gamma)V\right]dv = 0.$$

Now this relation will be identically satisfied, if we take as the equation to determine V

$$\frac{d}{dv}\{v(1-v)V\} = (\beta - v\gamma)V,$$

and assign, as the limits of the proposed integral, values of v such that

$$[Vv(1-v)(1-vx)^{-a-1}] = 0.$$

To solve the former equation, we have

$$\frac{d}{dv}\{v(1-v)V\} = v(1-v)V\frac{\beta - v\gamma}{v(1-v)}$$

$$= v(1-v)V\left\{\frac{\beta}{v} - \frac{\gamma - \beta}{1-v}\right\}.$$

Hence $v(1-v)V = Av^\beta(1-v)^{\gamma-\beta},$

where A is an arbitrary constant; and the equation determining the limits is

$$[v^\beta(1-v)^{\gamma-\beta}(1-vx)^{-a-1}] = 0,$$

which, on the supposition that β is positive and γ greater than β, is satisfied by $v = 0$ and $v = 1$. It therefore follows that the *equation of the hypergeometric series is satisfied by*

$$y = A\int_0^1 v^{\beta-1}(1-v)^{\gamma-\beta-1}(1-xv)^{-a}\,dv,$$

provided β be positive and γ greater than β.

It is easy to shew that, when $(1-xv)^{-a}$ is expanded and the coefficients of different powers of x are evaluated, the resulting series is a constant multiple of the hypergeometric series, this constant factor being

$$A\int_0^1 v^{\beta-1}(1-v)^{\gamma-\beta-1}\,dv.$$

144. If now we change the independent variable from x to $1-x$, the corresponding form of the differential equation is

$$x(1-x)\frac{d^2y}{dx^2} + \{\alpha+\beta+1-\gamma-(\alpha+\beta+1)x\}\frac{dy}{dx} - \alpha\beta y = 0.$$

A solution of this equation (and therefore of the original equation) is, from the foregoing analysis, given by

$$y = B \int_0^1 v^{\beta-1} (1-v)^{\alpha-\gamma} (1-xv)^{-\alpha} dv,$$

provided β is positive and $\alpha+1$ greater than γ. If the conditions of limitation of the parameters be satisfied, the primitive of the differential equation of the hypergeometric series is given by the sum of these two different solutions.

Ex. 1. Obtain, in terms of definite integrals, the complete solution of the equation

$$(A+Bx+Cx^2)\frac{d^2y}{dx^2}+(D+Ex)\frac{dy}{dx}+Fy=0;$$

(see Ex. 2, p. 240).

Ex. 2. Prove that,

(i) if β be positive and $\alpha+1$ greater than γ, then a solution is

$$y = \int_0^{-\infty} u^{\beta-1} (1-u)^{\gamma-\beta-1} (1-xu)^{-\alpha} du;$$

(ii) if γ be greater than β and less than $\alpha+1$, then a solution is

$$y = \int_1^{\infty} u^{\beta-1} (1-u)^{\gamma-\beta-1} (1-xu)^{-\alpha} du;$$

(iii) if γ be greater than β and α less than unity, then a solution is

$$y = \int_1^{\frac{1}{x}} u^{\beta-1} (1-u)^{\gamma-\beta-1} (1-xu)^{-\alpha} du.$$

(Jacobi.)

Ex. 3. Obtain the primitive of the equation

$$4xx'\frac{d^2y}{dx^2}+4(x'-x)\frac{dy}{dx}-y=0$$

(where $x'+x=1$) in the form

$$y = A \int_0^{\frac{\pi}{2}} (1-x\sin^2\phi)^{-\frac{1}{2}} d\phi + B \int_0^{\frac{\pi}{2}} (1-x'\sin^2\phi)^{-\frac{1}{2}} d\phi;$$

and of the equation

$$4xx'\frac{d^2y}{dx^2}=y$$

in the form

$$y = xx' \left[A \int_0^{\frac{\pi}{2}} \sin^2\phi \, (1-x\sin^2\phi)^{-\frac{3}{2}} d\phi + B \int_0^{\frac{\pi}{2}} \sin^2\phi \, (1-x'\sin^2\phi)^{-\frac{3}{2}} d\phi \right],$$

x' being the same as before.

Solve also

(i) $\quad 4xx'\dfrac{d^2y}{dx^2}+4x'\dfrac{dy}{dx}+y=0,$

(ii) $\quad 4xx'\dfrac{d^2y}{dx^2}-4x\dfrac{dy}{dx}+y=0,$

(iii) $\quad 4xx'\dfrac{d^2y}{dx^2}\pm 4\dfrac{dy}{dx}-y=0,$

(iv) $\quad 4xx'\dfrac{d^2y}{dx^2}+4\,(x'-x)\dfrac{dy}{dx}+3y=0.$

(Glaisher.)

Ex. 4. Prove that, if $n+1$ be positive, then

$$x^{-(n+1)}\int_0^1 t^{\frac{1}{2}n}(1-t)^{\frac{1}{2}(n-1)}\left(1-\frac{1}{x^2}\,t\right)^{-\frac{1}{2}(n+1)}dt$$

is a solution of Legendre's equation ; while, if n be negative, a solution is given by

$$x^n\int_0^1 t^{-\frac{1}{2}(n+1)}(1-t)^{-\frac{1}{2}(n+2)}\left(1-\frac{1}{x^2}\,t\right)^{\frac{1}{2}n}dt.$$

145. This chapter contains only a slight sketch of the method of solution of differential equations by means of definite integrals; the reader who wishes for fuller information on this part of the subject should consult two authorities in particular. By far the most important is PETZVAL, *Integration der linearen Differentialgleichungen*; the parts dealing with the method are §§ 2—5 of Section II.; §§ 19—22 of Section III.; §§ 10, 11 of Section V. The other authority is EULER, *Inst. Calc. Int.*, vol. II., c. X.; this work, however, labours under the disadvantage of assuming the form of the solution first and then of finding the differential equation satisfied by it. There are two other memoirs which might also with advantage be consulted; one by LOBATTO, *Crelle*, t. XVII., p. 363; and one by JACOBI, *Crelle*, t. LVI., p. 149.

A full discussion of the solution of linear differential equations by means of series and of definite integrals will be found, together with numerous examples, in a series of separately published memoirs by SPITZER.

MISCELLANEOUS EXAMPLES.

1. Integrate completely the equation

$$x\frac{d^ny}{dx^n}+ay=0.$$

2. Prove that the primitive of the equation

$$x\frac{d^2y}{dx^2}+a\frac{dy}{dx}+b^2xy=0$$

is given by

$$y = \int_0^b (b^2 - t^2)^{\frac{1}{2}a - 1} (A \sin xt + B \cos xt)\, dt$$

$$+ A \int_0^{\mp \infty} (t^2 + b^2)^{\frac{1}{2}a - 1}\, e^{xt}\, dt,$$

where the upper sign is to be taken if x be positive, and the lower if x be negative.

(Petzval.)

3. Prove that the equation

$$x \frac{d^3 y}{dx^3} - y = 0$$

has a solution given by

$$y = B \int_0^\infty \sin \frac{x}{v}\, e^{-\frac{1}{2}v^2}\, v\, dv\,;$$

and that a solution of

$$x \frac{d^3 y}{dx^3} + y = 0$$

is

$$y = C \int_0^{\mp \infty} e^{\frac{x}{v} - \frac{1}{2}v^2}\, v\, dv,$$

the minus or the plus sign being taken according as x is positive or is negative.

Obtain the primitive of each equation.

(Petzval.)

4. Investigate the primitive of the equation

$$\frac{d^2 y}{dx^2} + m^2 c^2 x^{2m - 2} y = 0$$

in the form

$$y = A \int_0^{\frac{\pi}{2}} \cos (cx^m \sin \phi) \cos^{-\frac{1}{m}} \phi\, d\phi$$

$$+ Bx \int_0^{\frac{\pi}{2}} \cos (cx^m \sin \phi) \cos^{\frac{1}{m}} \phi\, d\phi,$$

for values of m not included between -1 and $+1$.

(Kummer, and Lobatto.)

5. Shew that a particular solution of

$$\frac{d^2 y}{dx^2} + a^2 y = \frac{n(n+1)}{x^2} y$$

is

$$y = x^{n+1} \int_{-a}^{a} (v^2 - a^2)^n \cos xv\, dv\,;$$

and that a particular solution of

$$\frac{d^2y}{dx^2} - a^2y = \frac{n(n+1)}{x^2}y$$

is

$$y = x^{n+1}\int_0^\infty (x^2 + v^2)^{-n-1}\cos av\,dv.$$

6. Shew that the equation

$$\frac{d^{n+1}y}{dx^{n+1}} = x^m\frac{dy}{dx} + mx^{m-1}y$$

is satisfied by

$$y = \int_0^\infty z^{m-1}e^{-\frac{z^{m+n}}{m+n}}\psi(zx)\,dz,$$

where $\psi(x)$ is given by

$$\frac{d^{n+1}\psi(x)}{dx^{n+1}} = x^{m-1}\psi(x),$$

and m is positive. Hence from the solution of

$$\frac{d^3y}{dx^3} = y$$

deduce that of

$$\frac{d^2y}{dx^2} = xy.$$

7. Verify that

$$y = \int_{x^{\frac{1}{2}}}^x e^{-z^{2n} - x^{2n}}z^{-2n}\,dz$$

is a particular integral of

$$\frac{d^2y}{dx^2} - 4n^2x^{2n-2}y = \tfrac{1}{4}e^{-2x^n}x^{-\frac{3}{2}} - 2nx^{2n-1}e^{-x^{2n}-1}.$$

8. Shew that when the coefficients of the differential equation

$$(a_2 + b_2x)\frac{d^2y}{dx^2} + (a_1 + b_1x)\frac{dy}{dx} + (a_0 + b_0x)y = 0$$

satisfy the condition $a_1b_2 - a_2b_1 = b_2^2$, the solution will be

$$y = \int e^{ux}V\{A + B\log U_1(a_2 + b_2x)\}\,du,$$

where

$$U_1 = b_2u^2 + b_1u + b_0,$$

and

$$\log(VU_1) = \int\frac{a_2u^2 + a_1u + a_0}{U_1}\,du,$$

the limits being given by

$$e^{ux}U_1V = 0.$$

(Spitzer.)

9. Prove that equations of the form

$$x^2 \frac{d^2y}{dx^2} + (A_1 + B_1 x^m)\, x\, \frac{dy}{dx} + (A_0 + B_0 x^m + C_0 x^{2m})\, y = 0$$

may be reduced to the form

$$t\phi\left(\frac{d}{dt}\right) z + \psi\left(\frac{d}{dt}\right) z = 0$$

of § 136, by the substitutions $x^m = t$ and $y = t^k z$; and shew that k is determined by a quadratic equation.

(Petzval.)

10. Prove that the particular integral of

$$(\vartheta + a_1)(\vartheta + a_2) \ldots\ldots (\vartheta + a_n)\, y = f(x),$$

where ϑ denotes $x\,\dfrac{d}{dx}$, is

$$y = \int_0^1 \int_0^1 \int_0^1 \ldots\ldots f(\theta_1 \theta_2 \ldots\ldots \theta_n x)\, \theta_1^{a_1 - 1} \theta_2^{a_2 - 1} \ldots\ldots \theta_n^{a_n - 1} d\theta_1 d\theta_2 \ldots\ldots d\theta_n.$$

11. Prove that the definite integral

$$\int_0^1 \int_0^1 u^{\beta - 1}(1 - u)^{\theta - \beta - 1} v^{\gamma - 1}(1 - v)^{\epsilon - \gamma - 1}(1 - xuv)^{-a}\, du\, dv$$

is, when $\theta > \beta > 0$ and $\epsilon > \gamma > 0$, a solution of the differential equation

$$(1 - x)\, x^2 \frac{d^3y}{dx^3} + \{\theta + \epsilon + 1 - (a + \beta + \gamma + 3)\, x\}\, x\, \frac{d^2y}{dx^2}$$

$$+ \{\theta\epsilon - x\,(a\beta + \beta\gamma + \gamma a + a + \beta + \gamma + 1)\} \frac{dy}{dx} - a\beta\gamma y = 0.$$

Give, in the form of definite integrals, the primitive of this equation.

12. The primitive of the equation

$$\frac{d^3y}{dx^3} + 8\lambda x^3 y = bx$$

is $\quad y = A\int_0^a \dfrac{e^{ux^2}\, du}{(u^3 + \lambda)^{\frac{1}{2}}} + B\int_0^\beta \dfrac{e^{ux^2}\, du}{(u^3 + \lambda)^{\frac{1}{2}}} + C\int_0^\gamma \dfrac{e^{ux^2}\, du}{(u^3 + \lambda)^{\frac{1}{2}}} + D\int_{-\infty}^0 \dfrac{e^{ux^2}\, du}{(u^3 + \lambda)^{\frac{1}{2}}},$

where a, β, γ, are the roots of

$$u^3 + \lambda = 0,$$

and the arbitrary constants are connected by the single relation

$$A + B + C - D = -\tfrac{1}{8}b\lambda^{-\frac{1}{2}}.$$

(Petzval.)

13. Prove that the definite integral

$$y = \int_0^\infty e^{-x^m - bz^m x^{-m}} dx$$

satisfies the equation

$$\frac{d^2 y}{dz^2} = m^2 bz^{m-2} y.$$

(Poisson.)

14. Prove that

$$P_{-\frac{1}{2}}\left(\frac{1+x}{1-x}\right) = \frac{2}{\pi}(1-x)^{\frac{1}{2}} \int_0^{\frac{\pi}{2}} \frac{d\theta}{(1-x\sin^2\theta)^{\frac{1}{2}}},$$

P being Legendre's function.

(G. H. Stuart.)

15. Shew that, if β be positive and a less than unity,

$$\int_0^{\frac{1}{x}} u^{\beta-1}(1-u)^{\gamma-\beta-1}(1-xu)^{-a} du$$

is a solution of the differential equation of the hypergeometric series.

(Jacobi.)

16. Shew that three linearly independent integrals of the equation

$$2x\left(a_0 \frac{d^3 y}{dx^3} + 3a_1 \frac{d^2 y}{dx^2} + 3a_2 \frac{dy}{dx} + a_3 y\right) + k\left(3a_0 \frac{d^2 y}{dx^2} + 6a_1 \frac{dy}{dx} + 3a_2 y\right) = 0,$$

where k and the coefficients a are constants, are given by

$$\int_\beta^a e^{xu}\{(a-u)(u-\beta)(u-\gamma)\}^{\frac{1}{2}k-1} du,$$

$$\int_\gamma^\beta e^{xu}\{(a-u)(\beta-u)(u-\gamma)\}^{\frac{1}{2}k-1} du,$$

$$\int_{-\infty}^\gamma e^{xu}\{(a-u)(\beta-u)(\gamma-u)\}^{\frac{1}{2}k-1} du,$$

where a, β, γ, are the roots of

$$a_0 m^3 + 3a_1 m^2 + 3a_2 m + a_3 = 0,$$

which are supposed real and unequal, and $a > \beta > \gamma$.

Discuss the integrals of the differential equation (i) when $\beta = \gamma$, (ii) when $a = \beta = \gamma$.

(M. J. M. Hill.)

CHAPTER VIII

ORDINARY EQUATIONS WITH MORE THAN TWO VARIABLES

146. It has already appeared that in some cases, though the integration of separate terms of a differential equation would introduce new transcendental functions, the solution of the equation as a whole can be expressed in terms of purely algebraical functions. Thus, for instance, the equation

$$\frac{dx}{(1-x^2)^{\frac{1}{2}}} + \frac{dy}{(1-y^2)^{\frac{1}{2}}} = 0$$

can be integrated in terms of the transcendental functions arc sin x, arc sin y; but there is a solution of the form

$$x(1-y^2)^{\frac{1}{2}} + y(1-x^2)^{\frac{1}{2}} = C,$$

which is equivalent to the other. We are thus naturally led to enquire whether other cases exist in which such an algebraical relation between the variables of the integrals of functions can be obtained, when the integrals themselves cannot be evaluated without the introduction of new functions. The case next in point of simplicity, which furnishes a similar example, is that usually known as *Euler's equation*, in which the object is to find the integral algebraical relation between x and y which corresponds to the equation

$$X^{-\frac{1}{2}} dx + Y^{-\frac{1}{2}} dy = 0,$$

where $$X = a + bx + cx^2 + ex^3 + fx^4,$$

and $$Y = a + by + cy^2 + ey^3 + fy^4.$$

To integrate this, we assume

$$p = x + y,$$

and

$$\frac{dx}{dt} = \frac{X^{\frac{1}{2}}}{y - x},$$

so that

$$\frac{dy}{dt} = \frac{Y^{\frac{1}{2}}}{x - y},$$

and therefore

$$\frac{dp}{dt} = \frac{Y^{\frac{1}{2}} - X^{\frac{1}{2}}}{x - y}.$$

A second differentiation with regard to t gives

$$\frac{d^2p}{dt^2} = \frac{1}{x - y} \left\{ \frac{1}{2Y^{\frac{1}{2}}} \frac{dY}{dy} \frac{dy}{dt} - \frac{1}{2X^{\frac{1}{2}}} \frac{dX}{dx} \frac{dx}{dt} \right\} - \frac{Y^{\frac{1}{2}} - X^{\frac{1}{2}}}{(x - y)^2} \left(\frac{dx}{dt} - \frac{dy}{dt} \right)$$

$$= \frac{1}{(x - y)^2} \left\{ \frac{1}{2} \frac{dY}{dy} + \frac{1}{2} \frac{dX}{dx} - \frac{(Y^{\frac{1}{2}} - X^{\frac{1}{2}})(Y^{\frac{1}{2}} + X^{\frac{1}{2}})}{y - x} \right\}$$

$$= \frac{1}{(x - y)^2} \left\{ b + c(x + y) + \frac{3e}{2}(x^2 + y^2) + 2f(x^3 + y^3) \right.$$

$$\left. - b - c(x + y) - e(x^2 + xy + y^2) - f(x^3 + x^2y + xy^2 + y^3) \right\},$$

the last four terms inside the bracket being the value of $\dfrac{Y - X}{y - x}$.
Rearranging and collecting terms, we have

$$\frac{d^2p}{dt^2} = \frac{1}{(x - y)^2} \left\{ \frac{1}{2} e(x^2 - 2xy + y^2) + f(x^3 - x^2y - xy^2 + y^3) \right\}$$

$$= \frac{1}{2} e + f(x + y)$$

$$= \frac{1}{2} e + fp.$$

If we multiply by $2\dfrac{dp}{dt}$ and integrate, we obtain

$$\left(\frac{dp}{dt} \right)^2 = ep + fp^2 + C,$$

or substituting the value for $\dfrac{dp}{dt}$, we have

$$\frac{(Y^{\frac{1}{2}} - X^{\frac{1}{2}})^2}{(y - x)^2} = C + e(x + y) + f(x + y)^2,$$

an algebraical relation between x and y, though the separate integrals require for their expression elliptic functions.

Ex. 1. Prove that another integral of the equation

$$\frac{dx}{X^{\frac{1}{2}}} + \frac{dy}{Y^{\frac{1}{2}}} = 0$$

is

$$\left\{\frac{x^2 Y^{\frac{1}{2}} - y^2 X^{\frac{1}{2}}}{y-x}\right\}^2 = Cx^2 y^2 + bxy(x+y) + a(x+y)^2;$$

and verify the theorem of § 12 in this case by shewing that the two primitives are not independent.

Ex. 2. Prove that an integral of

$$\frac{dx}{X^{\frac{1}{2}}} = \frac{dy}{Y^{\frac{1}{2}}}$$

is

$$\frac{(Y^{\frac{1}{2}} + X^{\frac{1}{2}})^2}{(y-x)^2} = C + e(x+y) + f(x+y)^2.$$

Ex. 3. Express in an integral form the relation between y and x given by

$$\frac{dx}{(1-x^4)^{\frac{1}{2}}} + \frac{dy}{(1-y^4)^{\frac{1}{2}}} = 0.$$

Ex. 4. Shew that the primitive of

$$\frac{dx}{\{x(1-x)(1-\lambda x)\}^{\frac{1}{2}}} + \frac{dy}{\{y(1-y)(1-\lambda y)\}^{\frac{1}{2}}} = 0$$

may be exhibited in the form

$$\{x(1-y)(1-\lambda y)\}^{\frac{1}{2}} + \{y(1-x)(1-\lambda x)\}^{\frac{1}{2}} = A(1-\lambda xy),$$

where A is an arbitrary constant.

147. There is another method of proceeding, due to Cauchy; it is quite different from the former.

Consider a general equation between the two variables of the second degree of the form

$$u = X_0 y^2 + 2X_1 y + X_2$$
$$= Y_0 x^2 + 2Y_1 x + Y_2 = 0,$$

where $X_0, X_1, X_2, Y_0, Y_1, Y_2$, are all of the second degree, the first three in x, and the second three in y; thus if

$$X_0 = a_0 x^2 + 2a_1 x + a_2,$$
$$X_1 = b_0 x^2 + 2b_1 x + b_2,$$
$$X_2 = c_0 x^2 + 2c_1 x + c_2,$$

we should have

$$Y_0 = a_0 y^2 + 2b_0 y + c_0,$$
$$Y_1 = a_1 y^2 + 2b_1 y + c_1,$$
$$Y_2 = a_2 y^2 + 2b_2 y + c_2.$$

Then the ratio of $dy : dx$ is given by

$$\frac{\partial u}{\partial x} dx + \frac{\partial u}{\partial y} dy = 0.$$

But

$$\frac{\partial u}{\partial x} = 2 (Y_0 x + Y_1)$$
$$= 2 (Y_1{}^2 - Y_0 Y_2)^{\frac{1}{2}},$$

since $u = Y_0 x^2 + 2 Y_1 x + Y_2 = 0$; similarly

$$\frac{\partial u}{\partial y} = 2 (X_0 y + X_1)$$
$$= 2 (X_1{}^2 - X_0 X_2)^{\frac{1}{2}};$$

and therefore

$$\frac{dx}{(X_1{}^2 - X_0 X_2)^{\frac{1}{2}}} + \frac{dy}{(Y_1{}^2 - Y_0 Y_2)^{\frac{1}{2}}} = 0,$$

a differential equation the primitive of which is $u = 0$.

Now since Euler's differential equation is symmetrical with regard to x and y, it is necessary that its primitive $u = 0$ should be symmetrical with regard to x and y, in order that the preceding analysis may apply to the present case. In order that u may be symmetrical, we must have

$$b_0 = a_1, \quad c_0 = a_2, \quad c_1 = b_2,$$

and $X_1{}^2 - X_0 X_2$ is then the same function of x that $Y_1{}^2 - Y_0 Y_2$ is of y. In order to obtain the primitive of

$$\frac{dx}{X^{\frac{1}{2}}} + \frac{dy}{Y^{\frac{1}{2}}} = 0,$$

where

$$X = a + bx + cx^2 + ex^3 + fx^4,$$

and Y is the same function of y, we must make X and $X_1{}^2 - X_0 X_2$ the same. The comparison of their coefficients will give four equations to determine the coefficients of u; but in u there are five independent constants (there were originally eight as any one can be made unity, but three equations necessary for symmetry are satisfied) and therefore one will remain undetermined and so arbitrary. These equations giving the coefficients are

$$\frac{b_0{}^2 - a_0 c_0}{f} = \frac{4b_1 b_0 - 2\,(a_1 c_0 + a_0 c_1)}{e} = \frac{4b_1 b_2 - 2\,(a_1 c_2 + a_2 c_1)}{b} = \frac{b_2{}^2 - a_2 c_2}{a}$$

$$= \frac{4\,(b_1{}^2 - a_1 c_1) - (a_2 c_0 + a_0 c_2 - 2b_0 b_2)}{c};$$

when the values of the determined coefficients are substituted in u, the equation $u = 0$ contains one arbitrary constant and is thus the primitive.

Ex. 1. Prove that the primitive of

$$\frac{dx}{(Ax^2 + 2Bx + C)^{\frac{1}{2}}} + \frac{dy}{(Ay^2 + 2By + C)^{\frac{1}{2}}} = 0$$

is

$$a_0\,(x^2 + y^2) + 2b_0 xy + 2b_1\,(x + y) + c_2 = 0,$$

where

$$\frac{b_0{}^2 - a_0{}^2}{A} = \frac{b_1 b_0 - a_0 b_1}{B} = \frac{b_1{}^2 - a_0 c_2}{C}.$$

Ex. 2. Verify that the primitive of

$$\frac{dx}{(1 + a_2 x^2 + a_0 x^4)^{\frac{1}{2}}} + \frac{dy}{(1 + a_2 y^2 + a_0 y^4)^{\frac{1}{2}}} = 0$$

is

$$A_2\,(x^2 + y^2) + 2A_3 xy = 1 + a_0 x^2 y^2,$$

where

$$A_3{}^2 - a_0 = A_2{}^2 + a_2 A_2.$$

(Cauchy.)

Chap. XIV. of Cayley's *Elliptic Functions* may be consulted with advantage.

148. If instead of a single equation between two variables, the relation between which is expressible in an algebraical form, we have a system of $n - 1$ equations between n variables, we may without integration of each integrable expression represent in an integral form the dependence between the n variables in the shape of an algebraical equation; and as this equation is obtained by an integration, it must contain an arbitrary constant. The process made use of in order to derive it in the general case will be seen to differ materially from that adopted in the particular case of $n = 2$.

Let the differential equations be

$$\left.\begin{array}{l} \dfrac{dx_1}{X_1^{\frac{1}{2}}} + \dfrac{dx_2}{X_2^{\frac{1}{2}}} + \dots\dots\dots\dots\dots\dots\dots\dots + \dfrac{dx_n}{X_n^{\frac{1}{2}}} = 0 \\[3mm] \dfrac{x_1 dx_1}{X_1^{\frac{1}{2}}} + \dfrac{x_2 dx_2}{X_2^{\frac{1}{2}}} + \dots\dots\dots\dots\dots\dots\dots + \dfrac{x_n dx_n}{X_n^{\frac{1}{2}}} = 0 \\[3mm] \dots\dots\dots\dots\dots\dots\dots\dots\dots\dots\dots\dots\dots\dots \\[2mm] \dfrac{x_1^{n-2} dx_1}{X_1^{\frac{1}{2}}} + \dfrac{x_2^{n-2} dx_2}{X_2^{\frac{1}{2}}} + \dots\dots\dots\dots + \dfrac{x_n^{n-2} dx_n}{X_n^{\frac{1}{2}}} = 0 \end{array}\right\},$$

in which

$$X_\mu = A_0 + A_1 x_\mu + A_2 x_\mu^2 + \dots\dots\dots + A_{2n-1} x_\mu^{2n-1} + A_{2n} x_\mu^{2n},$$

for all the suffixes μ in the system. Let

$$f(x) = (x - x_1)(x - x_2)\dots\dots\dots(x - x_n);$$

and let $f'(x_\mu)$ denote the value of $\dfrac{df(x)}{dx}$ when in it, after the indicated differentiation has taken place, x_μ is substituted for x; the value of $f'(x_\mu)$ will therefore be

$$(x_\mu - x_1)(x_\mu - x_2)\dots\dots\dots(x_\mu - x_n),$$

the vanishing factor $x_\mu - x_\mu$ being absent. Solving now the above system of equations in order to obtain the algebraical ratios of the quantities $dx_1, dx_2, \dots\dots\dots, dx_n$, we find

$$\frac{f'(x_1)\,dx_1}{X_1^{\frac{1}{2}}} = \frac{f'(x_2)\,dx_2}{X_2^{\frac{1}{2}}} = \dots\dots\dots = \frac{f'(x_n)\,dx_n}{X_n^{\frac{1}{2}}}.$$

Let the common value of these equal fractions be denoted by dt, so that we have

$$\frac{dx_1}{dt} = \frac{X_1^{\frac{1}{2}}}{f'(x_1)}; \quad \frac{dx_2}{dt} = \frac{X_2^{\frac{1}{2}}}{f'(x_2)};$$

and so on.

The first of these gives

$$\left(\frac{dx_1}{dt}\right)^2 = \frac{X_1}{\{f'(x_1)\}^2},$$

and therefore, after differentiation with respect to t,

$$2\frac{dx_1}{dt}\frac{d^2x_1}{dt^2} = \frac{\partial}{\partial x_1}\left[\frac{X_1}{\{f'(x_1)\}^2}\right]\frac{dx_1}{dt} + \frac{\partial}{\partial x_2}\left[\frac{X_1}{\{f'(x_1)\}^2}\right]\frac{dx_2}{dt} + \dots\dots\dots$$

Now
$$\frac{\partial}{\partial x_\mu}\left[\frac{X_1}{\{f'(x_1)\}^2}\right] = -\frac{2X_1}{\{f'(x_1)\}^3}\frac{\partial}{\partial x_\mu}\{f'(x_1)\}.$$

But since
$$f'(x_1) = (x_1 - x_2)(x_1 - x_3)\ldots\ldots(x_1 - x_n),$$

we have
$$\frac{1}{f'(x_1)}\frac{\partial}{\partial x_\mu}\{f'(x_1)\} = \frac{-1}{x_1 - x_\mu},$$

and therefore
$$\frac{\partial}{\partial x_\mu}\left[\frac{X_1}{\{f'(x_1)\}^2}\right] = \frac{2X_1}{\{f'(x_1)\}^2}\frac{1}{x_1 - x_\mu},$$

provided μ be not unity. After substitution and division by the coefficient of $\dfrac{d^2 x_1}{dt^2}$ on the left-hand side, the equation becomes

$$\frac{d^2 x_1}{dt^2} = \frac{1}{2}\frac{\partial}{\partial x_1}\left[\frac{X_1}{\{f'(x_1)\}^2}\right] + \frac{X_1^{\frac{1}{2}}X_2^{\frac{1}{2}}}{f'(x_1)f'(x_2)}\frac{1}{x_1 - x_2} + \frac{X_1^{\frac{1}{2}}X_3^{\frac{1}{2}}}{f'(x_1)f'(x_3)}\frac{1}{x_1 - x_3}$$
$$+ \ldots\ldots + \frac{X_1^{\frac{1}{2}}X_n^{\frac{1}{2}}}{f'(x_1)f'(x_n)}\frac{1}{x_1 - x_n}.$$

Similarly
$$\frac{d^2 x_2}{dt^2} = \frac{1}{2}\frac{\partial}{\partial x_2}\left[\frac{X_2}{\{f'(x_2)\}^2}\right] + \frac{X_2^{\frac{1}{2}}X_1^{\frac{1}{2}}}{f'(x_2)f'(x_1)}\frac{1}{x_2 - x_1} + \frac{X_2^{\frac{1}{2}}X_3^{\frac{1}{2}}}{f'(x_2)f'(x_3)}\frac{1}{x_2 - x_3}$$
$$+ \ldots\ldots + \frac{X_2^{\frac{1}{2}}X_n^{\frac{1}{2}}}{f'(x_2)f'(x_n)}\frac{1}{x_2 - x_n};$$

and so for the others, making n in all. Now let the n left-hand sides of these equations be added together; the sum will be equal to that of the n right-hand sides. It will be seen that in the latter, when in the rth expression a term $\dfrac{X_r^{\frac{1}{2}}X_s^{\frac{1}{2}}}{f'(x_r)f'(x_s)}\dfrac{1}{x_r - x_s}$ enters, then in the sth expression a term $\dfrac{X_s^{\frac{1}{2}}X_r^{\frac{1}{2}}}{f'(x_s)f'(x_r)}\dfrac{1}{x_s - x_r}$ also enters, and the sum of the two terms is therefore zero. All the terms containing these fractions $\dfrac{1}{x_r - x_s}$ will thus disappear for all values of s and r; and so we have

$$2\frac{d^2}{dt^2}(x_1 + x_2 + \ldots\ldots + x_n) = \frac{\partial}{\partial x_1}\left[\frac{X_1}{\{f'(x_1)\}^2}\right] + \frac{\partial}{\partial x_2}\left[\frac{X_2}{\{f'(x_2)\}^2}\right] + \ldots\ldots$$
$$+ \frac{\partial}{\partial x_n}\left[\frac{X_n}{\{f'(x_n)\}^2}\right].$$

We denote $x_1 + x_2 + \ldots\ldots + x_n$ by p, so that the left-hand side of the last equation is $2\dfrac{d^2p}{dt^2}$.

149. We can obtain another value for the expression on the right-hand side of the equation. Let X denote the same function of x as X_1 of x_1; and let

$$\frac{X}{\{f(x)\}^2}$$

be expanded in partial fractions. Since X and $\{f(x)\}^2$ are both of the degree $2n$, there will be a term independent of x, which will be A_{2n}; and so we may write

$$\frac{X}{\{f(x)\}^2} = A_{2n} + \frac{B_1}{x-x_1} + \frac{B_2}{x-x_2} + \ldots\ldots + \frac{B_n}{x-x_n}$$
$$+ \frac{C_1}{(x-x_1)^2} + \frac{C_2}{(x-x_2)^2} + \ldots\ldots + \frac{C_n}{(x-x_n)^2}.$$

Multiplying up by $(x-x_1)^2$, we have

$$\frac{X(x-x_1)^2}{\{f(x)\}^2} = C_1 + B_1(x-x_1) + \text{terms multiplied by } (x-x_1)^2;$$

or, dividing out by the common factors in the numerator and the denominator on the left-hand side, we have $C_1 + B_1(x-x_1) +$ terms multiplied by $(x-x_1)^2 = \dfrac{X}{(x-x_2)^2 (x-x_3)^2 \ldots (x-x_n)^2}.$

If x be put equal to x_1, the left-hand side becomes C_1 and the right becomes $\dfrac{X_1}{\{f'(x_1)\}^2}$, so that

$$C_1 = \frac{X_1}{\{f'(x_1)\}^2}.$$

The right-hand side of the equation in the form last written does not involve x_1, and its partial differential coefficient with regard to x_1 is therefore zero; since the two sides of the equation are identically equal, zero must be the value of the partial differential coefficient of the left-hand side with regard to x_1, and so we have

$$\frac{\partial C_1}{\partial x_1} - B_1 + (x-x_1)\frac{\partial B_1}{\partial x_1} + \text{terms involving } (x-x_1) = 0.$$

This is true for all values of x, and therefore

$$B_1 = \frac{\partial C_1}{\partial x_1}$$

$$= \frac{\partial}{\partial x_1}\left[\frac{X_1}{\{f'(x_1)\}^2}\right].$$

Similarly

$$B_2 = \frac{\partial}{\partial x_2}\left[\frac{X_2}{\{f'(x_2)\}^2}\right],$$

with corresponding expressions for the other quantities B. Hence

$$2\frac{d^2p}{dt^2} = \frac{\partial}{\partial x_1}\left[\frac{X_1}{\{f'(x_1)\}^2}\right] + \frac{\partial}{\partial x_2}\left[\frac{X_2}{\{f'(x_2)\}^2}\right] + \ldots\ldots$$

$$= B_1 + B_2 + \ldots\ldots + B_n.$$

Let the equation expressing the resolution into partial fractions of the expression considered be multiplied throughout by $\{f(x)\}^2$; and let the coefficients of x^{2n-1} on the two sides of the relation

$$X = A_{2n}\{f(x)\}^2 + \sum_{\mu=1}^{\mu=n}\frac{B_\mu}{x-x_\mu}\{f(x)\}^2 + \sum_{\mu=1}^{\mu=n}\frac{C_\mu}{(x-x_\mu)^2}\{f(x)\}^2$$

be equated. None of the terms involving the quantities C can furnish terms of so high a degree, since each begins with x^{2n-2}; each of the terms involving the quantities B begins with x^{2n-1}, and the whole coefficient from this series of terms is therefore

$$B_1 + B_2 + \ldots\ldots + B_n.$$

Since

$$f(x) = (x-x_1)(x-x_2)\ldots\ldots(x-x_n)$$

$$= x^n - x^{n-1}(x_1 + x_2 + \ldots\ldots + x_n) + \text{lower powers of } x$$

$$= x^n - px^{n-1} + \text{lower powers},$$

the coefficient of x^{2n-1} in $A_{2n}\{f(x)\}^2$ is $-2A_{2n}p$. That coefficient on the left-hand side is A_{2n-1}; and therefore

$$A_{2n-1} = -2A_{2n}p + B_1 + B_2 + \ldots\ldots + B_n$$

$$= -2A_{2n}p + 2\frac{d^2p}{dt^2}.$$

Multiplying by $\dfrac{dp}{dt}$ and integrating, we have

$$\left(\frac{dp}{dt}\right)^2 = A_{2n}\,p^2 + A_{2n-1}\,p + E,$$

where E is an arbitrary constant. But

$$\frac{dp}{dt} = \frac{dx_1}{dt} + \frac{dx_2}{dt} + \ldots\ldots + \frac{dx_n}{dt}$$

$$= \frac{X_1^{\frac{1}{2}}}{f'(x_1)} + \frac{X_2^{\frac{1}{2}}}{f'(x_2)} + \ldots\ldots + \frac{X_n^{\frac{1}{2}}}{f'(x_n)}\ ;$$

and therefore the integral becomes

$$\left\{\frac{X_1^{\frac{1}{2}}}{f'(x_1)} + \frac{X_2^{\frac{1}{2}}}{f'(x_2)} + \ldots\ldots + \frac{X_n^{\frac{1}{2}}}{f'(x_n)}\right\}^2 = E + A_{2n}\,(x_1 + x_2 + \ldots\ldots + x_n)^2$$

$$+ A_{2n-1}\,(x_1 + x_2 + \ldots\ldots + x_n).$$

Ex. 1. Prove that an integral of the equations

$$\frac{dx}{X^{\frac{1}{2}}} + \frac{dy}{Y^{\frac{1}{2}}} + \frac{dz}{Z^{\frac{1}{2}}} = 0,$$

$$\frac{x\,dx}{X^{\frac{1}{2}}} + \frac{y\,dy}{Y^{\frac{1}{2}}} + \frac{z\,dz}{Z^{\frac{1}{2}}} = 0,$$

where

$$X = a + bx + cx^2 + \theta x^3 + \gamma x^4 + \beta x^5 + \alpha x^6,$$

and Y and Z are similar functions of y and z, is

$$\left\{\frac{(y-z)\,X^{\frac{1}{2}} + (z-x)\,Y^{\frac{1}{2}} + (x-y)\,Z^{\frac{1}{2}}}{(x-y)\,(y-z)\,(z-x)}\right\}^2 = \beta\,(x+y+z) + \alpha\,(x+y+z)^2 + C,$$

where C is an arbitrary constant.

(Richelot.)

Ex. 2. Deduce a second integral of these equations in the form

$$\left\{\frac{y^2 z^2\,(y-z)\,X^{\frac{1}{2}} + z^2 x^2\,(z-x)\,Y^{\frac{1}{2}} + x^2 y^2\,(x-y)\,Z^{\frac{1}{2}}}{(x-y)\,(y-z)\,(z-x)}\right\}^2$$

$$= C'x^2 y^2 z^2 + bxyz\,(xy + yz + zx) + a\,(xy + yz + zx)^2.$$

(Richelot.)

Note. The theory of these and kindred equations will not here be carried further, as it soon ceases to belong exclusively to differential equations and merges into the general theory of transcendental functions (commonly called Abelian Functions). The reader, who wishes for a fuller development on the lines of differential equations than can be given here, will find a paper by

RICHELOT, *Crelle*, t. XXIII., pp. 354—369, very useful; and he would do well to consult the following papers by JACOBI,

Crelle, t. IX., pp. 394—403;

t. XIII., pp. 55—78;

t. XXIV., pp. 28—35;

t. XXXII., pp. 220—226,

all of which are contained in the second volume of his collected works.

The theory of these transcendental functions, and of others, involves issues beyond the construction of integrals of systems of differential equations. It is expounded in treatises on the theory of algebraic functions and their integrals.

TOTAL DIFFERENTIAL EQUATIONS.

150. The differential equations with which we have hitherto had to deal have been, except in §§ 148 and 149, such as include one dependent and one independent variable; for the future we shall consider those which include more than two variables. These may be divided into two classes, one in which only one dependent variable occurs, the other in which only one independent variable occurs. In equations of the former class, we shall have the partial differential coefficients of the single dependent variable relatively to the independent variables; these are called *partial* differential equations, and will afterwards be discussed. In equations of the latter class, we shall have the differential coefficients of the several dependent variables with reference to the single independent variable (which may be either expressed or implied); these are usually called *total* differential equations.

When we have an integral equation

$$\phi\,(x,\,y,\,z) = C,$$

where C is a constant, we may suppose that x, y, z, undergo slight variations dx, dy, dz, which we know will be connected by the relation

$$\frac{\partial \phi}{\partial x}\,dx + \frac{\partial \phi}{\partial y}\,dy + \frac{\partial \phi}{\partial z}\,dz = 0.$$

If $\dfrac{\partial \phi}{\partial x}$, $\dfrac{\partial \phi}{\partial y}$, $\dfrac{\partial \phi}{\partial z}$, have any common factor, the equation can be simplified by the removal of that common factor; and so we may consider the general form of a derived total equation in three variables as represented by

$$P\,dx + Q\,dy + R\,dz = 0.$$

Here P, Q, R, are given functions of x, y, z; they are proportional to the differential coefficients of ϕ, and they may actually be equal to those differential coefficients.

But, conversely, when any equation of the form

$$P\,dx + Q\,dy + R\,dz = 0$$

is given, it does not necessarily lead to an equation of the form

$$\phi\,(x,\,y,\,z) = C\,;$$

for the existence of the latter equation implies that the three quantities P, Q, R, are proportional to the differential coefficients of some one function, and this requirement is not satisfied while P, Q, R, are quite general. We must therefore discuss in what circumstances such a differential equation will lead to an integral of the given form; and, on the assumption that such an integral is possible, indicate a method of obtaining it.

There will remain the further problem of obtaining an equivalent of the equation, when the conditions necessary for the existence of such an integral as the above are not satisfied.

151. In the first place, then, we assume that such an integral exists; we must therefore have P, Q, R, respectively proportional to the partial differential coefficients of some function ϕ with regard to x, y, z, so that we may write

$$= \frac{\partial\phi}{\partial x}, \quad \mu Q = \frac{\partial\phi}{\partial y}, \quad \mu R = \frac{\partial\phi}{\partial z},$$

in which μ is some quantity the value of which is not specified by the equation. From the first two of these equations we have

$$\frac{\partial}{\partial y}\,(\mu P) = \frac{\partial^2\phi}{\partial x\partial y} = \frac{\partial}{\partial x}\,(\mu Q),$$

or

$$\mu\,\frac{\partial P}{\partial y} + P\frac{\partial\mu}{\partial y} = \mu\,\frac{\partial Q}{\partial x} + Q\,\frac{\partial\mu}{\partial x},$$

that is

$$\mu\left(\frac{\partial P}{\partial y} - \frac{\partial Q}{\partial x}\right) = Q\,\frac{\partial\mu}{\partial x} - P\,\frac{\partial\mu}{\partial y}.$$

Similarly
$$\mu \left(\frac{\partial Q}{\partial z} - \frac{\partial R}{\partial y} \right) = R \frac{\partial \mu}{\partial y} - Q \frac{\partial \mu}{\partial z},$$

and
$$\mu \left(\frac{\partial R}{\partial x} - \frac{\partial P}{\partial z} \right) = P \frac{\partial \mu}{\partial z} - R \frac{\partial \mu}{\partial x}.$$

Multiply the last three equations respectively by R, P, Q, and add; we have

$$P \left(\frac{\partial Q}{\partial z} - \frac{\partial R}{\partial y} \right) + Q \left(\frac{\partial R}{\partial x} - \frac{\partial P}{\partial z} \right) + R \left(\frac{\partial P}{\partial y} - \frac{\partial Q}{\partial x} \right) = 0,$$

which is an equation giving a relation between P, Q, and R; it is a relation which must be satisfied if the proposed differential equation possesses an integral of the form considered.

Further, if we take

$$P_1 = \lambda P, \quad Q_1 = \lambda Q, \quad R_1 = \lambda R,$$

where λ is any quantity, the same condition is satisfied by P_1, Q_1, R_1, as by P, Q, R. For

$$\frac{\partial Q_1}{\partial z} - \frac{\partial R_1}{\partial y} = \lambda \left(\frac{\partial Q}{\partial z} - \frac{\partial R}{\partial y} \right) + Q \frac{\partial \lambda}{\partial z} - R \frac{\partial \lambda}{\partial y},$$

$$\frac{\partial R_1}{\partial x} - \frac{\partial P_1}{\partial z} = \lambda \left(\frac{\partial R}{\partial x} - \frac{\partial P}{\partial z} \right) + R \frac{\partial \lambda}{\partial x} - P \frac{\partial \lambda}{\partial z},$$

$$\frac{\partial P_1}{\partial y} - \frac{\partial Q_1}{\partial x} = \lambda \left(\frac{\partial P}{\partial y} - \frac{\partial Q}{\partial x} \right) + P \frac{\partial \lambda}{\partial y} - Q \frac{\partial \lambda}{\partial x};$$

and therefore

$$P_1 \left(\frac{\partial Q_1}{\partial z} - \frac{\partial R_1}{\partial y} \right) + Q_1 \left(\frac{\partial R_1}{\partial x} - \frac{\partial P_1}{\partial z} \right) + R_1 \left(\frac{\partial P_1}{\partial y} - \frac{\partial Q_1}{\partial x} \right)$$

$$= \lambda^2 \left\{ P \left(\frac{\partial Q}{\partial z} - \frac{\partial R}{\partial y} \right) + Q \left(\frac{\partial R}{\partial x} - \frac{\partial P}{\partial z} \right) + R \left(\frac{\partial P}{\partial y} - \frac{\partial Q}{\partial x} \right) \right\}$$

$$= 0,$$

which proves the statement.

Conversely, *when the relation is satisfied, the differential equation leads to a primitive of the form*

$$\phi (x, y, z) = C.$$

We proceed to prove this proposition. We know that, when a differential equation

$$P\,dx - Q\,dy = 0$$

is given, a function $u\,(x, y)$ exists such that the differential equation is satisfied in virtue of the relation

$$u\,(x, y) = \text{constant,}$$

and consequently that P and Q are proportional to the derivatives of u with regard to x and y respectively. If then, in the equation

$$P\,dx + Q\,dy + R\,dz = 0,$$

we consider P and Q as functions of x and y, which involve z also, we infer that a function of x and y exists, say u, such that for some quantity λ we can write

$$P_1 = \lambda P = \frac{\partial u}{\partial x}, \qquad Q_1 = \lambda Q = \frac{\partial u}{\partial y};$$

and u will involve other quantities occurring in P and Q that do not affect the partial differentiation with regard to x and y, that is, u will involve z. But it cannot be at once assumed that R is proportional to the remaining derivative of u; and we therefore write

$$R_1 - \frac{\partial u}{\partial z} = \lambda R - \frac{\partial u}{\partial z} = S.$$

When we substitute

$$P_1 = \frac{\partial u}{\partial x}, \qquad Q_1 = \frac{\partial u}{\partial y}, \qquad R_1 = \frac{\partial u}{\partial z} + S,$$

in the relation

$$P_1 \left(\frac{\partial Q_1}{\partial z} - \frac{\partial R_1}{\partial y} \right) + Q_1 \left(\frac{\partial R_1}{\partial x} - \frac{\partial P_1}{\partial z} \right) + R_1 \left(\frac{\partial P_1}{\partial y} - \frac{\partial Q_1}{\partial x} \right) = 0,$$

which is satisfied because the relation between $P, Q, R,$ is satisfied, we have

$$\frac{\partial S}{\partial x} \frac{\partial u}{\partial y} - \frac{\partial S}{\partial y} \frac{\partial u}{\partial x} = 0.$$

The only integral equation which is possessed at present is $u = \text{constant}$; and the equation between S and u is manifestly not satisfied in virtue of $u = $ constant. Hence the equation between S and u is satisfied identically; and therefore (§ 10) S and u, regarded as functions of x and y alone, are expressible in terms of one another. But the variable z may, and usually does, occur in u and in S; so we can take the expression for S in the form

$$S = f(z, u),$$

that is, S is expressible in terms of z and u only.

Moreover, we have

$$\lambda (P dx + Q dy + R dz)$$
$$= \frac{\partial u}{\partial x} dx + \frac{\partial u}{\partial y} dy + \frac{\partial u}{\partial z} dz + S dz$$
$$= du + S dz;$$

and therefore the original equation can be replaced by

$$du + S dz = 0,$$

where S is expressible in terms of z and u alone.

This equation now involves two variables only. We know that a function $\psi(u, z)$ exists, such that the equation is satisfied in virtue of

$$\psi(u, z) = \text{constant};$$

and we have

$$\mu = \frac{\partial \psi}{\partial u}, \qquad \mu S = \frac{\partial \psi}{\partial z},$$

for some quantity μ. Hence

$$\lambda \mu (P dx + Q dy + R dz) = \mu (du + S dz)$$
$$= d\psi;$$

and therefore the original equation is satisfied in virtue of

$$\psi(u, z) = \text{constant},$$

or, on replacing u by its value in terms of x, y, z, the original equation has a primitive of the form

$$\phi(x, y, z) = C.$$

Note. In consequence of this property, the relation

$$P\left(\frac{\partial Q}{\partial z} - \frac{\partial R}{\partial y}\right) + Q\left(\frac{\partial R}{\partial x} - \frac{\partial P}{\partial z}\right) + R\left(\frac{\partial P}{\partial y} - \frac{\partial Q}{\partial x}\right) = 0$$

is frequently called the *condition of integrability*.

152. We shall assume that the condition is satisfied, and that consequently the primitive exists. The actual deduction of the primitive can be effected in accordance with the preceding analysis.

It is first necessary to obtain the quantity u. For this purpose, we consider an equation

$$P\,dx + Q\,dy = 0,$$

with the assumption that z does not vary; then if its integral is

$$\theta\,(x,\,y) = u = \text{constant},$$

where the coefficients in θ may involve the variable z, we have the required function u. Moreover, we have

$$\lambda P = \frac{\partial u}{\partial x}, \qquad \lambda Q = \frac{\partial u}{\partial y},$$

so that λ is known.

We then multiply the original equation by λ, so that it becomes

$$\lambda\,(P\,dx + Q\,dy + R\,dz) = 0,$$

and we express it in the form

$$du + S\,dz = 0.$$

We express S in terms of z and u alone, which we know is possible; then the new equation

$$du + S\,dz = 0$$

involves only u and z. Its integral is the primitive of the original equation. Hence we have the rule :—

To obtain the primitive of the equation

$$P\,dx + Q\,dy + R\,dz = 0,$$

when the condition of integrability is satisfied, we integrate the equation

$$P\,dx + Q\,dy = 0$$

as if z were invariable* : let this integral be

$$u = constant,$$

---

* If more convenient in any particular case, either of the other variables could be considered temporarily constant: the corresponding changes to be made in the process are obvious.

and find the quantity λ *such that*

$$\lambda P = \frac{\partial u}{\partial x}.$$

The equation

$$\lambda (Pdx + Qdy + Rdz) = 0$$

has the form

$$du + Sdz = 0;$$

we express S in terms of z and u alone, and then integrate the equation

$$du + Sdz = 0.$$

Its integral is the primitive of the original equation.

Note. Another method of proceeding to the construction of the integral, when the condition of integrability is satisfied, is given in § 164 (*post*).

Ex. 1. Integrate the equation

$$(ydx + xdy)(a - z) + xy\,dz = 0.$$

Here $P = y(a-z)$, $Q = x(a-z)$, $R = xy$; the condition of integrability is satisfied.

We take the equation

$$(ydx + xdy)(a - z) = 0;$$

here, the factor $a - z$ can be neglected. Omitting it, we have an integral in the form

$$xy = \text{constant},$$

so that we take

$$u = xy.$$

Thus

$$\lambda P = \frac{\partial u}{\partial x} = y,$$

so that

$$\lambda = \frac{1}{a - z}.$$

Thus

$$\lambda (Pdx + Qdy + Rdz) = 0$$

becomes

$$du + \frac{xy}{a - z}\,dz = 0,$$

that is,

$$du + \frac{u}{a - z}\,dz = 0.$$

The primitive of this equation is

$$\frac{u}{a-z}=C,$$

and therefore the primitive of the original equation is

$$xy = C(a-z).$$

Ex. 2. Verify that for each of the following equations the condition of integrability is satisfied, and obtain the primitives:

(i) $(y+z)\,dx+(z+x)\,dy+(x+y)\,dz=0$;

(ii) $zy\,dx=zx\,dy+y^2\,dz$;

(iii) $(y+a)^2\,dx+z\,dy-(y+a)\,dz=0$;

(iv) $(x-a)\,dx+(z-c)\,dz+\{h^2-(x-a)^2-(z-c)^2\}^{\frac{1}{2}}\,dy=0$;

(v) $(2y^2+4az^2x^2)\,x\,dx+\{3y+2x^2+(y^2+z^2)^{-\frac{1}{2}}\}\,y\,dy$
$$+\{4z^2+2ax^4+(y^2+z^2)^{-\frac{1}{2}}\}\,z\,dz=0\,;$$

(vi) $(y^2+yz)\,dx+(xz+z^2)\,dy+(y^2-xy)\,dz=0$;

(vii) $(x^2y-y^3-y^2z)\,dx+(xy^2-x^3-x^2z)\,dy+(xy^2+x^2y)\,dz=0$;

(viii) $(2x^2+2xy+2xz^2+1)\,dx+dy+2z\,dz=0$;

(ix) $(2x+y^2+2xz)\,dx+2xy\,dy+x^2\,dz=du.$

153. The preceding solution has been obtained on the supposition that the equation of condition among the coefficients of the differential elements dx, dy, dz, is satisfied. It remains now to consider the class of equations for which the condition is not satisfied, and for which there cannot therefore be a single general integral.

Let us now assume any arbitrary relation between x, y, z, of the form

$$\psi(x,\,y,\,z)=0\,;$$

this, on being differentiated, gives

$$\frac{\partial\psi}{\partial x}\,dx+\frac{\partial\psi}{\partial y}\,dy+\frac{\partial\psi}{\partial z}\,dz=0.$$

When the form ψ is specified, these two equations will determine z and dz in terms of x, y, dx and dy (or, generally, one of the variables and its differential in terms of the other two and their differentials); when they are substituted in the equation

$$P\,dx+Q\,dy+R\,dz=0,$$

they make it of the form

$$Mdx + Ndy = 0,$$

where M and N are functions of x and y, the values of which will depend upon the form of the chosen function ψ. Now this equation may be integrated and the integral, containing an arbitrary constant, will together with the relation

$$\psi(x, y, z) = 0$$

constitute a solution of the differential equation.

For it is evident from the method of derivation of the integral that, in combination with $\psi = 0$, it furnishes relations between x, y, and z, such that the differential equation is satisfied.

By giving all possible forms to ψ, every possible solution will be obtained. Each solution will be constituted by two equations.

Ex. 1. Solve

$$dz = ay\,dx + b\,dy.$$

The equation of condition is not satisfied. Some relation between x, y, z, must therefore be assumed, and it may be perfectly arbitrary: let it be

$$y = f(x).$$

A combination of this relation with the differential equation gives

$$dz = af(x)\,dx + bf'(x)\,dx,$$

the integral of which is

$$z = a\int f(x)\,dx + bf(x) + C.$$

This equation, with $f(x) = y$, forms a solution of the proposed equation.

Ex. 2. Obtain the most general solution of the equation

$$x\,dx + y\,dy + c\left(1 - \frac{x^2}{a^2} - \frac{y^2}{b^2}\right)^{\frac{1}{2}} dz = 0,$$

which is consistent with the relation

$$\frac{x^2}{a^2} + \frac{y^2}{b^2} + \frac{z^2}{c^2} = 1.$$

Ex. 3. Find the equation, which must be associated with $x^2 + y^2 = \phi(z)$, in order to give an integral of

$$\{x(x-a) + y(y-b)\}\,dz = (z-c)(x\,dx + y\,dy);$$

and the equation, which must be associated with

$$y + z\log x + \phi(z) = 0,$$

so as to satisfy

$$z\,dx + x\,dy + y\,dz = 0.$$

Ex. 4. Prove that, if μ be a quantity such that

$$\mu(Pdx + Qdy) = dV,$$

then a solution of the general equation may be represented by

$$V = \phi(z),$$

$$\frac{\partial V}{\partial z} - \mu R = \phi'(z).$$

This is Monge's form.

Ex. 5. Obtain general equations which constitute the solution of

$$ydx = (x - z)(dy - dz).$$

154. It is not at first sight clear how the equation of condition affects the above process and, in particular, why what has been given as the solution in the latter case is not the solution in the former case. But the relation between the two solutions can be seen as follows.

The elimination of the differential element dz between the two equations in which it occurs leads to the equation

$$\left(R\frac{\partial \psi}{\partial x} - P\frac{\partial \psi}{\partial z}\right)dx + \left(R\frac{\partial \psi}{\partial y} - Q\frac{\partial \psi}{\partial z}\right)dy = 0;$$

and, in order that this may be reduced to the form

$$Mdx + Ndy = 0,$$

the variable z, which occurs in it, must be replaced by its value derived from $\psi(x, y, z) = 0$. Now suppose the equation of condition is satisfied so that P, Q, R, are proportional to the differential coefficients with regard to x, y, z, of some function; if this function be $\psi(x, y, z)$, then we have

$$\frac{1}{R}\frac{\partial \psi}{\partial z} = \frac{1}{Q}\frac{\partial \psi}{\partial y} = \frac{1}{P}\frac{\partial \psi}{\partial x} \quad \dots\dots\dots\dots(A),$$

and the equation involving dx and dy is identically satisfied. There will thus, on this supposition, be no other equation necessarily associated with the equation $\psi = 0$, or, what is equivalent for this case, $\psi = C$; this by itself is sufficient for the solution of the differential equation, and any other equation associated with $\psi = C$ may be perfectly arbitrary (such as $\chi = 0$), for its expression will not enter into the differential equation when formed from these integral equations. If however the equation first written down be not that which leads to the particular properties (A), but be another such as $\chi = 0$, it will still be possible to derive the equation $\psi = C$, into the expression of which the form of χ does not enter; and we may therefore consider as the general solution of the differential equation the equation

$$\psi = C\,;$$

while, if we wish to determine y and z separately as functions of x, we associate with this any arbitrary relation between x, y, z.

If however the equation of condition between the quantities P, Q, R, be not satisfied, there is no function ψ such that the relations (A) hold; and thus

$$M\,dx + N\,dy = 0$$

is not an identity but leads to an integral, the form of which is affected by the form of the arbitrary equation first written down and which must be associated with that equation in order to constitute the integral.

It thus appears that the difference between the two cases is as follows. While we may consider that in both cases two equations are necessary to give the complete solution, in the case when the equation of condition is satisfied one of these integral equations (called $\psi = C$) is completely unaffected in form by the other (called $\chi = 0$), but in the case when the equation of condition is not satisfied one of these integral equations is affected in form by the other.

155. The difference between the results in the two classes having been indicated, it is now possible to adopt a method of integration which shews the point of separation between the processes applying to these classes. Let

$$\chi\,(x,\,y,\,z) = 0$$

be any relation between x, y, and z; then

$$\frac{\partial \chi}{\partial x}\,dx + \frac{\partial \chi}{\partial y}\,dy + \frac{\partial \chi}{\partial z}\,dz = 0.$$

We also have

$$P\,dx + Q\,dy + R\,dz = 0.$$

Let the former equation be multiplied by λ (a quantity to be determined afterwards) and added to the latter, so that

$$\left(P+\lambda\frac{\partial\chi}{\partial x}\right)dx+\left(Q+\lambda\frac{c\chi}{\partial y}\right)dy+\left(R+\lambda\frac{\partial\chi}{\partial z}\right)dz=0,$$

or, say, $$P_1\,dx+Q_1\,dy+R_1\,dz=0.$$

Let λ be so chosen as to make $P_1,\,Q_1,\,R_1$, proportional to the differential coefficients with regard to $x,\,y,\,z$, respectively of some function ψ; then the integral of the last equation is

$$\psi\,(x,\,y,\,z)=C,$$

where C is arbitrary, and the primitive of the differential equation is given by the two equations

$$\left.\begin{array}{l}\chi\,(x,\,y,\,z)=0\\ \psi\,(x,\,y,\,z)=C\end{array}\right\}.$$

Now since $P_1,\,Q_1,\,R_1$, are proportional to differential coefficients with regard to $x,\,y,\,z$, we have

$$P_1\left(\frac{\partial Q_1}{\partial z}-\frac{\partial R_1}{\partial y}\right)+Q_1\left(\frac{\partial R_1}{\partial x}-\frac{\partial P_1}{\partial z}\right)+R_1\left(\frac{\partial P_1}{\partial y}-\frac{\partial Q_1}{\partial x}\right)=0;$$

or substituting for $P_1,\,Q_1,\,R_1$, and reducing, we have

$$P\left(\frac{\partial Q}{\partial z}-\frac{\partial R}{\partial y}\right)+Q\left(\frac{\partial R}{\partial x}-\frac{\partial P}{\partial z}\right)+R\left(\frac{\partial P}{\partial y}-\frac{\partial Q}{\partial x}\right)$$

$$+\lambda\left\{\frac{\partial\chi}{\partial x}\left(\frac{\partial Q}{\partial z}-\frac{\partial R}{\partial y}\right)+\frac{\partial\chi}{\partial y}\left(\frac{\partial R}{\partial x}-\frac{\partial P}{\partial z}\right)+\frac{\partial\chi}{\partial z}\left(\frac{\partial P}{\partial y}-\frac{\partial Q}{\partial x}\right)\right\}$$

$$+\frac{\partial\lambda}{\partial x}\left(Q\frac{\partial\chi}{\partial z}-R\frac{\partial\chi}{\partial y}\right)+\frac{\partial\lambda}{\partial y}\left(R\frac{\partial\chi}{\partial x}-P\frac{\partial\chi}{\partial z}\right)+\frac{\partial\lambda}{\partial z}\left(P\frac{\partial\chi}{\partial y}-Q\frac{\partial\chi}{\partial x}\right)=0.$$

If $P,\,Q,\,R$, be themselves proportional to differential coefficients with regard to $x,\,y,\,z$, the first line in this equation vanishes and a solution of the equation is $\lambda=0$; $P_1,\,Q_1,\,R_1$, are then independent of χ, and therefore $\psi\,(x,\,y,\,z)$ is independent of χ.

If $P,\,Q,\,R$, be not such as to make the first line vanish, then λ is shewn by this equation to depend upon the form of χ and therefore ψ also will depend upon the form of χ. The form of ψ will in this case be determined by the method given in § 153; but the foregoing investigation is useful as a means of instituting the analytical comparison between the methods.

GEOMETRICAL INTERPRETATION.

156. A geometrical interpretation can be given to the differential equation and its integral, which will illustrate the difference between the two classes of equation explained in the last two paragraphs.

If as usual x, y, z, represent the coordinates of a point A, the equation will then represent some locus. Let A' be a point on the locus adjacent to A; then dx, dy, dz, are proportional to the direction cosines of AA', and the differential equation implies a relation between these direction cosines; the locus which it represents will therefore be some curve or family of curves, and not a surface or family of surfaces.

Consider now the two differential equations

$$\frac{dx'}{P'} = \frac{dy'}{Q'} = \frac{dz'}{R'}\dotfill(\text{i}),$$

P', Q', R', being the same functions of x', y', z', that P, Q, R, are of x, y, z; their integrals are of the form

$$\left.\begin{array}{l} u_1 = a_1 \\ u_2 = a_2 \end{array}\right\}\dotfill(\text{ii}),$$

where u_1 and u_2 are functions of x', y', z'; and as they coexist, these integrals really represent the intersection of two surfaces, each of which is one of a family. This intersection of any two particular surfaces is a curve; and we therefore have a doubly infinite system of curves. One curve of this system passes through A; it is determined by those values of a_1 and a_2 obtained by substituting in u_1 and u_2 the coordinates of A. Let A'' be the point on this curve which is consecutive to A; then the direction cosines of AA'' are proportional to dx', dy', dz', or to the values of P', Q', R', at A, that is, to P, Q, R. Now the condition that AA'' and AA' may be perpendicular is

$$P dx + Q dy + R dz = 0,$$

which is the given differential equation; hence it expresses the fact that AA' is perpendicular to that curve of (ii) which passes through A. The solution of the differential equation must therefore include all the curves which cut the system (ii) orthogonally.

If we start from A in any direction which is perpendicular to the tangent at A to that curve of the system (ii) which passes through A, we shall come at A' to an adjacent curve of this system; moving from A' in any direction at right angles to this we shall at another consecutive point in this path reach another adjacent curve; and so on. The path thus obtained must be included in the solution of the differential equation; and as, at each point A, we may move in any one of an infinite number of directions (i.e. in any direction lying in the normal plane at A to the curve of the system), it follows that the solution of the equation will contain an arbitrary function.

Let us, then, draw through A any surface we please, and limit our path so as to be in this surface. Starting from A at right angles to the curve of (ii), there will, in general, be only one direction possible in the surface; moving along this through a small arc we shall at its extremity A' come to another curve; at A', there will as before be usually only one direction possible in the surface, and it will lead to another point A''; and so on. We shall thus obtain on the arbitrary surface a single path passing through the point A. Had a different point B on the same surface (but not lying in the path through A) been the starting point, there would have been similarly obtained a single path through B different from the former; and so for any point.

We should therefore have on any arbitrary surface a singly infinite series of curves.

157. This is the exact geometrical process corresponding to the analytical process applying to the case when the equation of condition was not satisfied. For what was there done was to assume an arbitrary relation among the variables—this is the equation of the arbitrary surface; it was combined with the differential equation and, after integration, another equation was obtained containing an arbitrary constant which with the original arbitrary relation was considered the solution. The new equation containing one arbitrary constant represents a family of surfaces; and the combination of the two gives the system of curves which form their intersection. Each of these curves lies on the surface first taken, and so we have an infinite series of curves on this surface. The process therefore gives the system of lines which lie on any surface and which satisfy the differential equation.

158. Now it may happen that the complete system of curves (ii) can be cut orthogonally not merely by one surface but by a family of surfaces; thus, if the system were a series of straight lines all passing through one point, they would be cut orthogonally by any sphere which had that point for centre. In this case, any curve drawn upon an orthogonal surface would cut the system (ii) at right angles, since it is at every point perpendicular to some one of the system; and such a curve would therefore be included in the solution. Hence the general solution must include all curves that can possibly be drawn upon any one of these surfaces; and therefore, if we look upon a surface as the aggregate of all the curves that can be drawn on it, we may say that the surface is included in the system of curves. As the surface is one of a family all the members of which possess the same property, we consider that the equation of this family of surfaces is the solution of the equation; and what has been said shews it to be thereby implied that the equations of every curve that can be drawn upon one of the family constitute a solution.

159. This corresponds exactly with the process applicable to the case for which the equation of condition was satisfied; we there had (§ 154) an equation $\psi = C$ and any other arbitrary equation $\chi = 0$, the two representing one curve on each of the surfaces $\psi = C$; by taking all possible arbitrary equations $\chi = 0$, we obtained all possible curves on the surfaces $\psi = C$. Thus ultimately the surfaces themselves into the expression of which the form of χ did not enter.

160. It only remains to shew how the equation of condition is derivable from the geometrical considerations. The arguments are applicable on the supposition that the system of curves, represented by

$$\frac{dx'}{P'} = \frac{dy'}{Q'} = \frac{dz'}{R'},$$

can be cut orthogonally. If they can be cut orthogonally, as at any point A, the tangent to the particular curve passing through A must coincide with the normal at A to the orthogonal surface.

Now the direction cosines of the tangent at A are proportional to the values of P', Q', R', at A, that is, to P, Q, R; and if

$$\phi\,(x',\,y',\,z') = C$$

be the orthogonal surface, the direction cosines of the normal at the point x, y, z (which is A) are proportional to $\dfrac{\partial\phi}{\partial x}, \dfrac{\partial\phi}{\partial y}, \dfrac{\partial\phi}{\partial z}$; since the direction cosines must be the same for the two lines, we must have

$$\frac{1}{P}\frac{\partial\phi}{\partial x} = \frac{1}{Q}\frac{\partial\phi}{\partial y} = \frac{1}{R}\frac{\partial\phi}{\partial z}.$$

Let each of these quantities be equal to μ so that

$$\frac{\partial\phi}{\partial x} = \mu P, \qquad \frac{\partial\phi}{\partial y} = \mu Q, \qquad \frac{\partial\phi}{\partial z} = \mu R;$$

the elimination of ϕ and μ between these leads (as in § 151) to the equation considered, which is therefore the condition that the system of curves may be cut orthogonally.

CASE OF n VARIABLES.

161. In what has preceded, only three variables have been supposed to occur; but it is easy to pass to the case when there are more than three. In order that the equation

$$X_1 dx_1 + X_2 dx_2 + X_3 dx_3 + \ldots\ldots\ldots + X_n dx_n = 0,$$

where X_1, X_2, $\ldots\ldots$, X_n, are functions of x_1, x_2, $\ldots\ldots$, x_n, should have a complete integral of the form

$$\phi\,(x_1,\,x_2,\,\ldots\ldots\ldots,\,x_n) = A,$$

the quantities X_μ must be proportional to the partial differential coefficients $\dfrac{\partial\phi}{\partial x_\mu}$, so that we may write

$$v X_\mu = \frac{\partial\phi}{\partial x_\mu},$$

for all values 1, 2, ..., n, of μ. If now λ, μ, ν, be three different suffixes, we have

$$\frac{\partial}{\partial x_\lambda}(v X_\mu) = \frac{\partial^2\phi}{\partial x_\mu \partial x_\lambda} = \frac{\partial}{\partial x_\mu}(v X_\lambda),$$

or

$$v\left(\frac{\partial X_\mu}{\partial x_\lambda} - \frac{\partial X_\lambda}{\partial x_\mu}\right) = X_\lambda \frac{\partial v}{\partial x_\mu} - X_\mu \frac{\partial v}{\partial x_\lambda}.$$

Similarly

$$v\left(\frac{\partial X_\lambda}{\partial x_\nu} - \frac{\partial X_\nu}{\partial x_\lambda}\right) = X_\nu \frac{\partial v}{\partial x_\lambda} - X_\lambda \frac{\partial v}{\partial x_\nu},$$

and

$$v\left(\frac{\partial X_\nu}{\partial x_\mu} - \frac{\partial X_\mu}{\partial x_\nu}\right) = X_\mu \frac{\partial v}{\partial x_\nu} - X_\nu \frac{\partial v}{\partial x_\mu};$$

and therefore

$$X_\nu\left(\frac{\partial X_\mu}{\partial x_\lambda} - \frac{\partial X_\lambda}{\partial x_\mu}\right) + X_\mu\left(\frac{\partial X_\lambda}{\partial x_\nu} - \frac{\partial X_\nu}{\partial x_\lambda}\right) + X_\lambda\left(\frac{\partial X_\nu}{\partial x_\mu} - \frac{\partial X_\mu}{\partial x_\nu}\right) = 0.$$

If the set of equations derived from this by all possible combinations of three different suffixes from among 1, 2, 3, ..., n, be satisfied, then the differential equation has an integral of the proposed form. The total number of these equations of condition is $\frac{1}{6}n(n-1)(n-2)$. They are not all independent; for if there be written down the four equations which involve three out of the four quantities X_λ, X_μ, X_ν, X_ρ, any one of them will be found to be derivable from the other three.

These are the conditions that must be satisfied if the differential equation has a primitive constituted by a single equation.

The converse proposition also is true: *when the conditions are satisfied, then the differential equation possesses a primitive of the form*

$$\phi(x_1, x_2, \ldots\ldots, x_n) = A.$$

The proof of this proposition can be effected in stages, each similar to the proof of the corresponding proposition in § 151 for the case of three variables*. It is, however, long: and it will be omitted here.

The truth of the proposition will be assumed.

Ex. Prove that the total number of independent equations of condition is

$$\tfrac{1}{2}(n-1)(n-2).$$

When these equations of condition or the necessarily independent equations are identically satisfied, the primitive, which must therefore exist, can be obtained by an extension of the method

---

* A proof, arranged rather differently from the proof suggested in the text, is given in the author's *Theory of Differential Equations*, vol. I., §§ 7—11.

adopted for equations with three variables. We integrate as if all but two of the variables were constant; and we replace the arbitrary constant by an arbitrary function of all those variables which were supposed constant. The equation so obtained is differentiated with regard to all the variables, and the result is made to agree with the given equation; the conditions necessary for this agreement will serve to determine the arbitrary function which was introduced and so to determine the primitive*.

Ex. 1. It is easily verifiable that the coefficients of the differentials in the equation

$$(2x_1 + x_2{}^2 + 2x_1 x_4 - x_3)\,dx_1 + 2x_1 x_2 dx_2 - x_1 dx_3 + x_1{}^2 dx_4 = 0,$$

satisfy the equations of condition which are four in number, three being independent. Following the rule, we assume that only two of the variables may change. These may be taken to be x_3 and x_4; the integral derived is

$$- x_1 x_3 + x_1{}^2 x_4 = C = \phi,$$

where ϕ is a function of x_1 and x_2. Differentiating this, we have

$$(- x_3 + 2x_1 x_4)\,dx_1 - x_1 dx_3 + x_1{}^2 dx_4 = d\phi,$$

and a comparison of this with the given equations shews that

$$- d\phi = (2x_1 + x_2{}^2)\,dx_1 + 2x_1 x_2 dx_2.$$

We thus have an equation involving three differentials $d\phi$, dx_1, dx_2, instead of four (we should have, in the general case, an equation involving $n-1$ differentials instead of n). The rule is reapplied to this, and the number is again decreased by unity; and so on, until we can obtain a final integral. In the example specially considered, the integral is easily seen to be

$$- \phi + A = x_1{}^2 + x_1 x_2{}^2,$$

where A is now an arbitrary constant; and the primitive is

$$x_1{}^2 + x_1 x_2{}^2 - x_1 x_3 + x_1{}^2 x_4 = A.$$

Ex. 2. Shew that the following equations have a primitive of the form considered, and obtain it for each of them:

    (i) $yzu\,dx + zux\,dy + uxy\,dz + xyz\,du = 0$;

    (ii) $(y+z+u)\,dx + (z+u+x)\,dy + (u+x+y)\,dz + (x+y+z)\,du = 0$;

    (iii) $z(y+z)\,dx + z(u-x)\,dy + y(x-u)\,dz + y(y+z)\,du = 0$.

162. When only some of the equations of condition are satisfied, or when none of them are satisfied, the inference from the preceding analysis is that no single integral equation exists which is equivalent to the differential equation.

---

  * Other methods are given in the author's *Theory of Differential Equations*, vol. I., Chap. I.

In such a case, the question of determining the integral equivalent is often called *Pfaff's problem*, and the differential equation itself is often called a *Pfaffian equation*: the reason being that the first important contribution to the solution was made in a memoir by Pfaff[*]. He shewed that an integral equivalent of a total differential equation, containing $2n - 1$ or $2n$ variables, can always be constituted by a system of integral equations not more than n in number. For the establishment of this result, and for allied investigations, reference must be made to other sources[†]. Here we shall be content with giving the solution for the case of three variables and for the case of four variables.

163. When there are three variables, we take the total equation in the form

$$P\,dx + Q\,dy + R\,dz = 0,$$

where we shall assume that the condition of integrability is not now satisfied.

We proceed to find whether it is possible to obtain quantities u, v, w, functions of the variables x, y, z, such that we may have

$$P\,dx + Q\,dy + R\,dz = du + v\,dw$$

identically. Should it be possible, we must have

$$P = \frac{\partial u}{\partial x} + v\,\frac{\partial w}{\partial x}, \quad Q = \frac{\partial u}{\partial y} + v\,\frac{\partial w}{\partial y}, \quad R = \frac{\partial u}{\partial z} + v\,\frac{\partial w}{\partial z};$$

and therefore, writing

$$P' = \frac{\partial Q}{\partial z} - \frac{\partial R}{\partial y}, \quad Q' = \frac{\partial R}{\partial x} - \frac{\partial P}{\partial z}, \quad R' = \frac{\partial P}{\partial y} - \frac{\partial Q}{\partial x};$$

we have, on substitution,

$$P' = \frac{\partial v}{\partial z}\frac{\partial w}{\partial y} - \frac{\partial v}{\partial y}\frac{\partial w}{\partial z},$$

$$Q' = \frac{\partial v}{\partial x}\frac{\partial w}{\partial z} - \frac{\partial v}{\partial z}\frac{\partial w}{\partial x},$$

$$R' = \frac{\partial v}{\partial y}\frac{\partial w}{\partial x} - \frac{\partial v}{\partial x}\frac{\partial w}{\partial y}.$$

[*] *Berlin. Abhand.* (1814–5), pp. 76—136.

[†] A full discussion of Pfaff's Problem will be found in the author's *Theory of Differential Equations*, vol. I., which also contains a discussion of systems of total equations (see Chap. II., for systems of exact equations; Chap. XIII., for systems of Pfaffian equations).

Multiplying these equations, first by $\dfrac{\partial w}{\partial x}, \dfrac{\partial w}{\partial y}, \dfrac{\partial w}{\partial z}$, respectively, and adding, we have

$$P'\frac{\partial w}{\partial x} + Q'\frac{\partial w}{\partial y} + R'\frac{\partial w}{\partial z} = 0;$$

and then by $\dfrac{\partial v}{\partial x}, \dfrac{\partial v}{\partial y}, \dfrac{\partial v}{\partial z}$, respectively, and adding, we have

$$P'\frac{\partial v}{\partial x} + Q'\frac{\partial v}{\partial y} + R'\frac{\partial v}{\partial z} = 0.$$

We also have

$$\left(P - \frac{\partial u}{\partial x}\right)P' + \left(Q - \frac{\partial u}{\partial y}\right)Q' + \left(R - \frac{\partial u}{\partial z}\right)R'$$
$$= v\left(P'\frac{\partial w}{\partial x} + Q'\frac{\partial w}{\partial y} + R'\frac{\partial w}{\partial z}\right) = 0;$$

so that

$$P'\frac{\partial u}{\partial x} + Q'\frac{\partial u}{\partial y} + R'\frac{\partial u}{\partial z} = PP' + QQ' + RR'.$$

The right-hand side in this equation does not vanish, because the condition of integrability is not satisfied; otherwise, u obviously would satisfy the same equation as w and v.

It thus appears that both v and w satisfy the equation[*]

$$P'\frac{\partial \theta}{\partial x} + Q'\frac{\partial \theta}{\partial y} + R'\frac{\partial \theta}{\partial z} = 0.$$

Anticipating the results of Chapter IX., we can solve this equation by constructing the system of equations

$$\frac{dx}{P'} = \frac{dy}{Q'} = \frac{dz}{R'},$$

and integrating these equations; if their two independent integrals are

$$\alpha\,(x,\, y,\, z) = \text{constant}, \quad \beta\,(x,\, y,\, z) = \text{constant},$$

every integral of the equation in θ is a function of α and β. Accordingly, both v and w are any functions of α and β.

---

[*] The quantities v and w may be expected to satisfy the same equation; for we can write

$$du + v\,dw = d\,(U + vw) - w\,dv,$$

that is, $P\,dx + Q\,dy + R\,dz$ is unaltered when we interchange w and $-v$, and substitute $U + vw$ for u.

Suppose now that we take w equal to any function of α and β, the simpler the better for our purpose; let us choose $w = \alpha$. When we make

$$\alpha(x, y, z) = a, \text{ a constant,}$$

which therefore is a relation among the variables, we have

$$P\,dx + Q\,dy + R\,dz = du,$$

that is, for this relation among the variables, $P\,dx + Q\,dy + R\,dz$ is a perfect differential. Accordingly we use the relation $\alpha(x, y, z) = a$ to remove one of the variables and its differential element, say z and dz, from $P\,dx + Q\,dy + R\,dz$; the resulting expression is a perfect differential, say $d\phi(x, y, a)$. In $\phi(x, y, a)$ we reinsert the variable z and remove the constant a by substituting $a = \alpha(x, y, z)$; and then $\phi(x, y, a)$ becomes u.

We thus have u and w. The value of v is then given by any one of the equations

$$P - \frac{\partial u}{\partial x} = v\,\frac{\partial w}{\partial x}, \quad Q - \frac{\partial u}{\partial y} = v\,\frac{\partial w}{\partial y}, \quad R - \frac{\partial u}{\partial z} = v\,\frac{\partial w}{\partial z}.$$

Consequently, we can consider the quantities u, v, w, known.

Now our differential equation is

$$P\,dx + Q\,dy + R\,dz = 0;$$

and therefore we have

$$du + v\,dw = 0.$$

Manifestly, there are different kinds of integrals, as follows. First, we can have

$$u = \text{constant}, \qquad w = \text{constant}.$$

Again, we can have

$$u = \text{constant}, \qquad v = 0.$$

Again, we can have

$$\psi(u, w) = 0, \qquad v\,\frac{\partial \psi}{\partial u} - \frac{\partial \psi}{\partial w} = 0,$$

where ψ is any arbitrary function.

In each case, *the integral equivalent of the differential equation consists of two equations; it does not consist of one alone, as is the fact when the condition of integrability is satisfied.*

Ex. 1. Obtain an integral equivalent of the equation

$$y\,dx + z\,dy + x\,dz = 0.$$

Here $P = y$, $Q = z$, $R = x$, $P' = 1$, $Q' = 1$, $R' = 1$; the condition of integrability is not satisfied.

The equations

$$\frac{dx}{P'} = \frac{dy}{Q'} = \frac{dz}{R'}$$

become

$$dx = dy = dz,$$

so that we have

$$a = x - y, \quad \beta = y - z.$$

According to the rule, we take $w = a$, and we simplify the equation by taking

$$x - y = a.$$

Then

$$y\,dx + z\,dy + x\,dz = y\,dy + z\,dy + (y + a)\,dz,$$

so that

$$\phi = \tfrac{1}{2}y^2 + yz + az,$$

and therefore

$$u = \tfrac{1}{2}y^2 + yz + (x - y)\,z$$
$$= \tfrac{1}{2}y^2 + xz.$$

Lastly, we have v given by

$$P - \frac{\partial u}{\partial x} = v\,\frac{\partial w}{\partial x},$$

that is,

$$v = y - z.$$

Hence, taking

$$u = \tfrac{1}{2}y^2 + xz, \quad v = y - z, \quad w = x - y,$$

we have

$$y\,dx + z\,dy + x\,dz = du + v\,dw.$$

Integral equivalents of the equation are given as follows:—

    (i) $u = c, \quad w = a$;

    (ii) $u = c, \quad v = 0$;

    (iii) $\psi(u, w) = 0, \quad v\,\dfrac{\partial \psi}{\partial u} - \dfrac{\partial \psi}{\partial w} = 0,$

where a and c are arbitrary constants, and ψ is any arbitrary function.

(It will be noticed that $v = \beta$, thus verifying the earlier remark that v and w satisfy the same equation.)

Ex. 2. Obtain integral equivalents of the equation, in Ex. 1, in the form

    (i) $\tfrac{1}{2}z^2 + xy = c, \quad y - z = a$;

    (ii) $\tfrac{1}{2}z^2 + xy = c, \quad z - x = 0$;

    (iii) $\psi(u', w') = 0, \quad (z - x)\,\dfrac{\partial \psi}{\partial u'} - \dfrac{\partial \psi}{\partial w'} = 0,$

where $u' = \tfrac{1}{2}z^2 + xy$, $w' = y - z$.

Ex. 3. Obtain integral equivalents of the equations

    (i) $z\,dx + x\,dy + y\,dz = 0$;

    (ii) $(ay + bz)\,dx + (a'z + b'x)\,dy + (a''x + b''y)\,dz = 0$,

        where the coefficients are constants;

    (iii) $y^2\,dx + z^2\,dy + x^2\,dz = 0$;

    (iv) $xy\,dx + yz\,dy + xz\,dz = 0$.

164. We have seen that when the condition of integrability is satisfied, so that $PP' + QQ' + RR' = 0$, then u satisfies the same equation as v and w.

In that case, each of the quantities u, v, w, is (by the later theory of partial differential equations of the first order) a function of α and β only, where $\alpha(x, y, z) = $ constant and $\beta(x, y, z) = $ constant are two independent integrals of the equations

$$\frac{dx}{P'} = \frac{dy}{Q'} = \frac{dz}{R'}.$$

Accordingly, substituting such values for u, v, w, in $du + v\,dw$, we have

$$P\,dx + Q\,dy + R\,dz = A\,d\alpha + B\,d\beta,$$

where A and B are functions of α and β as yet unknown. When α and β are supposed known, then A and B can be determined from the equations

$$P = A\frac{\partial\alpha}{\partial x} + B\frac{\partial\beta}{\partial x}, \quad Q = A\frac{\partial\alpha}{\partial y} + B\frac{\partial\beta}{\partial y}, \quad R = A\frac{\partial\alpha}{\partial z} + B\frac{\partial\beta}{\partial z};$$

after being determined, they should be expressed in terms of α and β alone.

The integration of the given equation then depends upon the integration of the equation

$$A\,d\alpha + B\,d\beta = 0,$$

which is an equation in two variables only.

We thus obtain another method for integrating an equation $P\,dx + Q\,dy + R\,dz = 0$ when the condition of integrability is satisfied: viz. *we take the two independent integrals*

$$\alpha(x, y, z) = a, \qquad \beta(x, y, z) = b,$$

of the equations

$$\frac{dx}{\dfrac{\partial Q}{\partial z}-\dfrac{\partial R}{\partial y}}=\frac{dy}{\dfrac{\partial R}{\partial x}-\dfrac{\partial P}{\partial z}}=\frac{dz}{\dfrac{\partial P}{\partial y}-\dfrac{\partial Q}{\partial x}};$$

we express $Pdx+Qdy+Rdz$ in the form $Ad\alpha+Bd\beta$, where A and B are to be expressed as functions of α and β only; and we integrate the equation

$$Ad\alpha+Bd\beta=0.$$

Its integral is the integral of the original equation.

Ex. 1. Integrate the equation

$$E=-zy\,dx+zx\,dy+y^2dz=0.$$

We have

$$P'=x-2y,\quad Q'=y,\quad R'=-2z.$$

The condition of integrability is satisfied. Two independent integrals of the equations

$$\frac{dx}{x-2y}=\frac{dy}{y}=\frac{dz}{-2z}$$

are given by

$$\alpha=y^2z=\text{constant},\quad \beta=\frac{x}{y}+2\log y=\text{constant}.$$

For the quantities A and B, we have the equations

$$-zy=B\,\frac{1}{y},\quad zx=A\,.\,2yz+B\left(-\frac{x}{y^2}+\frac{2}{y}\right),$$

that is,

$$A=1,\quad B=-\alpha.$$

Hence

$$E=d\alpha-\alpha\,d\beta=0,$$

the integral of which is

$$\beta-\log\alpha=A.$$

Consequently the primitive of the equation is

$$\frac{x}{y}-\log z=A.$$

Ex. 2. Integrate, in this manner, the equations in Ex. 2, § 152.

165. When there are four variables, we denote them by x_1, x_2, x_3, x_4; and we write the total equation in the form

$$\Omega=X_1dx_1+X_2dx_2+X_3dx_3+X_4dx_4=0,$$

where X_1, X_2, X_3, X_4, are functions of all the variables. We first proceed to obtain an equivalent expression for Ω, such that it shall contain the smallest number of differential elements.

Take three independent functions of all the variables, denoting them by y_1, y_2, y_3; and express three of the original independent variables, say x_1, x_2, x_3, in terms of y_1, y_2, y_3, x_4, so that these now become the independent variables. Making this change, we have Ω in the form

$$\Omega = Y_1 dy_1 + Y_2 dy_2 + Y_3 dy_3 + Y_4 dx_4,$$

where Y_1, Y_2, Y_3, Y_4, are functions of the variables y_1, y_2, y_3, x_4. Now the three functions y_1, y_2, y_3, are at our disposal; and they will be determinable by three conditions or relations. As one of these, we shall require that $Y_4 = 0$. As the other two, we shall require that the ratios $Y_1 : Y_2 : Y_3$ are independent of x_4. Accordingly, we can write

$$Y_1 = MU_1, \quad Y_2 = MU_2, \quad Y_3 = MU_3, \quad Y_4 = 0,$$

where U_1, U_2, U_3 are functions of y_1, y_2, y_3, alone; and we then have

$$\Omega = M(U_1 dy_1 + U_2 dy_2 + U_3 dy_3),$$

where the factor M is, or may be, a function of y_1, y_2, y_3, x_4.

Now the quantity

$$U_1 dy_1 + U_2 dy_2 + U_3 dy_3$$

is an expression, involving only the three variables y_1, y_2, y_3; it has exactly the same form as the expression discussed in § 163; and therefore, even in the most general case when no condition holds between the coefficients U_1, U_2, U_3, it can be represented in the form

$$U_1 dy_1 + U_2 dy_2 + U_3 dy_3 = du + v dw.$$

But unless the condition of integrability is satisfied, we cannot transform the differential expression into another which contains only one differential element. When this form is substituted in Ω, we have

$$\Omega = M(du + v dw)$$
$$= M du + N dw.$$

From what precedes it is clear that, in general, we could not obtain for Ω an expression containing only one differential element; and we therefore regard this form as the reduced equivalent form for Ω.

The foregoing method of proceeding involves two distinct stages. In the first, y_1, y_2, y_3, are determined. In the second, u, v, w, are determined. Instead of carrying out these two stages, we shall accept $M\,du + N\,dw$ as the type of the reduced form for Ω, and shall obtain u, w, M, N, by a different method. Accordingly, we take

$$X_1 dx_1 + X_2 dx_2 + X_3 dx_3 + X_4 dx_4 = \Omega = M\,du + N\,dw;$$

then writing $\dfrac{\partial u}{\partial x_1} = u_1$, and so for other derivatives, we have

$$X_r = M u_r + N w_r,$$

for $r = 1, 2, 3, 4$. Now let

$$a_{mn} = \frac{\partial X_m}{\partial x_n} - \frac{\partial X_n}{\partial x_m},$$

for all the combinations m, $n = 1, 2, 3, 4$, so that

$$a_{mm} = 0, \quad a_{mn} = - a_{nm};$$

and let

$$\left.\begin{aligned}
W_1 &= \quad\quad\; X_2 a_{34} + X_3 a_{42} + X_4 a_{23} \\
W_2 &= X_1 a_{34} \quad\quad\quad + X_3 a_{41} + X_4 a_{13} \\
W_3 &= X_1 a_{24} + X_2 a_{41} \quad\quad\quad + X_4 a_{12} \\
W_4 &= X_1 a_{23} + X_2 a_{31} + X_3 a_{12}
\end{aligned}\right\},$$

where obviously

$$W_1 X_1 - W_2 X_2 + W_3 X_3 - W_4 X_4 = 0.$$

We have

$$a_{mn} = \frac{\partial}{\partial x_n}(M u_m + N w_m) - \frac{\partial}{\partial x_m}(M u_n + N w_n)$$

$$= \frac{\partial M}{\partial x_n} u_m - \frac{\partial M}{\partial x_m} u_n + \frac{\partial N}{\partial x_n} w_m - \frac{\partial N}{\partial x_m} w_n;$$

and therefore, if

$$J_r = M \frac{\partial N}{\partial x_r} - N \frac{\partial M}{\partial x_r},$$

for $r = 1, 2, 3, 4$, we find

$$W_1 = (M u_2 + N w_2)\,a_{34} + (M u_3 + N w_3)\,a_{42} + (M u_4 + N w_4)\,a_{23}$$

$$= J_2 (u_3 w_4 - u_4 w_3) + J_3 (u_4 w_2 - u_2 w_4) + J_4 (u_2 w_3 - u_3 w_2).$$

Similarly

$$W_2 = J_1\,(u_3w_4 - u_4w_3) + J_3\,(u_4w_1 - u_1w_4) + J_4\,(u_1w_3 - u_3w_1),$$

$$W_3 = J_1\,(u_2w_4 - u_4w_2) + J_2\,(u_4w_1 - u_1w_4) + J_4\,(u_1w_2 - u_2w_1),$$

$$W_4 = J_1\,(u_2w_3 - u_3w_2) + J_2\,(u_3w_1 - u_1w_3) + J_3\,(u_1w_2 - u_2w_1).$$

It follows immediately that

$$W_1u_1 - W_2u_2 + W_3u_3 - W_4u_4 = 0,$$

$$W_1w_1 - W_2w_2 + W_3w_3 - W_4w_4 = 0;$$

and therefore u and w are two integrals of the partial differential equation

$$W_1\frac{\partial f}{\partial x_1} - W_2\frac{\partial f}{\partial x_2} + W_3\frac{\partial f}{\partial x_3} - W_4\frac{\partial f}{\partial x_4} = 0.$$

Hence, again (as in § 163) anticipating the results of Chapter IX., we form the equations

$$\frac{dx_1}{W_1} = \frac{dx_2}{-W_2} = \frac{dx_3}{W_3} = \frac{dx_4}{-W_4},$$

where

$$W_k = X_l a_{mn} + X_m a_{nl} + X_n a_{lm},$$

(k, l, m, n, $= 1$, 2, 3, 4, in cyclical order), we know that $u = $ constant and $w = $ constant are two integrals of these subsidiary equations.

166. We now can proceed definitely to the construction of the reduced form of Ω. We take any integral, the simpler the better, of the subsidiary equations

$$\frac{dx_1}{W_1} = \frac{dx_2}{-W_2} = \frac{dx_3}{W_3} = \frac{dx_4}{-W_4};$$

let it be

$$u\,(x_1,\,x_2,\,x_3,\,x_4) = a,$$

where a is a constant. Now

$$\Omega = M\,du + N\,dw;$$

and therefore, when we postulate the relation $u = a$ among the variables, we have

$$\Omega = N\,dw,$$

that is, the equation $\Omega = 0$ is exactly integrable; and its integral, under that relation, is $w = c$, where c is a constant. Accordingly, *we use the relation*

$$u(x_1, x_2, x_3, x_4) = a$$

to eliminate one of the variables and its differential element, say x_1 and dx_1, from the original equation $\Omega = 0$; we integrate the resulting form of the equation, obtaining its integral in the form

$$\phi(x_2, x_3, x_4, a) = c;$$

and we then substitute u for a, obtaining the modified form

$$w(x_1, x_2, x_3, x_4) = c$$

of the last relation.

With these functions u and w, the original total equation becomes

$$\Omega = M\,du + N\,dw;$$

and the quantities M and N are given by the relations

$$X_r = Mu_r + Nw_r,$$

for $r = 1, 2, 3, 4$. We thus obtain a reduced form for the differential expression Ω.

167. Integrals of the equation $\Omega = 0$ are now obvious. We have several classes as follows:—

$$\text{(i)} \quad u = a, \quad w = c;$$

$$\text{(ii)} \quad u = a, \quad N = 0;$$

$$\text{(iii)} \quad M = 0, \quad w = c;$$

$$\text{(iv)} \quad M = 0, \quad N = 0;$$

$$\text{(v)} \quad f(u, w) = 0, \quad M\frac{\partial f}{\partial w} - N\frac{\partial f}{\partial u} = 0;$$

where a and c are arbitrary constants, and f is an arbitrary function.

In each of these cases, the integral equivalent of the single total differential equation in four independent variables consists of two integral equations free from differential elements.

For further developments, references may be made as indicated in § 162.

Ex. 1. Obtain an integral equivalent of the equation

$$\Omega = x_4\,dx_1 + x_1\,dx_2 + x_2\,dx_3 + x_3\,dx_4 = 0.$$

Here

$$X_1 = x_4, \quad X_2 = x_1, \quad X_3 = x_2, \quad X_4 = x_3,$$

$$a_{12} = -1, \quad a_{13} = 0, \quad a_{14} = 1, \quad a_{23} = -1, \quad a_{24} = 0, \quad a_{34} = -1;$$

and therefore

$$W_1 = -x_1 - x_3 = W_3, \quad W_2 = -x_4 - x_2 = -W_4.$$

One integral of the subsidiary equations

$$\frac{dx_1}{W_1} = \frac{dx_2}{-W_2} = \frac{dx_3}{W_3} = \frac{dx_4}{-W_4}$$

is evidently given by

$$x_1 - x_3 = \text{constant}$$

so we take

$$u = x_1 - x_3 = a,$$

and use this relation to modify Ω. We have

$$\Omega = x_4\,dx_3 + (x_3 + a)\,dx_2 + x_2\,dx_3 + x_3\,dx_4$$

$$= d\,(x_3 x_4 + x_2 x_3 + a x_2);$$

hence

$$\phi = x_3 x_4 + x_2 x_3 + a x_2,$$

and so

$$w = x_3 x_4 + x_1 x_2.$$

We easily find

$$M = x_4 - x_2, \quad N = 1;$$

and with these values

$$\Omega = (x_4 - x_1)\,du + dw.$$

Integrals are given by

(i) $x_1 - x_3 = a, \quad x_1 x_2 + x_3 x_4 = c$;

(ii) $x_4 - x_1 = 0, \quad x_1 x_2 + x_3 x_4 = c$;

(iii) $f(u, w) = 0, \quad \dfrac{\partial f}{\partial u} - (x_4 - x_1)\dfrac{\partial f}{\partial w} = 0.$

Ex. 2. Another integral of the subsidiary equations in the preceding example is given by

$$u = x_2 - x_4 = a.$$

Obtain the corresponding reduced form for Ω; and state the corresponding integral equivalents.

Ex. 3. Obtain integral equivalents of the equations:

(i) $x_2\,dx_1 + x_3\,dx_2 + x_4\,dx_3 + x_1\,dx_4 = 0$;

(ii) $x_4\,dx_1 + 2x_1\,dx_2 + x_2\,dx_3 + 2x_3\,dx_4 = 0$;

(iii) $2x_2\,dx_1 + x_3\,dx_2 + 2x_4\,dx_3 + x_1\,dx_4 = 0.$

Ex. 4. Shew that, if the equation

$$\Omega = X_1 dx_1 + X_2 dx_2 + X_3 dx_3 + X_4 dx_4 = 0$$

is reducible to one or other of the following forms, the respective conditions are necessary and sufficient:—

(i) if $\Omega = du$, then $a_{mn} = 0$, for m, $n = 1, 2, 3, 4$; being six conditions;

(ii) if $\Omega = Mdu$, then $W_r = 0$, for $r = 1, 2, 3, 4$; being three independent conditions;

(iii) if $\Omega = Mdu + dw$, then $a_{12}a_{34} + a_{13}a_{42} + a_{14}a_{23} = 0$; being one condition.

Ex. 5. Shew that, in the most general case when no condition is satisfied by the quantities X, and when one reduced form of Ω is $Mdu + Ndw$, the quantities U and W for reduction to any similar form are given by

$$\left. \begin{array}{c} f(U, W, u, w) = 0 \\ \dfrac{1}{M}\dfrac{\partial f}{\partial u} = \dfrac{1}{N}\dfrac{\partial f}{\partial w} \end{array} \right\},$$

where f is any function whatever.

Deduce the most general integrals of $\Omega = 0$, when one reduced form $Mdu + Ndw = 0$ is known.

EQUATIONS OF A DEGREE HIGHER THAN THE FIRST.

168. Equations may arise in which the differentials of the variables occur in a degree higher than the first; into their solution it is not proposed to enter fully but only to indicate a method of proceeding in some cases. The general equation of the second degree may be taken as

$$Xdx^2 + Ydy^2 + Zdz^2 + 2X'dy\,dz + 2Y'dz\,dx + 2Z'dx\,dy = 0,$$

in which X, Y, Z, X', Y', Z', are functions of x, y, and z. If the left-hand side can be resolved into two factors, then the equation may be replaced by two others each of the form

$$Pdx + Qdy + Rdz = 0,$$

obtained by equating separately to zero the two factors. The solution of either of these, obtained by previous methods, will be a particular solution of the differential equation proposed; and the two general solutions taken together will constitute the complete solution. In the case when each of the linear equations is satisfied, in the sense of the preceding paragraphs, by a single integral of the respective forms

$$\psi_1(x, y, z) - C_1 = 0, \quad \psi_2(x, y, z) - C_2 = 0,$$

the general solution will, as in § 19, be represented by

$$\{\psi_1 (x, y, z) - C\} \{\psi_2 (x, y, z) - C\} = 0 \quad \ldots\ldots\ldots (A).$$

In the case when two separate equations are needed for the solution, each corresponding pair must be looked upon as a solution.

Now the condition that these should be solutions is that the left-hand side of the original equation should be resoluble into factors. The left-hand side is equal to

$$\frac{1}{Z} [(Z dz + Y' dx + X' dy)^2 - \{(Y'^2 - XZ) dx^2 - 2 (ZZ' - X'Y') dx dy$$
$$+ (X'^2 - YZ) dy^2\}].$$

In order that this may resolve into two factors, we must have

$$(Y'^2 - XZ) dx^2 - 2 (ZZ' - X'Y') dx dy + (X'^2 - YZ) dy^2$$

a perfect square, which will be the case if

$$(Y'^2 - XZ)(X'^2 - YZ) - (ZZ' - X'Y')^2 = 0,$$

that is, if

$$Z(XYZ + 2X'Y'Z' - XX'^2 - YY'^2 - ZZ'^2) = 0;$$

or, since Z is not zero, we must have

$$XYZ + 2X'Y'Z' - XX'^2 - YY'^2 - ZZ'^2 = 0.$$

When this condition is satisfied, the general solution is obtained in the foregoing manner.

When this condition is not satisfied, the proposed equation does not admit of a single primitive of the form (A), nor does it admit of a set of separate primitives each given by a pair of equations; but it does in general admit of a solution expressed by a system of simultaneous equations.

Note. Another method of proceeding is indicated by Guldberg[*]. When the equation of the second degree which has been discussed is satisfied in virtue of a single integral equation, the latter may be regarded as the primitive. This primitive defines z as a function of x and y, and is equivalent to

$$dz = p dx + q dy,$$

where

$$p = \frac{\partial z}{\partial x}, \quad q = \frac{\partial z}{\partial y};$$

[*] *Crelle*, t. cxxii. (1900), p. 34.

and therefore this new differential relation is equivalent to the original equation. Consequently, when dz in the original equation is replaced by the value $p\,dx + q\,dy$, the result must be satisfied for all values of dx and dy. Now the result of the substitution is

$$(X + 2pY' + p^2Z)\,dx^2 + (Y + 2qX' + q^2Z)\,dy^2$$
$$+ 2(Z' + pX' + qY' + pqZ)\,dx\,dy = 0;$$

as this equation must be satisfied for all values of dx and dy, we must have

$$X + 2pY' + p^2Z = 0,$$
$$Y + 2qX' + q^2Z = 0,$$
$$Z' + pX' + qY' + pqZ = 0.$$

The first of these gives

$$(pZ + Y')^2 = Y'^2 - XZ;$$

the second gives

$$(qZ + X')^2 = X'^2 - YZ;$$

and the third gives

$$(pZ + Y')(qZ + X') = X'Y' - ZZ'.$$

Hence

$$(Y'^2 - XZ)(X'^2 - YZ) = (X'Y' - ZZ')^2,$$

which, after removal of the factor Z from the expanded expression, gives

$$XYZ + 2X'Y'Z' - XX'^2 - YY'^2 - ZZ'^2 = 0,$$

the former condition.

Moreover, we have

$$pZ + Y' = (Y'^2 - XZ)^{\frac{1}{2}},$$
$$qZ + X' = (X'Y' - ZZ')(Y^2 - XZ)^{-\frac{1}{2}},$$

so that

$$p = \frac{1}{Z}\{- Y' + (Y'^2 - XZ)^{\frac{1}{2}}\},$$

$$q = \frac{1}{Z}\{- X' + (X'Y' - ZZ')(Y^2 - XZ)^{-\frac{1}{2}}\}.$$

In order that

$$dz = p\,dx + q\,dy$$

may be represented by a single primitive, we must have

$$\frac{\partial p}{\partial y} = \frac{\partial q}{\partial x}, \text{ or } \frac{\partial p}{\partial y} + q\frac{\partial p}{\partial z} = \frac{\partial q}{\partial x} + p\frac{\partial q}{\partial z},$$

according as p and q do not, or do, involve z. Substituting the values of p and q, we obtain a further condition among the coefficients of the original equation.

If it is satisfied for one sign of the radical, but not for both signs, then the primitive is of the form

$$\psi_1(x, y, z) - C = 0.$$

If it is satisfied for both signs of the radical, the primitive is of the form

$$\{\psi_1(x, y, z) - C\}\{\psi_2(x, y, z) - C\} = 0.$$

Ex. 1. The equation

$$x^2 dx^2 + y^2 dy^2 - z^2 dz^2 + 2xy\, dx\, dy = 0$$

satisfies the condition; and the equivalent equations are

$$x\, dx + y\, dy + z\, dz = 0, \quad x\, dx + y\, dy - z\, dz = 0,$$

which lead to the integrals

$$x^2 + y^2 + z^2 - a_1 = 0, \quad x^2 + y^2 - z^2 - a_2 = 0,$$

and therefore a general solution will be

$$(x^2 + y^2 + z^2 - a)(x^2 + y^2 - z^2 - a) = 0,$$

i.e.

$$(x^2 + y^2 - a)^2 = z^4,$$

in which a is an arbitrary constant.

Ex. 2. Solve

(i) $ll'dx^2 + mm'dy^2 + nn'dz^2 + (lm' + l'm)\, dx\, dy + (ln' + l'n)\, dx\, dz + (mn' + m'n)\, dz\, dy = 0;$

(ii) $(x\, dx + y\, dy + z\, dz)^2 z = (z^2 - x^2 - y^2)(x\, dx + y\, dy + z\, dz)\, dz;$

(iii) $dx\, dy\, dz = 0;$

(iv) $\begin{vmatrix} dx, & dy, & dz \\ x, & y, & mz \\ dx, & dy, & m\, dz \end{vmatrix} = 0$, where m is a constant.

Ex. 3. Obtain a solution of the equation

$$a(b - c)\, x\, dy\, dz + b(c - a)\, y\, dz\, dx + c(a - b)\, z\, dx\, dy = 0,$$

consistent with the equation

$$ax^2 + by^2 + cz^2 = 1.$$

(The former is the differential equation of the lines of curvature upon the surface represented by the latter.)

Ex. 4. Also of the equation

$$\begin{vmatrix} x^2 dx, & y^2 dy, & z^2 dz \\ dx, & dy, & dz \\ x, & y, & z \end{vmatrix} = 0,$$

consistent with the equation

$$xyz = 1.$$

SIMULTANEOUS EQUATIONS WITH CONSTANT COEFFICIENTS.

169. We have hitherto considered only single differential equations; we proceed now to consider systems of equations. The simplest, and at the same time the most frequently occurring, class is that containing only one independent variable of which all other variables are functions; for the separate and complete determination of each of these dependent variables, the number of equations in the system must be equal to the number of dependent variables. In this class are included most of the differential equations of dynamics. Thus in the case of the chief problem of physical astronomy—that of determining the motion of a system of material bodies under the influence of their mutual attractions—there is a single independent variable, the time elapsed from some definite epoch, while the dependent variables are the coordinates of the several bodies; these coordinates vary with the time and so furnish the varying positions of the bodies; and they are potentially determinate, since the number of equations is equal to the total number of coordinates. All equations dealing with the small oscillations in a steadily moving system of bodies are also included; in them there is the additional simplification that all the equations are linear, the quantities multiplying the differential coefficients being constants.

The general theory of the latter will be first considered.

170. Let t denote the independent variable and D stand for d/dt; taking the simplest possible general case, we have two equations involving two dependent variables denoted by x and y. As the equations are supposed linear, all the terms involving differential coefficients of x can be gathered together, and so also for all those involving differential coefficients of y; and the equations may therefore be written in the form

$$f_1(D)\,x + \phi_1(D)\,y = T_1 \brace f_2(D)\,x + \phi_2(D)\,y = T_2 \quad\cdots\cdots\cdots\cdots(\text{I}),$$

where f_1, f_2, ϕ_1, ϕ_2, are polynomial functions with constant coefficients, and T_1 and T_2 are explicit functions of t alone, a constant or a zero value not being excluded. Operate on both the sides of the first equation with $\phi_2(D)$ and on both the sides of the second with $\phi_1(D)$; then they become

$$\phi_2(D)\,f_1(D)\,x + \phi_2(D)\,\phi_1(D)\,y = \phi_2(D)\,T_1 \brace \phi_1(D)\,f_1(D)\,x + \phi_1(D)\,\phi_2(D)\,y = \phi_1(D)\,T_2 \cdot$$

Since the functions ϕ have only constants in their coefficients, it follows that

$$\phi_2(D)\,\phi_1(D)\,y = \phi_1(D)\,\phi_2(D)\,y,$$

and therefore the above equations give

$$\{\phi_2(D)\,f_1(D) - \phi_1(D)\,f_2(D)\}\,x = \phi_2(D)\,T_1 - \phi_1(D)\,T_2 \quad\cdots\cdots(\text{II}).$$

Now let l_1, l_2, m_1, m_2, be the indices of the highest differential coefficients in f_1, f_2, ϕ_1, ϕ_2, respectively; then the index of the highest differential in $\phi_2(D)\,f_1(D)$ is $m_2 + l_1$, and in $\phi_1(D)\,f_2(D)$ is $m_1 + l_2$; of these two numbers let n denote that which is not less than the other, so that n is generally* the order of the highest differential coefficient of x in the linear equation determining x. To solve it, we adopt the method of Chapter III. applicable to an ordinary single equation; if P be any value of x which satisfies the equation (there called the Particular Integral), and $\lambda_1, \lambda_2, \ldots, \lambda_n$, denote the n roots of the equation

$$\phi_2(\lambda)\,f_1(\lambda) - \phi_1(\lambda)\,f_2(\lambda) = 0 \quad\cdots\cdots\cdots\cdots(\text{A}),$$

the complete value of x is

$$x = A_1 e^{\lambda_1 t} + A_2 e^{\lambda_2 t} + \cdots\cdots + A_n e^{\lambda_n t} + P,$$

where $A_1, A_2, \ldots\ldots, A_n$, are arbitrary constants.

Proceed in the same way to eliminate x from the two fundamental equations, by operating on the first with $f_2(D)$ and subtracting it from the second after this has been operated upon with $f_1(D)$; we then have

$$\{\phi_2(D)\,f_1(D) - \phi_1(D)\,f_2(D)\}\,y = f_1(D)\,T_2 - f_2(D)\,T_1 \cdots\cdots(\text{III}),$$

---

* But not universally ; see the Note at the end of § 172.

and so, as before for x,

$$y = B_1 e^{\lambda_1 t} + B_2 e^{\lambda_2 t} + \ldots\ldots + B_n e^{\lambda_n t} + Q,$$

where $B_1, B_2, \ldots\ldots, B_n$, are arbitrary constants, and Q is the Particular Integral of the differential equation (III).

171. We have, in the expressions for the two dependent variables, two sets of constants arising from the differential equations (II) and (III); they are both composed of arbitrary constants, but we do not know whether they are independent of one another; this dependence may exist and yet the constants may be arbitrary. Thus any one of the constants B might be a multiple of one of the constants A; the latter being arbitrary, the former would be so also. We therefore must determine the number of independent arbitrary constants. To do this, let the values of x and y be substituted in either of the equations (I), say in the first; then the terms arising from P and Q, which are particular integrals, give on the left-hand side a term T_1 cancelling with that on the right-hand side. The resulting equation is

$$\{A_1 f_1(\lambda_1) + B_1 \phi_1(\lambda_1)\} e^{\lambda_1 t} + \{A_2 f_1(\lambda_2) + B_2 \phi_1(\lambda_2)\} e^{\lambda_2 t} + \ldots\ldots$$
$$+ \{A_n f_1(\lambda_n) + B_n \phi_1(\lambda_n)\} e^{\lambda_n t} = 0.$$

Since this is to be satisfied for all values of t, the coefficient of each exponential must be zero, and therefore

$$\left.\begin{aligned} A_1 f_1(\lambda_1) + B_1 \phi_1(\lambda_1) &= 0 \\ A_2 f_1(\lambda_2) + B_2 \phi_1(\lambda_2) &= 0 \\ \ldots\ldots\ldots\ldots\ldots\ldots\ldots\ldots & \\ A_n f_1(\lambda_n) + B_n \phi_1(\lambda_n) &= 0 \end{aligned}\right\} \ldots\ldots\ldots\ldots\ldots(B),$$

so that each constant B can be derived from each constant A. The number of independent arbitrary constants in the complete solution of the simultaneous equations is therefore n, i.e. the exponent of the highest index in the operator

$$\phi_2(D) f_1(D) - \phi_1(D) f_2(D).$$

Hence the solution of the equations (I) is given by the foregoing values of x and y; the quantities λ occurring in the expressions are the roots of the equation (A); and the relations between the constants are given by equations (B).

172. There is a corresponding theorem that, if there be three dependent variables given by the three equations

$$f_1(D)\,x + \phi_1(D)\,y + \psi_1(D)\,z = T_1,$$
$$f_2(D)\,x + \phi_2(D)\,y + \psi_2(D)\,z = T_2,$$
$$f_3(D)\,x + \phi_3(D)\,y + \psi_3(D)\,z = T_3,$$

the number of independent arbitrary constants entering into the complete solution is the index of the highest power of D in the determinant

$$\begin{vmatrix} f_1(D), & \phi_1(D), & \psi_1(D) \\ f_2(D), & \phi_2(D), & \psi_2(D) \\ f_3(D), & \phi_3(D), & \psi_3(D) \end{vmatrix}.$$

The same result is also true when there are n dependent variables, given by n simultaneous linear differential equations having constant coefficients; the number of independent arbitrary constants, which enter into the complete solution, is equal to the index of the highest power of D in the corresponding determinant.

For a rigorous proof of these statements, reference may be made to a memoir by Chrystal, *Trans. R. S. Edin.*, vol. XXXVIII. (1897), pp. 163—178.

Note. It must not be assumed that the number of arbitrary independent constants occurring in the general solution, in accordance with the theorems in the last section and the present, is always at least as great as the order of any one of the equations; the number is strictly equal to the degree of the determinant.

The number of arbitrary constants may even be zero. An example is given by Chrystal, in the form

$$(D^2+1)\,x+(D^2+D+1)\,y=t,$$
$$Dx+(D+1)\,y=e^t.$$

The quantity $\phi_2(D)f_1(D)-\phi_1(D)f_2(D)$ here is unity: and the solution is given by

$$x=1+t-3e^t, \quad y=2e^t-1.$$

173. If the roots of the equation (A) which give the coefficients of t in the exponents be real and unequal, the solution given above is complete. It remains to consider the cases

(i)   when there is a pair of imaginary roots;

(ii)   when there is a pair of equal real roots;

the case of equal imaginary roots will follow from a combination of these two.

For the former, the solution obtained remains general; but often it is desirable to change the form so that it may be free from imaginary quantities. The two imaginary roots, say λ_1 and λ_2, may be denoted by $\alpha \pm \beta i$; hence the corresponding part of x is

$$e^{at}(A_1 e^{\beta ti} + A_2 e^{-\beta ti}),$$

that is, $e^{at}(L_1 \cos \beta t + L_2 \sin \beta t),$

on changing the arbitrary constants as in § 44. The part of y corresponding to the two imaginary roots is similarly

$$e^{at}(M_1 \cos \beta t + M_2 \sin \beta t).$$

Instead of making the necessary changes in the relations between A and B, it is better to substitute again these expressions in one or other of the fundamental equations and derive the corresponding relations as before.

For the latter case, the solution obtained ceases to be general because two constants, say A_1 and A_2, become merged into one; but it may be proved, exactly as in § 44, that the part of x depending upon this repeated root λ is

$$e^{\lambda t}(A + A't),$$

and the part of y is

$$e^{\lambda t}(B + B't).$$

Ex. 1. Prove that in the latter case the relations between the four constants, reducing them to two independent constants, are

$$A'f_1(\lambda) + B'\phi_1(\lambda) = 0,$$

$$Af_1(\lambda) + B\phi_1(\lambda) + A'\frac{df_1(\lambda)}{d\lambda} + B'\frac{d\phi_1(\lambda)}{d\lambda} = 0.$$

Ex. 2. If an imaginary root $a + \beta i$ be repeated, write down the corresponding parts of the complementary functions in x and y.

174. It may happen that the question, in connection with which the differential equations arise, will afford some indication of the form of the result. Thus in a problem relating to small oscillations, we should expect the values of the dependent variables to be expressed in terms of purely periodic functions; and it would then be proper to substitute for x and y respectively functions of the form

$$L_1 \cos \beta t + L_2 \sin \beta t,$$

$$M_1 \cos \beta t + M_2 \sin \beta t,$$

instead of $e^{\lambda t}$ in the equations (II) and (III). By equating to zero the coefficients of $\cos \beta t$ and of $\sin \beta t$ in each equation after these values have been substituted, there will arise four equations linear and homogeneous in the quantities L and M; and the eliminants of these will furnish the values of β. If on the other hand the problem indicate a motion not purely periodic in character, the form of value for x adopted would be

$$e^{\alpha t} (L_1 \cos \beta t + L_2 \sin \beta t),$$

and so for y. If there be no external information as to the character of the motion, then the ordinary method should be adopted.

Ex. 1. Solve the equations

$$\frac{dx}{dt} = -\omega y,$$

$$\frac{dy}{dt} = \omega x.$$

Here we have

$$\left. \begin{array}{l} Dx + \omega y = 0 \\ -\omega x + Dy = 0 \end{array} \right\};$$

and therefore the equation for x is

$$(D^2 + \omega^2) x = 0,$$

so that $x = A \cos \omega t + B \sin \omega t.$

Similarly $y = A' \cos \omega t + B' \sin \omega t.$

The relations between A, B, A', B', are at once derived by substituting in the first equation: we have

$$-\omega A \sin \omega t + \omega B \cos \omega t = -\omega A' \cos \omega t - \omega B' \sin \omega t,$$

or $A' = -B, \quad B' = A.$

The shortest method would have been to use the first equation to give y in terms of x, so that

$$y = -\frac{1}{\omega} \frac{dx}{dt}$$

$$= A \sin \omega t - B \cos \omega t.$$

This method is however applicable only in particular cases.

Ex. 2. Solve the equations

$$\left.\begin{array}{l} \dfrac{d^2x}{dt^2} - a\dfrac{dy}{dt} + \mu^2 x = 0 \\[2mm] \dfrac{d^2y}{dt^2} + a\dfrac{dx}{dt} + \mu^2 y = 0 \end{array}\right\}.$$

When we collect the terms which belong to the separate variables, the equations are

$$\left.\begin{array}{l} (D^2 + \mu^2)\,x - aDy = 0 \\[1mm] aDx + (D^2 + \mu^2)\,y = 0 \end{array}\right\}.$$

Here the equation for x is

$$\{(D^2 + \mu^2)^2 + a^2 D^2\}\,x = 0,$$

and the value of x is

$$x = L_1 \cos\beta_1 t + L_2 \sin\beta_1 t + L_3 \cos\beta_2 t + L_4 \sin\beta_2 t,$$

where $\beta_1{}^2$ and $\beta_2{}^2$ are the roots of the equation

$$(\mu^2 - p)^2 - a^2 p = 0;$$

and the value of y is

$$y = M_1 \sin\beta_1 t - M_2 \cos\beta_1 t + M_3 \sin\beta_2 t - M_4 \cos\beta_2 t.$$

It is easy to prove that the relation between the constants is

$$\frac{L_1}{M_1} = \frac{L_2}{M_2} = \frac{L_3}{M_3} = \frac{L_4}{M_4} = 1.$$

Ex. 3. Prove that the solution of the equations in the preceding example can be expressed in the form

$$x = A\sin\{\mu e^{a}\,(t - t_1)\} + B\sin\{\mu e^{-a}\,(t - t_2)\},$$

$$y = A\cos\{\mu e^{a}\,(t - t_1)\} + B\cos\{\mu e^{-a}\,(t - t_2)\},$$

where A, B, t_1, t_2, are arbitrary constants, and a is defined by the relation

$$2\mu \sinh a = a.$$

Ex. 4. Solve

$$\left.\begin{array}{l} \dfrac{dx}{dt} = ax + by + c \\[2mm] \dfrac{dy}{dt} = a'x + b'y + c' \end{array}\right\}.$$

They might be solved by adopting the ordinary rule; the following is another method applicable to this form.

Multiply the second equation by m and add to the first; then

$$\frac{d}{dt}(x + my) = x\,(a + ma') + y\,(b + mb') + c + mc'$$

$$= (a + ma')\,(x + my) + c + mc',$$

provided m be so chosen that

$$b + mb' = m(a + ma'),$$

that is, if m be a root of the equation

$$m^2 a' + (a - b')m - b = 0.$$

The foregoing differential equation, being

$$\frac{d(x + my)}{(a + ma')(x + my) + c + mc'} = dt,$$

has an integral

$$(a + ma')(x + my) + c + mc' = Ae^{t(a + ma')}.$$

Let m_1 and m_2 be the roots of the quadratic equation; the last equation is an integral, provided m is either m_1 or m_2. On substituting $m = m_1$, we have

$$(a + m_1 a')(x + m_1 y) + c + m_1 c' = A_1 e^{t(a + m_1 a')},$$

and on substituting $m = m_2$ we have

$$(a + m_2 a')(x + m_2 y) + c + m_2 c' = A_2 e^{t(a + m_2 a')},$$

where A_1 and A_2 are arbitrary constants. These two equations constitute the complete solution of the given pair of simultaneous equations when the roots of the quadratic are unequal.

To discuss the case, when the roots of the quadratic are equal, we write

$$m_2 = m_1 + \epsilon,$$

and keep ϵ small: and we take

$$A_2 = A_1 + \epsilon A_3.$$

The second equation then gives

$$(a + m_1 a' + \epsilon a')(x + m_1 y + \epsilon y) + c_2 + m_1 c' + \epsilon c' = (A_1 + \epsilon A_3) e^{t(a + m_1 a') + \epsilon a' t}.$$

When we expand in powers of ϵ, the terms in ϵ^0 cancel, owing to the first equation. We then divide out by a factor ϵ and, after division, we make ϵ vanish; then

$$a'(x + m_1 y) + (a + m_1 a')y + c' = (A_3 + a't A_1)e^{t(a + m_1 a')},$$

which is to be associated with

$$(a + m_1 a')(x + m_1 y) + c + m_1 c' = A_1 e^{t(a + m_1 a')}.$$

These two equations constitute the complete solution in the present case.

Ex. 5. Solve, in the same way as the last example, the equations

$$\left. \begin{array}{l} \dfrac{d^2 x}{dt^2} = ax + by \\[2mm] \dfrac{d^2 y}{dt^2} = a'x + b'y \end{array} \right\}.$$

Ex. 6. Solve the following equations:

(i) $\dfrac{dx}{dt} + 7x - y = 0, \quad \dfrac{dy}{dt} + 2x + 5y = 0;$

(ii) $\dfrac{dx}{dt} + 5x + y = e^t, \quad \dfrac{dy}{dt} + 3y - x = e^{2t};$

(iii) $4\dfrac{dx}{dt} + 9\dfrac{dy}{dt} + 44x + 49y = t, \quad 3\dfrac{dx}{dt} + 7\dfrac{dy}{dt} + 34x + 38y = e^t;$

(iv) $4\dfrac{dx}{dt} + 9\dfrac{dy}{dt} + 11x + 31y = e^t, \quad 3\dfrac{dx}{dt} + 7\dfrac{dy}{dt} + 8x + 24y = e^{2t};$

(v) $4\dfrac{dx}{dt} + 9\dfrac{dy}{dt} + 2x + 31y = e^t, \quad 3\dfrac{dx}{dt} + 7\dfrac{dy}{dt} + x + 24y = 3;$

(vi) $\dfrac{d^2x}{dt^2} + m^2 y = 0, \quad \dfrac{d^2y}{dt^2} - m^2 x = 0;$

(vii) $\dfrac{d^2x}{dt^2} - 3x - 4y + 3 = 0, \quad \dfrac{d^2y}{dt^2} + x + y + 5 = 0.$

Simultaneous Equations with Variable Coefficients.

175. It will be assumed, as before, that there is only one independent variable, and that therefore the coexistence of m simultaneous equations should suffice to determine the relations between the m dependent variables and that of which each is a function.

Further, it will be sufficient to consider systems of simultaneous equations which are only of the first order; for, to these, any other system can be reduced. Thus if into any one of a given system a differential coefficient of the nth order should enter, such as $\dfrac{d^n y}{dx^n}$, we could obtain an equivalent series of equations of the first order by making the substitutions

$$y_1 = \frac{dy}{dx}, \quad y_2 = \frac{dy_1}{dx}, \quad y_3 = \frac{dy_2}{dx}, \quad \ldots\ldots, \quad y_n = \frac{dy_{n-1}}{dx},$$

which are all of the order stated; and the corresponding substitutions for all differential coefficients of order higher than unity will transform any system of simultaneous equations of any order into an equivalent system of equations of the first order. If there be m dependent variables, this system must contain m equations each of the form

$$\phi\left(x, y_1, y_2, \ldots\ldots, y_m, \frac{dy_1}{dx}, \ldots\ldots, \frac{dy_m}{dx}\right) = 0.$$

176. The solution of this system of equations can be made to depend upon the solution of a single differential equation of the mth order connecting one of the dependent variables with the independent variable.

For let the m equations be solved so as to give the m differential coefficients as explicit functions of the variables; and suppose these relations to be

$$\frac{dy_1}{dx} = \psi_1(x, y_1, y_2, \ldots\ldots, y_m),$$

$$\frac{dy_2}{dx} = \psi_2(x, y_1, y_2, \ldots\ldots, y_m),$$

$$\ldots\ldots\ldots\ldots\ldots\ldots\ldots\ldots\ldots\ldots\ldots\ldots$$

$$\frac{dy_m}{dx} = \psi_m(x, y_1, y_2, \ldots\ldots, y_m).$$

Let the first of these be differentiated $m - 1$ times in succession with regard to x and, after each differentiation but before the next, let the values of $\dfrac{dy_2}{dx}, \ldots\ldots, \dfrac{dy_m}{dx}$ be substituted from the last $m - 1$ of these equations. There will thus be obtained, including the first equation, m equations connecting

$$\frac{dy_1}{dx}, \frac{d^2y_1}{dx^2}, \ldots\ldots, \frac{d^my_1}{dx^m},$$

with the variables $x, y_1, y_2, \ldots\ldots, y_m$; from these m equations let the $m - 1$ variables $y_2, y_3, \ldots\ldots, y_m$ be eliminated, so that there will result a single equation which may be represented by

$$f\left(x, y_1, \frac{dy_1}{dx}, \frac{d^2y_1}{dx^2}, \ldots\ldots, \frac{d^my_1}{dx^m}\right) = 0.$$

This equation, being of the mth order, has (§ 8) m independent first integrals each involving one arbitrary constant, all the m

constants being independent of one another; these integrals we
may represent by the equations

$$
\left.
\begin{aligned}
& F_1\left(x,\ y_1,\ \frac{dy_1}{dx},\ \ldots\ldots,\ \frac{d^{m-1}y_1}{dx^{m-1}},\ C_1\right)=0 \\
& F_2\left(x,\ y_1,\ \frac{dy_1}{dx},\ \ldots\ldots,\ \frac{d^{m-1}y_1}{dx^{m-1}},\ C_2\right)=0 \\
& \ldots\ldots\ldots\ldots\ldots\ldots\ldots\ldots\ldots\ldots\ldots\ldots\ldots\ldots \\
& F_m\left(x,\ y_1,\ \frac{dy_1}{dx},\ \ldots\ldots,\ \frac{d^{m-1}y_1}{dx^{m-1}},\ C_m\right)=0
\end{aligned}
\right\},
$$

in which the constants C are independent. But from the preceding equations we know the values of the differential coefficients
of y_1, in terms of all the variables; when these are substituted in
the set of equations F, the latter take the form

$$
\left.
\begin{aligned}
& \Phi_1\left(x,\ y_1,\ y_2,\ \ldots\ldots,\ y_m,\ C_1\right)=0 \\
& \Phi_2\left(x,\ y_1,\ y_2,\ \ldots\ldots,\ y_m,\ C_2\right)=0 \\
& \ldots\ldots\ldots\ldots\ldots\ldots\ldots\ldots\ldots\ldots\ldots \\
& \Phi_m\left(x,\ y_1,\ y_2,\ \ldots\ldots,\ y_m,\ C_m\right)=0
\end{aligned}
\right\},
$$

which are sufficient to determine each of the variables y as a
function of x; they are an integral system and contain m arbitrary
constants.

Hence we have as the general result:

*The complete solution of a system of m differential equations
of the first order between $m+1$ variables depends on that of an
ordinary differential equation of the mth order and consists of m
equations, connecting the $m+1$ variables and containing m independent arbitrary constants.*

The foregoing is the general theory; but, in particular cases,
simplifications arise enabling much of the labour indicated in the
general theory to be dispensed with. Thus, if the equations consist
of a set each of which is linear in the derivatives, it may happen
that an integral of each equation of the form

$$
P\,dx + P_1\,dy_1 + P_2\,dy_2 + \ldots\ldots + P_m\,dy_m = 0
$$

can be obtained in the form

$$
\phi\left(x,\ y_1,\ y_2,\ \ldots\ldots,\ y_m\right) = C,
$$

and the long process would not need to be gone through. Again, instead of determining the m independent first integrals, it would be sufficient to determine the primitive of the ordinary equation of the mth order; for from it could be derived other $m-1$ equations in which the values of the differential coefficients could be substituted, and an equivalent result would be so deduced. Again, in the case when the equations are all linear, we can solve them to obtain the ratios of the $m+1$ differentials in the form

$$\frac{dx}{X} = \frac{dy_1}{Y_1} = \frac{dy_2}{Y_2} = \ldots\ldots = \frac{dy_m}{Y_m},$$

which might be called the symmetrical form; the mode of treatment for these will sometimes (depending upon the form of the denominators in these fractions) differ very materially from, and be much more convenient than, the general process. Examples illustrative of this statement will be found appended.

Ex. 1. The general method can be avoided, if integrals of all but one equation can be obtained and, *à fortiori*, if all the integrals can be obtained. Thus the equations

$$l\,dx + m\,dy + n\,dz = 0,$$
$$x\,dx + y\,dy + z\,dz = 0,$$

lead at once to the integrals

$$lx + my + nz = c_1, \qquad x^2 + y^2 + z^2 = c_2,$$

which determine y and z in terms of x.

Ex. 2. Solve

(i) $\begin{cases} t\,dx = (t - 2x)\,dt, \\ t\,dy = (tx + ty + 2x - t)\,dt. \end{cases}$

(ii) $\begin{cases} t\,\dfrac{d^2x}{dt^2} + 2\,\dfrac{dx}{dt} + tx = 0, \\[2mm] \dfrac{dy}{dt} + \dfrac{2}{t}\,y = \dfrac{dx}{dt}. \end{cases}$

Ex. 3. Solve $\dfrac{dx}{X} = \dfrac{dy}{Y} = \dfrac{dz}{Z}$, where

$$\left. \begin{array}{l} X = ax + by + cz + d \\ Y = a'x + b'y + c'z + d' \\ Z = a''x + b''y + c''z + d'' \end{array} \right\}.$$

In equations of this form, it is convenient to introduce some new independent variable and to make all those variables, which already occur in the equations given, functions of this new variable. Calling the latter t we may assume, as an advantageous form,

$$\frac{dt}{t} = \frac{dx}{X} = \frac{dy}{Y} = \frac{dz}{Z}$$

$$= \frac{l\,dx + m\,dy + n\,dz}{lX + mY + nZ}$$

$$= \frac{l\,dx + m\,dy + n\,dz}{\lambda\,(lx + my + nz) + r},$$

provided l, m, n, λ, be so chosen that

$$\left.\begin{array}{l} al + a'm + a''n = \lambda l \\ bl + b'm + b''n = \lambda m \\ cl + c'm + c''n = \lambda n \end{array}\right\} ;$$

the value of r is

$$ld + md' + nd''.$$

Eliminating l, m, n, between these three equations, we have

$$\begin{vmatrix} a - \lambda, & a', & a'' \\ b, & b' - \lambda, & b'' \\ c, & c', & c'' - \lambda \end{vmatrix} = 0,$$

a cubic equation determining λ; let its roots be λ_1, λ_2, λ_3. When λ_1 is substituted in any two of the foregoing equations, the ratios of $l : m : n$ can be derived; let them be denoted by $l_1 : m_1 : n_1$, and suppose the corresponding value of r to be r_1; with similar expressions for the other values of λ. Then for the value λ_1 we have

$$\frac{dt}{t} = \frac{l_1\,dx + m_1\,dy + n_1\,dz}{\lambda_1\,(l_1 x + m_1 y + n_1 z) + r_1},$$

the integral of which is

$$c_1 t = (l_1 x + m_1 y + n_1 z + r_1 \lambda_1^{-1})^{\lambda_1^{-1}}.$$

Similarly $\qquad c_2 t = (l_2 x + m_2 y + n_2 z + r_2 \lambda_2^{-1})^{\lambda_2^{-1}},$

and $\qquad c_3 t = (l_3 x + m_3 y + n_3 z + r_3 \lambda_3^{-1})^{\lambda_3^{-1}}.$

In order to obtain the general solution of the system of equations as given, we must eliminate t between these equations; when we write $c_1 = Ac_2 = Bc_3$, where A and B are arbitrary constants, the general integral as required is given by the equations

$$(l_1 x + m_1 y + n_1 z + r_1 \lambda_1^{-1})^{\lambda_1^{-1}} = A\,(l_2 x + m_2 y + n_2 z + r_2 \lambda_2^{-1})^{\lambda_2^{-1}}$$

$$= B\,(l_3 x + m_3 y + n_3 z + r_3 \lambda_3^{-1})^{\lambda_3^{-1}}.$$

Ex. 4. Solve in this manner the equations

$$-dx = \frac{dy}{3y + 4z} = \frac{dz}{2y + 5z}.$$

Ex. 5. This method may also be applied to solve certain systems of equations in which the variables do not occur so simply as in Ex. 3. Thus let us consider

$$\left. \begin{array}{c} \dfrac{dx}{dt} + T(ax+by) = T_1 \\[2mm] \dfrac{dy}{dt} + T(a'x+b'y) = T_2 \end{array} \right\},$$

where T, T_1, T_2, are functions of t. Multiplying the second equation by l and adding it to the first, we have

$$\frac{d}{dt}(x+ly) + \lambda T(x+ly) = T_1 + lT_2,$$

provided l and λ are determined so as to satisfy the equations

$$a + la' = \lambda,$$
$$b + lb' = l\lambda,$$

so that the values of λ are λ_1 and λ_2, the two roots of

$$(a-\lambda)(b'-\lambda) - a'b = 0.$$

The integral of the foregoing equation being

$$(x+ly)e^{\lambda \int Tdt} = A + \int (T_1 + lT_2) e^{\lambda \int Tdt} dt,$$

the complete solution of the original equations is given by

$$(x+l_1y) e^{\lambda_1 \int Tdt} = A_1 + \int (T_1 + l_1 T_2) e^{\lambda_1 \int Tdt} dt,$$
$$(x+l_2y) e^{\lambda_2 \int Tdt} = A_2 + \int (T_1 + l_2 T_2) e^{\lambda_2 \int Tdt} dt.$$

Ex. 6. Solve the systems of equations

$$(a) \quad \left. \begin{array}{c} \dfrac{dx}{dt} + \dfrac{2}{t}(x-y) = 1 \\[2mm] \dfrac{dy}{dt} + \dfrac{1}{t}(x+5y) = t \end{array} \right\};$$

$$(\beta) \quad \left. \begin{array}{c} lt\dfrac{dx}{dt} = mn(y-z) \\[2mm] mt\dfrac{dy}{dt} = nl(z-x) \\[2mm] nt\dfrac{dz}{dt} = lm(x-y) \end{array} \right\};$$

$$(\gamma) \quad \left. \begin{array}{c} \dfrac{dx}{dt} = ny - mz \\[2mm] \dfrac{dy}{dt} = lz - nx \\[2mm] \dfrac{dz}{dt} = mx - ly \end{array} \right\}.$$

JACOBI'S MULTIPLIERS.

177. At the end of § 176, reference was made to a system of m differential equations of the first order and the first degree, arising in the form

$$\frac{dx}{X} = \frac{dy_1}{Y_1} = \frac{dy_2}{Y_2} = \ldots\ldots = \frac{dy_m}{Y_m},$$

where all the quantities X, Y_1, Y_2, ..., Y_m, are functions of x, y_1, y_2, ..., y_m. Many investigations of general properties of the integrals of the equation have been made; among these, one of the most important is what is known as the theory of the *last multiplier*. Only the simplest cases will be considered here; reference to other sources must be made if ampler discussion is desired*.

Consider the equations

$$\frac{dx}{X} = \frac{dy}{Y} = \frac{dz}{Z},$$

where each of the quantities X, Y, Z, is a function of x, y, z; and let their integrals be denoted by

$$f(x, y, z) = a, \quad g(x, y, z) = b.$$

Then any relation of the form

$$F(f, g) = c$$

is also an integral. Moreover, the equations

$$f_x dx + f_y dy + f_z dz = 0,$$
$$g_x dx + g_y dy + g_z dz = 0,$$

are satisfied in virtue of the given equations; so that

$$X f_x + Y f_y + Z f_z = 0,$$
$$X g_x + Y g_y + Z g_z = 0.$$

---

* The theory is originally due to Jacobi; his fullest exposition is given in a memoir, *Ges. Werke*, t. IV. pp. 319—509, and a simpler exposition is given in his *Vorlesungen über Dynamik*, pp. 71 et seq. Cayley's memoir, *Coll. Math. Papers*, vol. X. pp. 93—133, may also be consulted.

Let

$$A = \begin{vmatrix} f_y, & f_z \\ g_y, & g_z \end{vmatrix}, \qquad B = \begin{vmatrix} f_z, & f_x \\ g_z, & g_x \end{vmatrix}, \qquad C = \begin{vmatrix} f_x, & f_y \\ g_x, & g_y \end{vmatrix};$$

then a quantity M exists such that

$$\frac{A}{X} = \frac{B}{Y} = \frac{C}{Z} = M.$$

Now it is easy to verify that

$$\frac{\partial A}{\partial x} + \frac{\partial B}{\partial y} + \frac{\partial C}{\partial z}$$

vanishes identically; consequently

$$\frac{\partial}{\partial x}(MX) + \frac{\partial}{\partial y}(MY) + \frac{\partial}{\partial z}(MZ) = 0,$$

and therefore

$$X\frac{\partial M}{\partial x} + Y\frac{\partial M}{\partial y} + Z\frac{\partial M}{\partial z} + M\left(\frac{\partial X}{\partial x} + \frac{\partial Y}{\partial y} + \frac{\partial Z}{\partial z}\right) = 0,$$

say

$$E(M) = 0,$$

an equation satisfied by M. It no longer contains any explicit reference to f or g.

Suppose that two solutions of the equation

$$E(\phi) = 0$$

are known, say M and N, and assume that their ratio is not a pure constant; then

$$E(M) = 0, \quad E(N) = 0,$$

and therefore

$$NE(M) - ME(N) = 0.$$

When we divide by N^2, this equation can be written in the form

$$X\frac{\partial}{\partial x}\left(\frac{M}{N}\right) + Y\frac{\partial}{\partial y}\left(\frac{M}{N}\right) + Z\frac{\partial}{\partial z}\left(\frac{M}{N}\right) = 0.$$

Consequently, the equation

$$\begin{vmatrix} \frac{\partial}{\partial x}\left(\frac{M}{N}\right), & \frac{\partial}{\partial y}\left(\frac{M}{N}\right), & \frac{\partial}{\partial z}\left(\frac{M}{N}\right) \\ f_x, & f_y, & f_z \\ g_x, & g_y, & g_z \end{vmatrix} = 0$$

must be satisfied. It cannot be in virtue of $f = a$ or $g = b$ that the equation is satisfied, for a and b do not occur; hence the equation must be satisfied identically, and therefore (§ 9) there must be a functional relation between $M \div N$, f, g, free from all variables. Hence

$$\frac{M}{N} = F(f, g) = F(a, b) = c,$$

say, so that $\dfrac{M}{N}$ is an integral of the original differential equations. In other words, when two solutions of the equation $E(M) = 0$ are known, these solutions not being constant multiples of one another, their quotient equated to an arbitrary constant is an integral of the original system.

Quantities, such as M and N, are called *multipliers*.

One or two corollaries are practically obvious: they will be stated without proof.

I. If one multiplier M is known, then every other multiplier is of the form $Mh(a, b)$.

II. If three multipliers L, M, N, are known, such that $L \div N$ and $M \div N$ are not pure constants, and such also that $L \div N$ and $M \div N$ are not connected by any functional relation involving only pure constants, then

$$\frac{L}{N} = a, \quad \frac{M}{N} = b,$$

can be taken as integrals of the original differential equations.

III. If

$$\frac{\partial X}{\partial x} + \frac{\partial Y}{\partial y} + \frac{\partial Z}{\partial z} = 0,$$

then one multiplier is known: we can take $M = 1$, so far as the equation $E(M) = 0$ is concerned.

The importance of the multiplier lies in the property that, *when one integral of the equations has been obtained and when any multiplier is known, another integral can be obtained by quadratures only.*

Let the integral which has been obtained be

$$f(x, y, z) = a \,;$$

let M be any known multiplier; and let the other integral, as yet unknown, which would be associated with $f = a$ to give the multiplier M, be

$$g(x, y, z) = b.$$

By means of $f(x, y, z) = a$, let one of the variables (say z) be expressed in terms of a and the other two variables; and let the value thus formed be supposed substituted in the other integral, with the result

$$G(x, y, a) = b.$$

Now

$$g_x = \frac{\partial G}{\partial x} + \frac{\partial G}{\partial a} f_x,$$

$$g_y = \frac{\partial G}{\partial y} + \frac{\partial G}{\partial a} f_y,$$

$$g_z = \qquad \frac{\partial G}{\partial a} f_z;$$

and therefore

$$MX = \begin{vmatrix} f_y, f_z \\ g_y, g_z \end{vmatrix} = -\frac{\partial G}{\partial y} f_z,$$

$$MY = \begin{vmatrix} f_z, f_x \\ g_z, g_x \end{vmatrix} = \quad \frac{\partial G}{\partial x} f_z.$$

Now

$$dG = \frac{\partial G}{\partial x} dx + \frac{\partial G}{\partial y} dy,$$

where dG is a perfect differential: substituting on the right-hand side, we have

$$dG = \frac{M}{f_z} (Y dx - X dy).$$

The left-hand side, regarded as a function of x, y, and the constant a, is a perfect differential; hence, when the quantities on the right-hand side (all of which are known) are expressed in terms of x, y, a, the expression is a perfect differential. Effecting the quadrature, we have

$$G(x, y, a) = \int \frac{M}{f_z} (Y dx - X dy).$$

The quadrature thus gives us $G(x, y, a)$. When the quantity a is replaced by $f(x, y, z)$, we have the second integral in the form

$$g(x, y, z) = b.$$

Ex. 1. Consider the equations

$$\frac{dx}{x(y-z)} = \frac{dy}{y(z-x)} = \frac{dz}{z(x-y)}.$$

Here, we have

$$\frac{\partial X}{\partial x} + \frac{\partial Y}{\partial y} + \frac{\partial Z}{\partial z} = 0;$$

hence we can take unity as a multiplier. Also $x+y+z=a$ is clearly an integral, so that

$$f = x+y+z,$$

and $f_z = 1$. To obtain another, we have to replace z by $a-x-y$ in

$$\int (Y dx - X dy)$$

and effect the quadrature: that is, we have

$$\int \{ y(a-2x-y)\, dx - x(x+2y-a)\, dy \}$$

$$= axy - x^2 y - y^2$$

$$= xy(a-x-y).$$

The other integral, by our theorem, therefore is

$$xyz = b.$$

Ex. 2. Integrate the equations:

$$\text{(i)} \quad \frac{dx}{x(y^2-z^2)} = \frac{dy}{y(z^2-x^2)} = \frac{dz}{z(x^2-y^2)};$$

$$\text{(ii)} \quad \frac{dx}{x(y^2-z^2)} = \frac{dy}{-y(z^2+x^2)} = \frac{dz}{z(x^2+y^2)};$$

$$\text{(iii)} \quad \frac{-dx}{x(x+y)} = \frac{dy}{y(x+y)} = \frac{dz}{(x-y)(2x+2y+z)};$$

$$\text{(iv)} \quad \frac{dx}{y^2+yz+z^2} = \frac{dy}{z^2+zx+x^2} = \frac{dz}{x^2+xy+y^2}.$$

Ex. 3. The method can be applied to a differential equation of the second order

$$\frac{d^2 y}{dx^2} = h(x, y),$$

when a first integral is known.

We have
$$\frac{dx}{1} = \frac{dy}{p} = \frac{dp}{h},$$
where $p = \frac{dy}{dx}$; so that an instance of the general system arises when we take
$$z = p,$$
$$X = 1, \quad Y = p = z, \quad Z = h(x, y).$$
Thus
$$\frac{\partial X}{\partial x} + \frac{\partial Y}{\partial y} + \frac{\partial Z}{\partial z} = 0;$$
and so unity can be taken as the value of M.

Let the first integral, supposed known, be
$$f(x, y, p) = a;$$
and let this be supposed resolved to give
$$p = \psi(x, y, a).$$
Then the general theorem shews that another integral is given by
$$\int \frac{dy - \psi dx}{\frac{\partial f}{\partial p}} = b.$$

Ex. 4. Consider the equation[*]
$$yy'' - \tfrac{2}{3}y'^2 = f(x),$$
where $f(x) = ax^2 + 2\beta x + \gamma$.

Writing $y' = z$, we have
$$\frac{dx}{1} = \frac{dy}{z} = \frac{dz}{\frac{1}{y}(f + \tfrac{2}{3}z^2)}.$$
The equation for M is
$$\frac{\partial M}{\partial x} + z\frac{\partial M}{\partial y} + \frac{1}{y}(f + \tfrac{2}{3}z^2)\frac{\partial M}{\partial z} + M\tfrac{4}{3}\frac{z}{y} = 0,$$
of which an integral manifestly is
$$M = y^{-\frac{4}{3}}.$$

If then we can obtain a first integral of the original equation, the primitive will be derivable by a quadrature.

Now
$$yy'' - \tfrac{2}{3}y'^2 = f(x),$$
$$yy''' - \tfrac{1}{3}y'y'' = f'(x),$$
$$yy^{\text{iv}} + \tfrac{2}{3}y'y''' - \tfrac{1}{3}y''^2 = f''(x),$$
$$yy^{\text{v}} + \tfrac{5}{3}y'y^{\text{iv}} = 0,$$

---

[*] Laguerre, *Œuvres*, t. I. p. 402.

because $f'''(x) = 0$. From the last we have

$$\frac{y^{\text{v}}}{y^{\text{iv}}} + \tfrac{5}{3}\frac{y'}{y} = 0,$$

and so

$$y^{\text{iv}} y^{\frac{5}{3}} = a.$$

Eliminating y^{iv}, y''', y'' between the last equation and the three above, we find

$$2y'f - 3yf' = 3\left(\Delta y^2 - 4f^3 - \tfrac{1}{9}ay^{\frac{4}{3}}f\right)^{\frac{1}{2}},$$

where

$$\Delta = f'^2 - 2ff'' = 4(\beta^2 - a\gamma).$$

Accordingly, we have a first integral of the equation in the form

$$p = y' = \tfrac{3}{2}\frac{f'}{f}y + \frac{3}{2f}\left(\Delta y^2 - 4f^3 - \tfrac{1}{9}ay^{\frac{4}{3}}f\right)^{\frac{1}{2}}$$

$$= \tfrac{3}{2}\frac{f'}{f}y + \frac{3}{2f}R.$$

To apply the theorem, we need $\dfrac{\partial a}{\partial p}$: we find

$$\frac{\partial a}{\partial p} = -12Ry^{-\frac{4}{3}}.$$

The integral, being

$$\int \frac{M}{\frac{\partial a}{\partial p}}(p\,dx - dy) = b,$$

becomes

$$\int 12R\,(dy - p\,dx) = b;$$

or multiplying up by -8, we have

$$\int \frac{dx}{f} + \int \frac{\frac{f'}{f}y\,dx - \tfrac{2}{3}dy}{R} = -8b.$$

To effect the last quadrature, write

$$y = \left(\frac{f}{u}\right)^{\frac{3}{2}};$$

then

$$\tfrac{2}{3}dy - \frac{f'}{f}y\,dx = -\frac{y}{u}\,du,$$

$$R = y\left(\Delta - \tfrac{1}{9}au - u^3\right)^{\frac{1}{2}},$$

so that the equation

$$\int \frac{dx}{f} + \int \frac{du}{u\left(\Delta - \tfrac{1}{9}au - u^3\right)^{\frac{1}{2}}} = c$$

gives the primitive.

Ex. 5. A first integral of the equation

$$\frac{1}{y}\frac{d^2y}{dx^2}=1+2\tan^2 x$$

is of the form

$$\frac{dy}{dx}=y\phi\,(x)+c\psi\,(x).$$

Determine the primitive; and compare the result, obtained by the process of the multiplier, with the result of immediate integration of the first integral.

There is one case in which a value of M can be obtained. Suppose that

$$\frac{1}{X}\left(\frac{\partial X}{\partial x}+\frac{\partial Y}{\partial y}+\frac{\partial Z}{\partial z}\right)$$

is the complete derivative of a function ξ with regard to x, account being taken of the relations

$$\frac{dx}{X}=\frac{dy}{Y}=\frac{dz}{Z}.$$

Then we have

$$\frac{\partial M}{\partial x}+\frac{Y}{X}\frac{\partial M}{\partial y}+\frac{Z}{X}\frac{\partial M}{\partial z}=\frac{\partial M}{\partial x}+\frac{\partial M}{\partial y}\frac{dy}{dx}+\frac{\partial M}{\partial z}\frac{dz}{dx}=\frac{dM}{dx};$$

and the equation for M becomes

$$\frac{dM}{dx}+M\frac{d\xi}{dx}=0,$$

so that we can take

$$M=e^{-\xi}.$$

A general instance of this type is given by Jacobi[*] in the form

$$\frac{d^2y}{dx^2}+\tfrac{1}{2}\frac{\partial\phi}{\partial y}\left(\frac{dy}{dx}\right)^2+\frac{\partial\phi}{\partial x}\frac{dy}{dx}+\psi=0,$$

where ϕ and ψ are any functions of x and y. As in some of the preceding examples, we take

$$\frac{dy}{dx}=p,$$

so that, writing

$$Z=-\tfrac{1}{2}p^2\frac{\partial\phi}{\partial y}-p\frac{\partial\phi}{\partial x}-\psi,$$

[*] *Ges. Werke,* t. IV. p. 408.

we may take the subsidiary system in the form

$$\frac{dx}{1} = \frac{dy}{p} = \frac{dz}{Z}.$$

Then

$$\frac{1}{X}\left(\frac{\partial X}{\partial x} + \frac{\partial Y}{\partial y} + \frac{\partial Z}{\partial z}\right) = -p\frac{\partial \phi}{\partial y} - \frac{\partial \phi}{\partial x}$$

$$= -\frac{d\phi}{dx};$$

and therefore

$$M = e^{\phi}.$$

If then a first integral of the differential equation can be obtained, the primitive can be constructed by means of a quadrature.

Ex. 1. Integrate the equations

(i) $\dfrac{dx}{y-z} = \dfrac{dy}{x+y} = \dfrac{dz}{x+z} = dt$;

(ii) $\dfrac{dx}{y-z} = \dfrac{dy}{x^2+y} = \dfrac{dz}{x^2+z} = dt$;

(iii) $\dfrac{dx}{y-z} = \dfrac{dy}{x+y+t} = \dfrac{dz}{x+z+t} = dt$.

Ex. 2. Obtain the primitive of the equation

$$y^2\frac{d^2y}{dx^2} + y\left(\frac{dy}{dx}\right)^2 + by - cx = 0,$$

where b and c are constants, having given that a first integral is

$$y^3p^3 + bxy^2p^2 + (by - 3cx)\,y^2p + cy^3 + by^2x - 2bcyx^2 + c^2x^3 = A,$$

or, what is the equivalent,

$$\left(yp + \lambda y + \frac{c}{\lambda}\,x\right)\left(yp + \lambda'y + \frac{c}{\lambda'}\,x\right)\left(yp + \lambda''y + \frac{c}{\lambda''}\,x\right) = A,$$

where λ, λ', λ'', are the roots of the equation

$$\lambda^3 + b\lambda - c = 0.$$

The preceding investigations have been restricted to the case of three variables and two equations. Corresponding results hold for the case of $n+1$ variables and n equations; but they will merely be stated here, and the proofs will be found in the works cited (p. 356) at the beginning of this section.

Let the equations be

$$\frac{dx_1}{X_1} = \frac{dx_2}{X_2} = \ldots\ldots = \frac{dx_n}{X_n} = \frac{dx_{n+1}}{X_{n+1}},$$

where each of the quantities X can be a function of all the variables.

I. The equation for the multiplier M is

$$\sum_{r=1}^{n+1} X_r \frac{\partial M}{\partial x_r} + M \sum_{r=1}^{n+1} \frac{\partial X_r}{\partial x_r} = 0.$$

If any two solutions of this equation, M_1 and M_2, are known, their ratio not being a pure constant, then

$$\frac{M_1}{M_2} = \alpha,$$

where α is an arbitrary constant, is an integral of the differential equations. Let $u_1 = \alpha_1$, $u_2 = \alpha_2$, ..., $u_n = \alpha_n$, be a complete system of integrals, and let M be any multiplier; then every other multiplier is of the form $M\phi(\alpha_1, \alpha_2, \ldots\ldots, \alpha_n)$. And if

$$\frac{\partial X_1}{\partial x_1} + \frac{\partial X_2}{\partial x_2} + \ldots\ldots + \frac{\partial X_{n+1}}{\partial x_{n+1}} = 0,$$

manifestly we can take $M = 1$.

II. Let

$$u_1 = \alpha_1, \quad u_2 = \alpha_2, \quad \ldots\ldots, \quad u_{n-1} = \alpha_{n-1},$$

be $n-1$ integrals of the equations, so that all the integrals save one are known; and let M be any multiplier. Then the last integral of the equations is given by

$$\int \frac{M}{\dfrac{\partial (u_1, u_2, \ldots\ldots, u_{n-1})}{\partial (x_3, x_4, \ldots\ldots, x_{n+1})}} (X_1 dx_2 - X_2 dx_1) = \alpha_n,$$

where α_n is an arbitrary constant, and all the quantities in the subject of integration are to be expressed in terms of

$$x_1, \quad x_2, \quad \alpha_1, \quad \ldots\ldots, \quad \alpha_{n-1},$$

by means of the given $n-1$ integrals of the system.

It is on account of the theorem last stated that the name *last multiplier* was attached to the quantity M by Jacobi.

Ex. Shew that, if two integrals of the differential equation

$$F(x, y, y', y''') = 0$$

be known in the form

$$f(x, y, y', y'') = a, \quad g(x, y, y', y'') = \beta,$$

then a third integral is given by the relation

$$\int \frac{dy - y'dx}{\dfrac{\partial(f, g)}{\partial(y', y'')}} = \gamma,$$

where y' and $\dfrac{\partial(f, g)}{\partial(y', y'')}$ are to be expressed in terms of x and y by means of the equations $f = a$ and $g = \beta$: the quantities a, β, γ, being constants.

A Special System of Equations in Dynamics.

178. There are two classes of simultaneous equations which are extremely important; one is the class already considered in §§ 148, 149 as the generalisation of Euler's equations leading to one class of higher transcendental functions ordinarily called Abelian functions; the other is the system of equations which determine the motion of a particle attracted to a centre of force that acts according to the gravitational law. The latter may be represented by the simultaneous equations

$$\frac{d^2x}{dt^2} = \frac{\partial R}{\partial x}, \quad \frac{d^2y}{dt^2} = \frac{\partial R}{\partial y}, \quad \frac{d^2z}{dt^2} = \frac{\partial R}{\partial z}, \quad \ldots\ldots\ldots\ldots(\text{i}),$$

in which R is a rational function of r or $(x^2 + y^2 + z^2)^{\frac{1}{2}}$, the distance of the point x, y, z, from the origin. To express the complete integral, three independent equations (or their equivalent) will be necessary. Since each equation may be replaced by two of the form

$$x_1 = \frac{dx}{dt}, \quad \frac{dx_1}{dt} = \frac{\partial R}{\partial x},$$

giving six equations in all to determine the six quantities, the investigation of § 176 shews that we must have six arbitrary constants in the solution.

If we multiply the equations (i) by $\frac{dx}{dt}$, $\frac{dy}{dt}$, $\frac{dz}{dt}$, respectively, add and integrate, we have

$$\tfrac{1}{2}\left\{\left(\frac{dx}{dt}\right)^2 + \left(\frac{dy}{dt}\right)^2 + \left(\frac{dz}{dt}\right)^2\right\} = R + B,$$

in which B is an arbitrary constant.

Another form may be given to the equations (i). Since R is a function of r, we have

$$\frac{\partial R}{\partial x} = \frac{dR}{dr}\frac{\partial r}{\partial x} = \frac{x}{r}\frac{dR}{dr},$$

and so for the others; and thus (i) becomes

$$\frac{d^2x}{dt^2} = \frac{x}{r}\frac{dR}{dr}; \quad \frac{d^2y}{dt^2} = \frac{y}{r}\frac{dR}{dr}; \quad \frac{d^2z}{dt^2} = \frac{z}{r}\frac{dR}{dr}.$$

Therefore

$$x\frac{d^2y}{dt^2} - y\frac{d^2x}{dt^2} = 0,$$

$$y\frac{d^2z}{dt^2} - z\frac{d^2y}{dt^2} = 0,$$

$$z\frac{d^2x}{dt^2} - x\frac{d^2z}{dt^2} = 0,$$

of which two only are independent; the integrals of these are respectively

$$x\frac{dy}{dt} - y\frac{dx}{dt} = C_1,$$

$$y\frac{dz}{dt} - z\frac{dy}{dt} = C_2,$$

$$z\frac{dx}{dt} - x\frac{dz}{dt} = C_3.$$

Squaring and adding these, we have

$$(x^2 + y^2 + z^2)\left\{\left(\frac{dx}{dt}\right)^2 + \left(\frac{dy}{dt}\right)^2 + \left(\frac{dz}{dt}\right)^2\right\} - \left(x\frac{dx}{dt} + y\frac{dy}{dt} + z\frac{dz}{dt}\right)^2$$

$$= C_1^2 + C_2^2 + C_3^2 = A^2,$$

where A is an arbitrary constant; this is equivalent to

$$2r^2(R+B) - \left(r\frac{dr}{dt}\right)^2 = A^2,$$

that is, to

$$dt = \frac{r\,dr}{\{2r^2(R+B) - A^2\}^{\frac{1}{2}}},$$

and therefore

$$t + \alpha = \int \frac{r\,dr}{\{2r^2(R+B) - A^2\}^{\frac{1}{2}}} \quad \dots\dots\dots\dots(ii).$$

From the equation just obtained, we have

$$2R = -2B + \frac{A^2}{r^2} + \left(\frac{dr}{dt}\right)^2,$$

and therefore

$$2\frac{dR}{dr}\frac{dr}{dt} = -2\frac{A^2}{r^3}\frac{dr}{dt} + 2\frac{dr}{dt}\frac{d^2r}{dt^2},$$

that is,

$$\frac{dR}{dr} = -\frac{A^2}{r^3} + \frac{d^2r}{dt^2}.$$

When this value is substituted in the modified form of the original equations, the first of them is

$$r\frac{d^2x}{dt^2} = x\frac{d^2r}{dt^2} - A^2\frac{x}{r^3},$$

or

$$\frac{d}{dt}\left(r\frac{dx}{dt} - x\frac{dr}{dt}\right) + A^2\frac{x}{r^3} = 0,$$

or

$$r^2\frac{d}{dt}\left\{r^2\frac{d}{dt}\left(\frac{x}{r}\right)\right\} + A^2\frac{x}{r} = 0.$$

Let

$$d\phi = A\frac{dt}{r^2}$$

$$= \frac{A\,dr}{r\{2r^2(R+B) - A^2\}^{\frac{1}{2}}};$$

then the foregoing equation for $\frac{x}{r}$ is

$$\frac{d^2}{d\phi^2}\left(\frac{x}{r}\right) + \frac{x}{r} = 0,$$

and therefore

$$\frac{x}{r} = a_1 \cos \phi + a_2 \sin \phi \quad \dots\dots\dots\dots\dots\text{(iii)}.$$

The second and third equations, similarly treated, lead to

$$\frac{y}{r} = b_1 \cos \phi + b_2 \sin \phi \quad \dots\dots\dots\dots\dots\text{(iv)},$$

$$\frac{z}{r} = c_1 \cos \phi + c_2 \sin \phi \quad \dots\dots\dots\dots\dots\text{(v)};$$

and in these the constants a, b, c, are arbitrary. But they are not independent; for we have always

$$x^2 + y^2 + z^2 = r^2,$$

whatever be the value of ϕ, and therefore the relation

$$(a_1{}^2 + b_1{}^2 + c_1{}^2) \cos^2 \phi + 2 (a_1 a_2 + b_1 b_2 + c_1 c_2) \cos \phi \sin \phi$$
$$+ (a_2{}^2 + b_2{}^2 + c_2{}^2) \sin^2 \phi = 1 = \cos^2 \phi + \sin^2 \phi$$

is satisfied for all values of ϕ, so that

$$\left. \begin{array}{l} a_1{}^2 + b_1{}^2 + c_1{}^2 = 1 \\ a_2{}^2 + b_2{}^2 + c_2{}^2 = 1 \\ a_1 a_2 + b_1 b_2 + c_1 c_2 = 0 \end{array} \right\} \quad \dots\dots\dots\dots\dots\text{(vi)}.$$

The six constants are thus equivalent to three independent constants. Further, we may put (iii) into the form

$$\frac{x}{r} = \rho_1 \cos (\phi + \beta_1),$$

where ρ_1 and β_1 are arbitrary constants; thus an arbitrary constant is associated with ϕ, and one will not require to be added in the equation

$$\phi = \int \frac{A \, dr}{r \{2r^2 (R + B) - A^2\}^{\frac{1}{2}}} \quad \dots\dots\dots\dots\text{(vii)}.$$

We have now sufficient equations to determine the general integral. By means of (vii) ϕ is given as a function of r, and therefore by (ii) as a function of t; hence (iii), (iv), (v), give x, y, z, as functions of t. Moreover, we have six independent arbitrary constants, viz., A^2, B, α, and the six quantities a_1, a_2, b_1, b_2, c_1, c_2, connected by the three relations (vi). These therefore constitute the general integral of the differential equations.

Ex. Solve in this way

$$\left.\begin{array}{l}(x^2+y^2)^{\frac{3}{2}}\dfrac{d^2x}{dt^2}+\mu x=0\\[2mm](x^2+y^2)^{\frac{3}{2}}\dfrac{d^2y}{dt^2}+\mu y=0\end{array}\right\}.$$

Also solve by transforming to polar coordinates.

MISCELLANEOUS EXAMPLES.

1. Prove that, if

$$d\theta\,(m-n\cos\phi)^{\frac{1}{2}}=d\phi\,(m-n\cos\theta)^{\frac{1}{2}},$$

then $\quad 2m-n\left(c^2+\dfrac{1}{c^2}\right)+n\left(c\cos\dfrac{\theta+\phi}{2}-\dfrac{1}{c}\cos\dfrac{\theta-\phi}{2}\right)^2=0,$

c being an arbitrary constant.

2. Let $F(x)$ denote the integral

$$\int_0^x\frac{dx}{\{(1-x^2)(1-k^2x^2)\}^{\frac{1}{2}}};$$

prove that an algebraical relation equivalent to

$$F(x_1)+F(x_2)+F(x_3)=0$$

is $\quad 4(1-x_1^2)(1-x_2^2)(1-x_3^2)=(2-x_1^2-x_2^2-x_3^2+k^2x_1^2x_2^2x_3^2)^2.$

3. Let $E(x)$ denote the integral

$$\int_0^x\left\{\frac{1-k^2x^2}{1-x^2}\right\}^{\frac{1}{2}}dx;$$

verify that

$$E(x_1)+E(x_2)+E(x_3)=-k^2x_1x_2x_3,$$

where x_1,x_2,x_3, are related as in the previous example.

4. Verify that

$$\begin{vmatrix}x_1,&y_1,&1\\x_2,&y_2,&1\\x_3,&y_3,&1\end{vmatrix}=0$$

is an integral of

$$\frac{dx_1}{(1-x_1^3)^{\frac{2}{3}}}+\frac{dx_2}{(1-x_2^3)^{\frac{2}{3}}}+\frac{dx_3}{(1-x_3^3)^{\frac{2}{3}}}=0,$$

y being given by the relation $x^3+y^3=1$.

Interpret the result geometrically.

(Cayley.)

5. Prove that the integral of

$$\frac{dx}{(1+x^3)^{\frac{2}{3}}} + \frac{dy}{(1+y^3)^{\frac{2}{3}}} = 0$$

may be exhibited in the form

$$(1+x^3)(1+y^3)(1+a^3) = (1+xya)^3,$$

where a is an arbitrary constant; and that the integral of

$$\frac{dx}{(4x^3 - Ix + J)^{\frac{2}{3}}} + \frac{dy}{(4y^3 - Iy + J)^{\frac{2}{3}}} = 0$$

may be exhibited in the form

$$(4x^3 - Ix + J)(4y^3 - Iy + J)(4a^3 - Ia + J) = \{4xya - \tfrac{1}{3}I(x+y+a) + J\}^3,$$

where I and J are definite constants, and a is an arbitrary constant.

Shew that a general integral of

$$X^{-\frac{2}{3}}dx + Y^{-\frac{2}{3}}dy = 0,$$

where

$$X = (k, l, m, n\Sigma x, 1)^3,$$

$$Y = (k, l, m, n\Sigma y, 1)^3,$$

is

$$XYZ = \{k + l(x+y+z) + m(xy+yz+zx) + nxyz\}^3,$$

where

$$Z = (k, l, m, n\Sigma z, 1)^3,$$

and z is an arbitrary constant.

(MacMahon and Russell.)

6. Prove that integral relations equivalent to

$$\left. \begin{array}{c} \dfrac{d\theta}{\Delta\theta} + \dfrac{d\phi}{\Delta\phi} + \dfrac{d\psi}{\Delta\psi} = 0 \\[2mm] \dfrac{\sin^2\theta\, d\theta}{\Delta\theta} + \dfrac{\sin^2\phi\, d\phi}{\Delta\phi} + \dfrac{\sin^2\psi\, d\psi}{\Delta\psi} = 0 \end{array} \right\},$$

where

$$\Delta\chi = \{(1 - \kappa\sin^2\chi)(1 - \lambda\sin^2\chi)(1 - \mu\sin^2\chi)\}^{\frac{1}{2}},$$

are

$$\frac{\sin\psi\sin\phi\cos\theta\,\Delta\theta}{(\sin^2\theta - \sin^2\phi)(\sin^2\theta - \sin^2\psi)} + \frac{\sin\theta\sin\psi\cos\phi\,\Delta\phi}{(\sin^2\phi - \sin^2\theta)(\sin^2\phi - \sin^2\psi)}$$

$$+ \frac{\sin\phi\sin\theta\cos\psi\,\Delta\psi}{(\sin^2\psi - \sin^2\theta)(\sin^2\psi - \sin^2\phi)} = A,$$

and

$$\frac{\cos\psi\cos\phi\sin\theta\,\Delta\theta}{(\sin^2\theta - \sin^2\phi)(\sin^2\theta - \sin^2\psi)} + \frac{\cos\theta\cos\psi\sin\phi\,\Delta\phi}{(\sin^2\phi - \sin^2\theta)(\sin^2\phi - \sin^2\psi)}$$

$$+ \frac{\cos\phi\cos\theta\sin\psi\,\Delta\psi}{(\sin^2\psi - \sin^2\theta)(\sin^2\psi - \sin^2\phi)} = B.$$

Determine A and B from the conditions that $\phi = a$ and $\psi = \beta$ when $\theta = 0$.

7. Find the primitives of the equations

(i) $(ay - bz) dx + (cz - ax) dy + (bx - cy) dz = 0$;

(ii) $\dfrac{dx\,(y + z - 2x)}{(y - x)\,(z - x)} + \dfrac{dy\,(z + x - 2y)}{(z - y)\,(x - y)} + \dfrac{dz\,(x + y - 2z)}{(x - z)\,(y - z)} = 0$;

(iii) $(y^2 + yz + z^2) dx + (z^2 + zx + x^2) dy + (x^2 + xy + y^2) dz = 0$.

8. Shew that, if the equation $P dx + Q dy + R dz = 0$ satisfies the condition of integrability (§ 151), and if P, Q, R, are homogeneous functions of x, y, z, of n dimensions, the integral can be obtained by introducing new variables u and v such that $x = uz$, $y = vz$.

Hence obtain the primitive of the equation

$$(x^2 - y^2 + z^2) dx + z dz (y - x) = z^2 dy - \frac{x\,dz}{z} (y^2 - x^2)$$

in the form

$$\int e^{-u^2} du + C = e^{-u^2} \frac{z}{y - x},$$

where $x = uz$.

(Euler.)

9. Solve the simultaneous equations

(i) $a\dfrac{dx}{dt} = (b - c) yz, \quad b\dfrac{dy}{dt} = (c - a) zx, \quad c\dfrac{dz}{dt} = (a - b) xy,$

expressing each of the quantities x, y, z, as elliptic functions;

(ii) $\dfrac{dx}{y^2 + yz + z^2} = \dfrac{dy}{z^2 + zx + x^2} = \dfrac{dz}{x^2 + xy + y^2} = dt.$

10. Integrate the system of equations

$$\frac{d\omega}{dt} + ax + by \cos nt + bz \sin nt = 0,$$

$$a\omega - \frac{dx}{dt} - by \sin nt + bz \cos nt = 0,$$

$$b\omega \cos nt + bx \sin nt - \frac{dy}{dt} - az = 0,$$

$$b\omega \sin nt - bx \cos nt + ay - \frac{dz}{dt} = 0.$$

11. Integrate the simultaneous equations

$$\frac{d^2u}{dt^2} + n^2 \{u - 3\xi\,(u\xi + v\eta)\} = 0,$$

$$\frac{d^2v}{dt^2} + n^2 \{v - 3\eta\,(u\xi + v\eta)\} = 0,$$

where ξ is written for $\cos(at + b)$ and η for $\sin(at + b)$.

(Liouville.)

12. Solve the simultaneous equations

$$\frac{d^4y}{dx^4} - a\frac{d^3z}{dx^3} + b\frac{d^2y}{dx^2} + cy = 0,$$

$$\frac{d^4z}{dx^4} + a\frac{d^3y}{dx^3} + b\frac{d^2z}{dx^2} + cz = 0.$$

13. Shew that any system of lines, described on the surface of the sphere $x^2 + y^2 + z^2 = r^2$ and satisfying the equation

$$(1 + 2m)\,x\,dx + y\,(1-x)\,dy + z\,dz = 0,$$

would be projected on the plane of xy into parabolas.

Find the equation of the projections of the same system of curves on the plane of yz.

14. Shew that Monge's method (Ex. 4, § 153) would, if we integrate first with respect to x and z, present the solution of the equation in the preceding example in the form

$$(1 + 2m)\,x^2 + z^2 = \phi\,(y), \quad 2y\,(1-x) = -\,\phi'\,(y).$$

Apply this to solve the problem of the preceding example; and identify the results.

15. Integrate the simultaneous equations

$$\frac{d^2x_1}{dt^2} = \frac{\partial R}{\partial x_1}, \quad \frac{d^2x_2}{dt^2} = \frac{\partial R}{\partial x_2}, \quad \ldots\ldots, \quad \frac{d^2x_n}{dt^2} = \frac{\partial R}{\partial x_n},$$

where R is a function of $(x_1^2 + x_2^2 + \ldots + x_n^2)^{\frac{1}{2}}$.

(Binet.)

16. Shew that, if

$$f\,(x + \omega y + \omega^2 z) = X + \omega Y + \omega^2 Z,$$

where ω is any cube root of unity, then

$$X\,dx + Z\,dy + Y\,dz,$$

$$Z\,dx + Y\,dy + X\,dz,$$

$$Y\,dx + X\,dy + Z\,dz,$$

are perfect differentials dP, dQ, dR; and prove that the surfaces

$$P = a, \quad Q = b, \quad R = c,$$

where they intersect at a common point, intersect in pairs at the same angle.

(Laurent.)

CHAPTER IX

PARTIAL DIFFERENTIAL EQUATIONS OF THE FIRST ORDER

179. HITHERTO we have been considering for the most part differential equations in which the dependent variable or, in the case of a set of simultaneous equations, the dependent variables are supposed to be functions of only a single independent variable. We shall now consider equations in which the number of independent variables is greater than unity; and we shall suppose that there is only a single dependent variable. The latter is usually denoted by z; if it be a function of only two variables, these are usually denoted by x and y; if z be a function of more than two, say of n, then it is convenient to denote the latter by $x_1, x_2, x_3, \ldots\ldots, x_n$. The first partial differential coefficients in the former case, viz., $\dfrac{\partial z}{\partial x}$ and $\dfrac{\partial z}{\partial y}$, are represented by p and q respectively; in the latter case, the partial differential coefficients $\dfrac{\partial z}{\partial x_1}, \dfrac{\partial z}{\partial x_2}, \ldots\ldots, \dfrac{\partial z}{\partial x_n}$, are represented by $p_1, p_2, \ldots\ldots, p_n$, respectively.

An equation in partial differential coefficients is a relation between the independent variables, the dependent variable (which is an unknown function of those variables) and its partial differential coefficients with regard to them. It is of the first order when all the partial differential coefficients which occur are of order not higher than unity, of the second order when the partial differential coefficients of highest order which occur are of order two; and so on. In this chapter, we shall consider only equations of the first order.

It may happen that we have more than a single differential equation relating to the same set of variables; for instance, we might have two equations between z, x, y, p, q. In this case the two equations could be solved and from them values of p and q in terms of x, y, and z, could be deduced; these could be substituted in the equation

$$dz = p\,dx + q\,dy,$$

and we should thus obtain a total differential equation. Similarly in the case of n independent variables, n equations would be sufficient and necessary to determine p_1, p_2, ..., p_n; these n equations would then be considered as furnishing a total differential equation. When the number of equations is less than the number of partial differential coefficients and therefore of course less than the number of independent variables, we are not able to deduce solely from them a total differential equation. When only a single equation is given, we call it a *partial differential equation*; when several equations are given, we call them a *simultaneous system*.

As in the case of ordinary differential equations, the integration of an equation is the derivation of all the values of z which, when substituted in the differential equation, render it an identity.

Classification of Integrals.

180. Before indicating methods of integration and giving such classes of equations as are easily integrable, it is necessary to classify the different kinds of integrals of a partial differential equation and to prove that the classes include all possible integrals of the equation. For perfect generality the propositions should be proved for an equation involving n variables, but the proofs are given for an equation involving only three variables; this limitation has the advantage of shortening the equations and of lessening their number, while very slight consideration will shew that it is possible to pass to the general case without any essential difficulties of analysis.

181. Suppose that we have between z, x_1, x_2, x_3, a relation of the form

$$f(z, x_1, x_2, x_3, a_1, a_2, a_3) = 0 \quad \ldots\ldots\ldots\ldots(1),$$

in which a_1, a_2, a_3, are arbitrary constants and which contains no differential coefficients of z. To obtain p_1, p_2, p_3, we have the equations

$$\left. \begin{array}{l} \dfrac{\partial f}{\partial z} p_1 + \dfrac{\partial f}{\partial x_1} = 0 \\[2ex] \dfrac{\partial f}{\partial z} p_2 + \dfrac{\partial f}{\partial x_2} = 0 \\[2ex] \dfrac{\partial f}{\partial z} p_3 + \dfrac{\partial f}{\partial x_3} = 0 \end{array} \right\} \quad \ldots\ldots\ldots\ldots\ldots\ldots(2).$$

Between equations (1) and (2) the three arbitrary constants can be eliminated; and usually the result of the elimination is to provide a single equation involving the variables z, x_1, x_2, x_3, and the derivatives p_1, p_2, p_3. If in (1) there were more than three arbitrary constants these equations would not be sufficient for the elimination, while if there were fewer than three there would be more than sufficient equations. Let the result of the elimination in the present case be denoted by

$$F(p_1, p_2, p_3, z, x_1, x_2, x_3) = 0 \quad \ldots\ldots\ldots\ldots(A),$$

which will be the partial differential equation corresponding to the integral relation (1).

Conversely, this integral relation (1) is a solution of (A), and it contains three arbitrary constants. We cannot expect more than three arbitrary constants in a solution of (A); for, on passing from such a solution to the differential equation by the method in which (A) has been obtained from (1), only three constants could be eliminated. Hence (1) contains the greatest number of arbitrary constants that we can expect in a solution of (A).

The name *Complete Integral* of an equation is given to a relation between the variables which includes as many arbitrary constants as there are independent variables.

Note. It must not however be assumed that an equation, having the form of a complete integral, always leads to one (and to only one) partial differential equation.

Thus the equation

$$z = \{(x_1 - a_1)^2 + (x_2 - a_2)^2\}^{\frac{1}{2}} + x_3 - a_3,$$

when treated so as to lead to a partial differential equation by the elimination of the three arbitrary constants a_1, a_2, a_3, leads to the two equations

$$p_1^2 + p_2^2 = 1, \quad p_3 = 1,$$

182. The supposition has been made that a_1, a_2, a_3, are constants; and we have deduced equation (A) from (1) and (2). But we may suppose that a_1, a_2, a_3, are functions of the independent variables. If these functions are chosen as to leave unaltered the forms of p_1, p_2, p_3, then the differential equation obtained by the elimination of the functions will be the same as in the case when the quantities a were arbitrary constants, for algebraical elimination takes no cognisance of the value of the quantity eliminated but only of its form. Now with the new supposition that the quantities a are functions of the variables x_1, x_2, x_3, the values of the partial differential coefficients are given by the equations

$$\frac{\partial f}{\partial z} p_1 + \frac{\partial f}{\partial x_1} + \frac{\partial f}{\partial a_1}\frac{\partial a_1}{\partial x_1} + \frac{\partial f}{\partial a_2}\frac{\partial a_2}{\partial x_1} + \frac{\partial f}{\partial a_3}\frac{\partial a_3}{\partial x_1} = 0,$$

$$\frac{\partial f}{\partial z} p_2 + \frac{\partial f}{\partial x_2} + \frac{\partial f}{\partial a_1}\frac{\partial a_1}{\partial x_2} + \frac{\partial f}{\partial a_2}\frac{\partial a_2}{\partial x_2} + \frac{\partial f}{\partial a_3}\frac{\partial a_3}{\partial x_2} = 0,$$

$$\frac{\partial f}{\partial z} p_3 + \frac{\partial f}{\partial x_3} + \frac{\partial f}{\partial a_1}\frac{\partial a_1}{\partial x_3} + \frac{\partial f}{\partial a_2}\frac{\partial a_2}{\partial x_3} + \frac{\partial f}{\partial a_3}\frac{\partial a_3}{\partial x_3} = 0.$$

But the forms of p_1, p_2, p_3, are to be the same as before, when they were given by equations (2); in order that this may be the case, we must have

$$\left.\begin{aligned}
\frac{\partial f}{\partial a_1}\frac{\partial a_1}{\partial x_1} + \frac{\partial f}{\partial a_2}\frac{\partial a_2}{\partial x_1} + \frac{\partial f}{\partial a_3}\frac{\partial a_3}{\partial x_1} = 0\\
\frac{\partial f}{\partial a_1}\frac{\partial a_1}{\partial x_2} + \frac{\partial f}{\partial a_2}\frac{\partial a_2}{\partial x_2} + \frac{\partial f}{\partial a_3}\frac{\partial a_3}{\partial x_2} = 0\\
\frac{\partial f}{\partial a_1}\frac{\partial a_1}{\partial x_3} + \frac{\partial f}{\partial a_2}\frac{\partial a_2}{\partial x_3} + \frac{\partial f}{\partial a_3}\frac{\partial a_3}{\partial x_3} = 0
\end{aligned}\right\} \quad\ldots\ldots\ldots\ldots(3).$$

Let R denote the value of the determinant

$$\begin{vmatrix}
\dfrac{\partial a_1}{\partial x_1}, & \dfrac{\partial a_2}{\partial x_1}, & \dfrac{\partial a_3}{\partial x_1} \\[2ex]
\dfrac{\partial a_1}{\partial x_2}, & \dfrac{\partial a_2}{\partial x_2}, & \dfrac{\partial a_3}{\partial x_2} \\[2ex]
\dfrac{\partial a_1}{\partial x_3}, & \dfrac{\partial a_2}{\partial x_3}, & \dfrac{\partial a_3}{\partial x_3}
\end{vmatrix},$$

so that the foregoing equations are equivalent to

$$R\frac{\partial f}{\partial a_1} = 0, \quad R\frac{\partial f}{\partial a_2} = 0, \quad R\frac{\partial f}{\partial a_3} = 0 \quad\ldots\ldots\ldots(4).$$

Now, if R do not vanish, these can only be satisfied by

$$\frac{\partial f}{\partial a_1} = 0, \quad \frac{\partial f}{\partial a_2} = 0, \quad \frac{\partial f}{\partial a_3} = 0 \quad \dots\dots\dots(B);$$

and these are three equations which may determine values of a_1, a_2, a_3, in terms of the variables. If they do determine values of a_1, a_2, a_3, in any of the possible forms (whether as constants or as functions of variables), the relation (1) is still a solution with the change in the quantities a; when the values thus found are substituted for them, we have a solution of (A) which contains no arbitrary constant. This solution moreover will generally differ from any solution, which contains no arbitrary constant but is derived from (1) by assigning particular constant values to a_1, a_2, a_3, in (1). The result of eliminating the arbitrary constants between (1) and (B) then gives a new solution.

This solution is called a *Singular Integral*; it is a relation between the variables involving no arbitrary constant. Sometimes it occurs also as a particular instance of a Complete Integral, when special values are given to the arbitrary constants; but this is usually not the case, and the Singular Integral (when it exists) is usually distinct from a Complete Integral.

183. The equations (4) will all be satisfied if $R = 0$; and as we are now assuming that a_1, a_2, a_3, are not arbitrary constants but functions of the variables, this equation will be satisfied (§ 9) by a functional relation between a_1, a_2, a_3. This functional relation may be arbitrary, so that we may write

$$a_3 = \phi(a_1, a_2) \quad \dots\dots\dots\dots\dots(C),$$

in which ϕ denotes an arbitrary function. Multiplying now the equations (3) by dx_1, dx_2, dx_3, respectively and adding, we obtain

$$\frac{\partial f}{\partial a_1} da_1 + \frac{\partial f}{\partial a_2} da_2 + \frac{\partial f}{\partial a_3} da_3 = 0.$$

But from equation (C) we have

$$da_3 = \frac{\partial \phi}{\partial a_1} da_1 + \frac{\partial \phi}{\partial a_2} da_2,$$

so that $\left(\dfrac{\partial f}{\partial a_1} + \dfrac{\partial f}{\partial a_3} \dfrac{\partial \phi}{\partial a_1} \right) da_1 + \left(\dfrac{\partial f}{\partial a_2} + \dfrac{\partial f}{\partial a_3} \dfrac{\partial \phi}{\partial a_2} \right) da_2 = 0.$

Since a_1 and a_2 are independent, their variations da_1 and da_2 are also independent; in order that this equation may be satisfied, we must therefore have

$$
\left.\begin{array}{l}
\dfrac{\partial f}{\partial a_1} + \dfrac{\partial f}{\partial a_2}\dfrac{\partial \phi}{\partial a_1} = 0 \\[2ex]
\dfrac{\partial f}{\partial a_2} + \dfrac{\partial f}{\partial a_3}\dfrac{\partial \phi}{\partial a_2} = 0
\end{array}\right\} \quad \dots\dots\dots\dots\dots(D).
$$

These equations (D) are sufficient to determine a_1, a_2, a_3, in terms of the variables, and the expressions so obtained will involve the arbitrary function ϕ; when they are substituted in (A), the solution takes a new form which is different from both of the other two.

This solution is called a *General Integral*; it is a relation between the variables involving two (or, in the case of n variables, $n-1$) independent functions of those variables, together with an arbitrary function of those two (or $n-1$) functions.

The equation $R = 0$ could also be satisfied by making a_3 an arbitrary function of a_2 alone or of a_1 alone, and also by making both a_2 and a_3 functions of a_1 alone. We should thus arrive at different classes of General Integrals; but these are all less general than the former, in which only a single arbitrary relation between all the quantities a occurs. This is easily seen from the consideration that if, in equation (C), a_3 be expanded in powers of a_1 the coefficients are arbitrary functions of a_2, while if $\psi(a_1)$, an arbitrary function of a_1, be expanded in powers of a_1 the coefficients are merely arbitrary constants; and the latter is obviously included in the former.

184. It is thus manifest that we have three fundamentally distinct classes of solutions of partial differential equations.

The question arises as to whether these three classes are completely comprehensive, in the sense that every integral can be included in one or other of the classes. Usually, but not universally, it is the case that the three classes do include all integrals. Exceptions may arise for partial differential equations of particular forms, and, when they do occur, the unincluded integrals will be called *special integrals*; but special integrals do not occur for equations of general form. It is, of course, necessary to have a test as to whether an integral is, or is not, special; the test will be obtained in the course of establishing the following theorem:

Every solution (which is not special) of the differential equation is included in one or other of the three classes of solutions of the equation which are constituted by a Complete Integral, the Singular Integral, and a General Integral.

Let (A) represent the differential equation, and (1) a Complete Integral of this equation; then the equations (B) and (C) will give the Singular Integral and the General Integral. Let any other solution of the equation be represented by

$$\psi (z, x_1, x_2, x_3) = 0 \quad \dots\dots\dots\dots\dots\dots(5).$$

As it is convenient to speak of z as explicitly expressed in terms of the independent variables, we shall use Z to represent the value of the dependent variable derived from (1) and ζ to represent the value derived from (5). Now suppose it possible to choose values for a_1, a_2, a_3, whether variable or constant, so that Z still continues to satisfy the partial equation, and so also that the value of Z in terms of the variables becomes the same as the value of ζ. In that case, the values of p_1, p_2, p_3, derived from Z become the same as the values derived from ζ, which are given by

$$\left.\begin{aligned} \frac{\partial \psi}{\partial z} p_1 + \frac{\partial \psi}{\partial x_1} &= 0 \\[4pt] \frac{\partial \psi}{\partial z} p_2 + \frac{\partial \psi}{\partial x_2} &= 0 \\[4pt] \frac{\partial \psi}{\partial z} p_3 + \frac{\partial \psi}{\partial x_3} &= 0 \end{aligned}\right\}.$$

Consequently, when we make these values of the differential coefficients of ζ agree with those of Z given by equations (2), we have the three equations

$$\left.\begin{aligned} \frac{\partial f}{\partial x_1} \frac{\partial \psi}{\partial z} - \frac{\partial \psi}{\partial x_1} \frac{\partial f}{\partial z} &= 0 \\[4pt] \frac{\partial f}{\partial x_2} \frac{\partial \psi}{\partial z} - \frac{\partial \psi}{\partial x_2} \frac{\partial f}{\partial z} &= 0 \\[4pt] \frac{\partial f}{\partial x_3} \frac{\partial \psi}{\partial z} - \frac{\partial \psi}{\partial x_3} \frac{\partial f}{\partial z} &= 0 \end{aligned}\right\} \quad \dots\dots\dots\dots\dots\dots(6).$$

These equations may, or may not, determine the values of a_1, a_2, a_3, in terms of x_1, x_2, x_3, and the dependent variable.

When the three equations do not determine values of a_1, a_2, a_3, we cannot proceed further in the attempt to modify Z so that it shall acquire the value ζ. In that event, the integral ζ is called *special*. It is sufficiently obvious that the failure of the three equations to determine the three quantities a_1, a_2, a_3, must arise through some peculiarities in the form of the integral equations.

When the three equations do determine values of a_1, a_2, a_3, we proceed as follows. Since (5) is a solution of the differential equation, we have

$$F(p_1, p_2, p_3, \zeta, x_1, x_2, x_3) = 0;$$

and since (1) is a solution, we have

$$F(p_1, p_2, p_3, Z, x_1, x_2, x_3) = 0$$

satisfied, when the quantities a are arbitrary. The last equation is also satisfied when the quantities a, instead of being arbitrary constants, become functions of the variables, provided these functions are such as to leave the forms of p_1, p_2, p_3, unaltered; and we may therefore replace them by the functions of x_1, x_2, x_3, obtained as their values from the equations (6), provided the necessary conditions be satisfied. When this is the case, the values of p_1, p_2, p_3, are the same for the two forms of the equation (A); and then, from a comparison of these two forms, we have the necessary equation

$$\zeta = Z,$$

where, in the integral equation for Z, the constants a_1, a_2, a_3, are changed into the values that have been derived for them.

In order that the forms of p_1, p_2, p_3, for the new values of the quantities a should be unchanged, the three equations of the form

$$\frac{\partial f}{\partial z}\frac{\partial \psi}{\partial x_1} = -\frac{\partial f}{\partial z}\frac{\partial \psi}{\partial z} p_1$$

$$= \frac{\partial \psi}{\partial z}\left(\frac{\partial f}{\partial x_1} + \frac{\partial f}{\partial a_1}\frac{\partial a_1}{\partial x_1} + \frac{\partial f}{\partial a_2}\frac{\partial a_2}{\partial x_1} + \frac{\partial f}{\partial a_3}\frac{\partial a_3}{\partial x_1}\right)$$

must be satisfied at the same time as (6); and therefore the values of a_1, a_2, a_3, are such as to satisfy the equations

$$\left. \begin{aligned} \frac{\partial f}{\partial a_1}\frac{\partial a_1}{\partial x_1} + \frac{\partial f}{\partial a_2}\frac{\partial a_2}{\partial x_1} + \frac{\partial f}{\partial a_3}\frac{\partial a_3}{\partial x_1} = 0 \\ \frac{\partial f}{\partial a_1}\frac{\partial a_1}{\partial x_2} + \frac{\partial f}{\partial a_2}\frac{\partial a_2}{\partial x_2} + \frac{\partial f}{\partial a_3}\frac{\partial a_3}{\partial x_2} = 0 \\ \frac{\partial f}{\partial a_1}\frac{\partial a_1}{\partial x_3} + \frac{\partial f}{\partial a_2}\frac{\partial a_2}{\partial x_3} + \frac{\partial f}{\partial a_3}\frac{\partial a_3}{\partial x_3} = 0 \end{aligned} \right\}.$$

But these are of the form of the equations (3) which enable us to pass, from the Complete Integral considered, to the other two Integrals; hence the values of a are included among those which give either the Complete Integral, the Singular Integral, or the General Integral of the equation. And as the necessary conditions have been satisfied, we have

$$\zeta = Z,$$

or the value of z derived from the given solution coincides with the value derived from one or other of the three principal integrals.

This proves the theorem and shews that, save for special integrals, the three classes adopted include all possible solutions.

If, on solving the equations (6), the quantities a be found to be all constant, then the given solution will be a particular case of the Complete Integral. If they be found to be functions of the variables and there exist a functional relation between them of the form

$$a_3 = \phi(a_2, a_1),$$

then the given solution will be a particular case of the General Integral. If they be found to be functions of the variables and there be no such functional relation between them, then the given solution is the Singular Integral.

Ex. 1. The equation

$$pq = 4xy$$

has a complete integral in the form

$$z = \frac{1}{a} x^2 + ay^2 + b,$$

where a and b are arbitrary constants; and

$$z = 2xy + c$$

is another integral, c being an arbitrary constant. What is the character of the latter integral?

The general integral, derived from the complete integral, is given by

$$z = \frac{1}{a} x^2 + ay^2 + \phi(a), \quad 0 = -\frac{x^2}{a^2} + y^2 + \phi'(a);$$

and there is no singular integral. Hence the second integral must be a particular case either of the complete integral or of the general integral: or it must be a special integral.

Forming the derivatives of the two values of z, and equating them as in the text, we find

$$\frac{2x}{a} = 2y, \quad 2ay = 2x,$$

both of which are satisfied by

$$a = \frac{x}{y};$$

with this value of a, the complete integral becomes the second integral.

The other equations

$$\frac{\partial f}{\partial a}\frac{\partial a}{\partial x} + \frac{\partial f}{\partial b}\frac{\partial b}{\partial x} = 0, \quad \frac{\partial f}{\partial a}\frac{\partial a}{\partial y} + \frac{\partial f}{\partial b}\frac{\partial b}{\partial y} = 0,$$

are satisfied by

$$\frac{\partial f}{\partial a} = 0, \quad b = \text{constant}.$$

The second integral is therefore a particular case of the general integral, obtained by taking

$$\phi(a) = c,$$

where c is the arbitrary constant in that second integral.

Ex. 2. The equation

$$px - qy = (z - xy)^2$$

has a complete integral in the form

$$x e^{\frac{1}{z - xy}} = \frac{xy + a}{xy + b}.$$

It is also satisfied by

$$\psi = z - xy = 0;$$

what is the character of the latter integral?

If the latter integral can be derived from the complete integral, which may be taken in the form

$$f = \log x + \frac{1}{z - xy} - \log(xy + a) + \log(xy + b) = 0,$$

the values of a and b must be determined by the equations

$$\frac{\partial f}{\partial x}\frac{\partial \psi}{\partial z} - \frac{\partial f}{\partial z}\frac{\partial \psi}{\partial x} = 0, \quad \frac{\partial f}{\partial y}\frac{\partial \psi}{\partial z} - \frac{\partial f}{\partial z}\frac{\partial \psi}{\partial y} = 0,$$

that is, by the equations

$$\frac{1}{x} - \frac{y}{xy + a} + \frac{y}{xy + b} = 0,$$

$$-\frac{x}{xy + a} + \frac{x}{xy + b} = 0,$$

which, being incompatible with one another, do not determine a and b. Thus values of a and b, suitable for the required modification of the complete integral, cannot be determined; the second integral is therefore of the type called special.

Ex. 3. Assuming that a Complete Integral of $z = pq$ is

$$4z = \left(ax + \frac{y}{a} + b\right)^2,$$

investigate the nature of the solution

$$4z - 2xy = (x^2 + y^2)\sec a + (x^2 - y^2)\tan a.$$

Ex. 4. Assuming that a Complete Integral of $z = px + qy$ is

$$\log z = a \log x + (1 - a)\log y + b,$$

investigate the nature of the solution

$$z = y\phi\left(\frac{y}{x}\right).$$

Ex. 5. Assuming that a Complete Integral of $z = px + qy + pq$ is

$$z = ax + by + ab,$$

investigate the nature of the solution

$$z + xy = 0.$$

Ex. 6. A complete integral of the equation

$$xp + 2yq = 2\left(z - \frac{x^2}{y}\right)^2$$

is given by

$$e^{\frac{1}{z - \frac{x^2}{y}}} = a\frac{x^2}{y^2} + \frac{b}{y},$$

where a and b are arbitrary constants; discuss the character of the integral

$$yz = x^2.$$

Ex. 7. The equation $3px + qy + q^3x^2 = 0$ possesses two complete integrals

$$z = a - \tfrac{1}{3}b^3x + byx^{-\frac{1}{3}},$$

$$z = a' + \tfrac{2}{3}y^{\frac{3}{2}}(x^2 + 2b'x)^{-\frac{1}{2}},$$

where a, b, a', b', are arbitrary constants. Shew that each of these integrals is a particular form of the general integral derivable from the other.

185. In the case when there are two independent variables and one dependent variable, the three may be taken as the coordinates of a point in space; and the relations between the separate integrals can be interpreted geometrically.

A Complete Integral, being a relation between x, y, and z, is the equation of a surface and it includes two arbitrary parameters. Thus the Complete Integral belongs to a doubly infinite system of surfaces, or to a singly infinite system of families of surfaces. This integral is of the form

$$\phi(x, y, z, a, b) = 0.$$

In order to obtain a General Integral, we make one of the parameters an arbitrary function of the other, say $b = \theta(a)$, and eliminate a between

$$\left. \begin{aligned} \phi(x, y, z, a, b) &= 0 \\ b &= \theta(a) \\ \frac{\partial \phi}{\partial a} + \frac{\partial \phi}{\partial b}\, \theta'(a) &= 0 \end{aligned} \right\}.$$

This operation is really equivalent to selecting from the system of families of surfaces a representative family and finding its envelope. If a particular family be taken (which occurs when b is made a *definite* function of a instead of an *arbitrary* function), then the equation of its envelope is a particular case of the General Integral. The foregoing equations, as they stand, represent a curve drawn on the surface of the family whose parameter is a, while the equation resulting from the elimination of a between them is the envelope of the family; hence the envelope touches the surface represented by the first two equations along the curve represented by the three equations. This curve is called the *characteristic* of the envelope; and the General Integral thus represents the envelope of a family of surfaces, considered as composed of its characteristics.

In order to obtain the Singular Integral, we eliminate the parameters between the equations

$$\left. \begin{aligned} \phi(x, y, z, a, b) &= 0 \\ \frac{\partial \phi}{\partial a} &= 0 \\ \frac{\partial \phi}{\partial b} &= 0 \end{aligned} \right\}.$$

This operation is the same as finding the envelope of all the surfaces included in the Complete Integral; the three foregoing equations give the point of contact of the particular surface represented by the first of them with the general envelope. The Singular Integral thus represents the general envelope of all the surfaces included in the Complete Integral.

But when the elimination has taken place so as to leave a relation between x, y, and z, it is necessary to ensure that the resulting equation is that of the envelope and not that of any of the loci which are included in the same equations*. Such loci are, for instance, the locus of conical points and the locus of double lines, neither of which satisfies the differential equation. It is therefore desirable to substitute the result (when it cannot at once be recognised as the equation of an envelope) in the differential equation; it is to be retained only when it is a solution.

It may happen that the entire system of surfaces does not admit of a general envelope; in such a case the Singular Integral does not exist for the corresponding differential equation, and its non-existence will be indicated by the equations ordinarily used to obtain it. Examples will hereafter occur.

As an example to illustrate the preceding discussion of the geometrical relations between the integrals, consider the equation

$$ax + by + cz = (a^2 + b^2 + c^2)^{\frac{1}{2}} = 1 \quad \dots\dots\dots\dots\dots\text{(i)},$$

which contains two independent constants. It is easy to prove that the corresponding differential equation is

$$(xp + yq - z)^2 = 1 + p^2 + q^2 \dots\dots\dots\dots\dots\text{(A)},$$

and that the general envelope of all the planes contained in (i) is the sphere

$$x^2 + y^2 + z^2 = 1 \quad \dots\dots\dots\dots\dots\dots\text{(ii)}.$$

Hence (ii) is the Singular Integral of (A), and the sphere represented by (ii) touches each of the planes represented by (i) in a point.

To obtain the General Integral we eliminate a between

$$\left.\begin{array}{l} ax + yf(a) + z\left[1 - a^2 - \{f(a)\}^2\right]^{\frac{1}{2}} = 1 \\[2mm] x + yf'(a) - z\,\dfrac{a + f(a)f'(a)}{\left[1 - a^2 - \{f(a)\}^2\right]^{\frac{1}{2}}} = 0 \end{array}\right\},$$

---

* In this connection a memoir by M. J. M. Hill, *Phil. Trans.* (A), 1892, pp. 141—278, may be consulted; it contains a full discussion of the occurrence of the various loci, which do not provide solutions of the partial differential equation and yet arise as parts of the eliminant.

in which $f(a)$ is an arbitrary function. This is clearly the envelope of a family of planes the equation of which contains only one parameter; and it is therefore a developable surface. The equation of any developable surface, which envelopes the sphere, is thus included in the above General Integral. The process of making b a function of a is equivalent to drawing on the sphere some definite curve; and the developable surface is the envelope of the tangent planes to the sphere at points which lie on this line.

186. The explanation of § 182 shews how the Singular Integral may be derived from a Complete Integral; it is, however, possible to derive it directly from the differential equation, as is the case (§ 27) in ordinary differential equations.

For the sake of brevity, suppose that there are only two independent variables. Let the equation be

$$\psi(x, y, z, p, q) = 0,$$

of which a Complete Integral is

$$F(x, y, z, a, b) = 0,$$

where a and b are arbitrary constants; the Singular Integral is obtained by combining the equation $F = 0$ with

$$\frac{\partial F}{\partial a} = 0, \quad \frac{\partial F}{\partial b} = 0 \quad \dots\dots\dots\dots\dots(A).$$

Since $F = 0$ is the integral of the differential equation the values of z, p, q, derived from the integral will render $\psi = 0$ an identity; and the substitution of the values of p and q (but not that of z) derived from $F = 0$ will in general render $\psi = 0$ equivalent to the integral equation. Let this latter substitution be made, so that p and q are replaced by functions of x, y, z, a, b; then, in order to find the Singular Integral, we must form the equations analogous to (A), which equations are

$$\frac{\partial \psi}{\partial p}\frac{\partial p}{\partial a} + \frac{\partial \psi}{\partial q}\frac{\partial q}{\partial a} = 0,$$

$$\frac{\partial \psi}{\partial p}\frac{\partial p}{\partial b} + \frac{\partial \psi}{\partial q}\frac{\partial q}{\partial b} = 0.$$

These equations may be satisfied in two ways: firstly, by writing

$$\frac{\partial \psi}{\partial p} = 0 = \frac{\partial \psi}{\partial q};$$

secondly, if $\dfrac{\partial \psi}{\partial p}$ and $\dfrac{\partial \psi}{\partial q}$ do not vanish, then

$$\frac{\partial p}{\partial a}\frac{\partial q}{\partial b} - \frac{\partial p}{\partial b}\frac{\partial q}{\partial a} = 0.$$

The latter equation implies a relation of the form

$$\phi(p, q) = 0,$$

which does not involve either a or b, but may involve quantities multiplying a and b in the expressions for p and q; that is, quantities depending on x, y, and z. If both the arbitrary constants occur in p and q (which does not always happen) the equation $\phi = 0$ would imply that they are effectively only **one**, or that one of them is a function of the other; the equations used then give the General Integral, with which we are not now concerned.

We thus return to

$$\frac{\partial \psi}{\partial p} = 0, \quad \frac{\partial \psi}{\partial q} = 0;$$

the elimination of p and q, between these and $\psi = 0$, will furnish a relation between x, y, z, which is independent of any arbitrary constant. *If this relation satisfy the differential equation, it is the Singular Integral.* When the relation is found by this method, it is necessary to see whether the differential equation is satisfied.

The reason that this precaution is necessary is similar to that which renders the corresponding precaution necessary in the case of ordinary differential equations. When the surfaces represented have an envelope, this envelope will be given by the equations

$$\psi = 0; \quad \frac{\partial \psi}{\partial p} = 0; \quad \frac{\partial \psi}{\partial q} = 0.$$

But these same equations will be satisfied by the coordinates of any pinch-point on one of the surfaces represented by the complete integral; the locus of these pinch-points, however, is easily seen not to be a solution of the equation. The equations will also be satisfied by the coordinates of any point P at which two different surfaces of the system touch, and therefore by the equation of the surface which is the locus of these points. But this surface has not necessarily for its tangent plane at P that tangent plane which is common to the two surfaces, and therefore the values of p and q (which give the direction cosines of the tangent plane) derived from this new locus are not the values of p and q which satisfy the given equation $\psi = 0$. Such a locus corresponds to what was before called the tac-locus (§ 27); and, while it may not be the only locus (other than the envelope) which is introduced, the possibility of its presence renders necessary an enquiry whether the equation between x, y, z, satisfies the differential equation.

As in the case of ordinary differential equations of the first order (§ 28), an analytical test of the form of the equation can be assigned when it possesses a singular integral. When the latter exists, it arises through the elimination of p and q between

$$\psi = 0, \quad \frac{\partial \psi}{\partial p} = 0, \quad \frac{\partial \psi}{\partial q} = 0.$$

Imagine the last two equations solved, so as to express p and q in terms of x, y, z; and let their values be substituted in ψ. Denoting the new value of ψ by ψ', we have

$$\psi' = 0;$$

and this equation is then to provide a solution of the differential equation. In order that the requirement may be satisfied, the proper values of p and q must be given by $\psi' = 0$: and they must be the values deduced from

$$\frac{\partial \psi}{\partial p} = 0, \quad \frac{\partial \psi}{\partial q} = 0.$$

The values of p and q are given by

$$\frac{\partial \psi'}{\partial x} + p \frac{\partial \psi'}{\partial z} = 0, \quad \frac{\partial \psi'}{\partial y} + q \frac{\partial \psi'}{\partial z} = 0.$$

But $\psi = \psi'$; and therefore

$$\frac{\partial \psi}{\partial x} + p \frac{\partial \psi}{\partial z} + \frac{\partial p}{\partial x} \frac{\partial \psi}{\partial p} + \frac{\partial q}{\partial x} \frac{\partial \psi}{\partial q} = \frac{\partial \psi'}{\partial x} + p \frac{\partial \psi'}{\partial z} = 0,$$

$$\frac{\partial \psi}{\partial y} + q \frac{\partial \psi}{\partial z} + \frac{\partial p}{\partial y} \frac{\partial \psi}{\partial p} + \frac{\partial q}{\partial y} \frac{\partial \psi}{\partial q} = \frac{\partial \psi'}{\partial y} + q \frac{\partial \psi'}{\partial z} = 0;$$

or, since $\dfrac{\partial \psi}{\partial p} = 0$, $\dfrac{\partial \psi}{\partial q} = 0$, the values of p and q satisfy the equations

$$\frac{\partial \psi}{\partial x} + p \frac{\partial \psi}{\partial z} = 0, \quad \frac{\partial \psi}{\partial y} + q \frac{\partial \psi}{\partial z} = 0.$$

Accordingly we infer that, *if a singular integral of the equation* $\psi(x, y, z, p, q) = 0$ *exists, it must simultaneously satisfy the equations*

$$\psi = 0, \quad \frac{\partial \psi}{\partial p} = 0, \quad \frac{\partial \psi}{\partial q} = 0,$$

$$\frac{\partial \psi}{\partial x} + p \frac{\partial \psi}{\partial z} = 0, \quad \frac{\partial \psi}{\partial y} + q \frac{\partial \psi}{\partial z} = 0.$$

We have seen that an integral satisfying the first three of these equations is a singular integral; we have just proved that then it must satisfy the remaining two; and thus we infer that an integral satisfying all five equations is a singular integral of the original differential equation.

Ex. 1. The differential equation

$$z^2(1 + p^2 + q^2) = \lambda^2\{(x + pz)^2 + (y + qz)^2\}$$

has for its complete integral

$$(x - a \cos a)^2 + (y - a \sin a)^2 + z^2 = \lambda^2 a^2,$$

λ being supposed a determinate constant. Forming the envelope of this sphere by taking

$$F = (x - a \cos a)^2 + (y - a \sin a)^2 + z^2 - \lambda^2 a^2 = 0,$$

$$\frac{\partial F}{\partial a} = 0, \quad \frac{\partial F}{\partial a} = 0,$$

we easily find it to be

$$\lambda^2(x^2 + y^2 + z^2) = z^2.$$

Now taking

$$\psi = z^2(1 + p^2 + q^2) - \lambda^2\{(x + pz)^2 + (y + qz)^2\}$$

and following the rule for deriving the Singular Integral from the differential equation, we have

$$\frac{\partial \psi}{\partial p} = 2pz^2 - 2\lambda^2 z(x + pz) = 0,$$

$$\frac{\partial \psi}{\partial q} = 2qz^2 - 2\lambda^2 z(y + qz) = 0.$$

The last two equations are satisfied by $z=0$ which, though free from p and q, is not a solution of the differential equation. In fact, by drawing a figure it is easily seen that $z=0$ is a tac-locus, being the plane which contains the points of contact of the different non-consecutive spheres with one another obtained by giving all possible values to a and a.

As regards the analytical test at the end of § 186, the two additional equations are

$$z(1+p^2+q^2)p=\lambda^2(x+pz)+\lambda^2(x+pz)p^2+\lambda^2(y+qz)qp,$$

$$z(1+p^2+q^2)q=\lambda^2(y+qz)+\lambda^2(x+pz)pq+\lambda^2(y+qz)q^2.$$

They are not satisfied by $z=0$, which accordingly is not a singular integral.

They are satisfied by values of p and q given by

$$pz-\lambda^2(x+pz)=0,\quad qz-\lambda^2(y+qz)=0.$$

Accordingly the integral, obtained by eliminating p and q between these various equations, is a singular integral. It is

$$\lambda^2(x^2+y^2+z^2)=z^2,$$

which therefore is the singular integral.

Ex. 2. Consider the system of cones

$$(x-a\cos\theta)^2+(y-a\sin\theta)^2=\left(\frac{z}{m}-\frac{a}{m^2}\right)^2,$$

in which m, θ, are arbitrary constants; the corresponding differential equation is easily obtained. The equations, which give the envelope, are

$$\sin\theta(x-a\cos\theta)-\cos\theta(y-a\sin\theta)=0,$$

$$\left(\frac{z}{m}-\frac{a}{m^2}\right)\left(-\frac{z}{m^2}+\frac{2a}{m^3}\right)=0.$$

These are satisfied by

$$x=a\cos\theta,\quad y=a\sin\theta,\quad z=\frac{a}{m},$$

which give

$$x^2+y^2=a^2,$$

but z is arbitrary.

The equations are also satisfied by

$$z=\frac{2a}{m},\quad x\sin\theta=y\cos\theta,$$

and the corresponding eliminant is

$$x^2+y^2=\left(a+\frac{z^2}{4a}\right)^2.$$

The last equation represents the envelope; the doubly infinite system of cones is generated by the revolution, round the directrix of a parabola, of all the right circular cones whose vertices lie on the tangent at the vertex to the parabola, and one slant side of any one of which coincides with the tangent to the parabola drawn through the vertex of the cone. The equation

$$x^2 + y^2 = a^2$$

is that of the cylinder on which lie all the (singular) circles which are the loci of the vertices of the cones in the revolution round the directrix.

For fuller information on the subject of the Singular Integrals of partial differential equations of the first order, a memoir of DARBOUX, *Mémoires de l'Institut de France*, t. XXVII. (1880), should be consulted. Reference may also be made to the author's *Theory of Differential Equations*, vol. v., ch. vii.

LAGRANGE'S LINEAR EQUATION.

187. We have seen that among the integrals of a differential equation there is one—the General Integral—into the expression of which an arbitrary function enters; the deduction of the differential equation from the integral implies the elimination of this arbitrary function. The simplest form possible for an integral of this nature, when there are two independent variables, is the equation

$$\phi(u, v) = 0 \quad \dots \dots \dots \dots \dots \text{(i)},$$

in which ϕ is an arbitrary functional symbol, and u and v are definite functions of x, y, and z. In order to eliminate ϕ, we differentiate with respect to each of the independent variables. Then

$$\frac{\partial \phi}{\partial u}\left(\frac{\partial u}{\partial x} + p\frac{\partial u}{\partial z}\right) + \frac{\partial \phi}{\partial v}\left(\frac{\partial v}{\partial x} + p\frac{\partial v}{\partial z}\right) = 0,$$

$$\frac{\partial \phi}{\partial u}\left(\frac{\partial u}{\partial y} + q\frac{\partial u}{\partial z}\right) + \frac{\partial \phi}{\partial v}\left(\frac{\partial v}{\partial y} + q\frac{\partial v}{\partial z}\right) = 0.$$

Therefore

$$\left(\frac{\partial u}{\partial x} + p\frac{\partial u}{\partial z}\right)\left(\frac{\partial v}{\partial y} + q\frac{\partial v}{\partial z}\right) = \left(\frac{\partial u}{\partial y} + q\frac{\partial u}{\partial z}\right)\left(\frac{\partial v}{\partial x} + p\frac{\partial v}{\partial z}\right);$$

which, on rearrangement, gives

$$Pp + Qq = R \quad \dots \dots \dots \dots \dots \text{(ii)},$$

where

$$\frac{P}{\left|\begin{matrix} \dfrac{\partial u}{\partial y}, & \dfrac{\partial u}{\partial z} \\[2mm] \dfrac{\partial v}{\partial y}, & \dfrac{\partial v}{\partial z} \end{matrix}\right|} = \frac{Q}{\left|\begin{matrix} \dfrac{\partial u}{\partial z}, & \dfrac{\partial u}{\partial x} \\[2mm] \dfrac{\partial v}{\partial z}, & \dfrac{\partial v}{\partial x} \end{matrix}\right|} = \frac{R}{\left|\begin{matrix} \dfrac{\partial u}{\partial x}, & \dfrac{\partial u}{\partial y} \\[2mm] \dfrac{\partial v}{\partial x}, & \dfrac{\partial v}{\partial y} \end{matrix}\right|} ;$$

or, what are the equivalents of these,

$$\left. \begin{aligned} P\frac{\partial u}{\partial x} + Q\frac{\partial u}{\partial y} + R\frac{\partial u}{\partial z} = 0 \\[2mm] P\frac{\partial v}{\partial x} + Q\frac{\partial v}{\partial y} + R\frac{\partial v}{\partial z} = 0 \end{aligned} \right\} \quad \ldots\ldots\ldots\ldots\ldots\text{(iii)}.$$

Hence, when we have a differential equation of the form (ii), into which the differential coefficients enter linearly while the quantities multiplying these may be any functions of x, y, z, we have a corresponding integral given by (i), provided we can obtain u and v in order to insert them in that integral equation. A differential equation of this form is said to be *linear*. The difficulty in constructing the solution is the derivation of the functions u and v.

188. Now let us consider the equations $u = a$ and $v = b$, where a and b are arbitrary constants, and let us form the differential equations corresponding to them. We have

$$\frac{\partial u}{\partial x} dx + \frac{\partial u}{\partial y} dy + \frac{\partial u}{\partial z} dz = 0,$$

$$\frac{\partial v}{\partial x} dx + \frac{\partial v}{\partial y} dy + \frac{\partial v}{\partial z} dz = 0,$$

and therefore

$$\frac{dx}{\left|\begin{matrix} \dfrac{\partial u}{\partial y}, & \dfrac{\partial u}{\partial z} \\[2mm] \dfrac{\partial v}{\partial y}, & \dfrac{\partial v}{\partial z} \end{matrix}\right|} = \frac{dy}{\left|\begin{matrix} \dfrac{\partial u}{\partial z}, & \dfrac{\partial u}{\partial x} \\[2mm] \dfrac{\partial v}{\partial z}, & \dfrac{\partial v}{\partial x} \end{matrix}\right|} = \frac{dz}{\left|\begin{matrix} \dfrac{\partial u}{\partial x}, & \dfrac{\partial u}{\partial y} \\[2mm] \dfrac{\partial v}{\partial x}, & \dfrac{\partial v}{\partial y} \end{matrix}\right|}$$

or

$$\frac{dx}{P} = \frac{dy}{Q} = \frac{dz}{R} \quad \ldots\ldots\ldots\ldots\ldots\ldots\text{(iv)}.$$

These are the differential equations which have for their integrals $u = a$ and $v = b$; they can be formed at once from the coefficients in the differential equation. We thus have the following rule*:—

To obtain an integral of the linear equation

$$Pp + Qq = R,$$

write down the subsidiary equations

$$\frac{dx}{P} = \frac{dy}{Q} = \frac{dz}{R},$$

and obtain two independent integrals of the latter; let these be

$$u = a \ \text{and} \ v = b.$$

Then an integral of the partial differential equation is given by

$$\phi(u, v) = 0,$$

where ϕ denotes an arbitrary function.

An arbitrary functional relation between u and v of any form will be satisfactory; thus we might have

$$u = \psi(v),$$

where ψ is an arbitrary function.

189. This rule enables us to obtain an integral involving an arbitrary function; it will now be shewn to provide all solutions of the equation which are not of the type called special.

Let $u = a$, $v = b$, be two independent integrals of the subsidiary equations

$$\frac{dx}{P} = \frac{dy}{Q} = \frac{dz}{R};$$

so that, as

$$u_x dx + u_y dy + u_z dz = 0,$$

$$v_x dx + v_y dy + v_z dz = 0,$$

are each of them satisfied in virtue of these subsidiary equations, we have

$$u_x P + u_y Q + u_z R = 0,$$

$$v_x P + v_y Q + v_z R = 0,$$

---

* The theory of linear partial differential equations was first given by Lagrange, as well as the classification of the integrals of equations of the first order. The subsidiary equations (iv) are sometimes called Lagrange's equations.

and therefore

$$\frac{P}{\begin{vmatrix} u_y, & u_z \\ v_y, & v_z \end{vmatrix}} = \frac{Q}{\begin{vmatrix} u_z, & u_x \\ v_z, & v_x \end{vmatrix}} = \frac{R}{\begin{vmatrix} u_x, & u_y \\ v_x, & v_y \end{vmatrix}} = M,$$

say, where M is the common value of the three fractions.

Consider now any integral of the partial differential equation

$$Pp + Qq = R.$$

The value of z which it gives, and the derived values of p and q, satisfy the equation identically. Let this value of z be substituted in u and v, making them functions of x and y only; in this form, let their values be denoted by u' and v' so that, in association with the integral of the equation, we have

$$u' = u, \quad v' = v.$$

Thus

$$\frac{\partial u'}{\partial x} = u_x + p u_z, \quad \frac{\partial v'}{\partial x} = v_x + p v_z,$$

$$\frac{\partial u'}{\partial y} = u_y + q u_z, \quad \frac{\partial v'}{\partial y} = v_y + q v_z;$$

and therefore, if J denote the Jacobian of u' and v' with regard to x and y,

$$J = \frac{\partial u'}{\partial x}\frac{\partial v'}{\partial y} - \frac{\partial u'}{\partial y}\frac{\partial v'}{\partial x}$$

$$= (u_x + p u_z)(v_y + q v_z) - (u_y + q u_z)(v_x + p v_z)$$

$$= \begin{vmatrix} u_x, & u_y \\ v_x, & v_y \end{vmatrix} + p \begin{vmatrix} u_z, & u_y \\ v_z, & v_y \end{vmatrix} + q \begin{vmatrix} u_x, & u_z \\ v_x, & v_z \end{vmatrix}$$

$$= \frac{1}{M}(R - Pp - Qq).$$

If then M is not zero, the right-hand side vanishes because the differential equation is satisfied by the integral; and therefore

$$J = 0,$$

an equation that is satisfied identically, because it is not satisfied in virtue of $u = a$ and $v = b$.

Now u' and v' are functions of x and y; as their Jacobian vanishes, there is some functional relation between them, and therefore there is some relation of the form

$$\phi(u', v') = 0.$$

But, in association with the integral of the partial differential equation, we have $u' = u$, $v' = v$; and therefore that integral is included in the equation

$$\phi(u, v) = 0.$$

But M might be zero. For instance, this would be the case if

$$P = 0, \quad Q = 0, \quad R = 0;$$

for as u and v are independent integrals of the subsidiary equations, not more than one of the minors at the utmost in the fractional expressions for M can vanish. In that particular case, we could not infer that J must vanish; and therefore we cannot assert that a singular integral must be given by an equation of the type

$$\phi(u, v) = 0.$$

Such an integral would however arise by equating to zero some common factor of P, Q, R; the vanishing of this would make P, Q, R, all vanish; and its existence could be determined by initial inspection of the equation.

The question arises as to whether M can vanish in any other circumstance. To test this, take any particular value of x, say c, and any arbitrary values of y and z, say $y = \alpha$ and $z = \beta$, such that not more than one of the three quantities P, Q, R, vanishes when $x = c$, $y = \alpha$, $z = \beta$; in particular, suppose that P does not vanish for these values.

Consider the subsidiary equations in the form

$$\frac{dy}{dx} = \frac{Q}{P}, \quad \frac{dz}{dx} = \frac{R}{P}.$$

It is known, by a theorem due to Cauchy*, that integrals of these equations exist under particular conditions, as follows:—

Let the functions P, Q, R, be regular for values of x, y, z,

---

* See the author's *Theory of Differential Equations*, vol. II. §§ 10—12.

in the vicinity of $x = c$, $y = \alpha$, $z = \beta$, so that they are expansible as series of powers of $x - c$, $y - \alpha$, $z - \beta$, having positive whole numbers for their indices; and suppose that P does not vanish for $x = c$, $y = \alpha$, $z = \beta$. Then solutions of the equations exist, uniquely determined by the condition that y and z acquire the values α and β respectively when $x = c$.

Let these integrals in the present case be denoted by

$$y - \alpha = \alpha_1 (x - c) + \alpha_2 (x - c)^2 + \ldots\ldots = P(x - c),$$

$$z - \beta = \beta_1 (x - c) + \beta_2 (x - c)^2 + \ldots\ldots = Q(x - c);$$

then as

$$\frac{Q}{P} = \alpha_1 + 2\alpha_2 (x - c) + \ldots,$$

$$\frac{R}{P} = \beta_1 + 2\beta_2 (x - c) + \ldots,$$

and as Q and R do not both vanish when $x = c$, $y = \alpha$, $z = \beta$, the two quantities α_1 and β_1 are not zero together.

Now the integrals $u = a$, $v = b$, of the subsidiary equations must be consistent with these integrals, for the latter are the only integrals of the equations that satisfy the assigned conditions. Let them be written in the forms

$$u(x, y, z) = a, \quad v(x, y, z) = b;$$

then each of the equations

$$u(x, \alpha + P, \beta + Q) = a,$$

$$v(x, \alpha + P, \beta + Q) = b,$$

is an identity, so that when the left-hand sides are expanded in powers of $x - c$, the coefficients of the various powers must vanish. Taking, in each of them, the coefficient of the first power, we have

$$u_c + \alpha_1 u_\alpha + \beta_1 u_\beta = 0,$$

$$v_c + \alpha_1 v_\alpha + \beta_1 v_\beta = 0,$$

where u_c, u_α, u_β, are the values of u_x, u_y, u_z, for the set of values $x = c$, $y = \alpha$, $z = \beta$, substituted in all of them, and similarly for

v_c, v_a, v_β. The quantities α_1 and β_1 are determinate, and they are not both zero; hence not more than one of the three minors

$$\left\| \begin{array}{ccc} u_c, & u_a, & u_\beta \\ v_c, & v_a, & v_\beta \end{array} \right\|$$

can be zero; and, in particular, $u_a v_\beta - v_a u_\beta$ cannot be zero.

Consequently the minor

$$\left| \begin{array}{cc} u_y, & u_z \\ v_y, & v_z \end{array} \right|$$

cannot vanish identically.

Nor can this minor vanish in virtue of any integral

$$f(x, y, z) = 0$$

of the original partial differential equation. The quantities c, α, β, have been arbitrarily chosen, subject only to the limitation that, at the utmost, not more than one of the magnitudes P, Q, R, shall vanish—in particular, that P does not vanish—when $x = c$, $y = \alpha$, $z = \beta$. Now let them be chosen so that they satisfy the equation $f = 0$, so that

$$f(c, \alpha, \beta) = 0.$$

Then the integral, defined by the original equation $f = 0$, acquires the value β when $x = c$ and $y = \alpha$.

If the minor $u_y v_z - v_y u_z$ were to vanish for all values of x, y, z, subject to the relation $f(x, y, z) = 0$, it would be zero for any particular set of values subject to that relation. But $u_a v_\beta - v_a u_\beta$ is not zero; and therefore $u_y v_z - v_y u_z$ does not vanish for all values of x, y, z, subject to the relation $f = 0$: in other words, $u_y v_z - v_y u_z$ does not vanish when for z we substitute its value as given by the integral under consideration. Hence, as we have

$$M(u_y v_z - u_z v_y) = P,$$

it follows that M does not vanish. Therefore the integral is included in the equation

$$\phi(u, v) = 0.$$

In order to establish the preceding result[*], the quantities c, α, β, have been chosen so as to satisfy the equation

$$f(x, y, z) = 0,$$

which gives the integral, and at the same time so that not more than one at most of the coefficients P, Q, R, should become zero when $x = c$, $y = \alpha$, $z = \beta$. This of course is possible and imposes no limitation. But it has been assumed that each of the quantities P, Q, R, is expressible as a series of positive integral powers of $x - c$, $y - \alpha$, $z - \beta$; if the assumption is not justified by the form of those quantities, Cauchy's theorem is not applicable, and the inferences cannot be maintained. We are not then in a position to assert that the integral is included in the equation

$$\phi(u, v) = 0.$$

190. In the preceding argument, it has been assumed that P does not vanish in connection with the equations

$$\frac{dy}{dx} = \frac{Q}{P}, \quad \frac{dz}{dx} = \frac{R}{P},$$

as associated with the integral

$$f(x, y, z) = 0$$

of the original partial differential equation.

In the same way, as connected with the equations

$$\frac{dx}{dy} = \frac{P}{Q}, \quad \frac{dz}{dy} = \frac{R}{Q},$$

having $u' = a$ and $v' = b$ for integrals, Q must not vanish in connection with that integral $f = 0$ if the latter is necessarily to be included in the equation $\phi(u', v') = 0$; and as connected with the equations

$$\frac{dx}{dz} = \frac{P}{R}, \quad \frac{dy}{dz} = \frac{Q}{R},$$

---

[*] The proof that has just been given is based partly upon the proof given by Goursat, *Leçons sur l'intégration des équations aux dérivées partielles du premier ordre*, § 18. A different proof is given by Chrystal, *Trans. Roy. Soc. Edin.*, vol. xxxvi. (1892), pp. 551—562; he also indicates fallacies that occur in several of the ordinarily accepted proofs.

having $u'' = a$ and $v'' = b$ for integrals, R must not vanish in connection with an integral $f = 0$ of the original equation if the latter is necessarily to be included in the equation $\phi(u'', v'') = 0$.

It therefore appears that exceptions to the complete comprehensiveness of the equation

$$\phi(u, v) = 0$$

may arise in the case of an integral $f = 0$ if, in connection with that integral, either $P = 0$, or $Q = 0$, or $R = 0$, or if P, Q, R, are not expressible as regular functions of $x - c$, $y - \alpha$, $z - \beta$, where c, α, β, are particular values of the variables satisfying $f = 0$.

All integrals, which are not included in the equation $\phi(u, v) = 0$, will be called *special*.

191. COROLLARY. *When either of the equations $u - a = 0$ and $v - b = 0$ involves z, it is an integral of the differential equation.* For the general solution may be written

$$u = \psi(v),$$

where ψ is an arbitrary function. Take then $\psi(v) = av^0$, where a is an arbitrary constant; the equation becomes $u - a = 0$, which is the first of the stated integrals. Similarly for the second.

These results can be obtained independently. The foregoing article shews that, in order that $\psi(x, y, z) = 0$ may be an integral, we must have

$$P \frac{\partial \psi}{\partial x} + Q \frac{\partial \psi}{\partial y} + R \frac{\partial \psi}{\partial z} = 0.$$

But the equations

$$P \frac{\partial u}{\partial x} + Q \frac{\partial u}{\partial y} + R \frac{\partial u}{\partial z} = 0,$$

$$P \frac{\partial v}{\partial x} + Q \frac{\partial v}{\partial y} + R \frac{\partial v}{\partial z} = 0,$$

are actually satisfied; hence $u - a = 0$ and $v - b = 0$ are integrals, with the limitations indicated.

To equations of the form

$$P \frac{\partial \psi}{\partial x} + Q \frac{\partial \psi}{\partial y} + R \frac{\partial \psi}{\partial z} = 0$$

we shall return in § 194.

192. We thus see that, when there is a single arbitrary function entering simply (that is, without any derivatives) into an integral equation, the corresponding differential equation is necessarily linear; and that the linear differential equation has, for its most general integral, a relation into which an arbitrary function enters. We therefore infer that, in the case of a differential equation which is not linear, the arbitrary function which is essential to the General Primitive cannot enter in a manner similar to that in which the arbitrary function enters in the foregoing equation; in fact, with it will be associated in the General Primitive its first differential coefficient.

193. In the foregoing discussion, we have limited ourselves to the case of two independent variables; the proof of the method when there are n independent variables follows the former on exactly the same lines, and the corresponding rule is :—

To obtain the most general integral of the linear equation

$$P_1 p_1 + P_2 p_2 + P_3 p_3 + \ldots\ldots + P_n p_n = R,$$

form the subsidiary equations

$$\frac{dx_1}{P_1} = \frac{dx_2}{P_2} = \ldots\ldots = \frac{dx_n}{P_n} = \frac{dz}{R},$$

and obtain n independent integrals of these equations; let them be

$$u_1 = a_1, \ \ u_2 = a_2, \ \ \ldots\ldots, \ \ u_n = a_n.$$

Connect these quantities u by an arbitrary functional relation

$$\phi(u_1, \ u_2, \ \ldots\ldots, \ u_n) = 0 \, ;$$

the resulting equation is the integral required.

The proof of this, as well as that of the corresponding corollaries, viz. *that, when any of the equations $u_1 = a_1$, $u_2 = a_2$, $\ldots\ldots$, $u_n = a_n$, involve z, they are integrals of the equation,* is not difficult.

Further, it can be proved that *the equation*

$$\phi(u_1, \ u_2, \ \ldots\ldots, \ u_n) = 0$$

contains all integrals of the partial differential equation which are not of the type called special : the proof can be made to follow the lines of the proof given in § 189 for the case of two variables. The instances of exception from the theorem arise in a similar manner to the corresponding instances of exception from the theorem as stated in § 190,

Ex. 1. Solve the equation $xp + yq = z$.

Lagrange's subsidiary equations are

$$\frac{dx}{x} = \frac{dy}{y} = \frac{dz}{z},$$

of which two integrals are $z = ay$, $z = bx$; hence the solution of the equation is

$$\phi\left(\frac{z}{y}, \frac{z}{x}\right) = 0.$$

It can be exhibited in the forms

$$\frac{z}{y} = \psi\left(\frac{z}{x}\right), \quad \frac{z}{y} = \chi\left(\frac{x}{y}\right),$$

which three are easily seen to be equivalent to one another.

Ex. 2. Solve the equation

$$(mz - ny)\, p + (nx - lz)\, q = ly - mx.$$

Lagrange's subsidiary equations are

$$\frac{dx}{mz - ny} = \frac{dy}{nx - lz} = \frac{dz}{ly - mx}.$$

Hence $x\,dx + y\,dy + z\,dz = 0$, whence $x^2 + y^2 + z^2 = a$;

and $l\,dx + m\,dy + n\,dz = 0$, whence $lx + my + nz = b$;

thus the integral of the equation is

$$lx + my + nz = \phi\,(x^2 + y^2 + z^2).$$

Ex. 3. Solve the equations:

    (i) $x^2 p - xyq + y^2 = 0$;

    (ii) $xzp + yzq = xy$;

    (iii) $(y^2 + z^2 - x^2)\, p - 2xyq + 2xz = 0$;

    (iv) $z - xp - yq = a\,(x^2 + y^2 + z^2)^{\frac{1}{2}}$;

    (v) $(a - x)\, p + (b - y)\, q = c - z$;

    (vi) $(y^3 x - 2x^4)\, p + (2y^4 - x^3 y)\, q = 9z\,(x^3 - y^3)$;

    (vii) $p \tan x + q \tan y = \tan z$;

    (viii) $(11x - 6y + 2z)\, p - (6x - 10y + 4z)\, q = 2x - 4y + 6z$;

    (ix) $x_1 p_1 + (z + x_3)\, p_2 + (z + x_2)\, p_3 = x_2 + x_3$.

Ex. 4. Solve the equation

$$(x_2+x_3+z)\, p_1 + (x_3+x_1+z)\, p_2 + (x_1+x_2+z)\, p_3 = x_1+x_2+x_3.$$

Lagrange's subsidiary equations are

$$\frac{dx_1}{x_2+x_3+z} = \frac{dx_2}{x_3+x_1+z} = \frac{dx_3}{x_1+x_2+z} = \frac{dz}{x_1+x_2+x_3}.$$

Each of these equal fractions

$$= \frac{dz-dx_1}{-(z-x_1)} = \frac{dz-dx_2}{-(z-x_2)} = \frac{dz-dx_3}{-(z-x_3)} = \frac{dz+dx_1+dx_2+dx_3}{3\,(z+x_1+x_2+x_3)}.$$

Integrals of these are

$$\frac{c_1}{z-x_1} = \frac{c_2}{z-x_2} = \frac{c_3}{z-x_3} = (z+x_1+x_2+x_3)^{\frac{1}{3}};$$

and therefore the integral of the equation is

$$\Phi\{(z-x_1)\, S^{\frac{1}{3}},\quad (z-x_2)\, S^{\frac{1}{3}},\quad (z-x_3)\, S^{\frac{1}{3}}\}=0,$$

where S stands for $\qquad z+x_1+x_2+x_3.$

Ex. 5. Prove that, if the variables in the last example be connected by the relation

$$x_1{}^3+x_2{}^3+x_3{}^3=1$$

when $z=0$, the integral is

$$\{(x_1-z)^3+(x_2-z)^3+(x_3-z)^3\}^4\,(x_1+x_2+x_3+z)^3=(x_1+x_2+x_3-3z)^3.$$

(Mansion.)

Ex. 6. Solve the equations:

(i) $p_1 x_1+p_2 x_2+p_3 x_3 = nz;$

(ii) $p_1 x_1+p_2 x_2+p_3 x_3 = az+\dfrac{x_1 x_2}{x_3};$

(iii) $x_2 x_3 z p_1 + x_3 x_1 z p_2 + x_1 x_2 z p_3 = x_1 x_2 x_3.$

Ex. 7. In order to illustrate the general theory in § 189, consider the equation

$$yz-x^2=0,$$

which provides an integral of the equation

$$xp+yq=z,$$

already discussed in Ex. 1. The integrals of the subsidiary equations are

$$\frac{z}{y}=a,\quad \frac{z}{x}=b.$$

With the notation of § 189, we choose constants c, a, β, such that

$$a\beta-c^2=0;$$

and then

$$z - \beta = -\beta + aa + a\,(y - a),$$
$$z - \beta = -\beta + bc + b\,(x - c).$$

We take

$$a = \frac{\beta}{a}, \quad b = \frac{\beta}{c};$$

and then

$$z - \beta = \frac{\beta}{c}\,(x - c), \quad y - a = \frac{a}{c}\,(x - c).$$

Then, still with the notation of § 189, we have

$$u_a = -\frac{\beta}{a^2}, \quad u_\beta = \frac{1}{a}; \quad v_a = 0, \quad v_\beta = \frac{1}{c};$$

thus $u_a v_\beta - v_a u_\beta$ is not zero. The coefficients P, Q, R, being x, y, z, are regular in the vicinity of the values c, a, β, of the variables. Hence $yz - x^2 = 0$ is included in the general solution

$$\phi\left(\frac{z}{y}, \ \frac{z}{x}\right) = 0:$$

in particular, it is included in the equation

$$\left(\frac{z}{x}\right)^2 - \frac{z}{y} = 0.$$

Ex. 8. Consider the equation

$$\{1 + (z - x - y)^{\frac{1}{2}}\}\,p + q = 2,$$

which is discussed by Chrystal in the memoir already quoted. The subsidiary equations are

$$\frac{dx}{1 + (z - x - y)^{\frac{1}{2}}} = \frac{dy}{1} = \frac{dz}{2};$$

and we can take

$$u = 2y - z, \quad v = y + 2\,(z - x - y)^{\frac{1}{2}}.$$

Now the relation

$$z = x + y$$

is easily seen to be an integral of the equation: we see equally easily that it cannot be included in any form

$$\phi\,(u,\ v) = 0.$$

As a matter of fact, making the quantities c, a, β, satisfy the condition

$$\beta = c + a,$$

we notice that the coefficient P, which is $1 + (z - x - y)^{\frac{1}{2}}$, is not uniform in the vicinity of $x = c$, $y = a$, $z = \beta$, and cannot be expanded in integral powers of $x - c$, $y - a$, $z - \beta$. Hence, by the general theory in § 189, we are not entitled to expect that the integral can be included in the form

$$\phi\,(u,\ v) = 0.$$

Ex. 9. Obtain the general integral of the equation

$$(x^2+z^2-1)\,p+\{xy+(1-z^2)^{\frac{1}{2}}\,(x^2+y^2+z^2-1)^{\frac{1}{2}}\}\,q=0\,;$$

and discuss the relation (if any) borne to it by the integral

$$x^2+y^2+z^2=1.$$

(Goursat.)

194. In the course of § 191, we had the equation

$$P\frac{\partial\psi}{\partial x}+Q\frac{\partial\psi}{\partial y}+R\frac{\partial\psi}{\partial z}=0,$$

where P, Q, R, are functions of x, y, z. This is a specialised form of the more general equation

$$R_1p_1+R_2p_2+\ldots+R_np_n=0,$$

where $R_1, \ldots, R_n$, are functions of the n independent variables $x_1, \ldots, x_n$, and do not involve the dependent variable z.

As in § 193, we form a subsidiary system

$$\frac{dx_1}{R_1}=\frac{dx_2}{R_2}=\ldots=\frac{dx_n}{R_n}$$

of $n-1$ ordinary simultaneous equations; and we construct $n-1$ independent integrals (the simpler the better) of these $n-1$ equations. Let these integrals be

$$u_m=u_m\,(x_1, \ldots, x_n)=a_m, \quad (m=1, \ldots, n-1),$$

where $a_1, \ldots, a_{n-1}$, are arbitrary constants. Then, for each of these integrals, we have

$$\frac{\partial u_m}{\partial x_1}dx_1+\ldots+\frac{\partial u_m}{\partial x_n}dx_n=0,$$

and therefore

$$R_1\frac{\partial u_m}{\partial x_1}+\ldots+R_n\frac{\partial u_m}{\partial x_n}=0.$$

Apparently, this is a new integral equation among the variables in the subsidiary system. Every integral equation connected with this subsidiary system, if not evanescent, must be expressible in terms of $u_1=a_1, \ldots, u_{n-1}=a_{n-1}$. The new equation manifestly is not satisfied in virtue of the integral equivalent of the subsidiary system, because it does not involve any of the arbitrary constants a. Hence it must be satisfied identically; that is, all the $n-1$ relations

$$R_1 \frac{\partial u_m}{\partial x_1} + \ldots + R_n \frac{\partial u_m}{\partial x_n} = 0,$$

for $m = 1, \ldots, n-1$, are satisfied identically.

Now let

$$z = f(x_1, \ldots, x_n)$$

be any integral of the differential equation

$$R_1 p_1 + \ldots + R_n p_n = 0;$$

then the relation

$$R_1 \frac{\partial f}{\partial x_1} + \ldots + R_n \frac{\partial f}{\partial x_n} = 0$$

is satisfied, obviously identically because it does not involve z. Consequently, we have

$$J\left(\frac{f, u_1, \ldots, u_{n-1}}{x_1, x_2, \ldots, x_n}\right) = 0.$$

This relation is not satisfied in virtue of any, or all, of the equations $u_1 = a_1, \ldots, u_{n-1} = a_{n-1}, z = f$. It therefore is satisfied identically; and consequently (§ 9) we have a functional relation of the form

$$\phi(f, u_1, \ldots, u_{n-1}) = 0,$$

where ϕ can be perfectly general. Resolving this functional relation so as to express f in terms of $u_1, \ldots, u_{n-1}$, we have

$$f = F(u_1, \ldots, u_{n-1}),$$

where, as ϕ can be perfectly general, so F can be perfectly general. But

$$z = f(x_1, \ldots, x_n) = f$$

is *any* integral of the differential equation; consequently, *every integral of the equation*

$$R_1 p_1 + \ldots + R_n p_n = 0$$

can be expressed in the form

$$z = F(u_1, \ldots, u_{n-1}),$$

where $u_1 = a_1, \ldots, u_{n-1} = a_{n-1},$ *are a complete system of* $n-1$ *distinct and independent integrals of the* $n-1$ *subsidiary equations*

$$\frac{dx_1}{R_1} = \frac{dx_2}{R_2} = \ldots = \frac{dx_n}{R_n}.$$

It should be noted that the form

$$z = F(u_1, \ldots, u_{n-1})$$

has been proved to include *every* integral of the equation. There are no special integrals (in the former sense of the term) of the equation

$$R_1 p_1 + \ldots + R_n p_n = 0.$$

The equation is somewhat different in form from the equation in § 193. Owing to the facts that z does not occur, that the derivatives $p_1, \ldots, p_n$, occur in the first power only, and that there is no term not involving a derivative, the equation is often called a *homogeneous linear* partial differential equation.

STANDARD FORMS.

195. Before proceeding to indicate a method of integration which is applicable to the most general equation of the first order, it is advisable to notice a few standard forms of differential equations in two independent variables, which admit of integration by very short processes, and to one or other of which many equations can be reduced. As the general method is usually much longer than that which is effective for any of the standard forms, it is advantageous to see whether an equation is included under one of them.

The General Integral and the Singular Integral must in the case of every equation be indicated as well as the Complete Integral, or the equation is not considered to be fully solved.

196. STANDARD I.

Equations, in which the variables do not explicitly occur, may be written in the form

$$\psi(p, q) = 0.$$

A solution of this is evidently

$$z = ax + by + c,$$

provided a and b are such as to satisfy

$$\psi(a, b) = 0.$$

If the value of b derived from this equation be $b = f(a)$, the Complete Integral of the equation is

$$z = ax + yf(a) + c.$$

Note. Equations, which do not explicitly come under this standard, can often be included by changes of the variables; thus for instance functions of x which occur in the equation might admit of association with the p, and functions of y with the q. But the changes needed for any equation can be determined only for the particular circumstances of the equation; there is no general rule, since an equation cannot always be reduced to this form.

Ex. 1. Solve $\qquad\qquad pq = k.$

The foregoing shews that

$$z = ax + by + c$$

is a solution provided

$$ab = k;$$

the Complete Integral therefore is

$$z = ax + \frac{k}{a}y + c.$$

The General Integral is obtained by eliminating a between the equations

$$\left.\begin{array}{l} z = ax + \dfrac{k}{a}y + \phi(a) \\[2mm] 0 = x - \dfrac{k}{a^3}y + \phi'(a) \end{array}\right\},$$

where ϕ is arbitrary.

The Singular Integral, if it exist, is determined by the equations

$$\left.\begin{array}{l} z = ax + \dfrac{k}{a}y + c \\[2mm] 0 = x - \dfrac{k}{a^2}y \\[2mm] 0 = \qquad 1 \end{array}\right\};$$

the last equation shews that the Singular Integral does not exist.

Ex. 2. Solve $pq = x^m y^n z^l.$

This can be put into the form

$$\frac{z^{-\frac{1}{2}l} dz}{x^m dx} \frac{z^{-\frac{1}{2}l} dz}{y^n dy} = 1.$$

Let $dZ = z^{-\frac{1}{2}l} dz,$ so that $(1 - \frac{1}{2}l) Z = z^{1 - \frac{1}{2}l},$

$\qquad\qquad d\xi = x^m dx,$ .... $(m + 1) \xi = x^{m+1},$

$\qquad\qquad d\eta = y^n dy,$ .... $(n + 1) \eta = y^{n+1},$

and the equation becomes

$$\frac{\partial Z}{\partial \xi} \frac{\partial Z}{\partial \eta} = 1,$$

which is included under the last example.

Ex. 3. Solve the equations:

$\qquad$ (i) $p^2 + q^2 = m^2;$

$\qquad$ (ii) $a(p + q) = z;$

$\qquad$ (iii) $x^2 p^2 + y^2 q^2 = z;$

$\qquad$ (iv) $p^m \sec^{2m} x + z^l q^n \operatorname{cosec}^{2n} y = z^{\frac{lm}{m-n}};$

$\qquad$ (v) $p^2 + q^2 = npq;$

$\qquad$ (vi) $p_1^m + p_2^m + p_3^m = 1;$

$\qquad$ (vii) $z p_1 p_2 p_3 = x_1 x_2 x_3.$

197. The differential equations included under the form

$$\psi(p, q) = 0$$

have an important interpretation when viewed geometrically. We know that the equation of the tangent plane to the surface

$$z = F(x, y)$$

at the point $\xi, \eta, \zeta,$ is

$$z = (x - \xi) \frac{\partial F}{\partial \xi} + (y - \eta) \frac{\partial F}{\partial \eta} + F(\xi, \eta)$$

$$= x \frac{\partial F}{\partial \xi} + y \frac{\partial F}{\partial \eta} - \left(\xi \frac{\partial F}{\partial \xi} + \eta \frac{\partial F}{\partial \eta} - F \right);$$

and the surface is the envelope of the tangent planes. Now if between $\dfrac{\partial F}{\partial \xi}$ and $\dfrac{\partial F}{\partial \eta}$ there be a relation

$$\psi\left(\frac{\partial F}{\partial \xi}, \frac{\partial F}{\partial \eta} \right) = 0,$$

we can take $\dfrac{\partial F}{\partial \xi} = \alpha$, and then $\dfrac{\partial F}{\partial \eta}$ is a function of α alone, say

$$\frac{\partial F}{\partial \eta} = f(\alpha).$$

Also

$$J\left(\frac{\xi \dfrac{\partial F}{\partial \xi} + \eta \dfrac{\partial F}{\partial \eta} - F, \ \dfrac{\partial F}{\partial \xi}}{\xi, \ \eta}\right) = -\eta \left\{\frac{\partial^2 F}{\partial \xi^2}\frac{\partial^2 F}{\partial \eta^2} - \left(\frac{\partial^2 F}{\partial \xi \partial \eta}\right)^2\right\}$$

$$= -\eta J\left(\frac{\dfrac{\partial F}{\partial \xi}, \ \dfrac{\partial F}{\partial \eta}}{\xi, \ \eta}\right)$$

$$= 0,$$

so that $\xi \dfrac{\partial F}{\partial \xi} + \eta \dfrac{\partial F}{\partial \eta} - F$ is (§ 9) expressible as a function of $\dfrac{\partial F}{\partial \xi}$ alone, that is, it can be taken in the form

$$\xi \frac{\partial F}{\partial \xi} + \eta \frac{\partial F}{\partial \eta} - F = g(\alpha).$$

The equation of the tangent plane becomes

$$z = x\alpha + yf(\alpha) - g(\alpha),$$

and so it contains only one parameter.

The envelope of a plane whose equation is of this form is a developable surface, and hence the surface considered is a developable surface.

It therefore follows that

$$\psi(p, q) = 0$$

is the general differential equation of a family of developable surfaces; and the equivalent General Integral is the integral equation of the family.

198. Standard II.

In attempting to reduce an equation to the preceding standard we may find it possible to remove from the equation the independent variables, so that they no longer occur explicitly; but it may not be possible to remove the dependent variable, and the equation will then be of the form

$$\chi(z, p, q) = 0.$$

We assume as a tentative solution

$$z = f(x + ay) = f(\xi)$$

(ξ being written instead of $x + ay$), in which a is an arbitrary constant. We then have

$$p = \frac{dz}{d\xi}\frac{\partial \xi}{\partial x} = \frac{dz}{d\xi},$$

$$q = \frac{dz}{d\xi}\frac{\partial \xi}{\partial y} = a\frac{dz}{d\xi};$$

and the substitution of these in the equation gives

$$\chi\left(z, \frac{dz}{d\xi}, a\frac{dz}{d\xi}\right) = 0.$$

This is no longer a partial differential equation, as there is now only one independent variable. This independent variable does not explicitly occur, and thus the equation comes under Standard IV. (§ 18) of ordinary differential equations of the first order. Solving for $\frac{dz}{d\xi}$, we have an equation of the form

$$\frac{dz}{d\xi} = \phi(z, a),$$

the solution of which is

$$\xi + b = \int \frac{dz}{\phi(z, a)},$$

or $$x + ay + b = F(z, a).$$

This is the Complete Integral; the General and the Singular Integrals may be found by the ordinary method.

Ex. 1. Solve the equation

$$9(p^2 z + q^2) = 4.$$

If we make the substitutions as in the standard case, the equation becomes

$$9\left(\frac{dz}{d\xi}\right)^2 (a^2 + z) = 4,$$

or $$\tfrac{3}{2}dz\,(a^2 + z)^{\frac{1}{2}} = d\xi,$$

the integral of which is

$$(z + a^2)^{\frac{3}{2}} = \xi + c;$$

the Complete Integral of the equation therefore is

$$(z + a^2)^3 = (x + ay + c)^2.$$

The General Integral is obtained by the elimination of a between

$$(z+a^2)^3 = \{x+ay+\theta(a)\}^2$$

and $\qquad 3a(z+a^2)^2 = \{x+ay+\theta(a)\}\{y+\theta'(a)\}$,

where θ is an arbitrary function.

It is not difficult to prove that there is no Singular Integral.

Ex. 2. Solve the equations:

     (i) $p^2 = z^2(1-pq)$;

     (ii) $q^2y^2 = z(z-px)$;

     (iii) $p(1+q^2) = q(z-a)$;

     (iv) $1 = p_2p_3 + p_3p_1z + p_1p_2z^2$;

     (v) $p_1{}^2 + zp_2{}^2 + z^2p_3{}^2 = z^3p_1p_2p_3$.

199. The relation between the integral and the differential equation admits of a geometrical interpretation. The first step in the process of solution is writing ξ for $x + ay$, which is equivalent to turning the axes in the plane of xy through an angle equal to $\tan^{-1} a$ and magnifying the coordinates in that plane in the ratio of $(1+a^2)^{\frac{1}{2}} : 1$. It is then assumed that z is a function of ξ, but is independent of the coordinate parallel to the new axis of y. Now

$$z = f(\xi)$$

represents a cylinder whose axis is parallel to the new axis of y; and therefore the equation gives the cylinders satisfying this condition. But now, returning to our original axes, since a is an arbitrary constant, the axis of ξ is an arbitrary line in the plane, and therefore also is the line taken for the transformed axis of y. It thus follows that what we find by our process of integration will be all the cylindrical surfaces, which satisfy the given differential equation and have their axes in the plane of xy.

200. Standard III.

In attempting to reduce a given equation to the first standard, it may happen that z may be removed from explicit occurrence in the equation, while x and y remain, and that then the functions of p and x may be associated with one another, and likewise the functions of q and y; the equation will thus take the form

$$\phi(x, p) = \psi(y, q).$$

We assume, as a trial solution, each of these equal quantities to be equal to an arbitrary constant a; from the first of the two equations so obtained we have

$$p = \theta_1 (x, a),$$

and from the second

$$q = \theta_2 (y, a).$$

Integrating both of these we find that, by the first,

$$z = f_1 (x, a) + \text{a quantity independent of } x,$$

and that, by the second,

$$z = f_2 (y, a) + \text{a quantity independent of } y.$$

These are evidently included in, and are equivalent to, the equation

$$z = f_1 (x, a) + f_2 (y, a) + b,$$

where b is an arbitrary constant. This is a solution of the original equation; as it contains two arbitrary constants, it is the Complete Integral.

The General Integral and the Singular Integral, if it exist, are to be deduced in the usual way.

Ex. 1. Solve the equation

$$p^2 + q^2 = x + y.$$

The equation, rearranged in the form

$$p^2 - x = -(q^2 - y),$$

comes under the standard; and we therefore write

$$p^2 - x = y - q^2 = a.$$

Hence
$$p = (x + a)^{\frac{1}{2}},$$
$$q = (y - a)^{\frac{1}{2}},$$

and therefore

$$z = \tfrac{2}{3} (x + a)^{\frac{3}{2}} + \tfrac{2}{3} (y - a)^{\frac{3}{2}} + b,$$

which is the Complete Integral.

The General Integral is given by the elimination of a between

$$\left. \begin{array}{l} z = \tfrac{2}{3} (x + a)^{\frac{3}{2}} + \tfrac{2}{3} (y - a)^{\frac{3}{2}} + \chi (a) \\ 0 = (x + a)^{\frac{1}{2}} - (y - a)^{\frac{1}{2}} + \chi' (a) \end{array} \right\},$$

where χ is an arbitrary function. There is no Singular Integral.

Ex. 2. Solve the equations:

$$\text{(i)}\quad z^2(p^2+q^2)=x^2+y^2;$$

$$\text{(ii)}\quad q=xp+p^2;$$

$$\text{(iii)}\quad p=(qy+z)^2;$$

$$\text{(iv)}\quad p^{\frac{1}{2}}+q^{\frac{1}{2}}=2x;$$

$$\text{(v)}\quad p^2-y^3q=x^2-y^2.$$

Ex. 3. Shew that this method can be applied to the solution of equations of the form

$$f_1(p_1,\,x_1)+f_2(p_2,\,x_2)+f_3(p_3,\,x_3)=0.$$

Thus solve fully the equation

$$p_1{}^2+p_2{}^2+p_3{}^2=x_1{}^2+x_2{}^2+x_3{}^2.$$

201. Standard IV.

In this class are included those equations involving partial differential coefficients, which are analogous to the equations included under Clairaut's form (§ 20) in ordinary differential equations. For two independent variables, they are represented by

$$z=px+qy+\phi(p,\,q),$$

where ϕ is a definite function.

A solution of the equation is

$$z=ax+by+\phi(a,\,b),$$

which admits of immediate verification. As it contains two arbitrary constants, it is the Complete Integral; the General Integral is to be obtained in the usual way; and there is a Singular Integral, unless ϕ is linear in p and in q.

Ex. 1. Solve the equations:

$$\text{(i)}\quad z=px+qy+pq;$$

$$\text{(ii)}\quad z=px+qy+(1+p^2+q^2)^{\frac{1}{2}};$$

$$\text{(iii)}\quad z=px+qy+(\alpha p^2+\beta q^2+\gamma)^{\frac{1}{2}};$$

$$\text{(iv)}\quad z=px+qy+3p^{\frac{1}{3}}q^{\frac{1}{3}};$$

obtaining in each case the Singular Integral as well as the Complete Integral.

Ex. 2. Solve the equations:

$$\text{(i)}\quad z=p_1x_1+p_2x_2+p_3x_3+f(p_1,\,p_2,\,p_3);$$

$$\text{(ii)}\quad z=\sum_{\mu=n}^{\mu=1}p_\mu x_\mu+(n+1)(p_1p_2\dots p_n)^{\frac{1}{n+1}};$$

and obtain the Singular Integral in each case.

PRINCIPLE OF DUALITY.

202. There exists in partial differential equations a remarkable duality, in virtue of which each equation is connected with some other equation of the same order by relations of a perfectly reciprocal character. We shall consider here only equations of the first order.

Considering the case of two independent variables only, we write as our new dependent variable

$$Z = px + qy - z,$$

and therefore

$$dZ = x\,dp + y\,dq.$$

We take as our new independent variables p and q, which we write X and Y for symmetry, so that

$$X = p, \quad Y = q;$$

and then we have

$$x = \frac{\partial Z}{\partial p} = \frac{\partial Z}{\partial X} = P,$$

$$y = \frac{\partial Z}{\partial q} = \frac{\partial Z}{\partial Y} = Q;$$

then

$$z = PX + QY - Z,$$

so that the relations between the variables are, as stated above, reciprocal.

If now we have an equation of the form

$$\psi\,(x,\,y,\,z,\,p,\,q) = 0,$$

the above relations transform it into

$$\psi\,(P,\,Q,\,PX + QY - Z,\,X,\,Y) = 0.$$

The integral of either of these being known, that of the other is deducible by a process of algebraical elimination. Thus let a solution of the second be given, or be derivable, in the form

$$\phi\,(Z,\,X,\,Y) = 0.$$

Then we have

$$P\frac{\partial \phi}{\partial Z} + \frac{\partial \phi}{\partial X} = 0 = Q\frac{\partial \phi}{\partial Z} + \frac{\partial \phi}{\partial Y},$$

that is,
$$x\frac{\partial\phi}{\partial Z}+\frac{\partial\phi}{\partial X}=0,$$

$$y\frac{\partial\phi}{\partial Z}+\frac{\partial\phi}{\partial Y}=0;$$

and
$$-z\frac{\partial\phi}{\partial Z}=Z\frac{\partial\phi}{\partial Z}+X\frac{\partial\phi}{\partial X}+Y\frac{\partial\phi}{\partial Y}.$$

The elimination of X, Y, Z, between these four equations will leave an equation in x, y, z, which will be a solution of

$$\psi(x,y,z,p,q)=0.$$

Ex. 1. The simplest example of an equation which can be treated by this method is that which comes under Standard IV. (§ 201); the equation being

$$z=px+qy+f(p,q),$$

the transformed equation is not differential, but algebraical, being in fact

$$-Z=f(X,Y).$$

Thus, in particular, consider

$$z=px+qy+p^2+q^2;$$

the transformed equation is

$$-Z=X^2+Y^2.$$

Hence
$$x=\frac{\partial Z}{\partial X}=-2X \text{ and } y=\frac{\partial Z}{\partial Y}=-2Y,$$

where
$$z=X\frac{\partial Z}{\partial X}+Y\frac{\partial Z}{\partial Y}-Z=-(X^2+Y^2).$$

Hence, eliminating the quantities X, Y, Z, we have

$$-4z=x^2+y^2,$$

which is easily seen to be the Singular Integral of

$$z=px+qy+p^2+q^2.$$

Ex. 2. Solve the equations:

    (i) $(xp+yq)(z-px-qy)+pq=0$;

    (ii) $z+1-x(x+p)-y(y+q)=0$;

    (iii) $p^2(x^2-x)+2pqxy+q^2(y^2-y)-2pxz-2qyz+z^2=0$;

    (iv) $(px+qy-z)(p^2x+q^2y)^{\frac{1}{2}}=pq$.

Ex. 3. Prove that the equations

  (i) $xf_1(z-px-qy,p,q)+yf_2(z-px-qy,p,q)=f_3(z-px-qy,p,q)$,

  (ii) $F(z-px-qy,x,y)=0$,

are reducible, by the foregoing substitutions, to standard forms.

Ex. 4. Prove that the equation

$$xf_1(y, p, z - px) + qf_2(y, p, z - px) = f_3(y, p, z - px)$$

is reducible to Lagrange's form by changing the variables, so that p and y are the new independent variables and $z - px$ is the new dependent variable.

Hence solve the equation

$$q(y - b)^2 + 2pxz = z^2 + xp^2(x + 1).$$

Ex. 5. Solve $\qquad (z - px - qy)^2 = 1 + p^2 + q^2.$

203. The process of derivation of one differential equation from another as exhibited in the preceding article is really a translation into analysis of the geometrical principle of duality between surfaces. When we take a fixed quadric, which we may denote by Σ, then with every surface S there is associated another surface S', called its polar reciprocal, which is the envelope of the polar planes with regard to Σ of points on the surface S; and the surface S is the polar reciprocal of S', being the envelope of the polar planes with regard to Σ of points on S'.

The polar reciprocal of a surface depends on the subsidiary quadric, Σ, and is different for different quadrics; the quadric most commonly chosen (on account of the geometrical simplicity) is a sphere with its centre at the origin of reciprocation.

Let us select, as the subsidiary quadric, not a sphere but a paraboloid of revolution whose equation is

$$x^2 + y^2 = 2z.$$

To the tangent plane at a point A on the surface S corresponds a point A' on the surface S'; and to the point A corresponds the tangent plane at A' to S'. Let $x, y, z, p, q,$ be the quantities associated with A; and $X, Y, Z, P, Q,$ the corresponding quantities associated with A'.

The tangent plane at x, y, z, to the given surface S is

$$\zeta - z = p(\xi - x) + q(\eta - y),$$

ξ, η, ζ, being current coordinates; the polar plane of X, Y, Z, with regard to the quadric, is

$$X\xi + Y\eta - \zeta - Z = 0.$$

But, because the two surfaces S and S' are polar reciprocals, these two planes are the same; a comparison of their equations gives

$$X = p; \quad Y = q; \quad Z = px + qy - z.$$

Similarly, taking a tangent plane at X, Y, Z, to the surface S', and noting that it must be the polar plane of x, y, z, with regard to the quadric, we obtain the equations

$$x = P; \quad y = Q; \quad z = PX + QY - Z.$$

These are the two sets of relations used in the preceding method.

Other relations could be obtained by taking other subsidiary quadrics in reference to which reciprocation should take place; but the preceding seem the simplest that can be found.

Note. The transformation of variables just outlined is originally due to Lagrange; often it is associated with the name of Legendre, because he developed its properties. But it is only an individual example of a general theory developed by Lie, which is usually called the theory of *tangential transformations* or *contact transformations*, when the number of independent variables is n and not merely two.

All the contact transformations, in the case of two independent variables x and y, are given by the equations which make

$$dZ - PdX - QdY = \rho \, (dz - pdx - qdy),$$

where ρ is a non-vanishing quantity independent of differential elements.

The elements of Lie's theory are expounded in vol. I. of my *Theory of Differential Equations*, chapter IX., where references are given; also in vol. V., chapter IX. The full exposition is contained (vol. II.) in the great treatise *Theorie der Transformationsgruppen* by Lie and Engel.

It is obvious geometrically that two surfaces, which touch one another, are reciprocated with respect to any quadric into two other surfaces which also touch one another. Thus the Lagrange-Legendre transformation is a contact transformation.

204. The General Integral of a differential equation involves an arbitrary function. It may be necessary to obtain an integral satisfying certain conditions; the latter will then be obtained if the arbitrary function be rightly determined. The process is equivalent to that which occurs in ordinary differential equations, where the arbitrary constants are determined by some particular relation or relations between special values of the variables. In every particular problem, the arbitrary function is determined by means of the specified conditions.

Ex. 1. The equation

$$ap + bq = 1$$

implies that the normal to the surface represented by the integral equation is perpendicular to a given line whose direction cosines are proportional to a, b, 1; this is the property of a cylindrical surface whose axis is parallel to that line. The integral obtained, either by Lagrange's method or by the method applied to Standard I., is

$$x - az = \phi (y - bz),$$

where ϕ is arbitrary. Suppose that the equation of a cylinder, having its axis parallel to the line $(a, b, 1)$ and passing through the curve $x^2 - y^2 = 1$ in the plane of xy, is desired. The section of the above surface by the plane of xy is obtained by writing $z = 0$ therein, and thus it is

$$x = \phi (y).$$

According to the assigned conditions, it should be

$$x^2 = 1 + y^2.$$

A comparison of these equations shews that

$$\phi (y) = (1 + y^2)^{\frac{1}{2}},$$

and therefore also

$$\phi (y - bz) = \{1 + (y - bz)^2\}^{\frac{1}{2}}.$$

Hence the equation required is

$$x - az = \{1 + (y - bz)^2\}^{\frac{1}{2}},$$

or, freed from radicals, is

$$(x - az)^2 - (y - bz)^2 = 1.$$

Ex. 2. Prove that the equation

$$p (x - a) + q (y - b) = z - c$$

represents a family of cones having the fixed point (a, b, c) for vertex. Shew that the member of the family, which passes through the circle

$$x^2 + y^2 = 1$$

in the plane of xy, has for its equation

$$(az - cx)^2 + (bz - cy)^2 = (z - c)^2.$$

Ex. 3. Obtain the integral of the equation

$$p (ny - mz) + q (lz - nx) = mx - ly,$$

so that the section, by the plane of xy, of the represented surface is a conic section of eccentricity e with its centre on the line

$$e^2 + (1 - e^2) (lx + my) = 0.$$

CHARPIT'S GENERAL METHOD OF SOLUTION.

205. We now proceed to consider a more general method, due to Charpit*. It applies to the general equation, which may be denoted by

$$F(x, y, z, p, q) = 0 ;$$

and its success depends, as will be seen, upon the integration of some ordinary differential equations.

If, in addition to the foregoing relation, we have another between the variables and the differential coefficients, the two can be considered as a pair of simultaneous equations which, when solved, will give p and q as explicit functions of $x, y,$ and z. The values so derived, when substituted in the equation

$$dz = p\,dx + q\,dy,$$

will render it either immediately integrable or integrable on multiplication by some factor; and the integral will be a solution of the original equation, since the values of p and q derived from it have in the inverse process been obtained from that equation. Let then another relation between the quantities be denoted by

$$\Phi(x, y, z, p, q) = 0 ;$$

if we can find the form of Φ, we shall be in a position to use this method of solution.

206. Now the integral of the equation gives z (and therefore also p and q) as functions of x and y; whatever these functions may be, they will, if substituted in the equations $F = 0$ and $\Phi = 0$, render them both identities. Let these values of z, p, q (as yet unknown) be supposed substituted; then the partial differential

---

* The method was contained in a memoir, presented 30 June, 1784, to the Académie des Sciences, Paris. Charpit died soon afterwards, and his memoir was never printed; see Lacroix, *Traité du calcul différentiel et du calcul intégral*, 2ᵉ éd. (1814), t. II. p. 548.

coefficients of the left-hand members of both equations with regard to x and y vanish, and therefore

$$\frac{\partial F}{\partial x} + \frac{\partial F}{\partial z} p + \frac{\partial F}{\partial p} \frac{\partial p}{\partial x} + \frac{\partial F}{\partial q} \frac{\partial q}{\partial x} = 0,$$

$$\frac{\partial \Phi}{\partial x} + \frac{\partial \Phi}{\partial z} p + \frac{\partial \Phi}{\partial p} \frac{\partial p}{\partial x} + \frac{\partial \Phi}{\partial q} \frac{\partial q}{\partial x} = 0,$$

$$\frac{\partial F}{\partial y} + \frac{\partial F}{\partial z} q + \frac{\partial F}{\partial p} \frac{\partial p}{\partial y} + \frac{\partial F}{\partial q} \frac{\partial q}{\partial y} = 0,$$

$$\frac{\partial \Phi}{\partial y} + \frac{\partial \Phi}{\partial z} q + \frac{\partial \Phi}{\partial p} \frac{\partial p}{\partial y} + \frac{\partial \Phi}{\partial q} \frac{\partial q}{\partial y} = 0.$$

Eliminating $\dfrac{\partial p}{\partial x}$ between the first pair of these equations, we have

$$\left(\frac{\partial F}{\partial x} \frac{\partial \Phi}{\partial p} - \frac{\partial F}{\partial p} \frac{\partial \Phi}{\partial x} \right) + p \left(\frac{\partial F}{\partial z} \frac{\partial \Phi}{\partial p} - \frac{\partial F}{\partial p} \frac{\partial \Phi}{\partial z} \right) + \frac{\partial q}{\partial x} \left(\frac{\partial F}{\partial q} \frac{\partial \Phi}{\partial p} - \frac{\partial F}{\partial p} \frac{\partial \Phi}{\partial q} \right) = 0;$$

and eliminating $\dfrac{\partial q}{\partial y}$ between the second pair, we have

$$\left(\frac{\partial F}{\partial y} \frac{\partial \Phi}{\partial q} - \frac{\partial F}{\partial q} \frac{\partial \Phi}{\partial y} \right) + q \left(\frac{\partial F}{\partial z} \frac{\partial \Phi}{\partial q} - \frac{\partial F}{\partial q} \frac{\partial \Phi}{\partial z} \right) + \frac{\partial p}{\partial y} \left(\frac{\partial F}{\partial p} \frac{\partial \Phi}{\partial q} - \frac{\partial F}{\partial q} \frac{\partial \Phi}{\partial p} \right) = 0.$$

Now

$$\frac{\partial q}{\partial x} = \frac{\partial^2 z}{\partial x \partial y} = \frac{\partial p}{\partial y},$$

so that from the last two equations, when added together as they stand, the terms involving these quantities disappear; and the result may be rearranged and written in the form

$$\left(\frac{\partial F}{\partial x} + p \frac{\partial F}{\partial z} \right) \frac{\partial \Phi}{\partial p} + \left(\frac{\partial F}{\partial y} + q \frac{\partial F}{\partial z} \right) \frac{\partial \Phi}{\partial q} + \left(-p \frac{\partial F}{\partial p} - q \frac{\partial F}{\partial q} \right) \frac{\partial \Phi}{\partial z}$$

$$+ \left(-\frac{\partial F}{\partial p} \right) \frac{\partial \Phi}{\partial x} + \left(-\frac{\partial F}{\partial q} \right) \frac{\partial \Phi}{\partial y} = 0,$$

which we may look upon as a linear differential equation of the first order to determine Φ. The method applicable to this equation is therefore the one used in the case of Lagrange's equation; we form the equations (§§ 193, 194)

$$\frac{dp}{\dfrac{\partial F}{\partial x} + p \dfrac{\partial F}{\partial z}} = \frac{dq}{\dfrac{\partial F}{\partial y} + q \dfrac{\partial F}{\partial z}} = \frac{dz}{-p \dfrac{\partial F}{\partial p} - q \dfrac{\partial F}{\partial q}} = \frac{dx}{-\dfrac{\partial F}{\partial p}} = \frac{dy}{-\dfrac{\partial F}{\partial q}} = \frac{d\Phi}{0},$$

and obtain integrals of these. Now, in order that these equations may hold, we must have

$$dF = 0,$$

or

$$F = A,$$

an arbitrary constant, which must be zero. If another integral can be obtained by equating any two of the first five fractions, it may be written in the form

$$u = B.$$

By the corollary in § 191, $u = B$ is a solution of the differential equation determining Φ. Now $\Phi = 0$ is the relation we are seeking between x, y, z, p, q; and the simpler this relation is, the easier will be the deduction of p and q from $\Phi = 0$ and $F = 0$. We may therefore take as the relation required the equation

$$u = B,$$

that is, we may take any one integral whatever of the foregoing system of ordinary differential equations, provided either p or q or both occur in it; when this integral has been obtained, we combine it with $F = 0$ and carry out the process indicated in the preceding article.

207. The preceding result, which is of fundamental importance, may be obtained by another process of using the necessary analysis.

The given differential equation to be integrated is the equation $F = 0$; and we may assume that it has some integral equivalent, in the form of a relation between x, y, z. Suppose that the other equation $\Phi = 0$ is also satisfied by this same unknown relation and by the values of p and q derived from it. Then the values of p and q, given in terms of x, y, z, by $F = 0$ and $\Phi = 0$ as two simultaneous equations in p and q, must be such as to make the equation

$$- dz + p\, dx + q\, dy = 0$$

an exact equation. Consequently, the condition of integrability (§ 151) must be satisfied; that is, we must have

$$p \frac{\partial q}{\partial z} - q \frac{\partial p}{\partial z} + \frac{\partial q}{\partial x} - \frac{\partial p}{\partial y} = 0,$$

an equation which was proved to be sufficient as well as necessary to secure the exactness of the equation $-dz + p\,dx + q\,dy = 0$. The values of p and q, given in terms of x, y, z, by $F = 0$ and $\Phi = 0$ as simultaneous equations, are such that

$$\frac{\partial F}{\partial x} + \frac{\partial F}{\partial p}\frac{\partial p}{\partial x} + \frac{\partial F}{\partial q}\frac{\partial q}{\partial x} = 0,$$

$$\frac{\partial \Phi}{\partial x} + \frac{\partial \Phi}{\partial p}\frac{\partial p}{\partial x} + \frac{\partial \Phi}{\partial q}\frac{\partial q}{\partial x} = 0,$$

so that

$$\left(\frac{\partial F}{\partial p}\frac{\partial \Phi}{\partial q} - \frac{\partial F}{\partial q}\frac{\partial \Phi}{\partial p}\right)\frac{\partial q}{\partial x} = \frac{\partial F}{\partial x}\frac{\partial \Phi}{\partial p} - \frac{\partial F}{\partial p}\frac{\partial \Phi}{\partial x}.$$

Similarly

$$\left(\frac{\partial F}{\partial p}\frac{\partial \Phi}{\partial q} - \frac{\partial F}{\partial q}\frac{\partial \Phi}{\partial p}\right)\frac{\partial p}{\partial y} = -\frac{\partial F}{\partial y}\frac{\partial \Phi}{\partial q} + \frac{\partial F}{\partial q}\frac{\partial \Phi}{\partial y},$$

$$\left(\frac{\partial F}{\partial p}\frac{\partial \Phi}{\partial q} - \frac{\partial F}{\partial q}\frac{\partial \Phi}{\partial p}\right)\frac{\partial p}{\partial z} = -\frac{\partial F}{\partial z}\frac{\partial \Phi}{\partial q} + \frac{\partial F}{\partial q}\frac{\partial \Phi}{\partial z},$$

$$\left(\frac{\partial F}{\partial p}\frac{\partial \Phi}{\partial q} - \frac{\partial F}{\partial q}\frac{\partial \Phi}{\partial p}\right)\frac{\partial q}{\partial z} = \frac{\partial F}{\partial z}\frac{\partial \Phi}{\partial p} - \frac{\partial F}{\partial p}\frac{\partial \Phi}{\partial z}.$$

Now the equations $F = 0$ and $\Phi = 0$ are regarded as determining p and q definitely, so that (§ 9) the quantity

$$\frac{\partial F}{\partial p}\frac{\partial \Phi}{\partial q} - \frac{\partial F}{\partial q}\frac{\partial \Phi}{\partial p}$$

does not vanish. Hence, when we insert in the condition of integrability these values of $\frac{\partial q}{\partial x}, \frac{\partial p}{\partial y}, \frac{\partial p}{\partial z}, \frac{\partial q}{\partial z}$, the condition takes the form

$$\frac{\partial F}{\partial x}\frac{\partial \Phi}{\partial p} - \frac{\partial F}{\partial p}\frac{\partial \Phi}{\partial x} + \frac{\partial F}{\partial y}\frac{\partial \Phi}{\partial q} - \frac{\partial F}{\partial q}\frac{\partial \Phi}{\partial y}$$

$$+ p\left(\frac{\partial F}{\partial z}\frac{\partial \Phi}{\partial p} - \frac{\partial F}{\partial p}\frac{\partial \Phi}{\partial z}\right) + q\left(\frac{\partial F}{\partial z}\frac{\partial \Phi}{\partial q} - \frac{\partial F}{\partial q}\frac{\partial \Phi}{\partial z}\right) = 0,$$

which is easily seen to be the same equation as in the preceding section (§ 206).

From this stage we proceed as before, constructing the equations subsidiary to the determination of Φ in the form

$$\frac{dp}{\dfrac{\partial F}{\partial x}+p\dfrac{\partial F}{\partial z}} = \frac{dq}{\dfrac{\partial F}{\partial y}+q\dfrac{\partial F}{\partial z}} = \frac{dz}{-p\dfrac{\partial F}{\partial p}-q\dfrac{\partial F}{\partial q}} = \frac{dx}{-\dfrac{\partial F}{\partial p}} = \frac{dy}{-\dfrac{\partial F}{\partial q}}.$$

One integral of these, distinct from $F = 0$ and involving either p or q or both, is to be obtained, say $u = B$; this, as before, can be taken for the equation $\Phi = 0$. We resolve $F = 0$ and $u = B$ for p and q; we substitute the deduced values of p and q in

$$dz = p\,dx + q\,dy;$$

and we integrate the latter which now is exact. The result is a complete primitive of the equation $F = 0$.

The subsidiary equations are often called *Charpit's equations*; and the method of integration is, as already stated, usually called *Charpit's method*.

208. The following proposition is an immediate corollary from the process of the preceding articles, or it may be considered merely as a re-enunciation of the result there obtained:

When two equations of the first order represented by

$$F(x, y, z, p, q) = 0,$$

$$\Phi(x, y, z, p, q) = 0,$$

are such that they satisfy identically the relation

$$\frac{\partial F}{\partial x}\frac{\partial \Phi}{\partial p} - \frac{\partial F}{\partial p}\frac{\partial \Phi}{\partial x} + \frac{\partial F}{\partial y}\frac{\partial \Phi}{\partial q} - \frac{\partial F}{\partial q}\frac{\partial \Phi}{\partial y}$$

$$+ p\left(\frac{\partial F}{\partial z}\frac{\partial \Phi}{\partial p} - \frac{\partial F}{\partial p}\frac{\partial \Phi}{\partial z}\right) + q\left(\frac{\partial F}{\partial z}\frac{\partial \Phi}{\partial q} - \frac{\partial F}{\partial q}\frac{\partial \Phi}{\partial z}\right) = 0,$$

and are considered as two simultaneous equations giving p and q as functions of x, y, and z, then the values of p and q, derived from them and substituted in the equation

$$dz = p\,dx + q\,dy,$$

render it (when multiplied, if necessary, by some factor) an exact differential.

Another form may be given to the relation. Let

$$F_x = \frac{\partial F}{\partial x} + p\,\frac{\partial F}{\partial z},$$

$$F_y = \frac{\partial F}{\partial y} + q\,\frac{\partial F}{\partial z},$$

and similarly for Φ; then the equation is easily transformed into

$$F_x \frac{\partial \Phi}{\partial p} - \Phi_x \frac{\partial F}{\partial p} + F_y \frac{\partial \Phi}{\partial q} - \Phi_y \frac{\partial F}{\partial q} = 0.$$

Ex. 1. Solve the equation

$$p^2 + q^2 - 2px - 2qy + 2xy = 0.$$

Forming the subsidiary equations, we have (among others)

$$\frac{dp}{2y - 2p} = \frac{dq}{2x - 2q} = \frac{dx}{-2p + 2x} = \frac{dy}{-2q + 2y}.$$

Hence $\qquad\qquad dp + dq = dx + dy,$

so that $\qquad\qquad p - x + q - y = a.$

Combining this with the original equation, which may be written

$$(p - x)^2 + (q - y)^2 = (x - y)^2,$$

we find $\qquad 2(p - x) = a + \{2(x - y)^2 - a^2\}^{\frac{1}{2}},$

$$2(q - y) = a - \{2(x - y)^2 - a^2\}^{\frac{1}{2}}.$$

Hence $\qquad\qquad dz = p\,dx + q\,dy$

gives $\qquad 2dz = (2x + a)\,dx + (2y + a)\,dy + (dx - dy)\{2(x - y)^2 - a^2\}^{\frac{1}{2}},$

the integral of which is

$$2z - b = x^2 + ax + y^2 + ay + \frac{x - y}{2^{\frac{3}{2}}}\{2(x - y)^2 - a^2\}^{\frac{1}{2}}$$

$$- \frac{a^2}{2^{\frac{3}{2}}}\log\left[2^{\frac{1}{2}}(x - y) + \{2(x - y)^2 - a^2\}^{\frac{1}{2}}\right].$$

This is the Complete Integral. The General Integral is deducible in the ordinary way. There is no Singular Integral.

The above equation may, however, be solved without having recourse to this method; but some transformations and substitutions are necessary. Taking the equation in the form

$$(p - x)^2 + (q - y)^2 = (x - y)^2,$$

we write $\qquad Z = z - \tfrac{1}{2}x^2 - \tfrac{1}{2}y^2,$

so that
$$\frac{\partial Z}{\partial x} = p - x, \quad \frac{\partial Z}{\partial y} = q - y.$$

Let the independent variables be changed by the equations

$$x - y = 2^{\frac{1}{2}} X \text{ and } x + y = 2^{\frac{1}{2}} Y;$$

then
$$\frac{\partial Z}{\partial x} = \left(\frac{\partial Z}{\partial X} + \frac{\partial Z}{\partial Y} \right) 2^{-\frac{1}{2}} = 2^{-\frac{1}{2}} (P + Q),$$

$$\frac{\partial Z}{\partial y} = \left(-\frac{\partial Z}{\partial X} + \frac{\partial Z}{\partial Y} \right) 2^{-\frac{1}{2}} = 2^{-\frac{1}{2}} (Q - P),$$

and therefore
$$\left(\frac{\partial Z}{\partial x} \right)^2 + \left(\frac{\partial Z}{\partial y} \right)^2 = P^2 + Q^2.$$

The equation becomes
$$P^2 + Q^2 = 2X^2,$$

and is thus of the form of Standard III. When the integral is obtained under the rule (§ 200), and the old variables are substituted for the new variables, it will be found to agree with the integral already obtained.

Ex. 2. Solve the equations

(i) $p^2 + q^2 - 2px - 2qy + 1 = 0$;

(ii) $2(pq + py + qx) + x^2 + y^2 = 0$;

by Charpit's method.

Also reduce both of them to one or other of the Standard forms and so integrate them, shewing that the integrals obtained by the two methods agree.

209. In these particular examples, Charpit's method is less laborious than the other; but this is by no means always the case. It often happens that an equation which furnishes an easy example of the general method is integrable still more easily because included in some one or other of the special Standard forms; and so the method is less used than would otherwise be the case. But it is more general than any of them; and equations, integrable by any of the special methods, are integrable by the general method. It is moreover important in the general theory, as indicating a process of obtaining a solution of the differential equation without any restrictions on its form.

The limitations to success in practice are connected with the integration of the subsidiary equations. Now these particular limitations are just such as give rise to the methods adopted for the different Standards and really indicate the classification therein adopted; in fact, *all the Standards are included in Charpit's method, and integration is possible by Charpit's method whenever it is possible by any of the special methods.*

210. Thus consider Lagrange's form, which is

$$R - Pp - Qq = 0,$$

in which P, Q, R, are functions of x, y, z, and do not involve p or q. In this case

$$F = R - Pp - Qq,$$

so that
$$-\frac{\partial F}{\partial p} = P, \quad -\frac{\partial F}{\partial q} = Q,$$

$$-p\frac{\partial F}{\partial p} - q\frac{\partial F}{\partial q} = pP + qQ = R;$$

thus two of Charpit's equations are

$$\frac{dx}{P} = \frac{dy}{Q} = \frac{dz}{R},$$

the equations on which the integration of Lagrange's form depends.

But it should be noticed that this is not a proof of Lagrange's method for linear differential equations; the result has already been assumed in the derivation of Charpit's equations. Moreover, Lagrange's method is concerned with the construction of the general integral. If Charpit's method be adopted in detail, it would be necessary to include the other fractions in the subsidiary system, so as to obtain an integral of that system involving either p or q or both p and q.

211. Next, consider the typical equation of the First Standard, which is

$$\psi(p, q) = 0,$$

so that
$$F = \psi(p, q),$$

in which x, y, z, do not explicitly occur; then

$$\frac{\partial F}{\partial x} = 0, \quad \frac{\partial F}{\partial y} = 0, \quad \frac{\partial F}{\partial z} = 0.$$

The subsidiary equations now are

$$\frac{dp}{0} = \frac{dq}{0} = \frac{dx}{-\dfrac{\partial \psi}{\partial p}} = \ldots,$$

so that we have $p=a$ and $q=b$, both arbitrary constants apparently. But, according to the rule, we must combine some one integral with the original equation, and so we have

$$\psi(a, q) = 0;$$

and therefore, if $q = b$, we have

$$\psi(a, b) = 0.$$

Then
$$dz = p\,dx + q\,dy$$
$$= a\,dx + b\,dy,$$

of which the integral is

$$z = ax + by + c,$$

with the limitation between a and b.

212. Proceeding to the typical equation of the Second Standard, which is

$$\psi(z, p, q) = 0,$$

an equation into which x and y do not explicitly enter, we have

$$F = \psi(z, p, q),$$

and therefore

$$\frac{\partial F}{\partial x} = 0, \quad \frac{\partial F}{\partial y} = 0.$$

The equation derived from the first pair of Charpit's fractions gives

$$\frac{dp}{p\dfrac{\partial F}{\partial z}} = \frac{dq}{q\dfrac{\partial F}{\partial z}};$$

and therefore $q = ap$. Combining this with $\psi = 0$, we can find both p and q in terms of z; let the values be $f(z)$ for p and therefore $af(z)$ for q. Substituting in

$$dz = p\,dx + q\,dy,$$

we have
$$\frac{dz}{f(z)} = dx + a\,dy,$$

or
$$\int \frac{dz}{f(z)} + C = x + ay,$$

which agrees with the former result.

213. Passing now to the Third Standard, in which the equation is

$$F = \phi(x, p) - \psi(y, q) = 0,$$

so that
$$\frac{\partial F}{\partial x} = \frac{\partial \phi}{\partial x}; \quad \frac{\partial F}{\partial p} = \frac{\partial \phi}{\partial p};$$

$$\frac{\partial F}{\partial y} = -\frac{\partial \psi}{\partial y}; \quad \frac{\partial F}{\partial q} = -\frac{\partial \psi}{\partial q}; \quad \frac{\partial F}{\partial z} = 0;$$

we have from the subsidiary equations

$$\frac{dp}{\dfrac{\partial \phi}{\partial x}} = \frac{dx}{-\dfrac{\partial \phi}{\partial p}},$$

or
$$\frac{\partial \phi}{\partial p}\, dp + \frac{\partial \phi}{\partial x}\, dx = 0,$$

that is,
$$\phi(x, p) = a;$$

and therefore from the original equation,

$$\psi(y, q) = a.$$

Solving these respectively for p and q, we have

$$p = \theta_1(x, a), \qquad q = \theta_2(y, a);$$

hence
$$dz = \theta_1(x, a)\, dx + \theta_2(y, a)\, dy,$$

the integral of which is

$$z + c = \int \theta_1(x, a)\, dx + \int \theta_2(y, a)\, dy.$$

Ex. 3. Derive by Charpit's method the integral of the differential equation of the form analogous to Clairaut's form for ordinary equations.

Ex. 4. Obtain by Charpit's method a solution of the equation

$$px + qy = f(p, q),$$

where $f(p, q)$ is a homogeneous function of p and q of the degree n.

Solve also
$$xp^2 + yq^2 = 2pq.$$

JACOBI'S METHOD FOR THE GENERAL EQUATION WITH ANY NUMBER OF INDEPENDENT VARIABLES.

214. It has been indicated, in §§ 193, 194, that the method used for the linear partial differential equation in Lagrange's form can be applied to the case when the number of variables is n. We now proceed to indicate the method, due to Jacobi, of solving the general partial differential equation when there are n independent variables. This general equation may be represented by

$$\Phi(z, p_1, p_2, \ldots\ldots, p_n, x_1, x_2, \ldots\ldots, x_n) = 0,$$

where $x_1, x_2, \ldots\ldots, x_n$, are the independent variables, and the p's are the partial differential coefficients of z with respect to the x's.

215. We will prove that, if the dependent variable explicitly occur in this equation (which will usually be the case since the equation is perfectly general), the equation $\Phi = 0$ can be changed into another with a new dependent variable, in which that dependent variable does not explicitly occur and the number of independent variables is increased by unity.

The differential equation $\Phi = 0$ has some solution. Let it be represented by

$$u = f(z, x_1, x_2, \ldots\ldots, x_n) = 0,$$

where f is as yet an unknown function; then we have

$$\frac{\partial u}{\partial x_r} + \frac{\partial u}{\partial z} p_r = 0,$$

for all values of the suffix from $r = 1$ to $r = n$. Let these values of p be substituted in the original equation, which therefore becomes

$$\Phi\left(z, -\frac{\frac{\partial u}{\partial x_1}}{\frac{\partial u}{\partial z}}, -\frac{\frac{\partial u}{\partial x_2}}{\frac{\partial u}{\partial z}}, \ldots\ldots, -\frac{\frac{\partial u}{\partial x_n}}{\frac{\partial u}{\partial z}}, x_1, x_2, \ldots\ldots, x_n\right) = 0,$$

and may be written in the form

$$\Psi\left(x_1, x_2, \ldots\ldots, x_n, z, \frac{\partial u}{\partial x_1}, \frac{\partial u}{\partial x_2}, \ldots\ldots, \frac{\partial u}{\partial x_n}, \frac{\partial u}{\partial z}\right) = 0.$$

We thus still have a partial differential equation of the first order; the dependent variable u does not explicitly occur, and there are $n+1$ independent variables $z, x_1, x_2, \ldots\ldots, x_n$. Hence the proposition is proved.

The integral of the modified equation leads to the integral of the original equation; it will be proved to be possible to obtain the integral of $\Psi = 0$ in the form

$$u = f(x_1, x_2, \ldots\ldots, x_n, z, a_1, a_2, \ldots\ldots, a_n),$$

where $a_1, a_2, \ldots\ldots, a_n$, are arbitrary constants.

When this integral of $\Psi = 0$ is known, the complete integral of the equation $\Phi = 0$ is given by

$$f(x_1, x_2, \ldots\ldots, x_n, z, a_1, a_2, \ldots\ldots, a_n) = 0,$$

in which z is now the dependent variable and there remain the original n independent variables.

For $u = f$ is the integral of $\Psi = 0$ and Ψ is a modified form of $\Phi = 0$, so that the latter is satisfied by $u = f$, and therefore

$$\Phi\left(z, -\frac{\dfrac{\partial f}{\partial x_1}}{\dfrac{\partial f}{\partial z}}, -\frac{\dfrac{\partial f}{\partial x_2}}{\dfrac{\partial f}{\partial z}}, \ldots\ldots, -\frac{\dfrac{\partial f}{\partial x_n}}{\dfrac{\partial f}{\partial z}}, x_1, x_2, \ldots\ldots, x_n\right) = 0.$$

But, since $f = 0$, we have

$$\frac{\partial f}{\partial x_r} + \frac{\partial f}{\partial z} p_r = 0,$$

and therefore

$$p_r = -\frac{\dfrac{\partial f}{\partial x_r}}{\dfrac{\partial f}{\partial z}},$$

which is satisfied for all the suffixes r from $r = 1$ to $r = n$; hence we obtain

$$\Phi(z, p_1, p_2, \ldots\ldots, p_n, x_1, x_2, \ldots\ldots, x_n) = 0,$$

the original differential equation.

216. It is thus theoretically sufficient to consider differential equations from which the dependent variable is explicitly absent. If the dependent variable explicitly occur in any given equation, it

can be removed in the manner indicated; and a transformed
differential equation $\Psi = 0$ can be obtained, the integral of which
will lead to the required integral. We shall therefore write the
general differential equation in the form

$$F(p_1, p_2, \ldots\ldots, p_n, x_1, x_2, \ldots\ldots, x_n) = 0.$$

If, in addition to $F = 0$, we have other $n - 1$ equations of the
form

$$F_1 = a_1, \ F_2 = a_2, \ \ldots\ldots, \ F_r = a_r, \ \ldots\ldots, \ F_{n-1} = a_{n-1},$$

where $F_1, F_2, \ldots\ldots, F_{n-1}$, are functions of $p_1, p_2, \ldots\ldots, p_n$ (or of
some of them), and may be, and usually are, functions also
of $x_1, x_2, \ldots\ldots, x_n$, and where $a_1, a_2, \ldots\ldots, a_{n-1}$, are arbitrary
constants, then from these n equations we can obtain values
of $p_1, p_2, \ldots\ldots, p_n$, as functions of the x's and the a's. Let these
values be substituted in

$$dz = p_1 dx_1 + p_2 dx_2 + \ldots\ldots + p_n dx_n ;$$

if they be such as to render this an exact differential, its integral
will be the complete integral of $F = 0$. For it will be an integral,
since the values of $p_1, p_2, \ldots\ldots, p_n$, are derived from n equations,
one of which is $F = 0$; and its expression will involve n arbitrary
constants, viz. the constants $a_1, a_2, \ldots\ldots, a_{n-1}$, and the constant of
integration. Moreover, the integral is of the form

$$z = \chi(x_1, x_2, \ldots\ldots, x_n, a_1, a_2, \ldots\ldots, a_{n-1}) + a_n,$$

which gives the dependent variable explicitly, and therefore
justifies the assumption made as to the form of the integral of
$\Psi = 0$.

The $n - 1$ functions F must be such that the values of the
quantities p will render the foregoing an exact differential equation. The necessary conditions, which are

$$\frac{\partial p_r}{\partial x_s} = \frac{\partial p_s}{\partial x_r},$$

for all values of r and s, will serve to determine these functions.

217. Suppose that the n equations

$$F = 0, \ F_1 = a_1, \ F_2 = a_2, \ \ldots\ldots, \ F_{n-1} = a_{n-1},$$

are solved so as to give the values of $p_1, p_2, \ldots\ldots, p_n$, as functions
of the variables x; these values will, when substituted, make each

equation an identity. When this substitution takes place in any two such equations as $F_r = a_r$ and $F_s = a_s$, we have

$$\begin{cases} \dfrac{\partial F_r}{\partial x_1} + \dfrac{\partial F_r}{\partial p_1}\dfrac{\partial p_1}{\partial x_1} + \dfrac{\partial F_r}{\partial p_2}\dfrac{\partial p_2}{\partial x_1} + \cdots\cdots + \dfrac{\partial F_r}{\partial p_n}\dfrac{\partial p_n}{\partial x_1} = 0, \\[2mm] \dfrac{\partial F_s}{\partial x_1} + \dfrac{\partial F_s}{\partial p_1}\dfrac{\partial p_1}{\partial x_1} + \dfrac{\partial F_s}{\partial p_2}\dfrac{\partial p_2}{\partial x_1} + \cdots\cdots + \dfrac{\partial F_s}{\partial p_n}\dfrac{\partial p_n}{\partial x_1} = 0, \end{cases}$$

$$\begin{cases} \dfrac{\partial F_r}{\partial x_2} + \dfrac{\partial F_r}{\partial p_1}\dfrac{\partial p_1}{\partial x_2} + \dfrac{\partial F_r}{\partial p_2}\dfrac{\partial p_2}{\partial x_2} + \cdots\cdots + \dfrac{\partial F_r}{\partial p_n}\dfrac{\partial p_n}{\partial x_2} = 0, \\[2mm] \dfrac{\partial F_s}{\partial x_2} + \dfrac{\partial F_s}{\partial p_1}\dfrac{\partial p_1}{\partial x_2} + \dfrac{\partial F_s}{\partial p_2}\dfrac{\partial p_2}{\partial x_2} + \cdots\cdots + \dfrac{\partial F_s}{\partial p_n}\dfrac{\partial p_n}{\partial x_2} = 0, \end{cases}$$

$$\cdots\cdots\cdots\cdots\cdots\cdots\cdots\cdots\cdots\cdots\cdots\cdots\cdots$$

giving altogether n pairs of equations; each pair is made up of the differential coefficients, with regard to the same independent variable, of F_r and F_s when in these the values of the p's are substituted. Between the first pair let the value of $\dfrac{\partial p_1}{\partial x_1}$ be eliminated; the resulting equation is

$$\begin{bmatrix} F_r, F_s \\ x_1, p_1 \end{bmatrix} + \begin{bmatrix} F_r, F_s \\ p_2, p_1 \end{bmatrix}\dfrac{\partial p_2}{\partial x_1} + \begin{bmatrix} F_r, F_s \\ p_3, p_1 \end{bmatrix}\dfrac{\partial p_3}{\partial x_1} + \cdots\cdots + \begin{bmatrix} F_r, F_s \\ p_n, p_1 \end{bmatrix}\dfrac{\partial p_n}{\partial x_1} = 0,$$

where
$$\begin{bmatrix} F_r, F_s \\ u, v \end{bmatrix} = \dfrac{\partial F_r}{\partial u}\dfrac{\partial F_s}{\partial v} - \dfrac{\partial F_r}{\partial v}\dfrac{\partial F_s}{\partial u}$$

$$= -\begin{bmatrix} F_r, F_s \\ v, u \end{bmatrix} = -\begin{bmatrix} F_s, F_r \\ u, v \end{bmatrix} = \begin{bmatrix} F_s, F_r \\ v, u \end{bmatrix}.$$

Similarly, the elimination of $\dfrac{\partial p_2}{\partial x_2}$ from the second pair gives

$$\begin{bmatrix} F_r, F_s \\ x_2, p_2 \end{bmatrix} + \begin{bmatrix} F_r, F_s \\ p_1, p_2 \end{bmatrix}\dfrac{\partial p_1}{\partial x_2} + \begin{bmatrix} F_r, F_s \\ p_3, p_2 \end{bmatrix}\dfrac{\partial p_3}{\partial x_2} + \cdots\cdots + \begin{bmatrix} F_r, F_s \\ p_n, p_2 \end{bmatrix}\dfrac{\partial p_n}{\partial x_2} = 0;$$

and so on, each pair leading to an equation of this form.

Now let all the left-hand members of these equations be added together. The coefficient of $\dfrac{\partial p_s}{\partial x_r}$ $\left(\text{which is equal to } \dfrac{\partial p_r}{\partial x_s}\right)$ will consist of the sum of two terms, viz. the term

$$\begin{bmatrix} F_r, F_s \\ p_s, p_r \end{bmatrix}$$

from the r'^{th} equation, and the term

$$\left[\frac{F_r, F_s}{p_{r'}, p_{s'}}\right]$$

from the s'^{th} equation; the sum of these two is zero, and thus the term in $\dfrac{\partial p_{s'}}{\partial x_{r'}}$ disappears, whatever be the values of r' and s'. The resulting equation is therefore

$$\left[\frac{F_r, F_s}{x_1, p_1}\right] + \left[\frac{F_r, F_s}{x_2, p_2}\right] + \left[\frac{F_r, F_s}{x_3, p_3}\right] + \ldots\ldots + \left[\frac{F_r, F_s}{x_n, p_n}\right] = 0.$$

Let the left-hand side be denoted by

$$(F_r, F_s);$$

then the equation is

$$(F_r, F_s) = 0;$$

and this must be satisfied, whatever the suffixes r and s may be. Hence the aggregate of the equations which these functions must satisfy may be represented in the form

$$0 = (F_i, F) = (F_i, F_1) = (F_i, F_2) = \ldots\ldots = (F_i, F_{i-1}),$$

for all values of the index i from $i = 1$ to $i = n - 1$.

218. These conditions, which are necessary for the integrability of the equation $dz = \Sigma p\, dx$, must now be proved sufficient; this will be established by shewing that, when the functions F satisfy the foregoing equations, we have

$$\frac{\partial p_{r'}}{\partial x_{s'}} = \frac{\partial p_{s'}}{\partial x_{r'}},$$

for all values of r' and s'.

The n equations derived from the n pairs of equations connected with any two given functions F_r and F_s still hold; when they are all added together, we have

$$(F_r, F_s) + \Sigma\Sigma \left[\frac{F_r, F_s}{p_{r'}, p_{s'}}\right]\left(\frac{\partial p_{r'}}{\partial x_{s'}} - \frac{\partial p_{s'}}{\partial x_{r'}}\right) = 0,$$

the double summation extending to all integral values of r' and s' from 1 to n, but not including pairs of equal values since for every such pair of values the term vanishes. But by the necessary conditions satisfied by the functions, we have

$$(F_r, F_s) = 0,$$

and therefore $\Sigma\Sigma \left[\dfrac{F_r,\ F_s}{p_{r'},\ p_{s'}} \right] \left(\dfrac{\partial p_{r'}}{\partial x_{s'}} - \dfrac{\partial p_{s'}}{\partial x_{r'}} \right) = 0,$

which holds for all the values of r and s given by the different functions; and every combination of the functions will give such an equation. The total number of these combinations is $\frac{1}{2}n(n-1)$; and therefore the number of such equations is $\frac{1}{2}n(n-1)$.

Now each equation is linear in the quantities

$$\frac{\partial p_{r'}}{\partial x_{s'}} - \frac{\partial p_{s'}}{\partial x_{r'}},$$

which are in number $\frac{1}{2}n(n-1)$ in all, that is, the same as the number of the equations. Since each right-hand side is zero it follows, either that each of these quantities

$$\frac{\partial p_{r'}}{\partial x_{s'}} - \frac{\partial p_{s'}}{\partial x_{r'}}$$

is zero, or that the determinant formed by the coefficients of these quantities is zero.

That the latter cannot be the case appears as follows. Let Δ denote the determinant

$$\begin{vmatrix} \dfrac{\partial F}{\partial p_1}, & \dfrac{\partial F}{\partial p_2}, & \cdots\cdots, & \dfrac{\partial F}{\partial p_n} \\[2mm] \dfrac{\partial F_1}{\partial p_1}, & \dfrac{\partial F_1}{\partial p_2}, & \cdots\cdots, & \dfrac{\partial F_1}{\partial p_n} \\[2mm] \cdots\cdots\cdots\cdots\cdots\cdots\cdots\cdots \\[1mm] \dfrac{\partial F_{n-1}}{\partial p_1}, & \dfrac{\partial F_{n-1}}{\partial p_2}, & \cdots\cdots, & \dfrac{\partial F_{n-1}}{\partial p_n} \end{vmatrix};$$

then each of the expressions

$$\left[\frac{F_r,\ F_s}{p_{r'},\ p_{s'}} \right]$$

is the complement of a second minor of Δ and there are in all $\frac{1}{4}n^2(n-1)^2$ of them; let Θ denote the determinant formed by them, so that Θ is the determinant which is under consideration. Let Θ' be the determinant formed by the complements in Δ of the constituents in Θ; then we have, on multiplying Θ and Θ' together,

$$\Theta\Theta' = \Delta^{\frac{1}{2}n(n-1)}.$$

Now Θ' is not infinite; hence if Θ vanish, we must have

$$\Delta = 0.$$

But this would imply that, among the n equations of the type $F' = 0$, the n quantities p could be eliminated; that is, that these equations would not suffice to determine the quantities p as functions of the independent variables. This is contrary to what has been assumed as to the independence of the functions F; hence Θ is not zero.

It follows that each of the $\frac{1}{2}n(n-1)$ quantities

$$\frac{\partial p_{r'}}{\partial x_{s'}} - \frac{\partial p_{s'}}{\partial x_{r'}}$$

is zero, and therefore that the assigned conditions are sufficient to ensure that

$$dz = p_1 dx_1 + p_2 dx_2 + \ldots \ldots + p_n dx_n$$

is a perfect differential.

219. We may therefore sum up our results, so far obtained, as follows:

To obtain the Complete Integral of any given equation $F = 0$, we first determine an integral $F_1 = a_1$ of the equation

$$(F_1, F) = 0;$$

then we obtain a common integral $F_2 = a_2$ of the equations

$$(F_2, F) = (F_2, F_1) = 0;$$

then a common integral $F_3 = a_3$ of the equations

$$(F_3, F) = (F_3, F_1) = (F_3, F_2) = 0;$$

and so on, thus obtaining in all $n - 1$ new equations, each containing an arbitrary constant. The n equations, which involve the n quantities p, are then solved so as to furnish the values of the p's as functions of the independent variables and the arbitrary constants; and these values are substituted in

$$dz = p_1 dx_1 + p_2 dx_2 + \ldots \ldots + p_n dx_n.$$

This, when integrated, gives the Complete Integral of the equation

$$F = 0.$$

Each of the equations determining any one of the functions F_r is linear in the partial differential coefficients of F_r; we have therefore to investigate a method of obtaining the common integral of a set of simultaneous linear partial differential equations.

Ex. Prove that, if the equations

$$F_1(x_1, x_2, ..., x_n, z, p_1, p_2, ..., p_n) = 0,$$
$$F_2(x_1, x_2, ..., x_n, z, p_1, p_2, ..., p_n) = 0,$$
$$\dotfill$$
$$F_n(x_1, x_2, ..., x_n, z, p_1, p_2, ..., p_n) = 0,$$

be solved so as to give $p_1, p_2, ..., p_n$, as functions of $x_1, x_2, ..., x_n, z$, the necessary and sufficient conditions in order that

$$dz = p_1 dx_1 + p_2 dx_2 + ... + p_n dx_n$$

should be an exact differential are, that the aggregate of equations

$$0 = \left(\frac{F_i, F_1}{x, p}\right) + \left\{\frac{F_i, F_1}{z, p}\right\} = \left(\frac{F_i, F_2}{x, p}\right) + \left\{\frac{F_i, F_2}{z, p}\right\} = ... = \left(\frac{F_i, F_{i-1}}{x, p}\right) + \left\{\frac{F_i, F_{i-1}}{z, p}\right\},$$

where
$$\left(\frac{F_i, F_k}{x, p}\right) = \left[\frac{F_i, F_k}{x_1, p_1}\right] + \left[\frac{F_i, F_k}{x_2, p_2}\right] + ... + \left[\frac{F_i, F_k}{x_n, p_n}\right],$$

and
$$\left\{\frac{F_i, F_k}{z, p}\right\} = p_1\left[\frac{F_i, F_k}{z, p_1}\right] + p_2\left[\frac{F_i, F_k}{z, p_2}\right] + ... + p_n\left[\frac{F_i, F_k}{z, p_n}\right],$$

should be satisfied for all values of the index i from $i = 2$ to $i = n$.

220. It is convenient to prove here an important Lemma which will be of use when the integration of the simultaneous equations is being considered.

If A, B, C, be any three functions of $2n$ independent variables $x_1, x_2, ..., x_n, p_1, p_2, ..., p_n$, and if the function (B, C) be denoted by α, and the function (A, α) by

$$[A, (B, C)],$$

then the equation

$$[A, (B, C)] + [B, (C, A)] + [C, (A, B)] = 0$$

is identically satisfied.

Consider the left-hand member of this equation; it consists of the sum of a number of terms all of the same form, each of which is the product of two first differential coefficients of two of the quantities A, B, C, by one second differential coefficient of the third of them. It moreover is a cyclically symmetrical function of $A, B,$ and C; and therefore, if the terms involving the second differential coefficient of any one function, such as C, disappear, all the terms will disappear, and thus the equation will be satisfied.

Let the quantity

$$\frac{\partial B}{\partial x_r}\frac{\partial C}{\partial p_r} - \frac{\partial B}{\partial p_r}\frac{\partial C}{\partial x_r}$$

be denoted by $\Delta_r BC$, so that Δ_r may be considered as a symbolical operator; we may write

$$(B,\ C) = (\Delta_1 + \Delta_2 + \ldots\ldots + \Delta_n)\ BC,$$

the operators being obviously subject to the distributive law

$$(\Delta_r + \Delta_s)\ BC = \Delta_r BC + \Delta_s BC.$$

Then, in accordance with this notation,

$$[A,\ (B,\ C)] = (\Delta_1 + \Delta_2 + \ldots\ldots + \Delta_n)\ A\ (\Delta_1 + \Delta_2 + \ldots\ldots + \Delta_n)\ BC,$$

and therefore $[A,\ (B,\ C)]$ is the sum of a series of pairs of terms

$$\Delta_r A \Delta_s BC + \Delta_s A \Delta_r BC$$

for all the values of r and s from 1 to n inclusive; in the case when r and s have the same value, only a single term occurs for consideration.

Expanding the functions thus symbolically represented, we find that the terms depending upon the second differential coefficients of C are

$$\frac{\partial A}{\partial x_r}\frac{\partial B}{\partial x_s}\frac{\partial^2 C}{\partial p_r \partial p_s} - \frac{\partial A}{\partial x_r}\frac{\partial B}{\partial p_s}\frac{\partial^2 C}{\partial p_r \partial x_s} - \frac{\partial A}{\partial p_r}\frac{\partial B}{\partial x_s}\frac{\partial^2 C}{\partial p_s \partial x_r} + \frac{\partial A}{\partial p_r}\frac{\partial B}{\partial p_s}\frac{\partial^2 C}{\partial x_r \partial x_s},$$

from the first of the foregoing pair, and

$$\frac{\partial A}{\partial x_s}\frac{\partial B}{\partial x_r}\frac{\partial^2 C}{\partial p_r \partial p_s} - \frac{\partial A}{\partial x_s}\frac{\partial B}{\partial p_r}\frac{\partial^2 C}{\partial p_s \partial x_r} - \frac{\partial A}{\partial p_s}\frac{\partial B}{\partial x_r}\frac{\partial^2 C}{\partial p_r \partial x_s} + \frac{\partial A}{\partial p_s}\frac{\partial B}{\partial p_r}\frac{\partial^2 C}{\partial x_r \partial x_s},$$

from the second.

Selecting in the same way from $[B,\ (C,\ A)]$ the corresponding pair of symbolical terms and considering in them the terms which involve second differential coefficients of C, we find them to be respectively

$$\frac{\partial B}{\partial x_r}\frac{\partial A}{\partial p_s}\frac{\partial^2 C}{\partial p_r \partial x_s} - \frac{\partial B}{\partial x_r}\frac{\partial A}{\partial x_s}\frac{\partial^2 C}{\partial p_r \partial p_s} - \frac{\partial B}{\partial p_r}\frac{\partial A}{\partial p_s}\frac{\partial^2 C}{\partial x_r \partial x_s} + \frac{\partial B}{\partial p_r}\frac{\partial A}{\partial x_s}\frac{\partial^2 C}{\partial p_s \partial x_r},$$

and

$$\frac{\partial B}{\partial x_s}\frac{\partial A}{\partial p_r}\frac{\partial^2 C}{\partial p_s \partial x_r} - \frac{\partial B}{\partial x_s}\frac{\partial A}{\partial x_r}\frac{\partial^2 C}{\partial p_r \partial p_s} - \frac{\partial B}{\partial p_s}\frac{\partial A}{\partial p_r}\frac{\partial^2 C}{\partial x_r \partial x_s} + \frac{\partial B}{\partial p_s}\frac{\partial A}{\partial x_r}\frac{\partial^2 C}{\partial p_r \partial x_s}.$$

The expression $[C, (A, B)]$ does not contain any second differential coefficients of C.

Hence, in

$$[A, (B, C)] + [B, (C, A)] + [C, (A, B)],$$

the coefficient of the term which involves $\dfrac{\partial^2 C}{\partial p_r \partial p_s}$ is the sum of those in the foregoing, and is therefore zero; so also are the coefficients of those which involve $\dfrac{\partial^2 C}{\partial p_r \partial x_s}$, $\dfrac{\partial^2 C}{\partial p_s \partial x_r}$, $\dfrac{\partial^2 C}{\partial x_r \partial x_s}$.

If r and s be the same, we need only to consider the first and third of the above lines of terms when in them we write $s = r$; it will be seen immediately that all the terms in $\dfrac{\partial^2 C}{\partial p_r{}^2}$, $\dfrac{\partial^2 C}{\partial p_r \partial x_r}$, $\dfrac{\partial^2 C}{\partial x_r{}^2}$, vanish.

Since this is true whatever r and s may be, it follows that all the terms involving second differential coefficients of C vanish; and therefore, by the symmetry, the whole expression vanishes.

Solution of the Subsidiary Equations.

221. We now proceed to obtain the values of $F_1, F_2, \ldots\ldots, F_{n-1}$, from the various differential equations which they must satisfy. To determine F_1 we have

$$(F, F_1) = 0,$$

or, what is the same thing,

$$\frac{\partial F}{\partial x_1}\frac{\partial F_1}{\partial p_1} - \frac{\partial F}{\partial p_1}\frac{\partial F_1}{\partial x_1} + \frac{\partial F}{\partial x_2}\frac{\partial F_1}{\partial p_2} - \frac{\partial F}{\partial p_2}\frac{\partial F_1}{\partial x_2} + \ldots\ldots + \frac{\partial F}{\partial x_n}\frac{\partial F_1}{\partial p_n} - \frac{\partial F}{\partial p_n}\frac{\partial F_1}{\partial x_n} = 0.$$

Since this is linear in the differential coefficients of F_1, we may obtain an integral of it by using as subsidiary equations (§ 194) the generalised form of Lagrange's equations. Let any integral of the system

$$\frac{dx_1}{-\dfrac{\partial F}{\partial p_1}} = \frac{dx_2}{-\dfrac{\partial F}{\partial p_2}} = \ldots\ldots = \frac{dx_n}{-\dfrac{\partial F}{\partial p_n}}$$

$$= \frac{dp_1}{\dfrac{\partial F}{\partial x_1}} = \frac{dp_2}{\dfrac{\partial F}{\partial x_2}} = \ldots\ldots = \frac{dp_n}{\dfrac{\partial F}{\partial x_n}} \quad \ldots(A),$$

containing at least one of the variables p, be denoted by

$$f_1 (x_1, x_2, \ldots\ldots, x_n, p_1, p_2, \ldots\ldots, p_n) = a_1,$$

where a_1 is an arbitrary constant; then $F_1 = f_1 = a_1$ is an integral of the original equation $(F, F_1) = 0$.

222. We have now to find a function F_2, which will satisfy the equations

$$(F, F_2) = 0; \ (F_1, F_2) = (f_1, F_2) = 0.$$

The former of these, being an equation to determine F_2, is identical in form with that which determines F_1, and therefore we shall have the same subsidiary equations; let

$$\phi (x_1, x_2, \ldots\ldots, x_n, p_1, p_2, \ldots\ldots, p_n) = \text{constant}$$

be an integral of the equations (A) different from $f_1 = a_1$, and containing at least one of the variables p; then $(F, \phi) = 0$.

If ϕ be such a function as to satisfy

$$(f_1, \phi) = 0,$$

then we may take

$$F_2 = \phi = a_2$$

as a common integral of the two equations which determine F_2.

If ϕ do not satisfy the equation, then we shall have

$$(f_1, \phi) = \phi_1;$$

the substitution of ϕ_1 may be repeated and so on indefinitely, so that we shall have a series of functions ϕ given by

$$(f_1, \phi_1) = \phi_2; \ (f_1, \phi_2) = \phi_3; \ \ldots\ldots; \ (f_1, \phi_{i-1}) = \phi_i; \ \ldots\ldots$$

Now all these functions ϕ satisfy the equation

$$(F, F_2) = 0$$

when substituted for F_2. In the identity

$$[A, (B, C)] + [B, (C, A)] + [C, (A, B)] = 0,$$

let F be substituted for A and f_1 for B; then

$$[C, (A, B)] = [C, (F, f_1)] = (C, 0) = 0,$$

and therefore

$$[F, (f_1, C)] = [f_1, (F, C)],$$

whatever C may be.

First, let $C = \phi$; then this equation becomes

$$[F, (f_1, \phi)] = [f_1, (F, \phi)] = (f_1, 0) = 0;$$

so that

$$(f_1, \phi) = \phi_1 = F_2$$

is a solution of

$$(F, F_2) = 0.$$

Next, let $C = \phi_1$; then we have

$$[F, (f_1, \phi_1)] = [f_1, (F, \phi_1)] = (f_1, 0) = 0,$$

so that

$$(f_1, \phi_1) = \phi_2 = F_2$$

is also a solution of

$$(F, F_2) = 0;$$

and so on with the whole series of functions ϕ, each of which is a solution of the first of the two equations which determine F_2, and is therefore, when equated to a constant, also a solution of the subsidiary equations (A).

Now these subsidiary equations have only $2n - 1$ independent integrals at the utmost; the functions ϕ_i, which arise from the indefinitely repeated substitution in (f_1, ϕ_{i-1}), cannot all be independent of one another; and therefore if the series of functions do not cease, we must ultimately come to some one which is expressible in terms of those already found.

There are thus three alternatives to be considered:

(i), some function ϕ_i of the series may be identically zero;

(ii), some function ϕ_i of the series is variable but expressible in terms of the preceding functions of the series;

(iii), some function ϕ_i of the series may be a determinate constant c.

We will consider these in turn.

223. Firstly, let $\phi_i = 0$; then $\phi_{i-1} = a_2$ will be the desired integral; for it is one of the series of functions and is therefore a solution of $(F, F_2) = 0$; also

$$(F_1, \phi_{i-1}) = (f_1, \phi_{i-1}) = \phi_i = 0,$$

and it is therefore a solution of $(F_1, F_2) = 0$. Hence it is a common integral of the two equations which determine F_2, and it therefore gives the second of the equations desired, viz.

$$\phi_{i-1} = F_2 = a_2.$$

224. Secondly, let ϕ_i be expressible in terms of the preceding functions of the series; suppose

$$\phi_i = \theta (F, f_1, \phi, \phi_1, \phi_2, \ldots\ldots, \phi_{i-1}),$$

where θ is a definite functional symbol. Proceeding now to form ϕ_{i+1}, we have

$$\phi_{i+1} = (f_1, \phi_i)$$

$$= (f_1, F) \frac{\partial \theta}{\partial F} + (f_1, f_1) \frac{\partial \theta}{\partial f_1} + (f_1, \phi) \frac{\partial \theta}{\partial \phi} + (f_1, \phi_1) \frac{\partial \theta}{\partial \phi_1} + \ldots\ldots,$$

when the value of ϕ_i is substituted. But

$$(f_1, F) = - (F, f_1) = 0,$$

since f_1 is a solution of the equations; and (f_1, f_1) vanishes identically, so that this equation becomes

$$\phi_{i+1} = \phi_1 \frac{\partial \theta}{\partial \phi} + \phi_2 \frac{\partial \theta}{\partial \phi_1} + \ldots\ldots + \phi_{i-1} \frac{\partial \theta}{\partial \phi_{i-2}} + \phi_i \frac{\partial \theta}{\partial \phi_{i-1}}.$$

But each of the differential coefficients of θ is a function of the previously obtained quantities ϕ; hence ϕ_{i+1} is so also.

It follows therefore that ϕ_i and all the functions ϕ of the series after ϕ_i are expressible in terms of those which precede ϕ_i.

Let us then seek to obtain some function of these quantities which shall satisfy the equations

$$(F, F_2) = 0 \text{ and } (F_1, F_2) = (f_1, F_2) = 0 ;$$

let it be given by

$$F_2 = \psi (F, f_1, \phi, \phi_1, \ldots\ldots, \phi_{i-1}).$$

When this value is substituted, the former equation becomes

$$0 = (F, F) \frac{\partial \psi}{\partial F} + (F, f_1) \frac{\partial \psi}{\partial f_1} + (F, \phi) \frac{\partial \psi}{\partial \phi} + \ldots\ldots + (F, \phi_{i-1}) \frac{\partial \psi}{\partial \phi_{i-1}},$$

which is satisfied identically since every function ϕ is a solution of

$$(F, F_2) = 0 ;$$

and the second equation becomes, as before,

$$0 = (f_1, F_2) = \phi_1 \frac{\partial \psi}{\partial \phi} + \phi_2 \frac{\partial \psi}{\partial \phi_1} + \phi_3 \frac{\partial \psi}{\partial \phi_2} + \ldots\ldots + \phi_i \frac{\partial \psi}{\partial \phi_{i-1}}.$$

The last equation is thus the only one which must be satisfied by ψ; and as no differential coefficients with regard to F or f_1 occur in it, we may consider them as replaced by their respective values 0 and a_1. Any integral of the system

$$\frac{d\phi}{\phi_1} = \frac{d\phi_1}{\phi_2} = \frac{d\phi_2}{\phi_3} = \ldots\ldots = \frac{d\phi_{i-1}}{\phi_i}$$

$$= \frac{d\phi_{i-1}}{\theta}$$

of the form $\Phi = a_2$ will be a solution of the equation in ψ; and therefore we may write

$$F_2 = \Phi = a_2,$$

and so we shall have the required common integral of the two equations which determine F_2.

225. Thirdly, let ϕ_i be some determinate constant c which will merely depend upon the coefficients of the original differential equation; the series of functions thus terminates because there is no further function to substitute. We then proceed as in the last case to find some function of the preceding quantities ϕ which will be a common solution of the two equations; let

$$F_2 = \chi (F, f_1, \phi, \phi_1, \ldots\ldots, \phi_{i-1}).$$

When this is substituted in $(F, F_2) = 0$ the equation is identically satisfied; when it is substituted in $(f_1, F_2) = 0$, the resulting equation is, just as before,

$$0 = \phi_1 \frac{\partial \chi}{\partial \phi} + \phi_2 \frac{\partial \chi}{\partial \phi_1} + \ldots\ldots + \phi_{i-1} \frac{\partial \chi}{\partial \phi_{i-2}} + \phi_i \frac{\partial \chi}{\partial \phi_{i-1}},$$

in which we may replace ϕ_i by c. An integral of this is given by

$$\frac{d\phi_{i-2}}{\phi_{i-1}} = \frac{d\phi_{i-1}}{c},$$

which when integrated gives

$$\phi_{i-1}{}^2 - 2c\phi_{i-2} = \text{constant};$$

and therefore we may, as in the last case, write

$$F_2 = \phi_{i-1}{}^2 - 2c\phi_{i-2} = a_2,$$

as the common integral desired.

This solution is satisfactory provided $i > 1$.

Now i cannot be zero since ϕ is determined as a function of the variables; the only exception therefore to be considered is the case $i = 1$, when

$$\phi_1 \frac{\partial \chi}{\partial \phi} = c \frac{\partial \chi}{\partial \phi} = 0,$$

so that χ is independent of ϕ. Now

$$F_2 = \chi (F, f_1, \phi),$$

and F and f_1 are replaceable by 0 and a_1 respectively; if then χ be independent of ϕ, it ceases to be a function of the variables and there is thus no solution common to the two equations to be derived from these functions.

Should this be the case, we return to the subsidiary equations (A) and determine a new integral distinct from those already obtained, which are

$$F_1 = f_1 = a_1, \quad \phi = \text{constant};$$

let this be

$$\Im (x_1, x_2, \ldots\ldots, x_n, p_1, p_2, \ldots\ldots, p_n) = \text{constant}.$$

Next we perform with the function ϑ all the operations which have been performed with the function ϕ; then the desired common integral

$$F_2 = a_2$$

will be obtained, except in the single case when we have

$$(f_1, \vartheta) = \vartheta_1 = c',$$

where c' is a determinate constant.

From a combination of these respective exceptional cases, which are the only ones in each of which the common integral F_2 has not been obtained, we can construct a common integral F_2. For let

$$F_2 = f_2(\phi, \vartheta)$$

be substituted in $(F, F_2) = 0 = (f_1, F_2)$; then these equations become

$$0 = (F, \phi)\frac{\partial f_2}{\partial \phi} + (F, \vartheta)\frac{\partial f_2}{\partial \vartheta},$$

$$0 = (f_1, \phi)\frac{\partial f_2}{\partial \phi} + (f_1, \vartheta)\frac{\partial f_2}{\partial \vartheta}.$$

Now the former equation is satisfied identically since ϕ and ϑ are both integrals of the subsidiary equations (A); while since

$$(f_1, \phi) = \phi_1 = c,$$
and
$$(f_1, \vartheta) = \vartheta_1 = c',$$

the latter equation becomes

$$c\frac{\partial f_2}{\partial \phi} + c'\frac{\partial f_2}{\partial \vartheta} = 0.$$

This is satisfied by

$$f_2 = \Theta(c'\phi - c\vartheta),$$

and therefore

$$F_2 = \Theta(c'\phi - c\vartheta) = a_2,$$

where Θ is an arbitrary functional symbol (which may at will be chosen of a simple form), is the desired integral.

Hence *in every case* a common integral of the equations which determine F_2 has been found; for convenience we may denote it by

$$F_2 = f_2 = a_2.$$

226. We now proceed to obtain F_3; it must be a common integral of the equations

$$(F, F_3) = 0 = (f_1, F_3) = (f_2, F_3).$$

To obtain one we find, by the preceding method, an integral common to the two equations

$$(F, F_3) = 0 = (f_1, F_3),$$

which is different from $f_2 = a_2$; this we may denote by

$$\lambda (x_1, x_2, \ldots\ldots, x_n, p_1, p_2, \ldots\ldots, p_n) = \text{constant}.$$

We then form as before the series of functions

$$(f_2, \lambda) = \lambda_1; \quad (f_2, \lambda_1) = \lambda_2; \quad \ldots\ldots; \quad (f_2, \lambda_{i-1}) = \lambda_i; \quad \ldots\ldots;$$

then all the functions λ of this series are common integrals of the first two of the equations which determine λ. For, in the identity

$$[A, (B, C)] + [B, (C, A)] + [C, (A, B)] = 0,$$

let $A = F$ and $B = f_2$; then since $(F, f_2) = 0$, we have

$$[F, (f_2, C)] = [f_2, (F, C)].$$

And, substituting in the same identity $A = f_1$ and $B = f_2$, and remembering that $(f_1, f_2) = 0$, we have

$$[f_1, (f_2, C)] = [f_2, (f_1, C)].$$

These two equations are satisfied whatever C may be. Now let $C = \lambda$; then

$$[F, (f_2, \lambda)] = [f_2, (F, \lambda)],$$

or $\qquad\qquad (F, \lambda_1) = (f_2, 0) = 0;$

and $\qquad\qquad [f_1, (f_2, \lambda)] = [f_2, (f_1, \lambda)],$

or $\qquad\qquad (f_1, \lambda_1) = (f_2, 0) = 0.$

Thus λ_1 is a common integral of the equations

$$(F, F_3) = 0 = (f_1, F_3).$$

Similarly the substitution of λ_1 for C would shew that λ_2 is a common integral of these equations; and so on, through all the series of functions.

As in the former case, the number of common integrals being limited, we shall in the series come to some integral λ_k which is expressible, as well as those that follow it, in terms of those which precede it, viz., F, f_1, f_2, λ, λ_1, ..., λ_{k-1}. The same three alternatives are presented and the value of F_3, the common integral in each, is determined as before. Either the single case of failure is avoided by the choice of a new integral different from λ; or in the case of failure of the latter, these two cases of failure are combined so as to furnish a common integral. Thus we obtain our third common integral, which may be represented by

$$F_3 = f_3 = a_3.$$

227. The remaining functions F_4, ..., F_{n-1}, may be derived in the same way as the above; and thus, with $F = 0$, we shall have n equations to determine the values of the p's in terms of the independent variables and $n-1$ arbitrary constants, which, when substituted in

$$dz = p_1 dx_1 + p_2 dx_2 + ... + p_n dx_n,$$

will render it integrable; its integral is the complete integral of the original differential equation.

The associated integrals are derivable from the results of §§ 182, 183.

228. The foregoing is an exposition of Jacobi's method of integration in its simplest form; there are, however, developments and simplifications and, arising out of these, methods of avoiding the exceptional cases which cannot be dealt with here. For these, as well as for a fuller exposition of the whole theory of partial differential equations of the first order, reference may be made to the fifth volume of my *Theory of Differential Equations*, particularly chapters III., IV., V. so far as concerns the preceding subject-matter. Other methods are given in the later chapters of that volume, particularly Lie's method based on the theory of groups and the method of characteristics. Full references to the original authorities are given there; so here it will be sufficient to refer to Jacobi, "Vorlesungen über Dynamik" (*Ges. Werke*, Suppl. Bd. pp. 248—269); Jacobi, "Nova methodus...integrandi" (*Crelle*, t. LX. pp. 1—181); to a very valuable memoir by Imschenetsky, *Grunert's Archiv der Mathematik und Physik*, t. L. pp. 278—474; and to a memoir by Graindorge, *Mémoires de la Société Royale des Sciences de Liège*, IIme série, t. V., as well as to the treatise by Mansion, *Théorie des équations aux dérivées partielles*, and the treatise by Goursat, *Leçons sur l'intégration des équations aux dérivées partielles du premier ordre* (Paris, Hermann, 1891).

The equations (A) are, when each fraction is equated to dt, of the form

$$\frac{dx_r}{dt} = -\frac{\partial F}{\partial p_r}; \quad \frac{dp_r}{dt} = \frac{\partial F}{\partial x_r};$$

these are the canonical equations of motion of a system of rigid bodies. A discussion of them will be found in the memoir by Imschenetsky, which has just been quoted, in ROUTH's *Rigid Dynamics*, and in chapter x. of the fifth volume of my *Theory of Differential Equations*, quoted above.

We now proceed to consider some examples of the foregoing theory.

Ex. 1. To solve the equation

$$z = f(p_1, p_2, \ldots, p_n),$$

where f does not explicitly involve the independent variables. We must first transform the equation so that the dependent variable does not explicitly occur; let the solution of the equation be

$$\psi(x_1, x_2, \ldots, x_n, z) = 0,$$

where the form of ψ has yet to be determined. Denoting $\dfrac{\partial \psi}{\partial x_r}$ by P_r and $\dfrac{\partial \psi}{\partial z}$ by P_{n+1}, we have

$$P_r + P_{n+1} p_r = 0;$$

and thus the equation is

$$z = f\left(-\frac{P_1}{P_{r+1}}, \quad -\frac{P_2}{P_{n+1}}, \quad \ldots\ldots, \quad -\frac{P_n}{P_{n+1}}\right),$$

in which the dependent variable ψ does not occur. Hence we have for our general formula

$$F = f\left(-\frac{P_1}{P_{n+1}}, \quad -\frac{P_2}{P_{n+1}}, \quad \ldots\ldots, \quad -\frac{P_n}{P_{n+1}}\right) - z;$$

and the subsidiary equations give

$$\frac{dP_1}{0} = \frac{dP_2}{0} = \ldots = \frac{dP_n}{0} = \frac{dP_{n+1}}{-1}.$$

From these, we have

$$P_1 = a_1, \quad P_2 = a_2, \quad \ldots\ldots, \quad P_n = a_n,$$

which give n integrals, coexisting with one another and with $F = 0$. From the equation $F = 0$, we have

$$z = f\left(-\frac{a_1}{P_{n+1}}, \quad -\frac{a_2}{P_{n+1}}, \quad \ldots\ldots, \quad -\frac{a_n}{P_{n+1}}\right).$$

Solving this for P_{n+1}, we should have $P_{n+1} = \chi(z)$, where χ involves the n constants a; and therefore

$$d\psi = P_1 dx_1 + P_2 dx_2 + \ldots\ldots + P_n dx_n + P_{n+1} dz$$
$$= a_1 dx_1 + a_2 dx_2 + \ldots\ldots + a_n dx_n + \chi(z) dz.$$

The integral of this is

$$\psi + a = a_1 x_1 + a_2 x_2 + \ldots\ldots + a_n x_n + \int \chi(z) dz,$$

where a is arbitrary and may be assumed to be absorbed in ψ. But the integral of the given differential equation is $\psi = 0$; hence the integral of

$$z = f(p_1, p_2, \ldots, p_n)$$

is
$$a_1 x_1 + a_2 x_2 + \ldots + a_n x_n = \int \chi(z)\, dz,$$

where χ, as a function of z, is given by the equation

$$z = f\left(\frac{a_1}{\chi}, \frac{a_2}{\chi}, \ldots, \frac{a_n}{\chi}\right).$$

Ex. 2. The case, when f is a homogeneous function of order μ in the p's, is readily reduced to the form already considered in § 196. For we may change the dependent variable from z to ξ, where

$$\xi = \frac{\mu}{\mu - 1} z^{\frac{\mu - 1}{\mu}};$$

and the equation is then

$$1 = f(\xi_1, \xi_2, \ldots, \xi_n),$$

where $\xi_r = \frac{\partial \xi}{\partial x_r}$. The integral of this equation is

$$\frac{\mu}{\mu - 1} z^{\frac{\mu - 1}{\mu}} = \xi = c + a_1 x_1 + a_2 x_2 + \ldots + a_n x_n,$$

provided
$$f(a_1, a_2, \ldots, a_n) = 1.$$

Ex. 3. Solve

(i) $z^2 + z p_3 = p_1^2 + p_2^2$;

(ii) $z + 2 p_3 = (p_1 + p_2)^2$;

(iii) $(p_1 - z)(p_2 - z)(p_3 - z) = p_1 p_2 p_3$.

Ex. 4. Solve

$$F = (x_2 p_1 + x_1 p_2) x_3 + a p_3 (p_1 - p_2) - 1 = 0.$$

The subsidiary equations are

$$\frac{-dx_1}{x_2 x_3 + a p_3} = \frac{-dx_2}{x_1 x_3 - a p_3} = \frac{-dx_3}{a(p_1 - p_2)} = \frac{dp_1}{x_3 p_2} = \frac{dp_2}{x_3 p_1} = \frac{dp_3}{x_2 p_1 + x_1 p_2}.$$

From the equality of the 1st, 2nd, 4th, and 5th, fractions, we have

$$-\frac{dx_1 + dx_2}{(x_2 + x_1) x_3} = \frac{dp_1 + dp_2}{(p_1 + p_2) x_3},$$

which leads to
$$(p_1 + p_2)(x_1 + x_2) = a_1.$$

We therefore (adopting the notation of the previous articles) take

$$F_1 = (p_1 + p_2)(x_1 + x_2);$$

and we have to determine a solution of the subsidiary equations $F_2 = a_2$, which shall satisfy

$$(F_1, F_2) = 0.$$

From the equality of the 4th and 5th fractions, we have

$$p_1 dp_1 = p_2 dp_2;$$

and therefore we have

$$\phi = p_1^2 - p_2^2 = \text{constant}.$$

Now

$$(F_1, \phi) = (p_1 + p_2) 2p_1 + (p_1 + p_2)(-2p_2)$$

$$= 2\phi = \phi_1;$$

the continued substitution in the equation

$$(F_1, \phi_{i-1}) = \phi_i$$

would thus not lead to a function such as is required. We therefore return to the original subsidiary equations to obtain an integral different from $F_1 = a_1$ and $\phi = $ constant. Such an one is derivable from the equality of the 3rd, 4th, and 5th, fractions, which give

$$\frac{dp_1 - dp_2}{x_3(p_1 - p_2)} = \frac{dx_3}{a(p_1 - p_2)};$$

and therefore we have

$$\psi = a(p_1 - p_2) - \tfrac{1}{2} x_3^2 = \text{constant}.$$

Now

$$(F_1, \psi) = (p_1 + p_2) a + (p_1 + p_2)(-a) = 0,$$

and so ψ satisfies the two equations; we thus have

$$F_2 = a(p_1 - p_2) - \tfrac{1}{2} x_3^2 = a_2.$$

We now solve the equations

$$F = 0, \quad F_1 = a_1, \quad F_2 = a_2,$$

to find the values of p_1, p_2, p_3. These values are

$$p_1 = \tfrac{1}{2} \frac{a_1}{x_1 + x_2} + \frac{a_2}{2a} + \frac{1}{4a} x_3^2,$$

$$p_2 = \tfrac{1}{2} \frac{a_1}{x_1 + x_2} - \frac{a_2}{2a} - \frac{1}{4a} x_3^2,$$

$$p_3 = \frac{2 - a_1 x_3}{2a_2 + x_3^2} + \frac{1}{2a}(x_1 - x_2) x;$$

hence

$$dz = \tfrac{1}{2} a_1 d \log(x_1 + x_2) + \frac{1}{2a} \{(a_2 + \tfrac{1}{2} x_3^2)(dx_1 - dx_2) + (x_1 - x_2) x_3 dx_3\} + \frac{z - a_1 x_3}{2a_2 + x_3^2} dx_3,$$

so that the complete integral of the differential equation is

$$z+A=\tfrac{1}{2}a_1\log(x_1+x_2)+\frac{1}{2a}(x_1-x_2)(a_2+\tfrac{1}{2}x_3{}^2)-\tfrac{1}{2}a_1\log(x_3{}^2+2a_2)$$
$$+\left(\frac{2}{a_2}\right)^{\frac{1}{2}}\text{arc tan}\left\{\frac{x_3}{(2a_2)^{\frac{1}{2}}}\right\},$$

in which A, a_1, a_2, are the arbitrary constants.

(Imschenetsky.)

Ex. 5. Integrate the equations:

$$\text{(i)}\quad p_1x_1{}^2=p_2{}^2+ap_3{}^2;$$

$$\text{(ii)}\quad x_1p_1{}^2+x_2p_2{}^2+x_3p_3{}^2=p_1p_2p_3;$$

$$\text{(iii)}\quad p_1{}^2+p_2{}^2+p_3{}^2=x_1{}^2+x_1x_2+x_2{}^2+x_1x_3+x_2x_3+x_3{}^2;$$

$$\text{(iv)}\quad p_1+\tfrac{1}{2}p_2{}^2+p_2x_1x_3+p_3x_1x_2=0;$$

$$\text{(v)}\quad x_1+\tfrac{1}{2}x_2{}^2+x_2p_1p_3+x_3p_1p_2=0;$$

$$\text{(vi)}\quad p_1p_2p_3=p_1x_1+p_2x_2+p_3x_3.$$

It has already been indicated that several of the forms (in §§ 194—201) in two independent variables, which admit of immediate integration without the use of Charpit's subsidiary equations, can be generalised so as to include the cases where the number of independent variables is greater than two.

Ex. 6. When a given differential equation can be written in the form

$$f_1(x_1,x_2,\ldots\ldots,x_r,p_1,p_2,\ldots\ldots,p_r)=f_2(x_{r+1},\ldots\ldots,x_n,p_{r+1},\ldots\ldots,p_n),$$

the complete integral is the common integral of the equations

$$f_1=a=f_2,$$

where a is arbitrary. For the subsidiary equations are

$$\frac{dx_1}{-\dfrac{\partial f_1}{\partial p_1}}=\frac{dp_1}{\dfrac{\partial f_1}{\partial x_1}}=\ldots\ldots=\frac{dx_r}{-\dfrac{\partial f_1}{\partial p_r}}=\frac{dp_r}{\dfrac{\partial f_1}{\partial x_r}}=\frac{dx_{r+1}}{\dfrac{\partial f_2}{\partial p_{r+1}}}=\frac{dp_{r+1}}{-\dfrac{\partial f_2}{\partial x_{r+1}}}=\ldots\ldots;$$

from these, we have

$$\frac{\partial f_1}{\partial x_1}dx_1+\frac{\partial f_1}{\partial p_1}+\ldots\ldots+\frac{\partial f_1}{\partial x_r}dx_r+\frac{\partial f_1}{\partial p_r}dp_r=0,$$

and therefore

$$f_1=a$$
$$=f_2$$

by the given equation.

As an example, we may take

$$x_2p_1+x_1p_2+(p_1-p_2)(p_3+x_4)(p_4+x_3)=1.$$

Here we may write

$$(p_3+x_4)(p_4+x_3)=a,$$

$$x_2p_1+x_1p_2+a(p_1-p_2)=1,$$

where a is an arbitrary constant. The integral of the former equation is

$$z+x_3x_4=Ax_3+\frac{a}{A}x_4+C,$$

where A and C are arbitrary constants. The integral of the latter is obtainable by Charpit's method; the subsidiary equations are

$$\frac{-dx_1}{x_2+a}=\frac{-dx_2}{x_1-a}=\frac{dp_1}{p_2}=\frac{dp_2}{p_1}.$$

From these, we have

$$\frac{dp_1+dp_2}{p_1+p_2}+\frac{dx_1+dx_2}{x_1+x_2}=0,$$

and therefore

$$(p_1+p_2)(x_1+x_2)=A_1+1.$$

Hence by combining with the equation the integral of which is sought, we have

$$(x_2+a)p_1+p_2(x_1-a)=1,$$

$$(x_1-a)p_1+p_2(x_2+a)=A_1;$$

and these give

$$p_1\{(x_1-a)^2-(x_2+a)^2\}=A_1(x_1-a)-(x_2+a),$$

$$p_2\{(x_1-a)^2-(x_2+a)^2\}=x_1-a-A_1(x_2+a).$$

Thus

$$dz=p_1dx_1+p_2dx_2$$

$$=A_1d\log\{(x_1-a)^2-(x_2+a)^2\}+\tfrac{1}{2}d\log\frac{1+\dfrac{x_2+a}{x_1-a}}{1-\dfrac{x_2+a}{x_1-a}};$$

and therefore

$$z=A_1\log\{(x_1-a)^2-(x_2+a)^2\}+\tfrac{1}{2}\log\frac{(x_1-a)+(x_2+a)}{(x_1-a)-(x_2+a)}+C_1.$$

The complete integral of the original equation is therefore

$$z+x_3x_4=Ax_3+\frac{a}{A}x_4+A_1\log\{(x_1-a)^2+(x_2+a)^2\}+B+\tfrac{1}{2}\log\frac{(x_1-a)+(x_2+a)}{(x_1-a)-(x_2+a)},$$

where A, A_1, B, a, are arbitrary constants.

Ex. 7. Integrate

$$(x_2p_1+x_1p_2)x_3+p_3(p_1-p_2)\{p_4{}^2+(p_5+x_4)(p_5+x_6)p_6\}=a.$$

(Imschenetsky.)

SIMULTANEOUS PARTIAL DIFFERENTIAL EQUATIONS*.

229. Instead of there being given only a single equation to determine the dependent variable, there may be given a number of simultaneous equations. If the dependent variable explicitly occur in any of them they all can be transformed, as in § 215, so that it shall disappear. The equations may then be taken of the form

$$F_1(x_1, x_2, \ldots\ldots, x_n, p_1, p_2, \ldots\ldots, p_n) = 0,$$
$$F_2(x_1, x_2, \ldots\ldots, x_n, p_1, p_2, \ldots\ldots, p_n) = 0,$$
$$\ldots\ldots\ldots\ldots\ldots\ldots\ldots\ldots\ldots\ldots\ldots\ldots\ldots\ldots$$
$$F_m(x_1, x_2, \ldots\ldots, x_n, p_1, p_2, \ldots\ldots, p_n) = 0.$$

If m be greater than n, the equations cannot be independent; for the first n of the equations may be resolved so as to give values of the p's in terms of the variables x; and these values, when substituted in the remaining $m - n$, must reduce them to identities, since there would otherwise be relations between the independent variables. Thus in effect there may be given at most n simultaneous equations; and we may therefore take m either equal to n, or less than n.

230. I. Let $m = n$. We have thus n equations giving the values of the n quantities p in terms of the variables; these values, substituted in the equation

$$dz = p_1 dx_1 + p_2 dx_2 + \ldots\ldots + p_n dx_n,$$

must make it a perfect differential if the given system have a common solution. The conditions for this are that

$$\frac{\partial p_r}{\partial x_s} = \frac{\partial p_s}{\partial x_r},$$

for all pairs of indices; and these, as in § 217, lead to equations of the form

$$(F_r, F_s) = 0.$$

Hence the given functions must satisfy all the equations for all

---

* This theory is due initially to Bour; see authorities cited in § 228, p. 447.

possible combinations of the suffixes; and then the common complete integral is obtained by the integration of the equation

$$dz = p_1 dx_1 + p_2 dx_2 + \ldots\ldots + p_n dx_n.$$

It therefore contains one arbitrary constant.

It may happen however that the functions F are not independent of one another; then the determinant Δ, where

$$\Delta = \begin{vmatrix} \dfrac{\partial F_1}{\partial p_1}, & \dfrac{\partial F_1}{\partial p_2}, & \ldots\ldots, & \dfrac{\partial F_1}{\partial p_n} \\ \ldots\ldots\ldots\ldots\ldots\ldots\ldots\ldots \\ \dfrac{\partial F_n}{\partial p_1}, & \dfrac{\partial F_n}{\partial p_2}, & \ldots\ldots, & \dfrac{\partial F_n}{\partial p_n} \end{vmatrix},$$

is zero, and there will (§ 9) be an identical relation of the form

$$\Phi(F_1, F_2, \ldots\ldots, F_n, x_1, x_2, \ldots\ldots, x_n) = 0.$$

But for the purposes of integration $F_1 = F_2 = \ldots\ldots = F_n = 0$; and this therefore becomes

$$\Phi(0, 0, \ldots\ldots, 0, x_1, x_2, \ldots\ldots, x_n) = 0.$$

If $\Phi = 0$ be not an identity, a relation is implied between the independent variables, which is of course impossible; it then follows that the given equations are inconsistent, and that there is no common integral. If $\Phi = 0$ be an identity, the number of given equations independent of one another is less than the number of the quantities p, which therefore cannot be determined from the given equations alone; we must therefore have recourse to the method which applies when m is less than n.

Thus, if there be four independent variables and if four equations $F_1 = 0 = F_2 = F_3 = F_4$ be given, there can be no common integral in a case when there is a relation of the form

$$F_4 = (x_1 - x_2) F_1 + (x_2 - x_3) F_2 + x_1 x_2 x_3 x_4;$$

when there is a relation of the form

$$F_4 = (x_1 - x_2) F_3 + (x_2 - x_3) F_1 + (x_3 - x_1) F_2,$$

there are only three independent equations.

231. II. Let m be less than n. We may suppose the equations reduced to such a number m, that they are algebraically independent of one another, even though they were not so in the form in which they were first given. It will be assumed that there is a common integral, so far as the algebraic relations which give the dependent functions in terms of the others indicate; this will be the case if these relations become identically-null equations because of the equations $F_1 = 0, \ldots\ldots, F_m = 0$.

First Case. The equations $F_1 = 0 = \ldots\ldots = F_m$ may satisfy the relations

$$(F_r, F_s) = 0,$$

for all values $1, 2, \ldots\ldots, m$, of r and s; they are therefore simultaneously integrable. To determine the values of the quantities p, other $n - m$ equations must be obtained by Jacobi's method; these will involve $n - m$ arbitrary constants. From these equations and the given m equations, the values of p must be derived and be substituted in

$$dz = p_1 dx_1 + p_2 dx_2 + \ldots\ldots + p_n dx_n.$$

The integral is the common complete integral of the original equations, and it contains $n - m + 1$ arbitrary constants.

Second Case. It may happen that, for one or for several combinations of the indices in the series $1, 2, \ldots\ldots, m$, we find (F_r, F_s) a function of the independent variables only, or (F_r, F_s) a determinate constant. In neither case can (F_r, F_s) be zero; the conditions that the equations should be simultaneously integrable are not satisfied, and there is no common integral of the proposed equations.

Third Case. It may happen that, for one or for several combinations of the indices in the series $1, 2, \ldots\ldots, m$, we find results of the form

$$(F_r, F_s) = f(x_1, x_2, \ldots\ldots, x_n, p_1, p_2, \ldots\ldots, p_n),$$

where f does not become identically zero on combination with the given equations. Let there be l such combinations, so that $m + l$ must not be greater than n; then, for combinations other than these l, the equations

$$(F_r, F_s) = 0$$

are satisfied. We now take

$$0 = F_{m+1} = f_1, \ 0 = F_{m+2} = f_2, \ \ldots\ldots, \ 0 = F_{m+l} = f_l;$$

and substitute in the functions

$$(F_r, F_s),$$

where either r or s at least must be greater than m.

If then these functions all vanish, we have $m + l$ equations which are simultaneously integrable; and we determine, by Jacobi's method, the $n - m - l$ remaining equations necessary to give the complete integral, which will therefore contain $n - m - l + 1$ arbitrary constants.

If for any combination (F_{m-i}, f_k), or for any one combination (f_i, f_k), the function be a determinate constant or a function of the independent variables only, the equations are not simultaneously integrable; and then there is no common integral.

If for any combination (F_{m-i}, f_k), or for any one combination (f_i, f_k), we obtain a function $\phi(x_1, x_2, \ldots\ldots, x_n, p_1, p_2, \ldots\ldots, p_n)$ which does not vanish in virtue of the equations already obtained, we proceed with the functions ϕ as we did before with the functions f. Ultimately, either we shall arrive at a finite number, not greater than n, of independent equations which are simultaneously integrable, and then, in the ordinary way, obtain the common integral; or we shall obtain a result indicating impossibility of simultaneous existence, in which case there will be no common integral.

Ex. 1. Obtain a common integral (if one exist) of the simultaneous equations

$$\left.\begin{array}{l} F_1 = p_1 p_2 - x_3 x_4 = 0 \\ F_2 = p_3 p_4 - x_1 x_2 = 0 \end{array}\right\}.$$

We have

$$(F_1, F_2) = p_1 x_1 + p_2 x_2 - p_3 x_3 - p_4 x_4,$$

where the right-hand side does not vanish in virtue of $F_1 = 0 = F_2$; we therefore write

$$F_3 = p_1 x_1 + p_2 x_2 - p_3 x_3 - p_4 x_4 = 0.$$

Then $(F_1, F_2) = 0.$

Also $(F_1, F_3) = -2p_1 p_2 + 2x_3 x_4 = 0,$

$$(F_2, F_3) = \ \ 2p_3 p_4 - 2x_1 x_2 = 0;$$

the three equations are therefore compatible. Let F_4 be the other function required, so that it will be determined as a common integral of the equations

$$(F_4, F_3) = 0 = (F_4, F_2) = (F_4, F_1).$$

Considering it as an integral of

$$(F_4, F_3) = 0,$$

we form the equations

$$-\frac{dx_1}{x_1} = -\frac{dx_2}{x_2} = \frac{dx_3}{x_3} = \frac{dx_4}{x_4} = \frac{dp_1}{p_1} = \frac{dp_2}{p_2} = -\frac{dp_3}{p_3} = -\frac{dp_4}{p_4}.$$

One integral is

$$p_1 = ax_3,$$

where a is arbitrary; we therefore tentatively take

$$F_4 = \frac{p_1}{x_3} = a.$$

We then find

$$(F_4, F_1) = 0;$$

and

$$(F_4, F_2) = \frac{x_2}{x_3} - \frac{p_1}{x_3{}^2} p_4.$$

On solving the equations

$$F_1 = 0 = F_2 = F_3; \quad F_4 = a,$$

we find

$$p_1 = ax_3, \quad p_2 = \frac{1}{a} x_4, \quad p_3 = ax_1, \quad p_4 = \frac{1}{a} x_2,$$

and therefore

$$p_1 p_4 = x_2 x_3,$$

so that

$$(F_4, F_2) = 0.$$

Hence a common solution is

$$p_1 = ax_3.$$

To obtain a complete common integral, we have

$$dz = a(x_3 dx_1 + x_1 dx_3) + \frac{1}{a}(x_4 dx_2 + x_2 dx_4),$$

so that the integral is

$$z = ax_1 x_3 + \frac{1}{a} x_2 x_4 + b,$$

where a and b are arbitrary constants.

Ex. 2. Obtain integrals of the preceding equations in the forms:

$$\text{(i)} \quad z = ax_1 x_4 + \frac{1}{a} x_2 x_3 + b;$$

$$\text{(ii)} \quad z = 2\{x_2 x_4 (x_1 x_3 - a)\}^{\frac{1}{2}} + b;$$

$$\text{(iii)} \quad z = 2\{x_1 x_3 (x_2 x_4 - a)\}^{\frac{1}{2}} + b.$$

Ex. 3. Obtain common complete integrals of the simultaneous equations:

$$\left.\begin{array}{l} \text{I.} \quad p_1 + (x_3 + x_1 x_2 + x_1 x_4)\, p_4 + (x_2 + x_4 - 3x_1)\, p_3 = 0 \\ \quad\;\; p_2 + (x_1 x_3 x_4 + x_2 - x_1 x_2)\, p_4 + (x_3 x_4 - x_2)\, p_2 = 0 \end{array}\right\};$$

$$\left.\begin{array}{l} \text{II.} \qquad\qquad\qquad\qquad 2x_5 p_4 + x_1{}^2 p_5 = 0 \\ x_1{}^2 p_1 - 2x_5 p_2 + (x_1{}^2 x_4 - 2x_5)\, p_3 - 2x_1 x_4 p_4 = 0 \end{array}\right\}.$$

(Imschenetsky and Graindorge.)

Homogeneous Linear Systems.

232. One of the simplest classes of systems of simultaneous partial differential equations is composed of those in which each equation is homogeneous and linear, as in § 194. They are, of course, included in the systems which have just been discussed. But the theory can be set out differently; and the practical process of integration is arranged in a different form.

We take the systems of equations

$$A_1 = A_1(z) = a_{11} p_1 + a_{12} p_2 + \ldots + a_{1n} p_n = 0,$$
$$A_2 = A_2(z) = a_{21} p_1 + a_{22} p_2 + \ldots + a_{2n} p_n = 0,$$
$$\dots\dots\dots\dots\dots\dots\dots\dots\dots\dots\dots\dots\dots\dots\dots$$
$$A_r = A_r(z) = a_{r1} p_1 + a_{r2} p_2 + \ldots + a_{rn} p_n = 0.$$

These r equations, which occur initially, are linearly independent of one another; and the coefficients a_{ij} of the derivatives are functions of the independent variables alone, so that they do not contain z.

It is obvious that any function z of the independent variables, which satisfies all the equations of the system, must also satisfy all the equations

$$A_i\{A_j(z)\} - A_j\{A_i(z)\} = 0,$$

for all the combinations $i, j = 1, \ldots, r$. This last typical equation is

$$\sum_{s=1}^{n} [A_i(a_{js}) - A_j(a_{is})]\, p_s = 0.$$

It is precisely the Jacobian condition of coexistence of the equations $A_i = 0$ and $A_j = 0$, expressed by the earlier relation

$$(A_i, A_j) = 0;$$

and we therefore shall use it in the new form which, as will be observed on the completion of the operations A, leads to an equation of the same character as each of the equations in the original system.

As before, we have three alternatives. It may happen that the condition is satisfied identically, so that the relation

$$A_i(a_{js}) - A_j(a_{is}) = 0$$

is satisfied for all values of s; no new equation arises. It may happen that the condition is satisfied, not identically, but only as a linear combination of the original equations; still, no new equation arises. It may happen that the condition is not satisfied, either identically or in virtue of the original equations; it must be satisfied and therefore, in this event, it is a new equation which must be associated with the original system.

This result, in its alternative forms, holds for every combination $i, j, = 1, \ldots, r$. Should any new equations thus arise, we may denote them by

$$A_{r+1} = 0, \ldots, A_{r+t} = 0.$$

Then each of these equations must satisfy the conditions of coexistence, alike with the original equations and with one another. These conditions may be satisfied either identically, or in virtue of the now augmented system, or they may produce new equations. In the last event, we associate them with the augmented system; and we proceed to consider the further conditions as before.

Ultimately, we obtain a system, for which all the conditions of coexistence of its constituents are satisfied. It is then said to be *complete*; and every constituent is homogeneous and linear. When thus complete, we may denote it by

$$A_1 = 0, \ldots, A_m = 0,$$

where (under the construction of the complete system) the m equations are linearly independent of one another.

If $m > n$, or if $m = n$, the only possible inference from these linearly independent equations is

$$p_1 = 0, \ldots, p_n = 0;$$

that is,

$$z = \text{constant.}$$

The original equations then do not possess any effective common integral.

We therefore need only consider the case when $m < n$, an assumption which accordingly will now be made.

233. We can change the complete system, in any manner, by linear combinations which leave the equations linearly independent; we proceed to prove that the modified system still will be complete. For take such a modified system

$$B_s = \mu_{s1}A_1 + \mu_{s2}A_2 + \dots + \mu_{sm}A_m, \quad (s = 1, \dots, m),$$

where the determinant of the multipliers μ is different from zero; manifestly, our system is

$$B_1 = 0, \dots, B_m = 0.$$

As regards the modified system, we must have

$$B_r\{B_t(z)\} - B_t\{B_r(z)\} = 0,$$

for all the combinations $r, t, = 1, \dots, m$, either identically, or in virtue of the equations of the system, if it is complete. But

$$B_r\{B_t(z)\} - B_t\{B_r(z)\}$$

$$= \sum_{i=1}^{m} \left[\mu_{ri} A_i \left\{ \sum_{j=1}^{m} \mu_{tj} A_j(z) \right\} \right] - \sum_{j=1}^{m} \left[\mu_{tj} A_j \left\{ \sum_{i=1}^{m} \mu_{ri} A_i(z) \right\} \right]$$

$$= \sum_{i=1}^{m} \sum_{j=1}^{m} \left[\mu_{ri}\mu_{tj} \left(A_i\{A_j(z)\} - A_j\{A_i(z)\} \right) \right.$$

$$\left. + \mu_{ri} A_j A_i(\mu_{tj}) - \mu_{tj} A_i A_j(\mu_{ri}) \right];$$

the quantity $A_i\{A_j(z)\} - A_j\{A_i(z)\}$ is a linear combination of the quantities $A_1, \dots, A_m$, and so the right-hand side is a linear combination of these quantities A. It therefore is also a linear combination of the equivalent quantities B, and so it vanishes with these quantities; that is, the system of equations

$$B_1 = 0, \dots, B_m = 0,$$

also is complete.

Accordingly, we may resolve the original complete system $A_1 = 0, \dots, A_m = 0$, so as to express m of the n derivatives p in terms of the remainder; let the result be

$$C_1(z) = p_1 + c_{11}p_{m+1} + \dots + c_{1k}p_n = 0,$$

$$C_2(z) = p_2 + c_{21}p_{m+1} + \dots + c_{2k}p_n = 0,$$

$$\cdots\cdots\cdots\cdots\cdots\cdots\cdots\cdots\cdots\cdots\cdots\cdots\cdots\cdots\cdots$$

$$C_m(z) = p_m + c_{m1}p_{m+1} + \dots + c_{mk}p_n = 0,$$

where $k = n - m$, and the coefficients c are functions of the independent variables. In this form, the system of equations $C = 0$ is complete; and so all the conditions of coexistence

$$C_r \{C_s(z)\} - C_s \{C_r(z)\} = 0,$$

for the combinations $r, s, = 1, \ldots, m$ must be satisfied. Now

$$C_r \{C_s(z)\} - C_s \{C_r(z)\} = \overset{\mu}{\underset{t=1}{\Sigma}} \{C_r(c_{st}) - C_s(c_{rt})\} p_{m+t}.$$

The right-hand side is to vanish, and manifestly it does not vanish in virtue of the equations

$$C_1 = 0, \ldots, C_m = 0,$$

for it does not involve $p_1, \ldots, p_m$; it therefore must vanish identically, and so we have

$$C_r(c_{st}) - C_s(c_{rt}) = 0,$$

for all the combinations $r, s = 1, \ldots, m$, and for $t = 1, \ldots, n - m$.

When expressed in this form, the system is often called a *complete Jacobian system*. There are various ways of constructing the integral; one of them is as follows*.

234. We begin with the equation

$$p_1 + c_{11} p_{m+1} + \ldots + c_{1\mu} p_n = 0,$$

where $\mu = n - m$. The subsidiary equations (§ 194) are the set

$$\frac{dx_1}{1} = \frac{dx_2}{0} = \ldots = \frac{dx_m}{0} = \frac{dx_{m+1}}{c_{11}} = \ldots = \frac{dx_n}{c_{1\mu}}.$$

Let the complete set of the integrals of these equations be denoted by

$$x_2, \ldots, x_m, y_1, \ldots, y_\mu;$$

the first $m - 1$ of these obviously belong to the set, and the remaining μ are functionally independent of one another. Moreover, each of the quantities y is such that

$$\frac{\partial y_s}{\partial x_1} dx_1 + \frac{\partial y_s}{\partial x_2} dx_2 + \ldots + \frac{\partial y_s}{\partial x_m} dx_m + \frac{\partial y_s}{\partial x_{m+1}} dx_{m+1} + \ldots + \frac{\partial y_s}{\partial x_n} dx_n = 0,$$

---

* Others are expounded in chapter III. of the fifth volume of my *Theory of Differential Equations*.

that is,

$$\frac{\partial y_s}{\partial x_1} + c_{11} \frac{\partial y_s}{\partial x_{m+1}} + \ldots + c_{1\mu} \frac{\partial y_s}{\partial x_n} = 0,$$

for $s = 1, \ldots, \mu$.

Having obtained these integrals of the system of equations subsidiary to the first equation, we change the independent variables so as to make $y_1, \ldots, y_\mu$, take the place of $x_{m+1}, \ldots, x_n$. The first equation becomes

$$\frac{\partial z}{\partial x_1} + \sum_{s=1}^{\mu} \frac{\partial z}{\partial y_s} \frac{\partial y_s}{\partial x_1} + \sum_{t=1}^{\mu} \left\{ c_{1t} \left(\sum_{s=1}^{\mu} \frac{\partial z}{\partial y_s} \frac{\partial y_s}{\partial x_{m+t}} \right) \right\} = 0,$$

that is,

$$\frac{\partial z}{\partial x_1} + \sum_{s=1}^{\mu} \frac{\partial z}{\partial y_s} \left\{ \frac{\partial y_s}{\partial x_1} + \Sigma c_{1t} \frac{\partial y_s}{\partial x_{m+t}} \right\} = 0,$$

and therefore

$$\frac{\partial z}{\partial x_1} = 0,$$

when z is expressed in terms of $x_1, x_2, \ldots, x_m, y_1, \ldots, y_\mu$. Thus, in the changed form, z will not explicitly involve x_1.

Further, for the transformed expression of any of the later equations (say $C_r = 0$), we have

$$\frac{\partial z}{\partial x_r} + \sum_{t=1}^{\mu} \left\{ c_{rt} \left(\sum_{s=1}^{\mu} \frac{\partial z}{\partial y_s} \frac{\partial y_s}{\partial x_{m+t}} \right) \right\} = 0,$$

that is,

$$\frac{\partial z}{\partial x_r} + \sum_{s=1}^{\mu} k_{rs} \frac{\partial z}{\partial y_s} = 0,$$

where the actual value of k_{rs} is given by

$$k_{rs} = \sum_{t=1}^{\mu} c_{rt} \frac{\partial y_s}{\partial x_{m+t}};$$

and we suppose that all the coefficients k are expressed in terms of the new set of variables $x_1, x_2, \ldots, x_m, y_1, \ldots, y_\mu$. Thus the whole original system of equations, after the transformation of the variables, is

$$C_1' = 0 = \frac{\partial z}{\partial x_1},$$

$$C_r' = 0 = \frac{\partial z}{\partial x_1} + \sum_{s=1}^{\mu} k_{rs} \frac{\partial z}{\partial y_s}, \quad (r = 2, \ldots, m),$$

the system being still complete. Among the conditions for completeness are those which bind the first equation with each of the remaining equations in turn; and these require that, for all values of s in the equation $C_r' = 0$, as well as for all the values of r, we must have

$$\frac{\partial k_{rs}}{\partial x_1} = 0.$$

In other words, no one of the coefficients k_{rs} involves explicitly the variable x_1; and so we now have the system of equations

$$C_r' = 0, \quad (r = 2, \ldots, m),$$

which still is a complete system, the independent variables now being $x_2, \ldots, x_m, y_1, \ldots, y_\mu$. Every integral of the original system is an integral of this later system; and it does not explicitly involve the variable x_1, when $x_2, \ldots, x_m, y_1, \ldots, y_\mu$, are taken to be the independent variables.

235. Accordingly, we take the complete system

$$C_2' = 0, \ldots, C_m' = 0,$$

involving $n - 1$ independent variables and containing $m - 1$ constituent equations. We proceed as before and attain a stage at which every integral is an integral of a new complete system

$$C_3'' = 0, \ldots, C_m'' = 0,$$

involving $n - 2$ independent variables and containing $m - 2$ constituent equations.

And so on, from stage to stage. At the end, we have a single equation

$$\frac{\partial z}{\partial x_m} + \vartheta_1 \frac{\partial z}{\partial \theta_1} + \ldots + \vartheta_\mu \frac{\partial z}{\partial \theta_\mu} = 0,$$

in $\mu + 1$ independent variables. Every integral of the original system is an integral of this equation. We know (§ 194) that this equation has μ, ($= n - m$), independent integrals, and also that, if these be $z_1, \ldots, z_\mu$, every integral of this last equation is given by

$$z = \phi(z_1, \ldots, z_\mu),$$

where ϕ is an arbitrary function. We therefore infer the theorem:—

When a complete Jacobian system of homogeneous linear partial differential equations of the first order involves n independent *variables and contains m linearly independent constituents, it possesses* $n - m$ *functionally independent integrals; and, if these be denoted by* $z_1, \ldots, z_{n-m}$, *every integral of the system is inclusible in the equation*

$$z = \phi(z_1, \ldots, z_{n-m}),$$

for some form ϕ *of that otherwise arbitrary function.*

Moreover, after the preceding analysis, a process of constructing the most general integral (if any) of a given initial homogeneous linear system is obviously as follows:—

(i)   we make the system complete, and resolve it for a number of the derivatives $p_1, \ldots, p_m$, each in terms of p_{m+1}, $\ldots, p_n$;

(ii)   we obtain all the independent integrals, the simpler the better, of the first equation, and use these as independent variables to transform the original system, which then involves one variable fewer and contains one constituent fewer;

(iii) we proceed from stage to stage as in (ii), until there is only a single equation which is homogeneous and linear; and then

(iv)   the most general integral of the last and single equation is the completely comprehensive integral of the original system.

Ex. 1. Integrate the system

$$A_1 = p_1 - \frac{1+x_4}{x_5} p_2 - p_3 = 0,$$

$$A_2 = p_5 + \frac{x_4}{1+x_5} p_4 = 0.$$

We have

$$A_1(A_2) - A_2(A_1) = -\frac{1+x_4+x_5}{1+x_5}\left(-\frac{p_2}{x_5^2}\right),$$

which must be zero; it manifestly does not vanish in virtue of $A_1 = 0$ and $A_2 = 0$, so that it is a new equation. Hence we have

$$A_3 = p_2 = 0.$$

Thus

$$A_1(A_2) - A_2(A_1) = \frac{1+x_4+x_5}{x_5^2(1+x_5)} A_3 = 0,$$

and clearly

$$A_1(A_3) - A_3(A_1) = 0, \quad A_2(A_3) - A (A_2) = 0;$$

that is, the system is complete. An equivalent complete system is

$$p_2 = 0,$$

$$p_1 - p_3 = 0,$$

$$p_5 + \frac{x_4}{1+x_5} p_4 = 0.$$

The first equation is satisfied by making z any function of x_1, x_3, x_4, x_5.

The second equation is then satisfied by making z any function of x_4, x_5 and ξ, where

$$\xi = x_1 + x_3.$$

The third equation is then satisfied by making z any function of ξ and η, where η is an integral of the equation

$$\frac{dx_5}{1} = \frac{dx_4}{\dfrac{x_4}{1+x_5}},$$

that is, by taking

$$\eta = \frac{1+x_5}{x_4}.$$

Consequently the most general integral of the original system is

$$z = \phi(\xi, \eta)$$
$$= \phi\left(x_1 + x_3, \ \frac{1+x_5}{x_4}\right),$$

where ϕ is an arbitrary function.

Ex. 2. Shew that the system

$$(1 + x_4^2)p_4 + x_4 x_5 p_5 - x_3 x_4 p_3 - x_3 p_1 = 0,$$

$$x_4 x_5 p_4 + (1 + x_5^2)p_5 - x_3 x_5 p_3 - x_3 p_2 = 0,$$

is complete as expressed; and prove that its most general integral has the form

$$z = \phi(x_1 + x_3 x_4, \ x_2 + x_3 x_5, \ x_3^2 + x_3^2 x_4^2 + x_3^2 x_5^2),$$

where ϕ is an arbitrary function of its three arguments.

Ex. 3. Shew that the system

$$p_4 - p_5 = 0,$$

$$p_2 - p_1 + (x_4 - x_3)p_3 - 2\frac{x_4}{x_1}p_4 = 0,$$

has no integral involving x_3, x_4, or x_5.

Ex. 4. Obtain the most general integrals (if any) of the respective systems :

(i)
$$\left.\begin{array}{c} x_4 p_4 + (1+x_5)\, p_5 = 0 \\[2mm] p_1 - \dfrac{1+x_4}{x_5}\, p_2 - p_3 = 0 \end{array}\right\} ;$$

(ii)
$$\left.\begin{array}{c} (1+x_4)\, p_4 + x_5 p_5 = 0 \\[2mm] p_1 - \dfrac{x_4}{1+x_5}\, p_2 + \dfrac{x_4}{1+x_5}\, p_3 = 0 \end{array}\right\} ;$$

(iii)
$$\left.\begin{array}{c} x_4 p_4 - x_5 p_5 = 0 \\[1mm] x_4 x_5 p_4 p_5 + x_4 x_5 p_1 p_2 + x_5 x_1 p_1 p_5 + x_4 x_2 p_2 p_4 \\[1mm] + (x_1 + x_2 + x_3 + x_4 + x_5)\, p_3 = 0 \end{array}\right\} ;$$

(iv)
$$\left.\begin{array}{c} p_5 + \rho p_4 = 0 \\[1mm] p_1 - \rho p_2 + (x_4 - \rho x_5)\, p_3 = 0 \end{array}\right\} ,$$

where, in the last system,

$$\rho\,(1+x_5{}^2) = x_4 x_5 + i\,(1 + x_4{}^2 + x_5{}^2)^{\frac{1}{2}}.$$

MISCELLANEOUS EXAMPLES.

1. Integrate the equations :

(i) $\{m\,(x+y) - n\,(x+z)\}\, p + \{n\,(y+z) - l\,(y+x)\}\, q = l\,(z+x) - m\,(z+y);$

(ii) $p\,(z+e^x) + q\,(z+e^y) = z^2 - e^{x+y};$

(iii) $x^2\,(y-z)\, p + y^2\,(z-x)\, q = z^2\,(x-y).$

2. Form the differential equation whose complete integral is

$$x^2 + y^2 + z^2 = 2ax + 2\beta y + 2\gamma z,$$

where $a^2 + \beta^2 + \gamma^2 = a^2$, a being a given constant, and a, β, γ, otherwise arbitrary. From the differential equation, deduce the singular integral.

Illustrate the connection of the complete, general, and singular integrals by a geometrical interpretation.

3. Integrate

$$x^2 p + y^2 q = z^2 ;$$

and find the equation of the cone of the second degree, which satisfies this equation and passes through the point (1, 2, 3).

4. Obtain the primitive of the equation

$$p\,(x+y+z+q) = 1 ;$$

and discuss the nature of the integral

$$x+y+z=0.$$

5. Integrate the equation

$$(y-z) X^{\frac{1}{2}} \frac{\partial v}{\partial x} + (z-x) Y^{\frac{1}{2}} \frac{\partial v}{\partial y} + (x-y) Z^{\frac{1}{2}} \frac{\partial v}{\partial z} = 0,$$

where X, Y, Z, are the same quadratic functions of x, y, z, respectively.

Integrate it also, (i), when they are quartic functions; (ii) when they are sextic functions.

(Richelot.)

6. Prove that, if

$$u = \exp h \left(\frac{d^2}{dx^2} \right) . \exp (kx^2),$$

then

$$\frac{\partial u}{\partial h} = 4k^2 \frac{\partial u}{\partial k} + 2ku,$$

and hence that

$$u = (1 - 4hk)^{-\frac{1}{2}} \exp \left(\frac{kx^2}{1 - 4hk} \right).$$

Shew also that

$$\exp \left(h \frac{\partial^2}{\partial x \partial y} \right) . \exp (kxy) = (1 - hk)^{-1} \exp \left(\frac{kxy}{1 - hk} \right).$$

Similarly prove that

$$e^{h \frac{d^2}{dx^2}} \{xe^{-kx^2}\} = \frac{x}{(1 + 4hk)^{\frac{3}{2}}} e^{-\frac{kx^2}{1 + 4hk}}.$$

7. Integrate the equation

$$(X_1 - x_1 X_3) p_1 + (X_2 - x_2 X_3) p_2 = 1,$$

where

$$X_\mu = a_{\mu, 1} x_1 + a_{\mu\ 2} x_2 + a_{\mu, 3},$$

and the equation

(Hesse;)

$$a_1 (x_2 p_3 - x_3 p_2)^2 + a_2 (x_3 p_1 - x_1 p_3)^2 + a_3 (x_1 p_2 - x_2 p_1)^2 = 1.$$

(Schläfli.)

8. Solve the equations:

(i) $p^2 + q^2 = x^2 + xy + y^2$;

(ii) $pq = px + qy$;

(iii) $pq = py + qx$;

(iv) $p_1 p_2 p_3 + x_1 x_2 x_3 (x_1 p_1 + x_2 p_2 + x_3 p_3) = x_2 x_3 p_2 p_3 + x_3 x_1 p_3 p_1 + x_1 x_2 p_1 p_2.$

9. Find the equation of a surface which belongs at once to surfaces of revolution defined by the equation $py - qx = 0$, and to conical surfaces defined by the equation $px + qy = z$.

10. If $z = f(x, y)$ denote any solution of the equation

$$p^2 - q^2 + 2pqz = c^2 (1 + z^2)^{\frac{3}{2}},$$

the curves represented by the equation

$$\left(\frac{dy}{dx}\right)^2 - 2f(x, y)\frac{dy}{dx} - 1 = 0$$

are an orthogonal system such that the product of the curvatures at any point is constant.

If $f(x, y)$ do not contain y, the form of the function is determined by

$$f(x) = (2 + \tan^2 \theta)^{\frac{1}{2}} \tan \theta,$$

where

$$cx = 2^{\frac{3}{2}} E(2^{\frac{1}{2}} \sin \tfrac{1}{2}\theta) - 2^{\frac{1}{2}} F(2^{\frac{1}{2}} \sin \tfrac{1}{2}\theta),$$

F and E being the first and second elliptic integrals, of modulus $2^{-\frac{1}{2}}$.

11. Find the surface cutting at right angles all the spheres, which pass through a given point and have their centres on a given line passing through that point.

12. Find the surface in which the coordinates of the point, where the normal meets the plane of xy, are proportional to the corresponding coordinates of the surface.

13. Find the system of surfaces orthogonal to the curves

$$\cosh x : \cosh y : \cosh z = a : b : c.$$

14. Prove that a solution of the differential equation

$$\frac{\partial u}{\partial x} + \frac{\partial v}{\partial y} + \frac{\partial w}{\partial z} = 0$$

is

$$u = \begin{vmatrix} \phi_y, & \phi_z \\ \psi_y, & \psi_z \end{vmatrix}, \qquad v = \begin{vmatrix} \phi_z, & \phi_x \\ \psi_z, & \psi_x \end{vmatrix}, \qquad w = \begin{vmatrix} \phi_x, & \phi_y \\ \psi_x, & \psi_y \end{vmatrix},$$

where ϕ and ψ are arbitrary functions of x, y, and z.

Prove also that this is the general solution.

15. Shew that, if the simultaneous equations

$$X \frac{\partial u}{\partial x} + Y \frac{\partial u}{\partial y} + Z \frac{\partial u}{\partial z} = 0,$$

$$X' \frac{\partial u}{\partial x} + Y' \frac{\partial u}{\partial y} + Z' \frac{\partial u}{\partial z} = 0,$$

have a solution different from $u = \text{constant}$, then

$$(YZ' - Y'Z)\,dx + (ZX' - Z'X)\,dy + (XY' - X'Y)\,dz = 0$$

is reducible to an exact equation, from the integral of which such common solution may be derived.

Have the equations

$$yz\,\frac{\partial u}{\partial x}+zx\,\frac{\partial u}{\partial y}+xy\,\frac{\partial u}{\partial z}=0,$$

$$(y-z)\,\frac{\partial u}{\partial x}+(z-x)\,\frac{\partial u}{\partial y}+(x-y)\,\frac{\partial u}{\partial z}=0,$$

a common solution other than $u=$ constant?

16. Solve by Jacobi's method the equation

$$p_1+(3x_2+2x_3)\,p_2+(4x_2+5x_3)\,p_3+\{x_4+x_5\,(p_2-p_3)\}\,p_5+\frac{x_5}{p_4}\,p_5{}^2=0.$$

(Imschenetsky.)

Shew that, by generalisation of the formulæ which in the case of two independent variables are the analytical expression of the principle of duality, this equation can be transformed into one which is linear in the partial differential coefficients of the new variable; and hence integrate the above equation.

17. Solve by Jacobi's method the equation

$$x_1{}^2 p_1+x_2{}^2 p_2-2x_1 z-b\log p_2+2b\log x_1=a,$$

(Ampère, and Graindorge;)

also the equation

$$zx_1=p_1\,(p_1+p_2)+x_1 p_2\,(p_3+x_3 p_2).$$

(Imschenetsky.)

18. Obtain the complete common integral of the simultaneous equations:

$$\left.\begin{array}{l}2x_2 x_4{}^2 p_1+x_3{}^2 x_4 p_4-x_3{}^2=0\\[2pt]2x_2 p_2-x_4 p_4-1=0\\[2pt]x_2 x_4{}^2 p_3+x_1 x_3 x_4 p_4-x_1 x_3=0\end{array}\right\}.$$

(Collet.)

19. Obtain the complete common integral of the equations

$$\left.\begin{array}{l}(x_4{}^2-x_3{}^2)\,p_1-(x_1 x_3-x_2 x_4)\,p_3+(x_2 x_3-x_1 x_4)\,p_4=0\\[2pt](x_4{}^2-x_3{}^2)\,p_2+(x_2 x_3-x_1 x_4)\,p_3+(x_1 x_3-x_2 x_4)\,p_4=0\end{array}\right\};$$

and that of the equations

$$\left.\begin{array}{l}x_1 p_1-x_2 p_2+x_3 p_3-x_4 p_4=0\\[2pt]x_3 p_1+x_4 p_2-x_1 p_3-x_2 p_4=0\end{array}\right\}.$$

(Collet.)

CHAPTER X

PARTIAL DIFFERENTIAL EQUATIONS OF THE SECOND
AND HIGHER ORDERS

236. It will be assumed through practically the whole of this chapter that there are only two independent variables. The notation already used for the partial differential coefficients of the first order will be retained; and it will be convenient to introduce similar symbols r, s, t, to represent those of the second order, which are thus defined:

$$r = \frac{\partial^2 z}{\partial x^2}, \quad s = \frac{\partial^2 z}{\partial x \partial y}, \quad t = \frac{\partial^2 z}{\partial y^2}.$$

An equation is said to be of the second order*, when it includes one at least of these differential coefficients r, s, t, but none of a higher order; the quantities p and q may also enter into the equation, the general form of which will therefore be

$$F(x, y, z, p, q, r, s, t) = 0.$$

The *complete integral* of the equation is the most general relation possible between x, y, z, such that, when the value of z derived from it and the associated differential coefficients thence formed are substituted in the differential equation, the latter becomes an identity. No condition is annexed to the definition in regard to the form of the complete integral, which may involve in its expression either arbitrary constants or arbitrary functions or both.

---

* For the whole theory of partial differential equations of the second order, reference may be made to the sixth volume of my *Theory of Differential Equations*, and to the treatise by Goursat, *Leçons sur l'intégration des équations aux dérivées partielles du second ordre à deux variables indépendantes*, (Paris, Hermann ; t. i., 1896, t. ii., 1898).

An *intermediate integral* is a relation in the form of a partial differential equation of the first order such that the given differential equation can be deduced from it. It does not necessarily exist as one distinct from, and derivable immediately by mere differentiation of, the complete integral; when such an integral, however, has been obtained, the application of the method of the preceding chapter will give an integral which may actually be, or may only be a particular case of, the complete integral.

A question naturally arises as to the most general form of an intermediate integral and also as to the form, which may or must be possessed by an equation of the second order having an intermediate integral. The obvious elements of generality in any equation are arbitrary functions and arbitrary constants.

When an intermediate integral is general, through the occurrence of an arbitrary function, it can contain only a single arbitrary function. As will be seen in § 239, when such an equation has the form

$$u = f(v),$$

where f is an arbitrary function, while u and v are specific functions of x, y, z, p, q, the equivalent equation has the form

$$Rr + Ss + Tt + U(rt - s^2) = V,$$

where R, S, T, U, V, do not involve derivatives of the second order.

When an intermediate integral is general, through the occurrence of arbitrary constants which are not coefficients in an arbitrary function, it cannot contain more than two independent arbitrary constants. If it has a form

$$\phi(x, y, z, p, q, a_1, \ldots) = 0,$$

we construct the derived equations

$$\frac{d\phi}{dx} = 0, \quad \frac{d\phi}{dy} = 0;$$

among these three equations we usually could eliminate two (but not more than two) constants a. There is no limitation upon the form of the deduced equation of the second order, corresponding to the limitation in the preceding case.

237. Hitherto, it has been possible only in particular cases to integrate the general equation. The most important of these cases is that in which the differential coefficients of the second order occur only in the first degree, so that the equation is linear; its most general form is then

$$Rr + Ss + Tt = V,$$

in which R, S, T, V, are functions of x, y, z, p, and q. This equation will now be discussed; but before giving the methods which have been used for its integration, it is desirable to consider some special forms which are simple and can be solved immediately; it will then be possible to exclude these cases afterwards from the general discussion.

One of the simplest cases is

$$r = f(x),$$

so that

$$\frac{\partial z}{\partial x} = \int f(x)\, dx + \phi(y),$$

where ϕ is an arbitrary function; another integration gives

$$z = \iint f(x)\, dx^2 + x\phi(y) + \psi(y),$$

where both ϕ and ψ are arbitrary.

Similarly, the integral of the equation

$$s = 0$$

is

$$z = \phi(x) + \psi(y),$$

where ϕ and ψ are arbitrary functions.

Similarly, we may integrate the equation

$$r = Mp = N,$$

where M and N are functions of x and of y respectively; it may be written

$$\frac{dp}{dx} + Mp = N,$$

y being constant for purposes of differentiation and integration with regard to x. Thus

$$p = e^{-\int M dx} \left[\int e^{\int M dx} N\, dx + \phi(y) \right],$$

where ϕ is an arbitrary function; and therefore

$$z = \int dx\, e^{-\int M dx} \left[\int e^{\int M dx}\, N\, dx + \phi\,(y) \right] + \psi\,(y),$$

ψ being an arbitrary function.

Ex. Integrate

(i) $s + Mp = N$;   (ii) $s + Mq = N$;

where M is a function of x only and N of y only.

After these simple examples, we proceed to the integration of those equations of the second order which possess an intermediate integral. We shall begin with an exposition of the customary method, originally devised by Monge ; it proves effective for some equations of the form

$$Rr + Ss + Tt + U\,(rt - s^2) = V,$$

whether U be zero or not.

Later, in §§ 251, 252, we shall give a quite different method, which proves effective for all equations (whether of this form or not) possessing intermediate integrals, and which also enables us to decide in the case of any given equation whether it does or does not possess an intermediate integral.

MONGE'S METHOD OF INTEGRATION OF THE EQUATION
$$Rr + Ss + Tt = V.$$

238. Monge's method consists in a certain process for the discovery of either one or two intermediate integrals of the form

$$u = f\,(v),$$

where u and v are functions of x, y, z, p, q, and f is some arbitrary functional symbol; there is thus implied in the method a tacit assumption that the differential equation admits of such an integral. It is therefore in the first place proper to enquire whether this assumption is justifiable in the general case and, if it should prove not to be so, to indicate how the general equation must be limited so that the assumption may be fairly made. For this purpose, it will be sufficient to proceed from the supposed intermediate integral and obtain the corresponding differential equation.

239. Since $u = f(v)$, and u and v are functions of x, y, z, p, q, we have

$$\frac{\partial u}{\partial x} + p\frac{\partial u}{\partial z} + r\frac{\partial u}{\partial p} + s\frac{\partial u}{\partial q} = \frac{df}{dv}\left(\frac{\partial v}{\partial x} + p\frac{\partial v}{\partial z} + r\frac{\partial v}{\partial p} + s\frac{\partial v}{\partial q}\right),$$

and $$\frac{\partial u}{\partial y} + q\frac{\partial u}{\partial z} + s\frac{\partial u}{\partial p} + t\frac{\partial u}{\partial q} = \frac{df}{dv}\left(\frac{\partial v}{\partial y} + q\frac{\partial v}{\partial z} + s\frac{\partial v}{\partial p} + t\frac{\partial v}{\partial q}\right).$$

Eliminating the quantity $\frac{df}{dv}$ between these two equations we find, as the equivalent differential equation freed from the arbitrary function,

$$rR_1 + sS_1 + tT_1 + U_1(rt - s^2) = V_1 \quad\ldots\ldots\ldots\ldots(1),$$

where R_1, S_1, T_1, U_1, V_1, are given by the relations

$$R_1 = \left(\frac{u,\ v}{p,\ y}\right) + q\left(\frac{u,\ v}{p,\ z}\right),$$

$$S_1 = \left(\frac{u,\ v}{q,\ y}\right) + q\left(\frac{u,\ v}{q,\ z}\right) + \left(\frac{u,\ v}{x,\ p}\right) + p\left(\frac{u,\ v}{x,\ p}\right),$$

$$T_1 = \left(\frac{u,\ v}{x,\ p}\right) + p\left(\frac{u,\ v}{z,\ q}\right),$$

$$U_1 = \left(\frac{u,\ v}{p,\ q}\right),$$

$$V_1 = q\left(\frac{u,\ v}{z,\ x}\right) + p\left(\frac{u,\ v}{y,\ z}\right) + \left(\frac{u,\ v}{y,\ x}\right),$$

the symbols $\left(\dfrac{u,\ v}{x,\ y}\right)$, $\ldots\ldots$ denoting, as usual, $\dfrac{\partial u}{\partial x}\dfrac{\partial v}{\partial y} - \dfrac{\partial u}{\partial y}\dfrac{\partial v}{\partial x}$, $\ldots\ldots$

If this differential equation of the second order be the same as the original equation, we must have

$$U_1 = 0,$$

and $$\frac{R_1}{R} = \frac{S_1}{S} = \frac{T_1}{T} = \frac{V_1}{V},$$

which are four equations in all. Now when

$$Rr + Ss + Tt = V \ldots\ldots\ldots\ldots\ldots\ldots(2)$$

is looked upon as the equation to be solved, these four equations just obtained will be equations satisfied by the quantities u and v from which the intermediate integral of (2) may be constructed.

But only two equations are necessary to determine the dependent variables u and v as functions of their independent variables; they may be therefore considered as given by any two of the equations though, in practice, these might prove too difficult to solve. When these values are substituted in the remaining two equations, the latter must become identities; and they will in this state involve the functions R, S, T, and V, of the original differential equation. There *will thus be two relations among these functions of x, y, z, p, q, which must be identically satisfied in order that the differential equation* (2) *may have an intermediate integral of the form*

$$u = f(v).$$

240. There is an important deduction from this result to be noted, though not affecting our present aim: it would be useless to seek an integral of the assumed intermediate form for any differential equation which is not of the form

$$Rr + Ss + Tt + U(rt - s^2) = V.$$

Just as in the particular case when $U = 0$, which has been already considered, it may be proved that a differential equation of this form can have an intermediate integral of the proposed type only when two identical relations among the coefficients R, S, T, U, V are satisfied.

Ex. When there are three independent variables, these may be conveniently denoted by x_1, x_2, x_3, and the corresponding differential coefficients of z by p_1, p_2, p_3. Prove that, if every first minor of the determinant

$$\begin{vmatrix} \dfrac{\partial \phi}{\partial p_1}, & \dfrac{\partial \phi}{\partial p_2}, & \dfrac{\partial \phi}{\partial p_3} \\[2mm] \dfrac{\partial \psi}{\partial p_1}, & \dfrac{\partial \psi}{\partial p_2}, & \dfrac{\partial \psi}{\partial p_3} \\[2mm] \dfrac{\partial \chi}{\partial p_1}, & \dfrac{\partial \chi}{\partial p_2}, & \dfrac{\partial \chi}{\partial p_3} \end{vmatrix}$$

(ϕ, ψ, χ being functions of z, x_1, x_2, x_3, p_1, p_2, p_3) vanish, then the equation

$$F(\phi, \psi, \chi) = 0,$$

where F is an arbitrary function, will lead to a differential equation of the second order of the form

$$R_1 \frac{\partial^2 z}{\partial x_1^2} + R_2 \frac{\partial^2 z}{\partial x_2^2} + R_3 \frac{\partial^2 z}{\partial x_3^2} + R_{12} \frac{\partial^2 z}{\partial x_1 \partial x_2} + R_{23} \frac{\partial^2 z}{\partial x_2 \partial x_3} + R_{31} \frac{\partial^2 z}{\partial x_3 \partial x_1} = V,$$

where R_1, R_2, ..., R_{31}, V, are functions of the variables and the first differential coefficients of z only, and that the coefficients R satisfy the relation

$$R_1 R_{23}{}^2 + R_2 R_{31}{}^2 + R_3 R_{12}{}^2 - 4R_1 R_2 R_3 - R_{12} R_{23} R_{31} = 0.$$

Information on this class of equations will be found in Euler, *Inst. Calc. Int.*, t. III. p. 448, and Legendre, *Mémoires de l'Académie des Sciences*, 1787, p. 323.

241. It therefore follows that we may consider

$$Rr + Ss + Tt + U(rt - s^2) = V$$

as the most general equation having an intermediate integral of the specified type. The linear equation of § 238 is included in this form, being given by the particular case when $U = 0$.

We now assume that the relations between the quantities R, S, T, U, and V, necessary for the possession of an intermediate integral of the assumed form, are satisfied; and we proceed to deduce this integral. We have always

$$dp = r\,dx + s\,dy,$$
$$dq = s\,dx + t\,dy;$$

when we substitute in the above general equation the values of r and t derived from these equations, it takes the form

$$R\,dp\,dy + T\,dq\,dx + U\,dp\,dq - V\,dx\,dy$$
$$= s(R\,dy^2 - S\,dx\,dy + T\,dx^2 + U\,dp\,dx + U\,dq\,dy).$$

Now let $u = a$ and $v = b$

(where a and b are arbitrary constants) be two integrals of the equations

$$R\,dp\,dy + T\,dq\,dx + U\,dp\,dq - V\,dx\,dy = 0,$$
$$R\,dy^2 + T\,dx^2 + U\,dp\,dx + U\,dq\,dy = S\,dx\,dy,$$
$$dz = p\,dx + q\,dy,$$

u and v being therefore functions of x, y, z, p, and q.

Hence we have

$$\left(\frac{\partial u}{\partial x} + p\frac{\partial u}{\partial z}\right)dx + \left(\frac{\partial u}{\partial y} + q\frac{\partial u}{\partial z}\right)dy + \frac{\partial u}{\partial p}\,dp + \frac{\partial u}{\partial q}\,dq = 0,$$

and $$\left(\frac{\partial v}{\partial x} + p\frac{\partial v}{\partial z}\right)dx + \left(\frac{\partial v}{\partial y} + q\frac{\partial v}{\partial z}\right)dy + \frac{\partial v}{\partial p}\,dp + \frac{\partial v}{\partial q}\,dq = 0,$$

which must be equivalent to the equation of which $u = a$ and $v = b$ are the integrals. Now solving these for dp and dq, and using the symbols of § 239, we find

$$- U_1 dp = T_1 dx + \left\{ \left(\frac{u, v}{y, q} \right) + \left(\frac{u, v}{z, q} \right) q \right\} dy,$$

$$- U_1 dq = R_1 dy + \left\{ \left(\frac{u, v}{p, x} \right) + \left(\frac{u, v}{p, z} \right) p \right\} dx,$$

and therefore

$$- U_1 dp \, dx - U_1 dq \, dy$$

$$= T_1 dx^2 + R_1 dy^2 + \left\{ \left(\frac{u, v}{y, q} \right) + \left(\frac{u, v}{z, q} \right) q + \left(\frac{u, v}{p, x} \right) + \left(\frac{u, v}{p, z} \right) p \right\} dx \, dy$$

$$= T_1 dx^2 + R_1 dy^2 - S_1 dx \, dy \, ;$$

and similarly we obtain

$$(U_1 dp + T_1 dx)(U_1 dq + R_1 dy) = (U_1 V_1 + R_1 T_1) \, dx \, dy,$$

or

$$R_1 dp \, dy + T_1 dq \, dx + U_1 dp \, dq - V_1 dx \, dy = 0.$$

These being identical with the former equations, we have

$$\frac{R_1}{R} = \frac{T_1}{T} = \frac{U_1}{U} = \frac{V_1}{V} = \frac{S_1}{S} ,$$

and therefore the equation to be solved becomes

$$R_1 r + S_1 s + T_1 t + U_1 (rt - s^2) = V_1.$$

But we already know the solution of this equation because it was derived from an intermediate integral ; and this integral is

$$u = f(v),$$

which is therefore an intermediate integral as required.

We thus derive the integral by making one of the functions deduced from the two subsidiary equations an arbitrary function of the other.

242. Let us consider in particular the case of the linear equation when $U = 0$; the subsidiary equations are now

$$R dy^2 + T dx^2 - S dx \, dy = 0,$$

$$R dp \, dy + T dq \, dx = V dx \, dy.$$

The former of these is of the second degree; it can, in general, be resolved into two distinct equations of the first degree.

Since the necessary conditions for the existence of an intermediate integral are supposed to be satisfied, it follows that one at least of the equations of the first degree will, when combined with

$$R\,dp\,dy + T\,dq\,dx = V\,dx\,dy,$$

and with $dz = p\,dx + q\,dy$ if necessary, lead to an integral system which determines u and v; and there will thus be obtained an intermediate integral of the form

$$u = f(v).$$

It may happen that each of the two equations of the first degree, similarly treated, will lead to integral systems of the desired form: and there will then be obtained two intermediate integrals

$$u_1 = f(v_1), \quad u_2 = \phi(v_2).$$

If $S^2 = 4RT$, there will be only a single equation of the first degree equivalent to

$$R\,dy^2 + T\,dx^2 - S\,dx\,dy = 0;$$

since the necessary conditions are satisfied, this single equation will lead, by a similar process, to an intermediate integral.

243. Passing now to the more general case in which U is not zero, we may similarly prove that one intermediate integral will, and two intermediate integrals may, be derivable from the subsidiary equations, provided the conditions necessary for the existence of an intermediate integral are satisfied. Let the subsidiary equation which involves V be multiplied by a quantity λ, as yet indeterminate, and added to the other; the result is

$$R\,dy^2 + T\,dx^2 - (S + \lambda V)\,dx\,dy + U\,dp\,dx + U\,dq\,dy$$
$$+ \lambda R\,dp\,dy + \lambda T\,dq\,dx + \lambda U\,dp\,dq = 0.$$

Now this can be resolved into two linear factors so as to be equivalent to

$$(R\,dy + kT\,dx + mU\,dp)\left(dy + \frac{1}{k}\,dx + \frac{\lambda}{m}\,dq\right) = 0,$$

provided the quantities k, m, λ, be such as to make the coefficients of the several terms in the expanded product the same as before.

Applying this condition, we find that the relations to be satisfied by these quantities are

$$kT + \frac{1}{k}R = -(S + \lambda V), \quad \frac{1}{m}\lambda R = U,$$

$$kT\frac{\lambda}{m} = \lambda T, \quad mU = \lambda R, \quad \frac{m}{k}U = U;$$

these are all satisfied by

$$m = k = \lambda\frac{R}{U},$$

provided λ be determined by the equation

$$\lambda^2(RT + UV) + \lambda US + U^2 = 0.$$

Let the two values of λ furnished by this equation be λ_1 and λ_2, which will be unequal except when

$$S^2 = 4(RT + UV).$$

The two subsidiary equations may be replaced by the two equations each resoluble into linear factors when the values of k, m, λ, are therein substituted, which two equations, after a slight reduction, may be written:

$$(Udy + \lambda_1 Tdx + \lambda_1 Udp)(Udx + \lambda_1 Rdy + \lambda_1 Udq) = 0,$$
$$(Udy + \lambda_2 Tdx + \lambda_2 Udp)(Udx + \lambda_2 Rdy + \lambda_2 Udq) = 0.$$

To obtain the functions u and v, from which an intermediate integral may be constructed, we must combine in pairs a factor from the first with a factor from the second. But of the four possible combinations two must be excluded, viz., that obtained by combining the first factors in these equations, for it would lead to a result

$$Udy = 0,$$

which obviously would not furnish any solution: and that obtained by combining the second factors in these equations, for it would lead to a result

$$Udx = 0,$$

which obviously also would furnish no solution. Hence the equations may again be replaced by the two pairs of equations

$$\left.\begin{array}{l}Udy + \lambda_1 Tdx + \lambda_1 Udp = 0\\Udx + \lambda_2 Rdy + \lambda_2 Udq = 0\end{array}\right\},$$

and
$$U dx + \lambda_1 R dy + \lambda_1 U dq = 0 \\ U dy + \lambda_2 T dx + \lambda_2 U dp = 0 \Big\}.$$

From one of the pairs we shall have two integrals of the form $u = a$ and $v = b$; and therefore also through that pair we obtain an intermediate integral.

And it may happen, as in the simpler case of § 242, that we can obtain an intermediate integral through each of the pairs of equations of the first degree.

These two integrals, which may be denoted as before by

$$u_1 = f(v_1), \quad u_2 = \phi(v_2),$$

are intermediate integrals of the original differential equation, and are distinct except when

$$S^2 = 4(RT + UV),$$

when only a single intermediate integral is obtainable.

244. We may now proceed further in the integration for either the linear equation of § 242 or the more general form of § 243. Taking the intermediate integral obtained if there be only one, or either of the intermediate integrals if there be two, we have a differential equation of the first order; the complete integral (and the associated integrals) of this can be obtained by the methods of Chap. IX. This integral will be the final integral of the original equation.

245. In the case when there are two intermediate integrals, we may apply an important proposition (now to be proved) which will considerably shorten the further labour of deriving this final integral. This proposition may be enunciated as follows:

When we have obtained two intermediate integrals of the form

$$u_1 = f(v_1), \quad u_2 = \phi(v_2),$$

and we consider them as simultaneous equations to determine p and q as functions of x, y, and z, the values of p and q given by these equations will be such as to render

$$dz = p\, dx + q\, dy$$

integrable.

Assuming this proposition established, we have therefore merely to solve the two intermediate integrals as simultaneous equations in p and q, to substitute the values of p and q thence derived in

$$dz = p\,dx + q\,dy,$$

and to integrate. The result will be the final integral.

246. We now proceed to establish the proposition enunciated in §245. Let $F = 0$ and $\Phi = 0$ respectively denote these integrals, so that $F = u_1 - f(v_1)$, $\Phi = u_2 - \phi(v_2)$; and, first, let $F = 0$ be a solution of the equation

$$Rr + Ss + Tt + U(rt - s^2) = V.$$

We have only the single equation $F = 0$, which is not sufficient to enable us to express r, s, and t, each as functions of x, y, z, p, and q; we can express any two of them in terms of the third and of quantities explicitly independent of them. When these values are substituted in the differential equation, the latter will contain one set of terms involving this second differential coefficient of the dependent variable and another set not involving it; and the equation is to be satisfied identically without regard to this differential coefficient. Now since $F = 0$, we have

$$\frac{\partial F}{\partial x} + \frac{\partial F}{\partial z}\,p + \frac{\partial F}{\partial p}\,r + \frac{\partial F}{\partial q}\,s = 0,$$

$$\frac{\partial F}{\partial y} + \frac{\partial F}{\partial z}\,q + \frac{\partial F}{\partial p}\,s + \frac{\partial F}{\partial q}\,t = 0\,;$$

when for brevity we replace $\dfrac{\partial F}{\partial x} + \dfrac{\partial F}{\partial z}\,p$ by F_x and $\dfrac{\partial F}{\partial y} + \dfrac{\partial F}{\partial z}\,q$ by F_y, these equations give

$$\frac{\partial F}{\partial p}\,r = -\frac{\partial F}{\partial q}\,s - F_x,$$

$$\frac{\partial F}{\partial q}\,t = -\frac{\partial F}{\partial p}\,s - F_y.$$

Let these values of r and t be substituted in the differential equation; it becomes

$$RF_x\frac{\partial F}{\partial q} + TF_y\frac{\partial F}{\partial p} + V\frac{\partial F}{\partial p}\frac{\partial F}{\partial q} - UF_xF_y$$

$$+ \left\{ R\left(\frac{\partial F}{\partial q}\right)^2 - S\frac{\partial F}{\partial p}\frac{\partial F}{\partial q} + T\left(\frac{\partial F}{\partial p}\right)^2 - UF_y\frac{\partial F}{\partial q} - UF_x\frac{\partial F}{\partial p} \right\}s = 0.$$

This must be satisfied identically without regard to s; and therefore the coefficient of s, and the term independent of s, must both vanish. If this were not so, the equation would determine s (and therefore also r and t) as functions of x, y, z, p, and q. This result, as we know, cannot be deduced from the single equation $F = 0$.

Hence we have

$$RF_x \frac{\partial F}{\partial q} + TF_y \frac{\partial F}{\partial p} + V \frac{\partial F}{\partial p}\frac{\partial F}{\partial q} - UF_x F_y = 0,$$

$$R\left(\frac{\partial F}{\partial q}\right)^2 - S\frac{\partial F}{\partial p}\frac{\partial F}{\partial q} + T\left(\frac{\partial F}{\partial p}\right)^2 - UF_y \frac{\partial F}{\partial q} - UF_x \frac{\partial F}{\partial p} = 0.$$

The same equations will be satisfied when we replace F by Φ; and we may therefore consider F and Φ as the solutions of the equations

$$R\Theta_x \frac{\partial \Theta}{\partial q} + T\Theta_y \frac{\partial \Theta}{\partial p} + V \frac{\partial \Theta}{\partial p}\frac{\partial \Theta}{\partial q} - U\Theta_x \Theta_y = 0,$$

$$R\left(\frac{\partial \Theta}{\partial q}\right)^2 - S\frac{\partial \Theta}{\partial p}\frac{\partial \Theta}{\partial q} + T\left(\frac{\partial \Theta}{\partial p}\right)^2 - U\Theta_y \frac{\partial \Theta}{\partial q} - U\Theta_x \frac{\partial \Theta}{\partial p} = 0.$$

247. We must now consider two cases.

(i) The linear equation, when $U = 0$.

Let ξ_1 and ξ_2 be the roots of

$$R\xi^2 - S\xi + T = 0,$$

so that the second equation becomes

$$\left(\frac{\partial \Theta}{\partial q} - \xi_1 \frac{\partial \Theta}{\partial p}\right)\left(\frac{\partial \Theta}{\partial q} - \xi_2 \frac{\partial \Theta}{\partial p}\right) = 0.$$

We may therefore write

$$\frac{\partial F}{\partial q} - \xi_1 \frac{\partial F}{\partial p} = 0,$$

$$\frac{\partial \Phi}{\partial q} - \xi_2 \frac{\partial \Phi}{\partial p} = 0,$$

thus associating ξ_1 with F and ξ_2 with Φ. The first equation, on dividing out by $\frac{\partial \Phi}{\partial p}$, becomes

$$R\xi\Theta_x + T\Theta_y + V\xi \frac{\partial \Theta}{\partial p} = 0,$$

and therefore

$$R\xi_1 F_x + TF_y + V\xi_1 \frac{\partial F}{\partial p} = 0.$$

But $T = R\xi_1\xi_2$, and the last may therefore be written

$$F_x + \xi_2 F_y + \frac{V}{R}\frac{\partial F}{\partial p} = 0.$$

Similarly

$$\Phi_x + \xi_1 \Phi_y + \frac{V}{R}\frac{\partial \Phi}{\partial p} = 0.$$

From the last two equations we have

$$F_x \frac{\partial \Phi}{\partial p} - \Phi_x \frac{\partial F}{\partial p} = \xi_1 \Phi_y \frac{\partial F}{\partial p} - \xi_2 F_y \frac{\partial \Phi}{\partial p} = \Phi_y \frac{\partial F}{\partial q} - F_y \frac{\partial \Phi}{\partial q},$$

and therefore

$$F_x \frac{\partial \Phi}{\partial p} - \Phi_x \frac{\partial F}{\partial p} + F_y \frac{\partial \Phi}{\partial q} - \Phi_y \frac{\partial F}{\partial q} = 0,$$

which is the condition (§ 208) to be satisfied by the two functions F and Φ in order that the values of p and q, derived from $F = 0 = \Phi$ as simultaneous equations, should render

$$dz = p\,dx + q\,dy$$

integrable. This proves the proposition for the case of $U = 0$.

(ii) The general form, when U is not zero.

We now proceed as in § 243; the first equation in Θ is multiplied by a quantity λ, given by

$$\lambda^2(RT + UV) - \lambda US + U^2 = 0,$$

and is added to the second; the resulting equation is resolved into factors for each of the values of λ; and the linear factors are combined as before, giving two pairs that may be retained. These are, if λ_1 and λ_2 be the two roots,

$$\left.\begin{aligned}
\lambda_1 T \frac{\partial F}{\partial p} &= \lambda_1 U F_x + U \frac{\partial F}{\partial q} \\
\lambda_2 R \frac{\partial F}{\partial q} &= \lambda_2 U F_y + U \frac{\partial F}{\partial p}
\end{aligned}\right\},$$

and

$$\left.\begin{aligned}
\lambda_2 T \frac{\partial \Phi}{\partial p} &= \lambda_2 U \Phi_x + U \frac{\partial \Phi}{\partial q} \\
\lambda_1 R \frac{\partial \Phi}{\partial q} &= \lambda_1 U \Phi_y + U \frac{\partial \Phi}{\partial p}
\end{aligned}\right\}.$$

From the first and third of these equations, we have

$$F_x \frac{\partial \Phi}{\partial p} - \Phi_x \frac{\partial F}{\partial p} = \frac{1}{\lambda_2} \frac{\partial F}{\partial p} \frac{\partial \Phi}{\partial q} - \frac{1}{\lambda_1} \frac{\partial F}{\partial q} \frac{\partial \Phi}{\partial p},$$

and from the second and fourth,

$$F_y \frac{\partial \Phi}{\partial q} - \Phi_y \frac{\partial F}{\partial q} = -\frac{1}{\lambda_2} \frac{\partial F}{\partial p} \frac{\partial \Phi}{\partial q} + \frac{1}{\lambda_1} \frac{\partial F}{\partial q} \frac{\partial \Phi}{\partial p};$$

and therefore

$$F_x \frac{\partial \Phi}{\partial p} - \Phi_x \frac{\partial F}{\partial p} + F_y \frac{\partial \Phi}{\partial q} - \Phi_y \frac{\partial F}{\partial q} = 0.$$

This shews that, for the more general form of the equation, when $F = 0 = \Phi$ are treated as simultaneous equations, the values of p and q thence derived are such as to render

$$dz = p\, dx + q\, dy$$

integrable.

Hence the proposition is proved in general. When these values of p and q are substituted, the integral of the resulting equation is the final integral of the proposed differential equation; it will involve in its expression either implicitly or explicitly the two arbitrary functions which occur in the two intermediate integrals.

248. The statement of the method of solution, as derived from the preceding investigation, is contained in the following Rules.

RULE I. When the equation

$$Rr + Ss + Tt = V$$

is integrable by this rule, we transform it by the equations

$$dp = r\, dx + s\, dy,$$
$$dq = s\, dx + t\, dy,$$

into

$$Rdp\,dy + Tdq\,dx - Vdx\,dy = s\,(R\,dy^2 - S\,dx\,dy + T\,dx^2);$$

we resolve the equation

$$Rdy^2 - Sdx\,dy + Tdx^2 = 0$$

into the two equations

$$dy - \xi_1 dx = 0, \quad dy - \xi_2 dx = 0.$$

From one of these linear equations and from the equation

$$Rdp\,dy + Tdq\,dx - Vdx\,dy = 0,$$

combined if necessary with $dz = pdx + qdy$, we may be able to obtain two integrals $u_1 = a_1$, $v_1 = b_1$; then

$$u_1 = f_1(v_1),$$

where f_1 is an arbitrary function, is an intermediate integral. From the other linear equation, combined with the same equations, we may be able to obtain another pair of integrals $u_2 = a_2$, $v_2 = b_2$; in that case, $u_2 = f_2(v_2)$ is another intermediate integral, f_2 being arbitrary.

To deduce the final integral, we integrate the intermediate integral, if only one has been obtainable, by the methods which apply to differential equations of the first order. If there be two intermediate integrals, we resolve them as equations giving p and q, and we substitute in

$$dz = pdx + qdy;$$

when this is integrated, it gives the complete integral.

RULE II. When the equation

$$Rr + Ss + Tt + U(rt - s^2) = V$$

is integrable by this rule, we either may be able to obtain two integrals $u_1 = a_1$ and $v_1 = b_1$ of the equations

$$\left.\begin{array}{l} Udy + \lambda_1 Tdx + \lambda_1 Udp = 0 \\ Udx + \lambda_2 Rdy + \lambda_2 Udq = 0 \end{array}\right\},$$

or may be able to obtain two integrals $u_2 = a_2$ and $v_2 = b_2$ of

$$\left.\begin{array}{l} Udx + \lambda_1 Rdy + \lambda_1 Udq = 0 \\ Udy + \lambda_2 Tdx + \lambda_2 Udp = 0 \end{array}\right\},$$

where λ_1 and λ_2 are the roots of

$$\lambda^2(RT + UV) + \lambda US + U^2 = 0;$$

or we may be able to obtain both sets of integrals.

Then $u_1 = f_1(v_1)$ and $u_2 = f_2(v_2)$, where f_1 and f_2 are arbitrary, are intermediate integrals in the respective cases. We proceed from these exactly as in Rule I.

249. It may, however, prove not to be possible to obtain, from the two intermediate integrals, values of p and q suitable for insertion in

$$dz = p\,dx + q\,dy;$$

and in that case we may proceed to obtain the final integral by integrating one of the intermediate integrals, adopting for this purpose Charpit's method as indicated in § 206. But without actually going through the work necessary in that method to derive the additional relation between p, q, and the variables, it will be sufficient to take, as this additional relation, any particular first integral of the general system other than that which is being directly integrated; thus we may take

$$u_1 = f(v_1), \quad u_2 = a,$$

where a is an arbitrary constant. Since an arbitrary constant is a particular case of an arbitrary function, the values of p and q derived from these equations will be such as to render

$$dz = p\,dx + q\,dy$$

integrable; and the integral will involve one arbitrary function f and two arbitrary constants, viz., a and the constant of integration. This result constitutes the complete integral of the intermediate integral; the general integral may be derived by Lagrange's rule (§ 183), by converting one of the arbitrary constants into an arbitrary function of the other and eliminating this remaining constant between the equation so transformed and that deduced from it by differentiation with respect to that constant.

250. This process ceases to be effective in the case in which the roots of the quadratic in λ are equal; there is then only one system of integrals given by $u_1 = a$ and $v_1 = b$, and so there is only one intermediate integral given by

$$u_1 = f(v_1),$$

and this must be integrated. Just as before, we may avoid the use of the general method for the integration of an equation of the first order by combining the general and particular first integrals

$$u_1 = f(v_1), \quad v_1 = b.$$

The values of p and q hence derived will evidently satisfy the condition of § 208, and therefore, when substituted in the equation

$$dz = p\,dx + q\,dy,$$

will give another integral of the form

$$w_1 = c.$$

If p and q occur in w_1, they may be eliminated by means of the former equations $v_1 = b$ and $u_1 = f(b)$; so that

$$w_1 = c$$

is a complete integral of the equation, since it involves two arbitrary constants b and c. To obtain the general integral, we must make c an arbitrary function of b, and then must eliminate b between the resulting equation and that derived from it by differentiation with respect to b.

Thus in the cases, when the roots of the quadratic are unequal and when they are equal, we are led to a general integral, into the expression of which two arbitrary functions enter.

It may be noticed that the foregoing reasoning would apply equally if there had been taken, instead of the particular integral

$$u_2 = a,$$

some other particular integral such as

$$ku_2 + lv_2 = a,$$

k and l being disposable constants. This particular integral may, in fact, be taken so as to render the subsequent integration as easy as possible.

Some examples will now be given.

Ex. 1. Solve $\qquad\qquad\qquad r = a^2 t.$

Substituting for r and t in terms of s, we have

$$dp\,dy - a^2\,dx\,dq = s\,(dy^2 - a^2\,dx^2),$$

so that the subsidiary equations are

$$dy^2 - a^2\,dx^2 = 0,$$
$$dp\,dy - a^2\,dx\,dq = 0.$$

The former can be resolved into the two equations

$$dy - a\,dx = 0, \quad dy + a\,dx = 0,$$

the respective integrals of which are

$$y - ax = A, \quad y + ax = B.$$

Taking the first of these and combining it with the second of the subsidiary equations, we find that the latter becomes

$$dp - a\,dq = 0,$$

which, when integrated, gives

$$p - aq = A'.$$

Hence one intermediate integral is

$$p - aq = \phi_1\,(y - ax).$$

Taking the second equation $y + ax = B$, and proceeding in the same way, we find

$$dp + a\,dq = 0,$$

which leads to

$$p + aq = B';$$

and therefore a second intermediate integral is

$$p + aq = \phi_2\,(y + ax).$$

We now, in accordance with our rule, treat these as simultaneous equations giving the values of p and q; and we find

$$dz = \tfrac{1}{2}dx\,\{\phi_2\,(y + ax) + \phi_1\,(y - ax)\} + \frac{1}{2a}\,dy\,\{\phi_2\,(y + ax) - \phi_1\,(y - ax)\}$$

$$= \frac{(dy + a\,dx)\,\phi_2\,(y + ax)}{2a} - \frac{(dy - a\,dx)\,\phi_1\,(y - ax)}{2a},$$

which can be integrated. Let

$$\phi\,(t) = \frac{1}{2a}\int \phi_2\,(t)\,dt \text{ and } \psi\,(t) = -\frac{1}{2a}\int \phi_1\,(t)\,dt;$$

then the integral is

$$z = \phi\,(y + ax) + \psi\,(y - ax).$$

The arbitrary constant of integration may be considered as absorbed in either of the functions ϕ and ψ. Since ϕ_1 and ϕ_2 are arbitrary, ϕ and ψ are also arbitrary.

Ex. 2. Solve
$$(b+cq)^2 r - 2(b+cq)(a+cp) s + (a+cp)^2 t = 0.$$

Transforming this by the relations of § 248 we find that the subsidiary equations are

$$(b+cq)^2 dy^2 + 2(b+cq)(a+cp) dx\, dy + (a+cp)^2 dx^2 = 0,$$

$$(b+cq)^2 dp\, dy + (a+cp)^2 dq\, dx = 0.$$

The former of these gives only a single equation

$$(b+cq) dy + (a+cp) dx = 0,$$

so that only a single intermediate integral can be obtained for the equation, if it be integrable by the method. When this equation is combined with

$$dz = p\, dx + q\, dy,$$

it gives
$$a\, dx + b\, dy + c\, dz = 0,$$

so that one integral of the subsidiary equations is

$$ax + by + cz = A.$$

Eliminating the ratio $dy : dx$ between the second subsidiary equation and the modified form of the first, we have

$$(b+cq) dp = (a+cp) dq,$$

the integral of which is
$$a + cp = B(b+cq),$$

B being an arbitrary constant. Hence the intermediate integral is

$$a + cp = (b+cq)\, \phi\,(ax+by+cz).$$

This must now be integrated; Lagrange's process for linear equations may be adopted. Denoting $\phi\,(ax+by+cz)$ by ϕ, we have as the auxiliary equations

$$\frac{dx}{c} = \frac{dy}{-c\phi} = \frac{dz}{b\phi-a}.$$

From these we have
$$a\, dx + b\, dy + c\, dz = 0,$$

so that
$$ax + by + cz = C,$$

and $\phi = \phi\,(ax+by+cz) = \phi\,(C)$ is a constant.

Hence, for a second integral,

$$dy + dx\, \phi\,(C) = 0,$$

that is,
$$y + x\phi\,(C) = C'.$$

The final integral of the differential equation is therefore

$$y + x\phi(ax + by + cz) = \psi(ax + by + cz),$$

where ϕ and ψ are arbitrary functions.

It may also be exhibited in the form

$$z = x\theta(ax + by + cz) + y\chi(ax + by + cz),$$

where θ and χ are arbitrary functions.

Ex. 3. Integrate

  (i) $r + ka^2t = 2as$,

    (1) when k is not unity, (2) when k is unity;

  (ii) $x^2r + 2xys + y^2t = 0$;

  (iii) $q^2r - 2pqs + p^2t = 0$;

  (iv) $x^2r - y^2t = 0$;

  (v) $r - a^2t + 2ab(p + aq) = 0$.

Ex. 4. Integrate the equation

$$ar + bs + ct + e(rt - s^2) = h,$$

a, b, c, e, h, being constants.

The equation in λ is

$$\lambda^2(ac + eh) + \lambda eb + e^2 = 0,$$

or, if we write $\lambda m + e = 0$, the equation which determines m is

$$m^2 - bm + ac + eh = 0;$$

let m_1 and m_2 be its roots. The first system of integrals is

$$\left.\begin{array}{l} cdx + edp - m_1dy = 0 \\ ady + edq - m_2dx = 0 \end{array}\right\},$$

so that one intermediate integral is

$$cx + ep - m_1y = F(ay + eq - m_2x).$$

The second system of integrals is

$$\left.\begin{array}{l} ady + edq - m_1dx = 0 \\ cdx + edp - m_2dy = 0 \end{array}\right\},$$

and therefore a second intermediate integral is

$$cx + ep - m_2y = \Phi(ay + eq - m_1x).$$

If it were possible to resolve these intermediate equations so as to express p and q in terms of x and y, the final integral would be at once derivable; but, as this is not the case, we combine any particular integral of the second with the general integral of the first system. Thus we may take

$$cx + ep - m_2y = a,$$

and then $\qquad F\left(ay+eq-m_2 x\right)=\left(m_2-m_1\right) y+a,$

so that, if Ψ be the inverse function of F and therefore an arbitrary function, we have

$$ay+eq=m_2 x+\Psi\left\{\left(m_2-m_1\right) y+a\right\}.$$

Thus

$$e\,dz=-cx\,dx-ay\,dy+\left(m_2 y+a\right) dx+\left[m_2 x+\Psi\left\{\left(m_2-m_1\right) y+a\right\}\right] dy,$$

the integral of which is

$$ez+\tfrac{1}{2} cx^2+\tfrac{1}{2} ay^2=m_2 xy+ax+\Theta\left\{\left(m_2-m_1\right) y+a\right\}+\beta,$$

where Θ is an arbitrary function (since it is given by

$$\left(m_2-m_1\right) \Theta\left(z\right)=\int \Psi\left(z\right) dz,$$

and Ψ is arbitrary) and β is an arbitrary constant.

This is the Complete Integral; to obtain the General Integral we eliminate a between the equations

$$\left.\begin{array}{r}ez+\tfrac{1}{2}\left(cx^2+ay^2\right)=m_2 xy+ax+\Theta\left\{\left(m_2-m_1\right) y+a\right\}+\chi\left(a\right)\\[2mm]0=x+\dfrac{1}{m_2-m_1} \Psi\left\{\left(m_2-m_1\right) y+a\right\}+\chi'\left(a\right)\end{array}\right\},$$

χ denoting an arbitrary function.

Ex. 5. Solve

(i) $\quad s^2-rt=a^2$;

(ii) $\quad qr+\left(p+x\right) s+yt=-q+y\left(s^2-rt\right)$;

(iii) $\quad 2pqyr+\left(p^2 y+qx\right) s+xpt=p^2 q\left(rt-s^2\right)+xy.$

Ex. 6. Solve

$$z\left(1+q^2\right) r-2pqzs+z\left(1+p^2\right) t-z^2\left(s^2-rt\right)+1+p^2+q^2=0.$$

The equation which determines m is

$$m^2+2pqzm+p^2 q^2 z^2=0,$$

so that the two values of m are equal, the common value being $-pqz$; and the system of integrals reduces to one given by

$$z\left(1+p^2\right) dx+z^2 dp+pqz\,dy=0,$$

$$z\left(1+q^2\right) dy+z^2 dp+pqz\,dx=0.$$

The former by means of

$$dz=p\,dx+q\,dy$$

gives, after division by z,

$$dx+p\,dz+z\,dp=0,$$

the integral of which is

$$x+pz=a;$$

the second similarly leads to

$$dy + q\,dz + z\,dq = 0,$$

the integral of which is

$$y + qz = b,$$

so that the intermediate integral is

$$F(x + pz, \; y + qz) = 0,$$

where F is arbitrary.

Proceeding as indicated in § 250, we have

$$x + pz = a,$$
$$y + qz = b\,;$$

and therefore

$$z\,dz = pz\,dx + qz\,dy$$
$$= (a - x)\,dx + (b - y)\,dy$$

the integral of which is

$$(x - a)^2 + (y - b)^2 + z^2 = c^2.$$

A general integral is found, as there explained, by eliminating c between the equations

$$\{x - \phi(c)\}^2 + \{y - \psi(c)\}^2 + z^2 = c^2,$$

and

$$\{x - \phi(c)\}\,\phi'(c) + \{y - \psi(c)\}\,\psi'(c) + c = 0,$$

ψ and ϕ being arbitrary functions.

Ex. 7. Solve

(i) $xqr + ypt + xy(s^2 - rt) = pq\,;$

(ii) $q^2 r + 4pqs + p^2 t + p^2 q^2 (rt - s^2) = a^2\,;$

(iii) $(1 + q^2)\,r - 2pqs + (1 + p^2)\,t = (s^2 - rt)(1 + p^2 + q^2)^{-\frac{1}{2}} - (1 + p^2 + q^2)^{\frac{3}{2}}.$

Ex. 8. Prove the converse of the general result of § 250, viz., let the equation of a surface be

$$\phi(x, y, z, a, b, c) = 0,$$

where a, b, c, are connected by any two conditions of the form

$$\chi(a, b, c) = 0 = \psi(a, b, c)\,;$$

shew that the equation of its envelope will satisfy a partial differential equation of the form

$$Rr + Ss + Tt + U(rt - s^2) = V,$$

the coefficients of which satisfy the relation

$$S^2 = 4(RT + UV).$$

GENERAL METHOD OF CONSTRUCTING AN INTERMEDIATE INTEGRAL IF IT EXISTS.

251. We now come to a more general method (mentioned at the end of § 237) for determining an intermediate integral of an algebraic equation

$$f(x, y, z, p, q, r, s, t) = 0,$$

polynomial in r, s, t, if any such integral exists.

Suppose that, if possible, this equation possesses an intermediate integral

$$n(x, y, z, p, q) = 0,$$

where we are not concerned, at the moment, with the form of the equation $u = 0$, nor with the kind of arbitrary element—whether of function or of constant—which this equation may contain. The significance of the intermediate integral lies in the property that the equation of the second order is satisfied in virtue of the integral and of its two first derivatives. The equation of the second order may not be the only deduced equation which is thus satisfied; but, under the hypothesis, it is satisfied in this way. It follows, then, that the possession, by the equation $f = 0$, of an intermediate integral $u = 0$ implies that the equation $f = 0$ is satisfied unconditionally in virtue of the equations

$$u = 0,$$

$$\frac{du}{dx} = \frac{\partial u}{\partial x} + p \frac{\partial u}{\partial z} + r \frac{\partial u}{\partial p} + s \frac{\partial u}{\partial q} = u_x + r u_p + s u_q = 0,$$

$$\frac{du}{dy} = \frac{\partial u}{\partial y} + q \frac{\partial u}{\partial z} + s \frac{\partial u}{\partial p} + t \frac{\partial u}{\partial q} = u_y + s u_p + t u_q = 0.$$

Consequently, when we use the last two equations to express t and r in terms of s and of the other quantities, and when we substitute these values of t and r in the equation $f = 0$, this last equation (under the hypothesis) must be satisfied unconditionally. It then is not an independent equation, and so it cannot determine s; that is, the coefficients of the various powers of s in the modified equation, integral in s, must vanish. We thus obtain a set of equations, involving derivatives of u and also the quantities x, y, z, p, q; and as the relations, which lead to the change in the form of $f = 0$, are

$$r = -\frac{u_x}{u_p} - s\frac{u_q}{u_p}, \quad t = -\frac{u_y}{u_q} - s\frac{u_p}{u_q},$$

each of the equations in the deduced set involves only ratios of the first derivatives of u.

But there are only four such ratios; consequently the deduced set of equations may not contain more than four members. There always will be two at least, for otherwise all the coefficients of powers of s, from the first power upwards, in the modified form of the equation $f = 0$ would vanish identically. But the members may be not independent of one another; so we can say that usually there will be not fewer than two independent equations, and that there cannot be more than four independent equations, in the set thus deduced.

We thus have algebraical equations, either two or three or four in number when the process is possible, involving the ratios $u_x : u_y : u_z : u_p : u_q$. The first step is to resolve the algebraical equations so as to express two of these ratios in terms of the others; and the result is to give a number of differential equations, two or three or four in number, which are homogeneous of zero order in the derivatives of u. The effective cases in the simplest form occur when these equations are rational in the derivatives of u; we then can express each equation as an equation, linear and homogeneous in the derivatives of u, the derivatives of u being taken with respect to x, y, z, p, q, and the coefficients of the derivatives of u being functions of the five quantities x, y, z, p, q.

We thus have a simultaneous system of equations satisfied by the quantity u, when it exists. So we proceed as in §§ 232—234. The first stage is to make the linear simultaneous system a complete Jacobian system. When this stage is finished, we proceed to construct the most general integral u of that system, if any such integral exists. When that integral u (if any) has been obtained, the corresponding intermediate integral of the original equation $f = 0$ is given by

$$u\,(x, y, z, p, q) = 0.$$

If the complete Jacobian system contains fewer than five members, it does possess some common integral. If it contains more than four members, it does not possess any integral other than $u = $ constant, which has no significance for the present purpose.

252. We now proceed to give some examples in detailed illustration of the method. As it is always possible to settle whether a given simultaneous system of linear and homogeneous partial differential equations does or does not possess a common integral, the method enables us to settle whether a given equation of the second order, algebraic in r, s, t, does or does not possess an intermediate integral. Further, subject solely to difficulties of mere quadrature, the method leads in practice to the most general form of that intermediate integral, if it exists.

Ex. 1. Does the equation

$$q(1+q)r - (1+p+q+2pq)s + p(1+p)t = 0$$

possess an intermediate integral ?

Denoting the possible intermediate integral by $u(x, y, z, p, q) = 0$, we are to have the original equation satisfied unconditionally in virtue of the equations

$$u_x + ru_p + su_q = 0, \quad u_y + su_p + tu_q = 0.$$

Adopting the process in the text so as to eliminate r and t, we have

$$q(1+q)\left(\frac{u_x}{u_p} + s\frac{u_q}{u_p}\right) + (1+p+q+2pq)s + p(1+p)\left(\frac{u_y}{u_q} + s\frac{u_p}{u_q}\right) = 0.$$

We then make this equation in s evanescent as regards s, so that

$$q(1+q)u_q^2 + (1+p+q+2pq)u_pu_q + p(1+p)u_p^2 = 0,$$

$$q(1+q)u_xu_q + p(1+p)u_yu_p = 0,$$

which are two equations satisfied by u.

The former equation is

$$\{pu_p + (1+q)u_q\}\{(1+p)u_p + qu_q\} = 0,$$

so that either

$$pu_p + (1+q)u_q = 0,$$

or

$$(1+p)u_p + qu_q = 0.$$

When the former is used, the second equation gives

$$u_x - \frac{1+p}{q}u_y = 0,$$

that is,

$$\frac{\partial u}{\partial x} - \frac{1+p}{q}\frac{\partial u}{\partial y} - \frac{\partial u}{\partial z} = 0;$$

and when the latter is used, it gives

$$u_y - \frac{1+q}{p}u_x = 0,$$

that is,

$$\frac{\partial u}{\partial y} - \frac{1+q}{p}\frac{\partial u}{\partial x} - \frac{\partial u}{\partial z} = 0.$$

Consequently, the function u satisfies one of the systems

$$\left. \begin{array}{c} p\dfrac{\partial u}{\partial p} + (1+q)\dfrac{\partial u}{\partial q} = 0 \\[2mm] \dfrac{\partial u}{\partial x} - \dfrac{1+p}{q}\dfrac{\partial u}{\partial y} - \dfrac{\partial u}{\partial z} = 0 \end{array}\right\}, \qquad \left.\begin{array}{c} q\dfrac{\partial u}{\partial q} + (1+p)\dfrac{\partial u}{\partial p} = 0 \\[2mm] \dfrac{\partial u}{\partial y} - \dfrac{1+q}{p}\dfrac{\partial u}{\partial x} - \dfrac{\partial u}{\partial z} = 0 \end{array}\right\}.$$

The condition of coexistence of the two equations in the first of the systems is

$$\frac{1+p+q}{q^2}\frac{\partial u}{\partial y} = 0,$$

which is not satisfied in virtue of either equation in the system; it is a new equation which must be satisfied, and so the system is

$$\left.\begin{array}{c} p\dfrac{\partial u}{\partial p} + (1+q)\dfrac{\partial u}{\partial q} = 0 \\[2mm] \dfrac{\partial u}{\partial x} - \dfrac{\partial u}{\partial z} = 0 \\[2mm] \dfrac{\partial u}{\partial y} = 0 \end{array}\right\}.$$

This system is easily seen to be complete; so (§ 235) it possesses two independent integrals which, by the process given for complete Jacobian systems, are easily found to be

$$\frac{1+q}{p}, \quad x+z.$$

The most general integral of the system is

$$u = \Phi\left(\frac{1+q}{p}, \ x+z\right),$$

where Φ is an arbitrary function of its arguments. Now the intermediate integral is $u = 0$, that is,

$$\Phi\left(\frac{1+q}{p}, \ x+z\right) = 0;$$

and this relation can be expressed in the form

$$x+z = f\left(\frac{1+q}{p}\right),$$

where f is an arbitrary function of its argument. This equation accordingly provides an intermediate integral of the original equation.

We proceed similarly for the alternative system; and we obtain a relation

$$y+z=g\left(\frac{1+p}{q}\right),$$

where g is an arbitrary function of its argument. (In this special case, the new relation can be deduced from the former relation by interchanging x and y, p and q, and by taking another arbitrary function g instead of f.) This equation accordingly provides another intermediate integral of the original equation, which thus possesses two independent intermediate integrals.

If we are to proceed to a primitive, we have to settle whether these intermediate integrals coexist. The form of the original equation is the same as that of § 239; and therefore we can apply the theorem of § 245, that is, the intermediate integrals do coexist. But it is possible that, for another equation not of the particular form, two intermediate integrals might be found; it then would be necessary to apply the Jacobian condition of coexistence $[F, G]=0$ of § 208. The condition is satisfied in the present instance.

Consequently, to obtain the primitive, we use the two simultaneous equations

$$x+z=f\left(\frac{1+q}{p}\right),\quad y+z=g\left(\frac{1+p}{q}\right).$$

Let

$$\frac{1+q}{p}=u,\quad \frac{1+p}{q}=v,$$

so that

$$\frac{p}{1+p+q}=\frac{1}{1+u},\quad \frac{q}{1+p+q}=\frac{1}{1+v};$$

then, as

$$dz=p\,dx+q\,dy,$$

we have

$$(1+p+q)\,dz=p\,(dx+dz)+q\,(dy+dz),$$

that is,

$$dz=\frac{1}{1+u}f'(u)\,du+\frac{1}{1+v}g'(v)\,dv,$$

and therefore

$$z=\theta(u)+\vartheta(v),$$

where θ and ϑ are arbitrary functions, because f and g are arbitrary. But

$$x+z=f(u),\quad y+z=g(v);$$

and therefore we have the primitive of the original equation in the form

$$z=F(x+z)+G(y+z),$$

where F and G are arbitrary functions of their respective arguments.

Ex. 2. Shew that, when the method is applied to the equation

$$rt - s^2 + Rr + 2Ss + Tt = V,$$

it leads to the two systems

$$\left.\begin{aligned} u_x - Tu_p - \rho u_q = 0 \\ u_y - \sigma u_p - Ru_q = 0 \end{aligned}\right\}, \qquad \left.\begin{aligned} u_x - Tu_p - \sigma u_q = 0 \\ u_y - \rho u_p - Ru_q = 0 \end{aligned}\right\},$$

where ρ and σ are the roots of the equation

$$\mu^2 + 2\mu S + RT + V = 0,$$

unless $\rho = \sigma$, when it leads to only one system.

Ex. 3. Deduce the conditions that the equation in Ex. 2 should possess (i) a single intermediate integral, (ii) a couple of intermediate integrals, each involving an arbitrary function.

Ex. 4. Apply the method to the equation

$$Rr + 2Ss + Tt = U,$$

so as to obtain the results corresponding to those in Ex. 2 and Ex. 3.

Ex. 5. Integrate the equation

$$z^2 (rt - s^2) + z \{(1 + q^2) r - 2pqs + (1 + p^2) t\} + 1 + p^2 + q^2 = 0.$$

Using the equations

$$r = -\frac{u_x}{u_p} - \frac{u_q}{u_p} s, \quad t = -\frac{u_y}{u_q} - \frac{u_p}{u_q} s,$$

to substitute for r and t, and making the resulting equation evanescent as an equation in s, we have the two relations

$$\left(z\frac{u_x}{u_p} - 1 - p^2\right)\left(z\frac{u_y}{u_q} - 1 - q^2\right) = p^2 q^2,$$

$$\left(z\frac{u_x}{u_p} - 1 - p^2\right)\frac{u_p}{u_q} + \left(z\frac{u_y}{u_q} - 1 - q^2\right)\frac{u_q}{u_p} = 2pq.$$

These are uniquely equivalent to the set

$$zu_x - (1 + p^2) u_p = pqu_q,$$

$$zu_y - (1 + q^2) u_q = pqu_p,$$

which (in this instance) is the only set of equations determining an intermediate integral $u = 0$, if any such integral exists. These equations are

$$E_1 = (1 + p^2)\frac{\partial u}{\partial p} + \quad pq\frac{\partial u}{\partial q} - pz\frac{\partial u}{\partial z} - z\frac{\partial u}{\partial x} = 0,$$

$$E_2 = \quad pq\frac{\partial u}{\partial p} + (1 + q^2)\frac{\partial u}{\partial q} - qz\frac{\partial u}{\partial z} - z\frac{\partial u}{\partial y} = 0.$$

The condition of coexistence (§ 232) is

$$q\frac{\partial u}{\partial p}-p\frac{\partial u}{\partial q}-qz\frac{\partial u}{\partial x}+pz\frac{\partial u}{\partial y}=0,$$

that is,

$$qE_1-pE_2=0,$$

and therefore it is satisfied. The two equations $E_1=0$, $E_2=0$, are a complete Jacobian system as they stand; and therefore (§ 235), as they involve five variables x, y, z, p, q, they possess three independent integrals.

To obtain these integrals, we adopt the regular process. The equations subsidiary to the integration of $E_1=0$ are

$$\frac{dp}{1+p^2}=\frac{dq}{pq}=\frac{dz}{-pz}=\frac{dx}{-z}=\frac{dy}{0}.$$

A complete set of independent integrals of these equations is

$$y=a_1,\quad \xi=a_2,\quad \eta=a_3,\quad \zeta=a_4,$$

where

$$\xi=x+pz,\quad \eta=q^2,\quad \zeta=(1+p^2)/q^2;$$

and therefore the equation $E_1=0$ will be satisfied by making u any function of y, ξ, η, ζ.

We therefore now make y, ξ, η, ζ, the independent variables; and we transform the other equation $E_2=0$, so as to have derivatives with respect to these variables. It is easily found that (after removal of a factor z) the modified form of the equation is

$$\frac{\partial u}{\partial \eta}-\frac{\partial u}{\partial y}-2\frac{1+\zeta}{\eta}\frac{\partial u}{\partial \zeta}=0.$$

(The property that the system $E_1=0$ and $E_2=0$ is complete secures, after § 234, that all the coefficients in $E_2=0$ are expressible in terms of y, ξ, η, ζ.) The equations, subsidiary to the integration of $E_2=0$, are

$$\frac{d\xi}{0}=\frac{d\eta}{1}=\frac{dy}{-1}=\frac{d\zeta}{-2\dfrac{1+\zeta}{\eta}}.$$

A complete set of independent integrals of these equations is

$$\xi=c_1,\quad \eta+y=c_2,\quad \eta^2(1+\zeta)=c_3.$$

If then

$$a=x+pz,\quad \beta=y+qz,\quad \gamma=z^2(1+p^2+q^2),$$

u can be any function of a, β, γ; and so the most general solution of the complete Jacobian system is given by

$$u=F(a,\beta,\gamma),$$

where F is an entirely arbitrary function of its arguments. But the intermediate integral, if any, of the original equation is $u=0$: that is, the most
general intermediate integral is given by

$$F(a, \beta, \gamma)=0,$$

which is an equation of the first order.

To complete the integration, this equation of the first order must be
integrated; for the purpose, Charpit's method will be used. The subsidiary
equations are

$$\frac{dp}{P}=\frac{dq}{Q}=\frac{dx}{X}=\frac{dy}{Y}=\frac{dz}{pX+qY},$$

where, if F_1, F_2, F_3, denote $\partial F/\partial a$, $\partial F/\partial \beta$, $\partial F/\partial \gamma$, respectively,

$$P=F_1+p\{pF_1+qF_2+(1+p^2+q^2)\,2zF_3\},$$
$$Q=F_2+q\{pF_1+qF_2+(1+p^2+q^2)\,2zF_3\},$$
$$X=-zF_1-2z^2pF_3,$$
$$Y=-zF_2-2z^2qF_3;$$

and one integral is required. We have, with these values,

$$dx+p\,dz+z\,dp=0,$$

so that an integral is given by

$$a=x+pz=a,$$

where a is a constant. This equation has to be taken simultaneously with
$F=0$, so that the values of p and q which they determine may be substituted in

$$dz=p\,dx+q\,dy.$$

The remaining analysis may be set out otherwise. Because

$$x+pz=a,$$

it follows that

$$z^2+(x-a)^2=\text{function of } y \text{ only};$$

and therefore

$$qz=\text{function of } y \text{ only},$$

that is, β is a function of y alone. But

$$F(a, \beta, \gamma)=0,$$

a is equal to a, and β is a function of y alone, and therefore γ is a function of
y alone. Now

$$\gamma=z^2(1+p^2+q^2)$$
$$=z^2+(a-x)^2+(\beta-y)^2,$$

so that

$$\tfrac{1}{2}d\gamma=z\,dz-(a-x)\,dx+(\beta-y)(d\beta-dy)$$
$$=(\beta-y)\,d\beta,$$

because

$$dz = p\,dx + q\,dy$$

$$= \frac{1}{z}\{(a-x)\,dx + (\beta-y)\,dy\}.$$

Also

$$\frac{\partial F}{\partial \beta}\,d\beta + \frac{\partial F}{\partial \gamma}\,d\gamma = 0;$$

and therefore

$$\frac{\partial F}{\partial \beta} + 2(\beta-y)\frac{\partial F}{\partial \gamma} = 0.$$

Eliminating β, γ, p, q, between this equation and the equations

$$F(a, \beta, \gamma) = 0,$$

$$x + pz = a, \quad y + qz = \beta, \quad z^2(1+p^2+q^2) = \gamma,$$

we have the primitive of the equation.

Ex. 6. Shew how the preceding primitive must be modified, so that it may include the primitive of the same equation given in Ex. 6 of § 250.

Ex. 7. Shew that the equation

$$r - t = 2\frac{p}{x}$$

has no intermediate integral.

Ex. 8. Obtain an intermediate integral (if any) of the equation

$$q^2rs + pq(rt+s^2) + p^2st - qxr - (px+qy)s - pyt + pq = 0.$$

Proceeding as usual from the equations

$$r = -\frac{u_x}{u_p} - \frac{u_q}{u_p}s, \quad t = -\frac{u_y}{u_q} - \frac{u_p}{u_q}s,$$

we find that u must satisfy the three equations

$$-q^2\frac{u_q}{u_p} + 2pq - p^2\frac{u_p}{u_q} = 0,$$

$$-q^2\frac{u_x}{u_p} + pq\left(\frac{u_y}{u_p} + \frac{u_x}{u_q}\right) - p^2\frac{u_y}{u_q} + qx\frac{u_q}{u_p} - px - qy + py\frac{u_p}{u_q} = 0,$$

$$pq\frac{u_xu_y}{u_qu_p} + pq + qx\frac{u_x}{u_p} + py\frac{u_y}{u_q} = 0.$$

From the first of these we have

$$F = pu_p - qu_q = 0.$$

The second of them is satisfied identically in virtue of $F = 0$. The third is

$$G = pq\,u_pu_q + qx\,u_xu_q + py\,u_yu_p + pq\,u_xu_y = 0.$$

The Jacobian condition of coexistence of these equations, being the relation $(F, G) = 0$ of § 208, leads to

$$\frac{\partial u}{\partial z} = 0;$$

so that, if we write x, y, z, p, q, $= x_1$, x_2, x_3, x_4, x_5, and denote the derivatives of u by p_1, p_2, p_3, p_4, p_5, the three equations are

$$p_3 = 0, \quad x_4 p_4 - x_5 p_5 = 0,$$

$$x_4 x_5 p_4 p_5 + x_4 x_5 p_1 p_2 + x_1 x_5 p_1 p_5 + x_2 x_4 p_2 p_4 = 0.$$

The first of them is satisfied if u does not explicitly involve x_3, that is, z.

The second of them then is satisfied if u is any function of x_1, x_2, and x_6, where $x_6 = x_4 x_5$. Write p_6 for $\dfrac{\partial u}{\partial x_6}$, so that

$$p_4 = x_5 p_6, \quad p_5 = x_4 p_6;$$

then the third equation is

$$x_6 p_6{}^2 + x_1 p_1 p_6 + x_2 p_2 p_6 + p_1 p_2 = 0,$$

on the removal of a factor x_6. We need the integral of this equation of the first order; to obtain it, we adopt the Jacobian process of § 219, and we find that

$$u = \Phi \left(\frac{x_6 + a x_1}{x_2 + a} \right),$$

where Φ is any arbitrary function, is the most general integral of the system. But the intermediate integral is $u = 0$, that is,

$$\Phi \left(\frac{x_6 + a x_1}{x_2 + a} \right) = 0;$$

or, as Φ is an arbitrary function, it is

$$\frac{x_6 + a x_1}{x_2 + a} = -b,$$

where b is an arbitrary constant. We thus obtain the intermediate integral in the form

$$pq + ax + by + ab = 0,$$

where a and b are arbitrary constants.

Ex. 9. Prove that the equation

$$z (rt - s^2)^2 + (rt + s^2) pq - rsq^2 - stp^2 = (rt - s^2) (p^2 t - 2pqs + q^2 r)$$

possesses an intermediate integral

$$z + ap + bq + ab = 0,$$

where a and b are arbitrary constants.

Ex. 10. Obtain the most general intermediate integral of the equation

$$\{(1+q^2)\,r - 2pqs + (1+p^2)\,t\}^2 = 4\,(rt - s^2)\,(1 + p^2 + q^2);$$

and prove that the only real surfaces, which satisfy the equation, are spheres.

Ex. 11. Integrate, by the foregoing method, the equations in Ex. 3 and Ex. 7 of § 250.

Note. Further discussion of equations of the second order having an intermediate integral will be found in the author's *Theory of Differential Equations*, vol. VI., ch. XVI.

PRINCIPLE OF DUALITY.

253. This principle, which was shewn (§ 202) to be effective in deducing from the solution of one equation of the first order that of another associated with the former by relations of a perfectly reciprocal character, may be applied to equations of the second order. The analytical connection consisted in taking new variables defined by the equations

$$X = p, \quad Y = q, \quad Z = px + qy - z,$$

from which there were derived the reciprocal equations

$$x = P, \quad y = Q, \quad z = PX + QY - Z.$$

From these we have

$$dx = dP = RdX + SdY,$$
$$dy = dQ = SdX + TdY;$$

so that

$$dX = \frac{Tdx - Sdy}{RT - S^2},$$

$$dY = \frac{-Sdx + Rdy}{RT - S^2}.$$

But

$$rdx + sdy = dp = dX,$$
$$sdx + tdy = dq = dY;$$

we therefore obtain, by equating coefficients,

$$r = \frac{T}{RT - S^2}, \quad s = \frac{-S}{RT - S^2}, \quad t = \frac{R}{RT - S^2};$$

and also

$$rt - s^2 = \frac{1}{RT - S^2}.$$

Let these substitutions be applied to any equation of the form

$$\lambda r + \mu s + \nu t + \sigma (rt - s^2) = 0,$$

in which λ, μ, ν, σ, are functions of x, y, z, p, q. Let their values after the transformations have taken place be denoted by $\lambda', \mu', \nu', \sigma'$, respectively; then the result of the substitution gives

$$\lambda' T - \mu' S + \nu' R + \sigma' = 0.$$

If then the solution of the former equation be known, that of the latter can be obtained; and *vice versa*.

Thus, in particular, the solutions of the two equations

$$r\phi (p, q) + s\psi (p, q) + t\chi (p, q) = 0,$$

and

$$r\chi (x, y) - s\psi (x, y) + t\phi (x, y) = 0,$$

are derivable from one another.

Ex. 1. From the solution of

$$x^2 r + 2xys + y^2 t = 0,$$

derive that of

$$q^2 r - 2pqs + p^2 t = 0.$$

Ex. 2. Integrate the equations:

(i) $px + qy - sxy = z$;

(ii) $z(rt - s^2) = -pqs$;

(iii) $q^2 (z - px - qy) = (pt - qs) xz$;

(iv) $p^2 r + 2pqs + q^2 t = (xp + yq)(rt - s^2)$;

(v) $(1 + pq)(r - t) = (p^2 - q^2) s + p^2 t - q^2 r$.

LAPLACE'S METHOD FOR THE TRANSFORMATION OF THE LINEAR EQUATION.

254. The linear equation

$$Rr + Ss + Tt + Pp + Qq + Zz = U,$$

in which R, S, T, P, Q, Z, U, are functions of x and y only, can be reduced to simpler forms. The process consists in changing the variables.

Let the independent variables x and y be changed to ξ and η, as yet undetermined; then, when p', q', ... denote $\dfrac{\partial z}{\partial \xi}$, $\dfrac{\partial z}{\partial \eta}$, ... the equation becomes

$$r' \left\{ R \left(\frac{\partial \xi}{\partial x} \right)^2 + S \frac{\partial \xi}{\partial x} \frac{\partial \xi}{\partial y} + T \left(\frac{\partial \xi}{\partial y} \right)^2 \right\}$$

$$+ t' \left\{ R \left(\frac{\partial \eta}{\partial x} \right)^2 + S \frac{\partial \eta}{\partial x} \frac{\partial \eta}{\partial y} + T \left(\frac{\partial \eta}{\partial y} \right)^2 \right\}$$

$$+ s' \left\{ 2R \frac{\partial \xi}{\partial x} \frac{\partial \eta}{\partial x} + S \left(\frac{\partial \eta}{\partial x} \frac{\partial \xi}{\partial y} + \frac{\partial \eta}{\partial y} \frac{\partial \xi}{\partial x} \right) + 2T \frac{\partial \xi}{\partial y} \frac{\partial \eta}{\partial y} \right\}$$

$$+ p' \left\{ R \frac{\partial^2 \xi}{\partial x^2} + S \frac{\partial^2 \xi}{\partial x \partial y} + T \frac{\partial^2 \xi}{\partial y^2} + P \frac{\partial \xi}{\partial x} + Q \frac{\partial \xi}{\partial y} \right\}$$

$$+ q' \left\{ R \frac{\partial^2 \eta}{\partial x^2} + S \frac{\partial^2 \eta}{\partial x \partial y} + T \frac{\partial^2 \eta}{\partial y^2} + P \frac{\partial \eta}{\partial x} + Q \frac{\partial \eta}{\partial y} \right\} + Zz = U.$$

Let m and n be the roots of the quadratic equation in k

$$Rk^2 + Sk + T = 0.$$

First, suppose that these roots are unequal; then choose ξ and η so that

$$\frac{\partial \xi}{\partial x} = m \frac{\partial \xi}{\partial y},$$

$$\frac{\partial \eta}{\partial x} = n \frac{\partial \eta}{\partial y},$$

which determine ξ and η. The terms involving r' and t' now disappear; and the coefficient of s', being

$$\frac{\partial \xi}{\partial y} \frac{\partial \eta}{\partial y} \left(4T - \frac{S^2}{R} \right),$$

does not vanish since the roots of the quadratic are unequal. Let the equation be divided throughout by this coefficient; then it takes the form

$$\frac{\partial^2 z}{\partial \xi \partial \eta} + L \frac{\partial z}{\partial \xi} + M \frac{\partial z}{\partial \eta} + Nz = V.$$

255. In two cases, the integral of this equation can be obtained without further transformation. We may write it in the form

$$\frac{\partial}{\partial \xi}\left(\frac{\partial z}{\partial \eta} + Lz\right) + M\left(\frac{\partial z}{\partial \eta} + Lz\right) + z\left(N - LM - \frac{\partial L}{\partial \xi}\right) = V,$$

so that, if the condition

$$N - LM - \frac{\partial L}{\partial \xi} = 0$$

be satisfied, the equation becomes

$$\frac{\partial u}{\partial \xi} + Mu = V,$$

where u replaces $\frac{\partial z}{\partial \eta} + Lz$. A general value of u can be obtained, and thence a general value of z.

We may write the equation also in the form

$$\frac{\partial}{\partial \eta}\left(\frac{\partial z}{\partial \xi} + Mz\right) + L\left(\frac{\partial z}{\partial \xi} + Mz\right) + z\left(N - LM - \frac{\partial M}{\partial \eta}\right) = V,$$

so that, if the condition

$$N - LM - \frac{\partial M}{\partial \eta} = 0$$

be satisfied, the equation becomes

$$\frac{\partial v}{\partial \eta} + Lv = V,$$

where v replaces $\frac{\partial z}{\partial \xi} + Mz$. From this, through v, a general value of z can be obtained.

256. If however neither of the conditions in § 255 between the coefficients in the transformed equation be satisfied, it can still be transformed by changing the dependent variable. Thus when we write

$$\frac{\partial z}{\partial \eta} + Lz = \zeta,$$

we have

$$\frac{\partial \zeta}{\partial \xi} + M\zeta + z\left(N - LM - \frac{\partial L}{\partial \xi}\right) = V.$$

Denoting $LM + \dfrac{\partial L}{\partial \xi} - N$ by K, we may write

$$z = \frac{1}{K} \frac{\partial \zeta}{\partial \xi} + \frac{M}{K} \zeta - \frac{V}{K};$$

and therefore

$$\zeta = \frac{L}{K} \frac{\partial \zeta}{\partial \xi} + \frac{LM}{K} \zeta - \frac{LV}{K} + \frac{\partial}{\partial \eta} \left\{ \frac{1}{K} \frac{\partial \zeta}{\partial \xi} + \frac{M}{K} \zeta - \frac{V}{K} \right\},$$

which is equivalent to

$$\frac{\partial^2 \zeta}{\partial \xi \partial \eta} + L' \frac{\partial \zeta}{\partial \xi} + M' \frac{\partial \zeta}{\partial \eta} + N' \zeta = V',$$

where

$$L' = L - \frac{1}{K} \frac{\partial K}{\partial \eta},$$

$$M' = M,$$

$$N' = LM - K + K \frac{\partial}{\partial \eta} \left(\frac{M}{K} \right),$$

so that the same form is reproduced but with altered coefficients. The equation in its new form can be integrated, if the analogous relations between the new coefficients be satisfied. From the values of L', M', N', we have

$$L'M' - N' = K - \frac{\partial M}{\partial \eta}$$

$$= K - \frac{\partial M'}{\partial \eta},$$

so that, as K is not zero (by hypothesis), the relation

$$L'M' + \frac{\partial M'}{\partial \eta} - N' = 0$$

is not satisfied. The other condition, being that the equation

$$L'M' + \frac{\partial L'}{\partial \xi} - N' = 0$$

should be satisfied, is when expressed in terms of the original coefficients

$$K + \frac{\partial L}{\partial \xi} - \frac{\partial M}{\partial \eta} - \frac{1}{K} \frac{\partial^2 K}{\partial \eta \partial \xi} + \frac{1}{K^2} \frac{\partial K}{\partial \eta} \frac{\partial K}{\partial \xi} = 0.$$

If this be not satisfied, nor the corresponding relation derived by the consideration of the other expression

$$LM + \frac{\partial M}{\partial \eta} - N,$$

the process of transformation may be repeated indefinitely. If, at any step of the process, the requisite condition should be satisfied, the solution may then be found.

Ex. 1. Prove that for any substitution of the form

$$z = \lambda u,$$

where u is to be the new dependent variable and λ is a function of ξ and η,

$$LM - N + \frac{\partial M}{\partial \eta} \quad \text{and} \quad LM - N + \frac{\partial L}{\partial \xi}$$

are absolute invariants, and that therefore such a transformation is ineffective for the purpose of solution.

Ex. 2. Prove that, if

$$K_r = N_r - L_r M_r - \frac{\partial L_r}{\partial \xi}, \quad J_r = N_r - L_r M_r - \frac{\partial M_r}{\partial \eta},$$

being the functions of the coefficients after r transformations, then

$$K_{r+1} = \frac{\partial^2 (\log K_r)}{\partial \xi \partial \eta} + 2K_r - J_r,$$

$$J_{r+1} = K_r.$$

Hence solve the equation

$$s + xyp = 2yz.$$

(Imschenetsky.)

Note. A full discussion of these equations, subjected to Laplace transformations, will be found in the author's *Theory of Differential Equations*, vol. VI., chapters XIII. and XIV.

257. Next, consider the case when the roots of the quadratic are equal, so that

$$S^2 - 4RT = 0.$$

The two equations determining ξ and η now coincide, so that from them only one of these quantities can be obtained; let it be ξ, given by

$$\frac{\partial \xi}{\partial x} = m \frac{\partial \xi}{\partial y},$$

and suppose ξ and y to be the new independent variables; then we may write $\eta = y$. In the transformed equation the coefficient of r' now is zero, that of t' is T, and that of s' is

$$S \frac{\partial \xi}{\partial x} + 2T \frac{\partial \xi}{\partial y}.$$

But m being a repeated root of

$$Rk^2 + Sk + T = 0,$$

we have

$$m = -\frac{S}{2R} = -\frac{2T}{S},$$

so that the coefficient of s' is

$$\frac{\partial \xi}{\partial y} \left(2T - \frac{S^2}{2R} \right),$$

which is zero. Hence the transformed equation, on division throughout by T, becomes

$$\frac{\partial^2 z}{\partial y^2} + L \frac{\partial z}{\partial \xi} + M \frac{\partial z}{\partial y} + Nz = V.$$

The case suitable for treatment by this method is that in which L is zero; the equation may then be looked upon as an ordinary equation in y, the variable x being considered parametric; the arbitrary constants of integration should be replaced by arbitrary functions of x.

POISSON'S METHOD.

258. Poisson has shewn how to deduce a particular integral of any partial differential equation which is of the form

$$P = (rt - s^2)^n Q,$$

where P is a function of p, q, r, s, and t, homogeneous with respect to the last three quantities, and Q is any function of the variables x, y, z, and the differential coefficients of z, which remains finite when $rt - s^2 = 0$.

He assumes

$$q = \phi(p),$$

and therefore

$$s = r\phi'(p), \quad t = s\phi'(p) = r\{\phi'(p)\}^2.$$

These values make
$$rt - s^2 = 0,$$
and reduce the differential equation to
$$P = 0.$$

Now P being homogeneous with respect to r, s, and t, there will, when the foregoing values are substituted, occur a common factor throughout, being some power of r; this may be rejected and the remaining equation will involve only p, $\phi(p)$ and $\phi'(p)$ which when integrated will determine the value of $\phi(p)$ and so will lead to an integral of the original equation. This integral, being of the form
$$q = \phi(p),$$
can always be further integrated.

It may be noticed that Poisson's process is equivalent to obtaining the developable surfaces which are included under the given differential equation, for
$$q = \phi(p)$$
is the differential equation of developable surfaces.

Ex. 1. Solve $\qquad r^2 - t^2 = rt - s^2.$

Proceeding as above we find
$$1 - \{\phi'(p)\}^4 = 0,$$
so that, retaining only the real values, we have
$$\phi'(p) = \pm 1,$$
whence $\qquad q = \phi(p) = a \pm p,$
where a is an arbitrary constant. The complete integral of this, considered as a partial differential equation of the first order, is
$$z = ay + \lambda(x \pm y) + \nu,$$
where λ and ν are arbitrary constants; the general integral is
$$z = ay + \phi(x \pm y),$$
where ϕ is an arbitrary function.

Ex. 2. Solve
$$\text{(i)} \quad t + 2ps + (p^2 - a^2)r = 0;$$
$$\text{(ii)} \quad (1 + q^2)r - 2pqs + (1 + p^2)t = 0.$$

LINEAR EQUATIONS WITH CONSTANT COEFFICIENTS.

259. We now proceed to consider equations which are linear not merely with regard to the differential coefficients of highest order but also with regard to the dependent variable and all its differential coefficients, and in which the various terms are multiplied by constants only. Such an equation is

$$\Phi\left(\frac{\partial}{\partial x}, \ \frac{\partial}{\partial y}\right) z = V,$$

where Φ is a rational integral algebraical function all the coefficients of which are constant; V may be any function of the independent variables.

As in the case of ordinary differential equations the complete integral consists of the sum of two parts:

first, the most general integral of

$$\Phi\left(\frac{\partial}{\partial x}, \ \frac{\partial}{\partial y}\right) z = 0;$$

second, any particular solution of

$$\Phi\left(\frac{\partial}{\partial x}, \ \frac{\partial}{\partial y}\right) z = V.$$

These will be obtained separately. For convenience, let $\dfrac{\partial}{\partial x}$ and $\dfrac{\partial}{\partial y}$ be respectively denoted by D and D'.

260. The simplest case of the general equation is that in which only differential coefficients of the nth order occur, so that it may be written

$$(D^n + A_1 D^{n-1} D' + A_2 D^{n-2} D'^2 + \ldots\ldots + A_n D'^n) z = V.$$

Let $\alpha_1, \alpha_2, \ldots\ldots, \alpha_n$, be the n roots of

$$\xi^n + A_1 \xi^{n-1} + A_2 \xi^{n-2} + \ldots\ldots + A_{n-1}\xi + A_n = 0;$$

then the equation may be transformed into

$$(D - \alpha_1 D')(D - \alpha_2 D') \ldots\ldots (D - \alpha_n D') z = V.$$

To find the complementary function, we write $V = 0$; then a solution of

$$(D - \alpha_r D') z = 0$$

will be a term in the complementary function; and as there are n such factors there will be n such terms.

Now the solution of

$$(D - \lambda) z = 0,$$

where λ is independent of x, is given by

$$z = e^{\lambda x} C,$$

C being also independent of x. The quantity C may therefore, in the solution of

$$(D - \alpha D') z = 0,$$

be made an arbitrary function of y, and we then have

$$z = e^{\alpha x \frac{d}{dy}} \phi(y)$$
$$= \phi(y + \alpha x).$$

There is one such solution for every value of α; and the sum of these different solutions is also a solution, so that the complementary function is

$$z = \phi_1(y + \alpha_1 x) + \phi_2(y + \alpha_2 x) + \ldots\ldots + \phi_n(y + \alpha_n x),$$

where all the functions ϕ_1, ϕ_2, $\ldots\ldots$, ϕ_n, are arbitrary.

In the case, however, in which two roots α are equal, this value ceases to be general, as the sum of two arbitrary functions of the same argument is merely an arbitrary function of that argument; the corresponding terms are then obtained as follows.

The solution of

$$(D - \lambda)^2 z = 0$$

is

$$z = e^{\lambda x}(A + Bx),$$

where A and B are independent of x; hence the integral of

$$(D - \alpha D')^2 z = 0$$

is

$$z = e^{\alpha x \frac{d}{dy}} \{\phi(y) + x\psi(y)\}$$
$$= \phi(y + \alpha x) + x\psi(y + \alpha x),$$

where both ϕ and ψ are arbitrary; the sum of these two terms replaces the sum of the two terms, which had coalesced into one, and the general character of the solution is restored. Similarly, when any number of the roots α are equal, the corresponding terms of the complementary function, which coalesce into one, are replaced by a series of terms derived in the same manner as the above.

261. To obtain the particular integral, we may represent it symbolically by

$$z = \frac{1}{(D - \alpha_1 D')(D - \alpha_2 D')\dots(D - \alpha_n D')} V$$

$$= \frac{1}{D'^n} \frac{1}{\left(\dfrac{D}{D'} - \alpha_1\right)\left(\dfrac{D}{D'} - \alpha_2\right)\dots\left(\dfrac{D}{D'} - \alpha_n\right)} V.$$

To evaluate this, we resolve the second symbolical fraction into the sum of n symbolical partial fractions, into the denominator of each of which only one of the quantities $D/D' - \alpha$ enters. Thus, if

$$\frac{1}{(\xi - \alpha_1)(\xi - \alpha_2)\dots(\xi - \alpha_n)} = \sum_{r=1}^{r=n} \frac{N_r}{\xi - \alpha_r},$$

we have

$$z = \frac{1}{D'^n} \sum_{r=1}^{r=n} \frac{N_r}{\dfrac{D}{D'} - \alpha_r} V$$

$$= \frac{1}{D'^{n-1}} \sum_{r=1}^{r=n} \frac{N_r}{D - \alpha_r D'} V,$$

N_r being a constant and depending only upon the constants α. Let

$$V = \psi(x, y);$$

then, since

$$(D - \alpha D')^{-1} = e^{\alpha x \frac{\partial}{\partial y}} \int dx\, e^{-\alpha x \frac{\partial}{\partial y}},$$

we have

$$\frac{1}{D - \alpha_1 D'} V = e^{\alpha x \frac{\partial}{\partial y}} \int^x d\xi\, e^{-\alpha \xi \frac{\partial}{\partial y}} \psi(\xi, y)$$

$$= e^{\alpha x \frac{\partial}{\partial y}} \int^x d\xi\, \psi(\xi, y - \alpha \xi)$$

$$= \int^x d\xi\, \psi(\xi, y + \alpha x - \alpha \xi);$$

hence the particular integral of the equation is

$$z = \iiint \ldots dy^{n-1} \int^x d\xi \sum_{r=1}^{r=n} [N_r \psi\{\xi, y + \alpha_r(x - \xi)\}].$$

This is the value in the most general case possible; in particular cases, the actual evaluation becomes much more easy. Thus, if V be a function of x only, we may consider $\{\Phi(D, D')\}^{-1}$ as expanded in a series of ascending powers of D'; every term may then be neglected (so far as the particular integral is concerned) except that which does not contain D'. Corresponding simplifications arise in other examples.

Ex. 1. Solve
$$\frac{\partial^2 u}{\partial x^2} - a^2 \frac{\partial^2 u}{\partial y^2} = x.$$

(See Ex. 1, § 250.)

For the Complementary Function, we have

$$\left(\frac{\partial}{\partial x} - a\frac{\partial}{\partial y}\right)\left(\frac{\partial}{\partial x} + a\frac{\partial}{\partial y}\right) u = 0,$$

and therefore

$$u = e^{ax\frac{\partial}{\partial y}} \phi(y) + e^{-ax\frac{\partial}{\partial y}} \psi(y)$$
$$= \phi(y + ax) + \psi(y - ax),$$

ϕ and ψ denoting arbitrary functions.

For the Particular Integral, we have

$$u = \frac{1}{D^2 - a^2 D'^2} x$$
$$= \frac{1}{D^2}\left(1 + a^2 \frac{D'^2}{D^2} + \ldots\right) x$$
$$= \frac{1}{D^2} x$$
$$= \frac{x^3}{3!}.$$

Hence the Complete Integral is

$$u = \phi(y + ax) + \psi(y - ax) + \frac{x^3}{3!}.$$

Ex. 2. Obtain a solution of the equation

$$\frac{\partial^2 y}{\partial t^2} = a^2 \frac{\partial^2 y}{\partial x^2}$$

such that, when $t = 0$, $y = F(x)$ and $\frac{\partial y}{\partial t} = \frac{df(x)}{dx}$, $F(x)$ and $f(x)$ being known functions of x.

Ex. 3. Solve the equations:

$$\text{(i)} \quad \frac{\partial^2 z}{\partial x^2} + \frac{\partial^2 z}{\partial y^2} = \cos mx \cos ny;$$

$$\text{(ii)} \quad \frac{\partial^2 z}{\partial x^2} + 3 \frac{\partial^2 z}{\partial x \partial y} + 2 \frac{\partial^2 z}{\partial y^2} = x + y;$$

$$\text{(iii)} \quad \frac{\partial^2 z}{\partial x^2} - 2a \frac{\partial^2 z}{\partial x \partial y} + a^2 \frac{\partial^2 z}{\partial y^2} = f(y + ax);$$

$$\text{(iv)} \quad \frac{\partial^3 z}{\partial x^3} - \frac{\partial^3 z}{\partial y^3} = x^3 y^3;$$

$$\text{(v)} \quad (D - aD')^2 z = \phi(x) + \psi(y) + \chi(x + by);$$

$$\text{(vi)} \quad (D - D')^2 z = x + \phi(x + y).$$

Ex. 4. Solve the equation

$$\frac{\partial^3 u}{\partial x^3} + \frac{\partial^3 u}{\partial y^3} + \frac{\partial^3 u}{\partial z^3} - 3 \frac{\partial^3 u}{\partial x \partial y \partial z} = x^3 + y^3 + z^3 - 3xyz.$$

For the Complementary Function, we have

$$\left(\frac{\partial}{\partial x} + \frac{\partial}{\partial y} + \frac{\partial}{\partial z} \right) \left(\frac{\partial}{\partial x} + \omega \frac{\partial}{\partial y} + \omega^2 \frac{\partial}{\partial z} \right) \left(\frac{\partial}{\partial x} + \omega^2 \frac{\partial}{\partial y} + \omega \frac{\partial}{\partial z} \right) u = 0,$$

ω being a cube root of unity. The solution of

$$\left(\frac{\partial}{\partial x} + \lambda \frac{\partial}{\partial y} + \mu \frac{\partial}{\partial z} \right) u = 0$$

is

$$u = e^{-x \left(\lambda \frac{\partial}{\partial y} + \mu \frac{\partial}{\partial z} \right)} \phi(y, z)$$

$$= \phi(y - \lambda x, \quad z - \mu x);$$

hence the Complementary Function is

$$\phi_1(y - x, \ z - x) + \phi_2(y - \omega x, \ z - \omega^2 x) + \phi_3(y - \omega^2 x, \ z - \omega x),$$

where ϕ_1, ϕ_2, ϕ_3, are arbitrary functions.

The part of the Particular Integral corresponding to x^3 is

$$\frac{1}{\left(\frac{\partial}{\partial x} \right)^3 + \left(\frac{\partial}{\partial y} \right)^3 + \left(\frac{\partial}{\partial z} \right)^3 - 3 \frac{\partial^3}{\partial x \partial y \partial z}} x^3 = \frac{1}{\left(\frac{\partial}{\partial x} \right)^3} x^3 = \frac{x^6}{4.5.6},$$

and so for the other terms; the full value is

$$\frac{x^6 + y^6 + z^6}{4.5.6} + \frac{x^2 y^2 z^2}{8}.$$

The Complete Integral is the sum of the Complementary Function and the Particular Integral.

Ex. 5. Solve

(i) $\dfrac{\partial^2 u}{\partial x^2} + \dfrac{\partial^2 u}{\partial x \partial z} - \dfrac{\partial^2 u}{\partial y^2} - \dfrac{\partial^2 u}{\partial y \partial z} = xyz$;

(ii) $\dfrac{\partial^3 u}{\partial x^2 \partial y} - 2\dfrac{\partial^3 u}{\partial x \partial y^2} - 3\dfrac{\partial^3 u}{\partial x^2 \partial z} - 3\dfrac{\partial^3 u}{\partial x \partial z^2} - 2\dfrac{\partial^3 u}{\partial y^2 \partial z} + 6\dfrac{\partial^3 u}{\partial y \partial z^2} + 7\dfrac{\partial^3 u}{\partial x \partial y \partial z} = 0.$

262. Passing now to the general equation, we proceed to find the solution of

$$\Phi\left(\frac{\partial}{\partial x}, \ \frac{\partial}{\partial y}\right) z = 0,$$

where Φ is of the form

$$A_0 \frac{\partial^n}{\partial x^n} + A_1 \frac{\partial^n}{\partial x^{n-1} \partial y} + A_2 \frac{\partial^n}{\partial x^{n-2} \partial y^2} + \ldots\ldots$$

$$+ B_0 \frac{\partial^{n-1}}{\partial x^{n-1}} + B_1 \frac{\partial^{n-1}}{\partial x^{n-2} \partial y} + \ldots\ldots$$

$$+ \ldots\ldots$$

$$\ldots\ldots + K_0 \frac{\partial}{\partial x} + K_1 \frac{\partial}{\partial y} + L.$$

We assume as a trial solution

$$z = A e^{hx + ky},$$

where h and k are constants yet to be determined. With this value, we have

$$\frac{\partial z}{\partial x} = hz, \quad \frac{\partial z}{\partial y} = kz;$$

and therefore

$$\Phi(h, k) z = 0,$$

which will be satisfied, if h and k be determined so as to satisfy

$$\Phi(h, k) = 0.$$

This obviously makes one of the constants to depend on the other. Let the equation be resolved to determine k, so that we obtain results of the form

$$k = \theta(h),$$

n in number. Taking one of them, as $k = \theta_1(h)$, we have the solution in the form

$$z = A e^{hx + y\theta_1(h)},$$

for all values of A and h. The sum of any number of solutions is also a solution; hence another is given by

$$z = \Sigma A e^{hx+y\theta_1 (h)},$$

where Σ implies summation for all values of h; and A, an arbitrary constant, may be looked upon as an arbitrary function of h which may vary from term to term of the series.

Similarly another value of k, such as $\theta_2(h)$, will lead to another solution which may be represented by

$$z = \Sigma B e^{h'x+y\theta_2 (h')};$$

and, as each value of k will lead to a corresponding series, the general solution may be represented as the sum of n series in the form

$$z = \Sigma \{A e^{hx+y\theta_1(h)}\} + \Sigma \{B e^{h'x+y\theta_2(h')}\} + \ldots,$$

the summation in each series extending to terms arising from all possible values of the constants h.

The fact, that the coefficient belonging to any term may be considered as an arbitrary function of the constant which occurs in that term, shews that each series may be regarded as having in its expression one general arbitrary function. Thus, in the Complementary Function, we should expect n arbitrary functions.

263. This general result, in the form of the sum of n series each containing arbitrary elements, may appear to be of slight value. Sometimes, however, by the form of the differential equation, a simplification is introduced such as that indicated in the next paragraph. Sometimes, by conditions imposed on the dependent variable other than the satisfaction of the differential equation, the number of terms of the series is limited to those which contain particular values of the parametric constant.

For example, whenever a solution of the equation which determines k is of the form

$$k = \alpha h + \beta,$$

where α and β are determinate constants, the corresponding series may be expressed in a finite form. The solution is

$$e^{\beta y} \Sigma A e^{h(x+\alpha y)},$$

that is (save as to the factor outside Σ), it is the sum of any number of arbitrary powers of $e^{x+\alpha y}$ each multiplied by an arbitrary constant; such a sum is an arbitrary function of $e^{x+\alpha y}$ or, what is the equivalent, an arbitrary function of $x + \alpha y$; and the series may therefore be replaced by

$$e^{\beta y}\phi(x + \alpha y),$$

where ϕ is arbitrary. Corresponding to the conditions which in any particular case limit the number of terms included in the series, there will be analogous conditions which determine the form of the arbitrary function.

Ex. Prove that, if the root

$$k = ah + \beta$$

occur $r + 1$ times, the corresponding part of the Complementary Function is

$$e^{\beta y}\left[\phi_0(x+ay)+y\phi_1(x+ay)+\ldots\ldots+y^r\phi_r(x+ay)\right],$$

where $\phi_0, \phi_1, \ldots\ldots, \phi_r$, denote arbitrary functions.

264. To obtain the Particular Integral, we may represent it by

$$z = \frac{1}{\Phi(D, D')}\,V;$$

the evaluation of this expression will depend upon the form of V. Thus if

$$V = e^{ax+by},$$

we should have

$$\frac{1}{\Phi(a, b)}\,e^{ax+by},$$

as the value of z required. If V were a polynomial function of x and y, then it would be possible to evaluate the expression by expanding the inverse operator in a series of ascending powers of both D and D', if permissible, or in powers of either of them. The methods, applied to the particular forms considered in §46 in the case of ordinary differential equations, indicate the corresponding methods to be adopted for some of the varying forms of V.

Ex. Solve

$$\frac{\partial^2 z}{\partial x^2} - \frac{\partial^2 z}{\partial y^2} - 3 \frac{\partial z}{\partial x} + 3 \frac{\partial z}{\partial y} = xy + e^{x+2y}.$$

First, for the Complementary Function, we must solve

$$(D-D')(D+D'-3) z = 0.$$

Let
$$z = A e^{hx+ky}$$

be substituted. Then

$$(h-k)(h+k-3) = 0,$$

so that
$$k = h, \quad k = 3 - h,$$

are the relations between h and k. Hence

$$z = \Sigma A e^{h(x+y)} + e^{3y} \Sigma B e^{h'(x-y)}$$

$$= \phi(x+y) + e^{3y} \psi(x-y),$$

where ϕ and ψ are both arbitrary.

The part of the Particular Integral corresponding to e^{x+2y} is

$$z = \frac{1}{(D-D')(D+D'-3)} e^{x+2y}$$

$$= e^x \frac{1}{(1-D')(D'-2)} e^{2y}$$

$$= e^{x+2y} \frac{1}{(1-D'-2)(D'+2-2)} \cdot 1$$

$$= -e^{x+2y} \frac{1}{D'(D'+1)} \cdot 1$$

$$= -y e^{x+2y}.$$

The result indicates that a term of the form e^{x+2y} will arise in the Complementary Function, as is obvious from the identity

$$e^{x+2y} = e^{3y} e^{x-y}.$$

The part of the Particular Integral corresponding to the term xy is

$$z = \frac{1}{(D-D')(D+D'-3)} xy$$

$$= -\frac{1}{3} \frac{1}{D-D'} \left\{ 1 + \frac{D+D'}{3} + \left(\frac{D+D'}{3} \right)^2 \right\} xy$$

$$= -\frac{1}{3D} \left(1 + \frac{D'}{D} \right) (xy + \tfrac{1}{3}x + \tfrac{1}{3}y + \tfrac{2}{9})$$

$$= -\frac{1}{3D} (xy + \tfrac{1}{3}x + \tfrac{1}{3}y + \tfrac{2}{9} + \tfrac{1}{2}x^2 + \tfrac{1}{3}x)$$

$$= -(\tfrac{1}{6}x^2 y + \tfrac{1}{8}x^2 + \tfrac{1}{6}xy + \tfrac{2}{27}x + \tfrac{1}{18}x^3),$$

the expansions in each case being taken no further than is necessary to furnish non-evanescent terms.

The Complete Integral is, as usual, the sum of the foregoing three parts.

Note. It might happen that, by a different method of procedure such as expanding in powers of $\dfrac{D}{D'}$, a particular integral of apparently different form would be obtained; it would however be found that the two could be transformed into each other by means of the Complementary Function.

265. Any equation, such that the coefficient of a differential coefficient of any order is a constant multiple of the variables of the same degree, may be reduced to an equation of the foregoing form. Such an equation will be of the form

$$\Sigma A_r x^r \frac{\partial^r z}{\partial x^r} + \Sigma B_{pq} x^p y^q \frac{\partial^{p+q} z}{\partial x^p \partial y^q} + \Sigma C_s y^s \frac{\partial^s z}{\partial y^s} = V.$$

We may either change the independent variables to u and v, where $x = e^u$ and $y = e^v$; or we may represent $x\dfrac{\partial}{\partial x}$ by ϑ and $y\dfrac{\partial}{\partial y}$ by ϕ, and then we have

$$x^r \frac{\partial^r z}{\partial x^r} = \vartheta (\vartheta - 1)(\vartheta - 2)\dots(\vartheta - r + 1) z,$$

$$x^p y^q \frac{\partial^{p+q} z}{\partial x^p \partial y^q} = \vartheta (\vartheta - 1)\dots(\vartheta - p + 1)\phi(\phi - 1)\dots(\phi - q + 1) z.$$

In either case, the equation is reduced to the form already considered.

Ex. 1. Solve the equation

$$x^2 \frac{\partial^2 z}{\partial x^2} + 2xy \frac{\partial^2 z}{\partial x \partial y} + y^2 \frac{\partial^2 z}{\partial y^2} = x^m y^n.$$

We have, on assuming $u = \log x$ and $v = \log y$,

$$\left(\frac{\partial}{\partial u} + \frac{\partial}{\partial v}\right)\left(\frac{\partial}{\partial u} + \frac{\partial}{\partial v} - 1\right) z = e^{mu + nv}.$$

The integral of this is

$$z = e^u F_1 (v - u) + f_1 (v - u) + \frac{e^{mu + nv}}{(m+n)(m+n-1)}$$

$$= xF\left(\frac{y}{x}\right) + f\left(\frac{y}{x}\right) + \frac{x^m y^n}{(m+n)(m+n-1)}$$

where f and F are arbitrary.

Ex. 2. Solve the equations:

$$\text{(i)} \quad x^2 \frac{\partial^2 z}{\partial x^2} - y^2 \frac{\partial^2 z}{\partial y^2} = xy \; ;$$

$$\text{(ii)} \quad x^2 \frac{\partial^2 z}{\partial x^2} - y^2 \frac{\partial^2 z}{\partial y^2} = y \frac{\partial z}{\partial y} - x \frac{\partial z}{\partial x}.$$

Ex. 3. Solve the equations:

$$\text{(i)} \quad x^2 \frac{\partial^2 u}{\partial x^2} + 2xy \frac{\partial^2 u}{\partial x \partial y} + y^2 \frac{\partial^2 u}{\partial y^2} + nu = n \left(x \frac{\partial u}{\partial x} + y \frac{\partial u}{\partial y} \right) + x^2 + y^2 + x^3 \; ;$$

$$\text{(ii)} \quad x^2 \frac{\partial^2 u}{\partial x^2} + 2xy \frac{\partial^2 u}{\partial x \partial y} + y^2 \frac{\partial^2 u}{\partial y^2} = (x^2 + y^2)^{\frac{1}{2}n} \; ;$$

$$\text{(iii)} \quad \left(x \frac{\partial}{\partial x} + y \frac{\partial}{\partial y} + z \frac{\partial}{\partial z} \right)^2 u + n^2 u = 0.$$

Ex. 4. Solve

$$(1-x^2)^2 \frac{\partial^2 u}{\partial x^2} + 2(1-x^2)(1-xy) \frac{\partial^2 u}{\partial x \partial y} + (1-xy)^2 \frac{\partial^2 u}{\partial y^2} + n^2 u$$
$$= 2x(1-x^2) \frac{\partial u}{\partial x} + (x+y-2x^2 y) \frac{\partial u}{\partial y}.$$

Ex. 5. Solve the equations:

$$\text{(i)} \quad \frac{\partial^2 z}{\partial x^2} - a^2 \frac{\partial^2 z}{\partial y^2} + 2ab \frac{\partial z}{\partial x} + 2a^2 b \frac{\partial z}{\partial y} = 0 \; ;$$

$$\text{(ii)} \quad mn \left(\frac{\partial^2 z}{\partial x^2} + \frac{\partial^2 z}{\partial y^2} \right) - (m^2+n^2) \frac{\partial^2 z}{\partial x \partial y} + mn \left(n \frac{\partial z}{\partial x} - m \frac{\partial z}{\partial y} \right)$$
$$= \cos(mx+ny) + \cos(kx+ly) \; ;$$

$$\text{(iii)} \quad a \frac{\partial^2 z}{\partial x^2} + 2h \frac{\partial^2 z}{\partial x \partial y} + b \frac{\partial^2 z}{\partial y^2} + 2g \frac{\partial z}{\partial x} + 2f \frac{\partial z}{\partial y} + cz = 0.$$

Ex. 6. Solve
$$f(\varpi) z = H_n,$$

where ϖ denotes the operator $x_1 \frac{\partial}{\partial x_1} + x_2 \frac{\partial}{\partial x_2} + \dots + x_m \frac{\partial}{\partial x_m}$, f is a polynomial function of ϖ, and H_n is a homogeneous function of n dimensions of the quantities $x_1, x_2, \dots, x_m$.

MISCELLANEOUS METHODS.

266. There are several partial differential equations which are of frequent occurrence in physical investigations. Solutions of these have frequently been obtained by methods, the application of most of which to equations other than those connected with their discovery is very limited. The two chief methods are integration by means of definite integrals and integration in series; but as each method is of special application only, and as the variations which arise owe their origin to the conditions imposed upon the function whose value is sought and not to any variety in the differential equations to which it can be applied, it is not possible to give here a full discussion. The discussion will, accordingly, be limited to a few examples; for fuller investigations, recourse must be had to the treatises on those branches of mathematical physics in which the differential equations occur.

267. Consider first an equation which can be integrated by both methods.

Such an equation is

$$\frac{\partial u}{\partial t} = a^2 \frac{\partial^2 u}{\partial x^2},$$

which arises in investigations connected with the conduction of heat. It is not without interest to indicate some of the different methods which may be applied to obtain a solution.

By the method of § 260, we may write

$$u = e^{a^2 t \frac{d^2}{dx^2}} \phi(x),$$

where $\phi(x)$ is arbitrary; expanding the differential operator, we obtain

$$u = \phi(x) + a^2 t \frac{d^2\phi}{dx^2} + \frac{a^4 t^2}{2!} \frac{d^4\phi}{dx^4} + \frac{a^6 t^3}{3!} \frac{d^6\phi}{dx^6} + \cdots.$$

This solution contains *one* arbitrary function.

We may proceed otherwise thus: the solution of the equation

$$\frac{d^2 u}{dx^2} = \lambda^2 u$$

is

$$u = e^{\lambda x} A + e^{-\lambda x} B,$$

where A and B are independent of x; so that we may express the solution of the equation

$$\frac{\partial^2 u}{\partial x^2} = \frac{1}{a^2} \frac{\partial u}{\partial t}$$

in the form

$$u = e^{\frac{x}{a}\left(\frac{d}{dt}\right)^{\frac{1}{2}}} \psi(t) + e^{-\frac{x}{a}\left(\frac{d}{dt}\right)^{\frac{1}{2}}} \chi(t),$$

where ψ and χ are arbitrary functions. In order to free the result from symbolical operations, which would require interpretation if they remained, we change the arbitrary functions to f and F, where

$$f(t) = \psi(t) + \chi(t),$$

$$F(t) = \left(\frac{d}{dt}\right)^{\frac{1}{2}} \{\psi(t) - \chi(t)\};$$

then, since ψ and χ are arbitrary, both f and F will be arbitrary, whatever interpretation be assigned to $\left(\frac{d}{dt}\right)^{\frac{1}{2}}$. When the symbolical operators in the first form of solution involving ψ and χ are expanded, and the terms of the same order in differentiation are gathered together, the solution becomes

$$u = f(t) + \frac{x^2}{2! \, a^2} \frac{df}{dt} + \frac{x^4}{4! \, a^4} \frac{d^2f}{dt^2} + \cdots$$

$$+ \frac{x}{a} F(t) + \frac{x^3}{3! \, a^3} \frac{dF}{dt} + \frac{x^5}{5! \, a^5} \frac{d^2F}{dt^2} + \cdots.$$

This solution contains *two* arbitrary functions.

268. It may at first sight seem paradoxical that two perfectly general solutions of the same differential equation can be obtained of apparently so different a character. The difficulty will disappear if it be noticed that the equation is only of the first order in t while it is of the second order in x; the former solution contains only a single arbitrary function of x, which is all that can be expected in the case of an equation of the first order; the second solution contains two arbitrary functions of t, which is the number of arbitrary functions to be expected in the case of an equation of the second order.

If we assume that all the arbitrary functions can be expanded in positive integral powers of their arguments, we are able to transform one of these solutions into the other. For let

$$\phi(x) = \sum_{n=0}^{n=\infty} \frac{A_n}{n!} \left(\frac{x}{a}\right)^n,$$

where the coefficients A_n are arbitrary; and let this value be substituted in the first solution. Then the term independent of x is

$$A_0 + A_2 t + \frac{A_4}{2!} t^2 + \frac{A_6}{3!} t^3 + \ldots\ldots,$$

which is a series with arbitrary coefficients and so may be denoted by $f(t)$, where f is arbitrary; the coefficient of $\left(\dfrac{x}{a}\right)^2 \dfrac{1}{2!}$ is

$$A_2 + A_4 t + \frac{A_6}{2!} t^2 + \ldots\ldots,$$

that is, $\dfrac{df}{dt}$; and so for the other even powers of x. Thus the part of the solution depending upon the even powers of x is

$$f(t) + \frac{x^2}{2!\,a^2} \frac{df}{dt} + \frac{x^4}{4!\,a^4} \frac{d^2 f}{dt^2} + \ldots\ldots$$

Similarly collecting the terms which depend upon the odd powers of x, and writing

$$F(t) = A_1 + A_3 t + \frac{A_5}{2!} t^2 + \ldots\ldots,$$

which is another arbitrary function, we should obtain the second part of the second solution. It thus appears that the two algebraical expressions are equivalent, independently of the fact that they are both solutions of the differential equation.

SOLUTION BY DEFINITE INTEGRALS.

269. Now let the method of § 262 be applied. We substitute

$$u = e^{\alpha x + \beta t};$$

the necessary relation between the constants α and β is

$$\beta = a^2 \alpha^2,$$

so that
$$u = A e^{\alpha x + a^2 \alpha^2 t},$$

for all values of A and α, would be a solution. Instead of α write αi, so that solutions are given by

$$e^{-a^2 \alpha^2 t + \alpha x i}, \quad e^{-a^2 \alpha^2 t - \alpha x i},$$

and therefore by

$$A e^{-a^2 \alpha^2 t + \alpha (x - \lambda) i}, \quad B e^{-a^2 \alpha^2 t - \alpha (x - \lambda) i},$$

where λ is any constant, and A and B are arbitrary functions of λ. These solutions may be replaced by

$$A'e^{-a^2a^2t}\cos\alpha\,(x-\lambda),\quad B'e^{-a^2a^2t}\sin\alpha\,(x-\lambda),$$

where A' and B' are arbitrary functions of λ.

Further, the sum of any number of solutions is also a solution. Consider the solution obtained by summing any number of terms of the first form for all values of λ and α, and by assuming that, while A' is an arbitrary function of λ, the *form* of the arbitrary function is the same for different values of λ. (The corresponding terms which would arise from the second may be deemed included in this because, so far as the variable part is concerned, we need only to change λ into $\lambda - \dfrac{\pi}{2\alpha}$ to obtain the first.)

Let then $\qquad\qquad A' = \psi\,(\lambda)\,d\lambda,$

and suppose summation to take place for all values of λ between $-\infty$ and $+\infty$; the corresponding solution is

$$\int_{-\infty}^{\infty} e^{-a^2a^2t}\cos\alpha\,(x-\lambda)\,\psi\,(\lambda)\,d\lambda.$$

This again may be multiplied by any function of α and the summation taken for all values of α; as it stands, the function is an even one of α; and so, if the factor be taken as $d\alpha$, it will suffice to take 0 and ∞ as the limits of α. Consequently, we have as the solution

$$u = \int_0^{\infty} d\alpha \int_{-\infty}^{\infty} e^{-a^2a^2t}\cos\alpha\,(x-\lambda)\,\psi\,(\lambda)\,d\lambda.$$

The solution in this form is specially suitable for the case in which u is to satisfy some condition, for instance that

$$u = f(x)$$

when t is zero; we then are to have

$$f(x) = \int_0^{\infty} d\alpha \int_{-\infty}^{\infty} \cos a\,(x-\lambda)\,\psi\,(\lambda)\,d\lambda.$$

But, by Fourier's theorem, the value of the right-hand side is $\pi\psi\,(x)$, so that ψ is a determinate function; thus

$$u = \frac{1}{\pi} \int_0^{\infty} d\alpha \int_{-\infty}^{\infty} e^{-a^2a^2t}\cos a\,(x-\lambda)\,f(\lambda)\,d\lambda.$$

$$\text{(Riemann.)}$$

Ex. Obtain a solution of the equation

$$\frac{\partial^2 u}{\partial t^2} = a^2 \frac{\partial^2 u}{\partial x^2},$$

which is such that

$$u = f(x), \text{ and } \frac{\partial u}{\partial t} = F(x),$$

when $t = 0$.

The result is

$$u = \tfrac{1}{2}\{f(x+at) + f(x-at)\} + \frac{1}{2a}\int_{x-at}^{x+at} F(\lambda)\, d\lambda.$$

(Riemann.)

270. We may again solve the equation by a method, due originally to Laplace and extended by Poisson.

We have

$$\int_{-\infty}^{\infty} e^{-u^2}\, du = \pi^{\frac{1}{2}};$$

hence, writing $u - l$ for u, where l is independent of u,

$$\int_{-\infty}^{\infty} e^{-u^2 + 2ul}\, du = \pi^{\frac{1}{2}} e^{l^2}.$$

When l is any differential operation to be performed, this relation indicates that the symbolical operation e^{l^2} can be expressed provided e^{2ul} can be expressed.

This method may be applied to the equation

$$\frac{\partial u}{\partial t} = a^2 \frac{\partial^2 u}{\partial x^2};$$

for we have

$$u = e^{\left(t^{\frac{1}{2}} a \frac{d}{dx}\right)^2} f(x),$$

where $f(x)$ is an arbitrary function independent of t. The foregoing formula in equivalent operators may be applied, if l be replaced by $at^{\frac{1}{2}} \dfrac{d}{dx}$; and thus we have

$$u = \pi^{-\frac{1}{2}} \int_{-\infty}^{\infty} e^{-u^2 + 2uat^{\frac{1}{2}} \frac{d}{dx}} f(x)\, du$$

$$= \pi^{-\frac{1}{2}} \int_{-\infty}^{\infty} e^{-u^2} f(x + 2uat^{\frac{1}{2}})\, du.$$

Another form may be given to this result by substituting λ for $x + 2uat^{\frac{1}{2}}$. Then u becomes

$$\frac{1}{2a\,(\pi t)^{\frac{1}{2}}} \int_{-\infty}^{\infty} e^{-\frac{(x-\lambda)^2}{4a^2 t}} f(\lambda)\, d\lambda.$$

Now $f(\lambda)$ is an arbitrary function; if we choose to assume its value to be zero everywhere except when $\lambda = r$, and then write $f(\lambda)\, d\lambda = H$, we have

$$u = \frac{H}{2a\,(\pi t)^{\frac{1}{2}}} e^{-\frac{(x-r)^2}{4a^2 t}}.$$

Ex. 1. Prove that, if u satisfy the conditions

(i) $u = f(x)$ when $t = 0$,

(ii) $u = \phi(t)$ when $x = 0$,

then its value is

$$\frac{1}{2a\,(\pi t)^{\frac{1}{2}}} \int_0^{\infty} \left\{ e^{-\frac{(x-\lambda)^2}{4a^2 t}} - e^{-\frac{(x+\lambda)^2}{4a^2 t}} \right\} f(\lambda)\, d\lambda + \frac{x}{2a\pi^{\frac{1}{2}}} \int_0^t e^{-\frac{x^2}{4a^2(t-\lambda)}} \frac{\phi(\lambda)\, d\lambda}{(t-\lambda)^{\frac{3}{2}}}.$$

Ex. 2. Obtain a solution of the equation

$$\frac{\partial^2 \phi}{\partial t^2} + b^2 \left(\frac{\partial^4 \phi}{\partial x^4} + 2 \frac{\partial^4 \phi}{\partial x^2 \partial y^2} + \frac{\partial^4 \phi}{\partial y^4} \right) = 0$$

in the form

$$y = \iint f(x + 2utb^{\frac{1}{2}},\ y + 2vtb^{\frac{1}{2}}) \sin(u^2 + v^2)\, du\, dv$$

$$+ \iint F(x + 2utb^{\frac{1}{2}},\ y + 2vtb^{\frac{1}{2}}) \cos(u^2 + v^2)\, du\, dv.$$

Ex. 3. Verify that

$$u = \frac{1}{4\pi} \int_0^{2\pi} d\phi \int_0^{\pi} t f(x + at\sin\theta\cos\phi,\ y + at\sin\theta\sin\phi,\ z + at\cos\theta) \sin\theta\, d\theta$$

$$+ \frac{1}{4\pi} \frac{d}{dt} \int_0^{2\pi} d\phi \int_0^{\pi} t F(x + at\sin\theta\cos\phi,\ y + at\sin\theta\sin\phi,\ z + at\cos\theta) \sin\theta\, d\theta$$

satisfies the differential equation

$$\frac{\partial^2 u}{\partial t^2} = a^2 \left(\frac{\partial^2 u}{\partial x^2} + \frac{\partial^2 u}{\partial y^2} + \frac{\partial^2 u}{\partial z^2} \right),$$

and is such that $u = F(x, y, z)$ and $\dfrac{\partial u}{\partial t} = f(x, y, z)$, when $t = 0$.

Ex. 4. Obtain the value of the integral

$$\int\int e^{\alpha x + \beta y + \gamma z}\, dS,$$

taken over the surface of a sphere whose centre is the origin and radius R, in the form

$$4\pi \frac{R}{p} \sinh (Rp),$$

where

$$p^2 = a^2 + \beta^2 + \gamma^2.$$

Hence shew that over the surface of any sphere the mean value of a function, which satisfies the equation

$$\frac{\partial^2 u}{\partial x^2} + \frac{\partial^2 u}{\partial y^2} + \frac{\partial^2 u}{\partial z^2} = 0,$$

and which, for all points within the sphere, is expressible by a converging series, is equal to the value of the function at the centre of the sphere.

Further information on this part of the subject and, in particular, on the applications in physical investigations, will be found in H. Weber's edition of Riemann's *Die partielle Differentialgleichungen der mathematischen Physik.*

SOLUTION IN SERIES.

271. Consider now a case of integration by means of series.

The most important equation, to which this method is applied, is the equation

$$\frac{\partial^2 u}{\partial x^2} + \frac{\partial^2 u}{\partial y^2} + \frac{\partial^2 u}{\partial z^2} = 0,$$

which continually occurs in physical investigations. To solve it by the method under consideration, we change the independent variables from x, y, z, to r, θ, ϕ, given by the relations

$$x = r \sin \theta \cos \phi, \quad y = r \sin \theta \sin \phi, \quad z = r \cos \theta,$$

which will in effect be changing from the Cartesian to the polar coordinates of a point in space. The equation becomes

$$r \frac{\partial^2 (ru)}{\partial r^2} + \frac{1}{\sin \theta} \frac{\partial}{\partial \theta}\left(\sin \theta \frac{\partial u}{\partial \theta}\right) + \frac{1}{\sin^2 \theta}\frac{\partial^2 u}{\partial \phi^2} = 0;$$

and, if another change be made by writing μ instead of $\cos \theta$, the resulting form is

$$r \frac{\partial^2 (ru)}{\partial r^2} + \frac{\partial}{\partial \mu}\left\{ (1 - \mu^2) \frac{\partial u}{\partial \mu}\right\} + \frac{1}{1 - \mu^2}\frac{\partial^2 u}{\partial \phi^2} = 0.$$

272. First, let a solution be desired which is to be a function of r only, that is, of $(x^2 + y^2 + z^2)^{\frac{1}{2}}$, so that it will be a specially symmetrical solution; the equation then reduces to

$$\frac{\partial^2 (ru)}{\partial r^2} = 0,$$

and therefore

$$u = A + \frac{B}{r}.$$

In a similar way, a solution which would be a function of θ alone, and one which would be a function of ϕ alone, may be deduced; but they are not so useful as that just obtained.

Next, suppose that solutions which are not functions of r alone may be expanded in a series of integral powers of r; and in u let there be a term

$$r^n u_n,$$

where u_n is independent of r but may be a function of θ and ϕ, the form of which is still to be determined. Then, when the value of u is substituted, the term on the left-hand side of the differential equation corresponding to this particular term of u is

$$r^n \left[n (n + 1) u_n + \frac{\partial}{\partial \mu} \left\{ (1 - \mu^2) \frac{\partial u_n}{\partial \mu} \right\} + \frac{1}{1 - \mu^2} \frac{\partial^2 u_n}{\partial \phi^2} \right];$$

and the sum of all these terms is to be zero for all values of the independent variables. The foregoing is the only term which involves the nth power of r; in order to have the equation satisfied, the coefficient of the term must vanish for every value of n. Hence u_n is determined by

$$n (n + 1) u_n + \frac{\partial}{\partial \mu} \left\{ (1 - \mu^2) \frac{\partial u_n}{\partial \mu} \right\} + \frac{1}{1 - \mu^2} \frac{\partial^2 u_n}{\partial \phi^2} = 0,$$

and therefore $r^n u_n$ is a solution of the original differential equation. The coefficients of the terms involving the differential coefficients of u_n do not depend upon n; and the coefficient of u_n is unaltered if for n there be substituted $-(n + 1)$; hence $r^{-(n+1)} u_n$ is another solution of the original equation. These two solutions just obtained may be combined into one, so as to give a solution

$$\left(A_n r^n + \frac{B_n}{r^{n+1}} \right) u_n,$$

A_n and B_n being arbitrary constants. Thus the general value of u is

$$u = \sum_{n=0}^{n=\infty} \left(A_n r^n + \frac{B_n}{r^{n+1}} \right) u_n,$$

provided u_n be determined by the equation

$$\frac{\partial}{\partial \mu} \left\{ (1 - \mu^2) \frac{\partial u_n}{\partial \mu} \right\} + \frac{1}{1 - \mu^2} \frac{\partial^2 u_n}{\partial \phi^2} + n(n+1) u_n = 0.$$

273. The general solution of this equation would give u_n as a function of θ and ϕ; consider the case in which u_n is a function of θ only. The value is then determined by

$$\frac{d}{d\mu} \left\{ (1 - \mu^2) \frac{du_n}{d\mu} \right\} + n(n+1) u_n = 0,$$

the independent particular integrals of which (§§ 90, 91) are $P_n(\mu)$ and $Q_n(\mu)$; the corresponding terms in u are

$$\left(A_n r^n + \frac{B_n}{r^{n+1}} \right) P_n(\mu) + \left(A_n' r^n + \frac{B_n'}{r^{n+1}} \right) Q_n(\mu).$$

In most physical investigations, the term dependent upon $Q_n(\mu)$ has to be rejected; and then the general value of u, expressed as a function of r and θ, that is, of z and $(x^2 + y^2)^{\frac{1}{2}}$, is

$$u = \sum_{n=0}^{n=\infty} \left\{ \left(A_n r^n + \frac{B_n}{r^{n+1}} \right) P_n(\mu) \right\},$$

in which the A's and B's are arbitrary constants. It will be noticed that the solution formerly obtained, viz.,

$$A + \frac{B}{r},$$

is the particular case obtained, by making all these arbitrary constants zero except A_0 and B_0, and by remembering that $P_0(\mu)$ is a constant.

274. Consider now the general case in which u_n is a function of θ and ϕ; it may be expanded in a series of trigonometrical functions of multiples of ϕ, the coefficients of which are functions of μ. Any term of the series for u_n may be denoted by

$$v_n^{(\sigma)} \cos \sigma \phi,$$

where v is a function of μ only; and, just as in the case of the separate terms in u considered as involving different powers of r when each such term was a solution of the equation, this will be a solution of the equation giving u_n. Substituting and dividing out by $\cos \sigma\phi$, we find that $v_n^{(\sigma)}$ is determined by the equation

$$\frac{d}{d\mu}\left\{(1-\mu^2)\frac{dv_n^{(\sigma)}}{d\mu}\right\} + n(n+1)v_n^{(\sigma)} = \frac{\sigma^2}{1-\mu^2}v_n^{(\sigma)}.$$

This equation would also have been obtained by the substitution of $v_n^{(\sigma)} \sin \sigma\phi$ in the equation for u_n; and therefore the solution of the equation in u_n is

$$\sum_{\sigma=1}^{\sigma=n}\left\{E_\sigma \sin \sigma\phi + F_\sigma \cos \sigma\phi\right\}v_n^{(\sigma)},$$

the value $\sigma = 0$ not being here included, since it gives terms independent of ϕ which have already been found.

Now, by Ex. 12, Chap. v., p. 202, the solution of the equation giving $v_n^{(\sigma)}$ is

$$v_n^{(\sigma)} = (1-\mu^2)^{\frac{1}{2}\sigma}\frac{d^\sigma y_n}{d\mu^\sigma},$$

where y_n is a solution of the equation when σ is zero and thus may be either P_n or Q_n. Hence the corresponding term in u_n is

$$(E_\sigma \sin \sigma\phi + F_\sigma \cos \sigma\phi)(1-\mu^2)^{\frac{1}{2}\sigma}\frac{d^\sigma P_n}{d\mu^\sigma}$$
$$+ (E'_\sigma \sin \sigma\phi + F'_\sigma \cos \sigma\phi)(1-\mu^2)^{\frac{1}{2}\sigma}\frac{d^\sigma Q_n}{d\mu^\sigma}.$$

The term involving Q_n often has to be rejected in physical investigations; the suitable value of u_n then is

$$\sum_{\sigma=1}^{\sigma=n}(1-\mu^2)^{\frac{1}{2}\sigma}(E_\sigma \sin \sigma\phi + F_\sigma \cos \sigma\phi)\frac{d^\sigma P_n}{d\mu^\sigma},$$

it being obviously useless to include values of σ higher than n.

The sum of any number of solutions of the original equation is a solution; and therefore the most general value of u expressed in a series is

$$u = A + \frac{B}{r}$$
$$+ \sum_{n=1}^{n=\infty} \left\{ \left(A_n r^n + \frac{B_n}{r^{n+1}} \right) P_n \right\}$$
$$+ \sum_{n=1}^{n=\infty} \left[\sum_{\sigma=1}^{\sigma=n} (1-\mu^2)^{\frac{1}{2}\sigma} \frac{d^\sigma P_n}{d\mu^\sigma} \left\{ \left(A'_n r^n + \frac{B'_n}{r^{n+1}} \right) \sin \sigma\phi \right. \right.$$
$$\left. \left. + \left(A''_n r^n + \frac{B''_n}{r^{n+1}} \right) \cos \sigma\phi \right\} \right].$$

We have omitted from the foregoing general value, (i) the terms which would arise from the part of u independent of r and ϕ, which can easily be proved to be

$$C \log \frac{1+\mu}{1-\mu},$$

(ii) the term dependent upon ϕ alone which obviously is $M\phi$, and (iii) the terms usually rejected as unsuitable in physical investigations.

Any further investigations on the solution of the equation are connected either with other equivalent forms of solution or with the particular solutions obtained by a determination of the constants in accordance with imposed conditions. For them, recourse should be had to the authorities on the several subjects in applied mathematics in which this equation arises; in particular, those quoted on p. 187 will be found of great value.

Ex. 1. Solve the equation

$$\frac{\partial^2 u}{\partial x^2} + \frac{\partial^2 u}{\partial y^2} = 0$$

in series, by transforming the polar coordinates.

Ex. 2. Prove that the equation

$$\frac{\partial^2 u}{\partial t^2} = a^2 \left(\frac{\partial^2 u}{\partial x^2} + \frac{\partial^2 u}{\partial y^2} + \frac{\partial^2 u}{\partial z^2} \right)$$

has a solution of the form

$$u = e^{akti} \sum_{n=0}^{n=\infty} \frac{1}{r} P_n \{ A e^{-ikr} f_n(ikr) + B e^{ikr} f_n(-ikr) \},$$

where

$$f_n(z) = 1 + \frac{n(n+1)}{2z} + \frac{(n-1)n(n+1)(n+2)}{2 \cdot 4 \cdot z^2} + \frac{(n-2)\ldots(n+3)}{2 \cdot 4 \cdot 6 \cdot z^3} + \cdots$$

$$\cdots\cdots + \frac{1 \cdot 2 \cdot 3 \ldots 2n}{2 \cdot 4 \cdot 6 \ldots 2n \cdot z^n}.$$

Obtain a more general solution which is not independent of the spherical coordinate ϕ.

(Stokes.)

Ex. 3. Shew that the general solution of the equation

$$a^2 \left(\frac{\partial^2 u}{\partial x^2} + \frac{\partial^2 u}{\partial y^2} \right) = \frac{\partial^2 u}{\partial t^2},$$

or, by transformation to plane polar coordinates, its equivalent

$$a^2 \left(\frac{\partial^2 u}{\partial r^2} + \frac{1}{r} \frac{\partial u}{\partial r} + \frac{1}{r^2} \frac{\partial^2 u}{\partial \theta^2} \right) = \frac{\partial^2 u}{\partial t^2},$$

can be expressed in terms of Bessel's functions as the sum of two terms of the form

$$u = \cos akt \sum_{n=0}^{n=\infty} [\{AJ_n(kr) + BY_n(kr)\} \cos n\theta + \{A'J_n(kr) + B'Y_n(kr)\} \sin n\theta].$$

Ampère's Method for Equations of the Second Order.

275. The methods, given in §§ 238—252 for the construction of a primitive of a given equation of the second order, have proceeded upon the supposition that an intermediate integral exists; and they have been directed towards the construction of that intermediate integral. Usually, however, the supposition is not justified: that is to say, an equation must satisfy conditions in order that an intermediate integral may exist. Thus the coefficients R, S, T, U, V, in the equation

$$Rr + 2Ss + Tt + U(rt - s^2) = V,$$

must (§ 239) satisfy two conditions in order that an intermediate integral may exist. An equation of the second order, which is not of this form, may or may not possess an intermediate integral; at present, the discrimination can be made only by the process of §§ 251, 252, or by some similar process. In every case where there is no intermediate integral, the various methods are ineffective.

Other methods suggested are somewhat of a tentative character; the most general, in its bearing, is the method associated with the name of Ampère. We proceed to give a brief outline of it[*] in its application to equations, which possess a primitive of a specified type of a fairly general character.

276. In the various examples which have been considered, we have seen that the primitive involves two arbitrary functions in its expression. This character is not restricted to equations which have an intermediate integral; thus it is easy to prove that the equation

$$r - t = 2n\frac{p}{x},$$

where n is a positive integer, possesses no intermediate integral, and to verify that its primitive is

$$z = \sum_{m=0}^{n} (-1)^m \frac{2^m}{m!} \frac{\binom{n}{m}}{\binom{2n}{m}} x^m \{\phi^{(m)}(x-y) + \psi^{(m)}(x+y)\},$$

where ϕ and ψ denote arbitrary functions of their arguments. Moreover, it is a consequence of Cauchy's general theorem[†] as to the existence of an integral of a partial equation of any order, that the number of independent arbitrary functions involved in the primitive is equal to the order of the equation; so that, for our case in general, the primitive will contain two arbitrary functions.

Accordingly, we consider equations which have a primitive involving explicitly two arbitrary functions, free from all operations of quadrature; and we shall allow the primitive to contain derivatives of these up to any finite order. Thus the preceding equation

$$r - t = 2n\frac{p}{x}$$

would be admitted; but we should exclude the equation

$$r - t + \frac{4p}{x+y} = 0,$$

---

[*] For a fuller account, see the author's *Theory of Differential Equations*, vol. VI., ch. XVII. Ampère's two fundamental memoirs are contained in the *Journal de l'École Polytechnique*, Cah. XVII. (1815), pp. 549—611, Cah. XVIII. (1819), pp. 1—188.

[†] See the author's *Theory of Differential Equations*, vol. VI., § 184.

because its primitive, which can be expressed either by

$$(x+y)\,z + e^{\frac{2y}{x+y}}\,F(x+y) + e^{\frac{2y}{x+y}}\int e^{-\frac{2y}{a}}f(2y-a)\,dy = 0,$$

(where, after integration, $x+y$ is to be substituted for a), or by

$$z + e^{\frac{2y}{x+y}}\,G(x+y) = g(y-x) + (x+y)\,g'(y-x) + (x+y)^2 g''(y-x) + \dots,$$

is not of the specified form.

Each of the arbitrary functions in the primitive has its argument; the two arguments may be the same or may be different, according to the equation. Let α denote the sole argument when there is only one, and one of the arguments when they differ.

277. Now change the independent variables, choosing α as one of them and temporarily keeping x as the other*; and denote the changed partial derivation with respect to x and α by $\dfrac{\delta}{\delta x}$ and $\dfrac{\delta}{\delta \alpha}$. For any function u, we have

$$\frac{\delta u}{\delta x}\,dx + \frac{\delta u}{\delta \alpha}\,d\alpha = du$$
$$= \frac{\partial u}{\partial x}\,dx + \frac{\partial u}{\partial y}\left(\frac{\delta y}{\delta x}\,dx + \frac{\delta y}{\delta \alpha}\,d\alpha\right),$$

so that

$$\frac{\delta u}{\delta x} = \frac{\partial u}{\partial x} + \frac{\partial u}{\partial y}\frac{\delta y}{\delta x}, \qquad \frac{\delta u}{\delta \alpha} = \frac{\partial u}{\partial y}\frac{\delta y}{\delta \alpha}.$$

Denoting $\dfrac{\delta u}{\delta x}$ by u', for all functions u, when α is the other independent variable, we have

$$z' = p + q y', \qquad \frac{\delta z}{\delta \alpha} = q\,\frac{\delta y}{\delta \alpha},$$

$$p' = r + s y', \qquad \frac{\delta p}{\delta \alpha} = s\,\frac{\delta y}{\delta \alpha},$$

$$q' = s + t y', \qquad \frac{\delta q}{\delta \alpha} = t\,\frac{\delta y}{\delta \alpha};$$

---

* If the subsequently obtained subsidiary equations prove inconsistent with one another, then we should take a and y as the temporary set of independent variables.

and therefore, when we retain the relation

$$\frac{\delta q}{\delta \alpha} = t \frac{\delta y}{\delta \alpha}$$

as defining t, we have

$$s = q' - ty',$$
$$r = p' - q'y' + ty'^2.$$

Let these values of r and s be substituted in the equation

$$f(x, y, z, p, q, r, s, t) = 0,$$

which will be assumed to be polynomial in the derivatives r, s, t; and denote the result of the substitution in this equation by

$$f_0 + tf_1 + t^2f_2 + \dots + t^nf_n = 0,$$

where the quantities $f_0, f_1, f_2, \dots, f_n$, involve $x, y, z, p, q, y', p', q'$, together with z' as given by

$$z' = p + qy'.$$

In the primitive of the equation, the number of derivatives of the arbitrary function of α is finite; let the number be m. Then the number of derivatives of that arbitrary function, which occur in p or in q or in both, is $m + 1$.

When x and α are made the independent variables, so that we take α as non-varying when we form p' and q', these quantities p' and q' still will involve derivatives of the arbitrary function of order no higher than $m + 1$. But the operation $\frac{\delta}{\delta \alpha}$ does affect those derivatives; and therefore $\frac{\delta q}{\delta \alpha}$ will contain the derivative of the arbitrary function, which is of order $m + 2$; that is, the quantity t will involve this derivative of the arbitrary function, of order higher than the derivatives which can occur in $f_0, f_1, \dots, f_n$.

The differential equation, in any form, and therefore in the form

$$f_0 + tf_1 + t^2f_2 + \dots + t^nf_n = 0,$$

has to be satisfied in connection with the primitive; and t contains a derivative of the arbitrary function of order higher than any which occurs in the primitive or in $f_0, f_1, \ldots, f_n$. Hence the coefficient of the highest power of t must vanish; and so on, in the complete succession of the decreasing powers of t. Thus we have

$$f_n = 0, \; f_{n-1} = 0, \; \ldots, \; f_1 = 0, \; f_0 = 0.$$

These equations involve z, p, q, x, y, y', p', q'; there is also the equation

$$z' = p + qy';$$

and there is no derivative with respect to α. Consequently, these deduced equations are formally a system of ordinary equations for the determination of z, p, q, y, in terms of x, the arbitrary elements in their set of integrals being functions of α, as arbitrary as we please. As we have a permanent equation $z' = p + qy'$, which must be associated with every equation to be solved, it follows that we cannot have more than three algebraically independent equations equivalent to the set

$$f_n = 0, \; f_{n-1} = 0, \; \ldots, \; f_1 = 0, \; f_0 = 0.$$

278. Accordingly, we resolve these equations, so as to have their simplest equivalents. If there is only one resulting set, we associate it with the argument α which temporarily is non-varying. If there are two resulting distinct sets, we associate them with arguments α and β respectively. If there are more than two resulting sets, we infer that the propounded equation of the second order does not possess a primitive of the specified type.

When there is one set of equations, we require integrals of the set. The arbitrary element in one such integral can, without loss of generality, be made equal to α; and then the arbitrary elements in other integrals are made arbitrary functions of α.

When there are two sets of equations, it is desirable to have at least one integral of each set. For an integral of the one set, we make the arbitrary element equal to α; for an integral of the other set, we make it equal to β. We then make α and β to be the independent variables for the whole of the two sets of equations; the modified equations are to be treated as a system of equations, expressing x, y, z, p, q, in terms of α and β. The equations, which express x, y, z, in terms of α and β, constitute (either explicitly or implicitly) the primitive of the differential equation.

Some examples will illustrate the detailed process of the method.

Ex. 1. Required a primitive of the equation

$$st + x\,(rt - s^2)^2 = 0.$$

Following the process in the text, we take

$$s = q' - ty', \quad r = p' - q'y' + ty'^2;$$

and we make the resulting equation evanescent as an equation in t. Thus we find

$$xq'^4 = 0,$$
$$q' - 2xq'^2\,(p' + q'y') = 0,$$
$$-y' + x\,(p' + q'y')^2 = 0.$$

Consequently, we have

$$q' = 0, \quad xp'^2 - y' = 0,$$

a single system of two not inconsistent equations.

As $q' = 0$, that is, $\dfrac{\delta q}{\delta x} = 0$, we have

$$q = \text{any function of } a \text{ alone} = \phi'(a),$$

where ϕ is an arbitrary function. Also (p. 535)

$$\frac{\delta p}{\delta a} + y'\,\frac{\delta q}{\delta a} = (s + ty')\,\frac{\delta y}{\delta a} = q'\,\frac{\delta y}{\delta a} = 0;$$

and therefore, by the equation $xp'^2 - y' = 0$, we have

$$\frac{\delta p}{\delta a} + x\left(\frac{\delta p}{\delta x}\right)^2 \phi''(a) = 0,$$

an equation of the first order for p. Its complete integral is

$$p = l + 2kx^{\frac{1}{2}} - k^2\phi'(a),$$

where k and l are arbitrary constants; and so

$$y' = xp'^2 = k^2,$$

whence

$$y = k^2x + \theta\,(a),$$

where θ is an arbitrary constant, that is, a is an arbitrary function of $y - k^2x$.

Further,

$$\frac{\delta z}{\delta a} = q\,\frac{\delta y}{\delta a} = \phi'(a)\,\theta'(a),$$

$$\frac{\delta z}{\delta x} = p + qy' = 2kx^{\frac{1}{2}} + l,$$

on substitution; hence

$$z = lx + \tfrac{4}{3} kx^{\frac{3}{2}} + G(a)$$

$$= lx + \tfrac{4}{3} kx^{\frac{3}{2}} + F(y - k^2 x),$$

where F is an arbitrary function, while k and l are arbitrary constants.

We have taken the complete integral of

$$\frac{\delta p}{\delta a} + x \left(\frac{\delta p}{\delta x}\right)^2 \phi''(a) = 0.$$

When we take its general integral, the second arbitrary function arises for occurrence in the final primitive.

Ex. 2. Required a primitive of the equation

$$(1 + q^2) r - 2pqs + (1 + p^2) t = 0.$$

(It is the equation of minimal surfaces; and it expresses the property that the principal radii of curvature are equal and opposite.)

Using the equations

$$s = q' - ty', \quad r = p' - q'y' + ty'^2,$$

as before, we find

$$(1 + q^2) y'^2 + 2pqy' + 1 + p^2 = 0,$$

$$(1 + q^2)(p' - q'y') - 2pqq' = 0,$$

together with

$$z' - p - qy' = 0.$$

Let w denote $(1 + p^2 + q^2)^{\frac{1}{2}}$; the first of these equations gives

$$(1 + q^2) y' + pq = \pm iw.$$

We thus have two sets of equations, viz.

$$\left.\begin{aligned} y' + \frac{pq - iw}{1 + q^2} &= 0 \\[4pt] p' - \frac{pq + iw}{1 + q^2} q' &= 0 \\[4pt] z' - \frac{p + iqw}{1 + q^2} &= 0 \end{aligned}\right\}, \qquad \left.\begin{aligned} y'' + \frac{pq + iw}{1 + q^2} &= 0 \\[4pt] p'' - \frac{pq - iw}{1 + q^2} q'' &= 0 \\[4pt] z'' - \frac{p - iqw}{1 + q^2} &= 0 \end{aligned}\right\}.$$

For the first set, we denote the non-varying argument by a; for the second, we denote it by β; while $y' = \dfrac{\delta y}{\delta x}$ when a is constant, $y'' = \dfrac{\delta y}{\delta x}$ when β is constant, and so for p', q', z', p'', q'', z''.

The second equation in the first set of ordinary equations may be written

$$\frac{dp}{dq} = \frac{pq + iw}{1 + q^2}.$$

It is easily seen to be a Clairaut equation of the first order; and its primitive is

$$\frac{pq+iw}{1+q^2} = \text{constant.}$$

After the explanations in the text, we can take

$$\frac{pq+iw}{1+q^2} = a.$$

Similarly, an integral of the second set is given by

$$\frac{pq-iw}{1+q^2} = \beta.$$

We now proceed to make a and β the independent variables. In the first set of equations, y is a function of x and a; thus it comes to be a function of β and a, through the expression of x in terms of β and a, and therefore

$$\frac{\partial y}{\partial \beta} = y' \frac{\partial x}{\partial \beta} = -\beta \frac{\partial x}{\partial \beta}.$$

Similarly, in the second set, y is a function of x and β; thus it comes to be a function of a and β, through the expression of x in terms of β and a, and therefore

$$\frac{\partial y}{\partial a} = y'' \frac{\partial x}{\partial a} = -a \frac{\partial x}{\partial a}.$$

Consequently

$$\frac{\partial}{\partial a}\left(\beta \frac{\partial x}{\partial \beta}\right) = -\frac{\partial^2 y}{\partial a \partial \beta} = \frac{\partial}{\partial \beta}\left(a \frac{\partial x}{\partial a}\right),$$

and therefore

$$\frac{\partial^2 x}{\partial a \partial \beta} = 0,$$

so that

$$x = \phi'(a) + \psi'(\beta),$$

where ϕ and ψ are arbitrary functions. Then

$$dy = \frac{\partial y}{\partial a}\, da + \frac{\partial y}{\partial \beta}\, d\beta$$

$$= -a\phi''(a)\, da - \beta\psi''(\beta)\, d\beta;$$

hence

$$y = \phi(a) - a\phi'(a) + \psi(\beta) - \beta\psi'(\beta).$$

Finally, we have

$$dz = \left(p \frac{\partial x}{\partial a} + q \frac{\partial y}{\partial a}\right) da + \left(p \frac{\partial x}{\partial \beta} + q \frac{\partial y}{\partial \beta}\right) d\beta$$

$$= (p - aq)\phi''(a)\, da + (p - \beta q)\psi''(\beta)\, d\beta.$$

But

$$p - aq = i(1+a^2)^{\frac{1}{2}}, \quad p - \beta q = i(1+\beta^2)^{\frac{1}{2}};$$

and therefore

$$z = i \int (1+a^2)^{\frac{1}{2}} \phi''(a)\,da + i \int (1+\beta^2)^{\frac{1}{2}} \psi''(\beta)\,d\beta.$$

The three expressions for x, y, z, when combined, constitute the primitive of the equation as constructed by Ampère and Legendre.

When (as is permissible, without loss of generality) we change the variables a and β to v and u, by the equations

$$a = i\frac{v^2+1}{v^2-1}, \quad \beta = -i\frac{u^2+1}{u^2-1},$$

$$i(1+a^2)^{\frac{1}{2}} = \frac{2v}{1-v^2}, \quad i(1+\beta^2) = \frac{2u}{1-u^2},$$

and we change the arbitrary functions by the relations

$$\phi'(a) = (1-v^2)\,V'' + 2v\,V' - 2V,$$

$$\psi'(\beta) = (1-u^2)\,U'' + 2u\,U' - 2U,$$

where U and V are arbitrary functions of u and of v respectively, we find

$$\left.\begin{array}{l} x = (1-v^2)\,V'' + 2v\,V' - 2V + (1-u^2)\,U'' + 2u\,U' - 2U \\[4pt] y = i\{(1+v^2)\,V'' + 2v\,V' + 2V\} - i\{(1+u^2)\,U'' - 2u\,U' + 2U\} \\[4pt] z = 2v\,V'' - 2V' + 2u\,U'' - 2U' \end{array}\right\}.$$

These equations, expressing x, y, z, in terms of two parameters u and v, and of two arbitrary U and V functions of those parameters respectively, constitute a primitive of the equation. The primitive, in this form, is Weierstrass's set of equations for a minimal surface.

Note. For the developed inferences from this equation as made by Weierstrass, Schwarz, and others, reference may be made to the author's *Lectures on Differential Geometry*, chapter VIII. Above all, as regards these inferences, Darboux's *Théorie générale des surfaces*, vol. I., book III., should be consulted.

Ex. 3. Obtain the primitive of the equation

$$(r - pt)^2 = q^2 rt$$

in the form

$$x = \frac{1}{a}\psi'(\beta) + \phi''''(a),$$

$$y = a\psi'(\beta) - a^2\phi''''(a) + 2a\phi'''(a) - 2\phi''(a),$$

$$z = \tfrac{1}{3}(2\beta - a^3)\,\psi'(\beta) - \tfrac{2}{3}\psi(\beta) + a^4\phi''''(a) - 4a^3\phi'''(a) + 12a^2\phi''(a) - 24a\phi'(a) + 24\phi(a).$$

Ex. 4. Shew that, when the equation in Ex. 3 is resolved into two equations which are linear in r and t, each of these equations has an intermediate integral; and obtain these integrals.

Ex. 5. Shew that the primitive of the equation

$$qr + (zq - p)\, s - zpt = 0$$

can be represented in the form

$$\left.\begin{array}{l} x = \phi'(a) + \psi'(z) \\ y = a\phi'(a) - \phi(a) + z\psi'(z) - \psi(z) \end{array}\right\}.$$

Ex. 6. Integrate the equations :

   (i) $(q + yt)\,(r + 1) = s\,(ys - p - x)$;

   (ii) $2pqyr + (p^2 y + qx)\, s + xpt = p^2 q\,(rt - s^2) + xy$;

   (iii) $\{pqx^2 + (1 + q^2)\, xy\}\, r + \{(1 + q^2)\, y^2 - (1 + p^2)\, x^2\}\, s$

$$- \{pqy^2 + (1 + p^2)\, xy\}\, t = 0.$$

IMSCHENETSKY'S GENERALISATION OF A LIMITED PRIMITIVE.

279. It may be added that, when an equation (such as

$$Rr + 2Ss + Tt + U\,(rt - s^2) = V$$

or any other equation) possesses an intermediate integral, the analysis connected with Ampère's method is somewhat simplified. In actual practice, difficulty may arise in developing the method beyond the stage of constructing an intermediate integral or two independent intermediate integrals.

In that case, we proceed as before from the single intermediate integral (by Charpit's method) or from the combination of the two intermediate integrals (as in § 245) to the general integral of the equation; and this integral will usually involve either two arbitrary functions or three arbitrary constants. But it is not the most general integral possible. For if we had an original integral equation of the form

$$\phi\,(z,\, x,\, y,\, a_1,\, a_2,\, a_3,\, a_4,\, a_5) = 0,$$

and obtained thence five other equations giving the values of $p,\, q,\, r,\, s,\, t,$ we could between the six resulting equations eliminate the five constants a and have a differential equation of the second order ; and, according to the form of ϕ, the degree of this equation would vary. Conversely, in the integral, which is most general so far as the number of arbitrary constants that enter is concerned, we might expect more than three constants. But $\phi = 0$ will not necessarily be the most general integral; the only inference to be made is that the equation containing three arbitrary constants is not the most general integral.

The integral can be replaced however by one which is more general; the method of obtaining this, due to Imschenetsky, is similar to that employed by Lagrange for partial differential equations of the first order—viz., variation of the constants. The method will be sufficiently illustrated by considering an integral of the equation

$$Rr + 2Ss - Tt + U\,(rt - s^2) = V.$$

280. Let the integral obtained by any of the foregoing methods be represented by

$$z = f(x, y, a, b, c).$$

To obtain the general integral, we shall suppose c to be changed into a function of a and b the value of which is, as yet, undetermined; and we then consider a and b to be functions of x and y such that p and q preserve the same *forms* as when a, b, c, are constants. Denoting

$$\frac{\partial f}{\partial a} + \frac{\partial f}{\partial c}\frac{\partial c}{\partial a} \ \text{ and } \ \frac{\partial f}{\partial b} + \frac{\partial f}{\partial c}\frac{\partial c}{\partial b}$$

respectively by $\dfrac{df}{da}$ and $\dfrac{df}{db}$, we have

$$\frac{\partial z}{\partial x} = p + \frac{df}{da}\frac{\partial a}{\partial x} + \frac{df}{db}\frac{\partial b}{\partial x},$$

$$\frac{\partial z}{\partial y} = q + \frac{df}{da}\frac{\partial a}{\partial y} + \frac{df}{db}\frac{\partial b}{\partial y};$$

and therefore, since $\dfrac{\partial z}{\partial x} = p$ and $\dfrac{\partial z}{\partial y} = q$, we have

$$\frac{df}{da}\frac{\partial a}{\partial x} + \frac{df}{db}\frac{\partial b}{\partial x} = 0,$$

$$\frac{df}{da}\frac{\partial a}{\partial y} + \frac{df}{db}\frac{\partial b}{\partial y} = 0,$$

which will be satisfied if we write

$$\frac{df}{da} = 0 = \frac{df}{db}.$$

The second differential coefficients are

$$\frac{\partial^2 z}{\partial x^2} = r + \frac{dp}{da}\frac{\partial a}{\partial x} + \frac{dp}{db}\frac{\partial b}{\partial x} = r + h;$$

$$\frac{\partial^2 z}{\partial x \partial y} = s + \frac{dp}{da}\frac{\partial a}{\partial y} + \frac{dp}{db}\frac{\partial b}{\partial y} = s + \frac{dq}{da}\frac{\partial a}{\partial x} + \frac{dq}{db}\frac{\partial b}{\partial x} = s + k;$$

$$\frac{\partial^2 z}{\partial y^2} = t + \frac{dq}{da}\frac{\partial a}{\partial y} + \frac{dq}{db}\frac{\partial b}{\partial y} = t + l.$$

But since $\frac{df}{da}$ is identically zero when we suppose a and b replaced by their values in terms of x and y, we have

$$\frac{\partial}{\partial x}\left(\frac{df}{da}\right) + \frac{d^2 f}{da^2}\frac{\partial a}{\partial x} + \frac{d^2 f}{da\,db}\frac{\partial b}{\partial x} = 0;$$

and

$$\frac{\partial}{\partial x}\left(\frac{df}{da}\right) = \frac{d}{da}\left(\frac{\partial f}{\partial x}\right) = \frac{dp}{da},$$

so that

$$\frac{dp}{da} + \frac{d^2 f}{da^2}\frac{\partial a}{\partial x} + \frac{d^2 f}{da\,db}\frac{\partial b}{\partial x} = 0.$$

Similarly

$$\frac{dq}{da} + \frac{d^2 f}{da^2}\frac{\partial a}{\partial y} + \frac{d^2 f}{da\,db}\frac{\partial b}{\partial y} = 0;$$

$$\frac{dp}{db} + \frac{d^2 f}{da\,db}\frac{\partial a}{\partial x} + \frac{d^2 f}{db^2}\frac{\partial b}{\partial x} = 0;$$

$$\frac{dq}{db} + \frac{d^2 f}{da\,db}\frac{\partial a}{\partial y} + \frac{d^2 f}{db^2}\frac{\partial b}{\partial y} = 0.$$

These equations satisfy the condition

$$k = \frac{dp}{da}\frac{\partial a}{\partial y} + \frac{dp}{db}\frac{\partial b}{\partial y} = \frac{dq}{da}\frac{\partial a}{\partial x} + \frac{dq}{db}\frac{\partial b}{\partial x};$$

and from them there can be obtained the expressions

$$h\Delta = \frac{d^2 f}{da^2}\left(\frac{dp}{db}\right)^2 - 2\frac{d^2 f}{da\,db}\frac{dp}{da}\frac{dp}{db} + \frac{d^2 f}{db^2}\left(\frac{dp}{da}\right)^2,$$

$$l\Delta = \frac{d^2 f}{da^2}\left(\frac{dq}{db}\right)^2 - 2\frac{d^2 f}{da\,db}\frac{dq}{da}\frac{dq}{db} + \frac{d^2 f}{db^2}\left(\frac{dq}{da}\right)^2,$$

$$k\Delta = \frac{d^2 f}{da^2}\frac{dp}{db}\frac{dq}{db} - \frac{d^2 f}{da\,db}\left(\frac{dp}{da}\frac{dq}{db} + \frac{dq}{da}\frac{dp}{db}\right) + \frac{d^2 f}{db^2}\frac{dp}{da}\frac{dq}{da},$$

where
$$\Delta = \left(\frac{d^2f}{da\,db}\right)^2 - \frac{d^2f}{da^2}\frac{d^2f}{db^2}.$$

But with the modified forms of a, b, c, the equation

$$z = f(x, y, a, b, c)$$

is still to provide a solution of the equation

$$R\frac{\partial^2 z}{\partial x^2} + 2S\frac{\partial^2 z}{\partial x \partial y} + T\frac{\partial^2 z}{\partial y^2} + U\left\{\frac{\partial^2 z}{\partial x^2}\frac{\partial^2 z}{\partial y^2} - \left(\frac{\partial^2 z}{\partial x \partial y}\right)^2\right\} = V.$$

The coefficients of the second differential coefficients are unaltered in form, since we have retained the forms of the first differential coefficients; and therefore R, S, T, U, V, remain unmodified. Substituting now in this equation the values of $\frac{\partial^2 z}{\partial x^2}$, $\frac{\partial^2 z}{\partial x \partial y}$, $\frac{\partial^2 z}{\partial y^2}$, and remembering that the differential equation is satisfied when h, k, l, are zero, we find that it takes the form

$$(R + Ut)h + 2(S - Us)k + (T + Ur)l + U(lh - k^2) = 0,$$

where the quantities r, s, t, which explicitly occur, and the quantities p, q, z, which implicitly occur, are to be replaced by their respective values derived from the integral

$$z = f(x, y, a, b, c),$$

in which a, b, c, are considered constants. We must now substitute the expressions found for h, k, l; then the equation, after some reductions, is found to be of the form

$$R_1\frac{d^2f}{da^2} - 2S_1\frac{d^2f}{da\,db} + T_1\frac{d^2f}{db^2} = V_1,$$

where

$$R_1 = (R + Ut)\left(\frac{dp}{db}\right)^2 + 2(S - Us)\frac{dp}{db}\frac{dq}{db} + (T + Ur)\left(\frac{dq}{db}\right)^2,$$

$$T_1 = (R + Ut)\left(\frac{dp}{da}\right)^2 + 2(S - Us)\frac{dp}{da}\frac{dq}{da} + (T + Ur)\left(\frac{dq}{da}\right)^2,$$

$$S_1 = (R + Ut)\frac{dp}{da}\frac{dp}{db} + (S - Us)\left(\frac{dp}{da}\frac{dq}{db} + \frac{dp}{db}\frac{dq}{da}\right) + (T + Ur)\frac{dq}{da}\frac{dq}{db},$$

$$V_1 = U\left(\frac{dp}{da}\frac{dq}{db} - \frac{dp}{db}\frac{dq}{da}\right)^2.$$

In all these coefficients, the quantities z, p, q, r, s, t, are to be replaced by their values in terms of x and y as derived from the given integral equation.

This differential equation is linear in the second differential coefficients of f with regard to a and b; it is, moreover, the equation which is to determine the value of c as a function of a and b. Now

$$\frac{df}{da} = \frac{\partial f}{\partial a} + \frac{\partial f}{\partial c}\frac{\partial c}{\partial a},$$

$$\frac{d^2f}{da^2} = \frac{\partial^2 f}{\partial a^2} + 2\frac{\partial^2 f}{\partial a \partial c}\frac{\partial c}{\partial a} + \frac{\partial^2 f}{\partial c^2}\left(\frac{\partial c}{\partial a}\right)^2 + \frac{\partial f}{\partial c}\frac{\partial^2 c}{\partial a^2},$$

and also for the other coefficients; when these are substituted for $\frac{d^2f}{da^2}$, $\frac{d^2f}{da\,db}$, $\frac{d^2f}{db^2}$, the resulting equation is linear in the second differential coefficients of c with regard to a and b, and the quantities multiplying these are functions of x, y, a, b, c, $\frac{\partial c}{\partial a}$, $\frac{\partial c}{\partial b}$. But we also have

$$\frac{df}{da} = 0 = \frac{df}{db},$$

from which the values of x and y can be found as functions of a, b, c, $\frac{\partial c}{\partial a}$, $\frac{\partial c}{\partial b}$; and these, when substituted, will make the equation one which involves only the quantities a, b, c, and the differential coefficients of c. This equation is found to be of the form

$$A\frac{\partial^2 c}{\partial a^2} + 2C\frac{\partial^2 c}{\partial a \partial b} + B\frac{\partial^2 c}{\partial b^2} = F,$$

where A, B, C, F, are functions of a, b, c, $\frac{\partial c}{\partial a}$, $\frac{\partial c}{\partial b}$.

Now it may not be possible to integrate directly the original differential equation, while it may be possible to obtain, almost by inspection, a particular solution which involves three arbitrary constants; or it may be possible to derive such an integral not obtainable merely by inspection. In either case, such particular integral can be generalised provided the solution of the new equation satisfied by c can be obtained; and if this solution be represented by

$$\theta(a, b, c) = 0,$$

then the new integral of the original equation is obtained from

$$z = f(x, y, a, b, c)$$
$$0 = \theta(a, b, c)$$
$$0 = \frac{\partial f}{\partial a}\frac{\partial \theta}{\partial c} - \frac{\partial f}{\partial c}\frac{\partial \theta}{\partial a}$$
$$0 = \frac{\partial f}{\partial b}\frac{\partial \theta}{\partial c} - \frac{\partial f}{\partial c}\frac{\partial \theta}{\partial b}$$

by eliminating a, b, c, between them.

Ex. 1. Integrate the equation

$$r + 2(q-x)s + (q-x)^2 t = q.$$

Here $R=1$, $S=q-x$, $T=(q-x)^2$, $U=0$, $V=q$; thus $G=0$, and the equations determining W are only a single pair, viz.

$$\begin{cases} 0 = \dfrac{\partial W}{\partial q} - (q-x)\dfrac{\partial W}{\partial p}, \\ 0 = \dfrac{\partial W}{\partial x} + (q-x)\dfrac{\partial W}{\partial y} + (p+q^2-qx)\dfrac{\partial W}{\partial z} + q\dfrac{\partial W}{\partial p}. \end{cases}$$

We proceed as in §§ 232—235. We denote the equations by

$$0 = F_1 = Q - (q-x)P,$$
$$0 = F_2 = X + (q-x)Y + (p+q^2-qx)Z + qP.$$

As a condition that these equations may coexist, we must have

$$0 = (F_1, F_2) = -qZ - Y.$$

Hence we write

$$0 = F_3 = -qZ - Y;$$

then $(F_2, F_3) = 0$; $(F_1, F_3) = Z$;

and so we take $0 = F_4 = Z$,

and then $0 = (F_1, F_2) = \ldots = (F_3, F_4).$

Hence $Y = 0 = Z$; $X + qP = 0$; $Q - (q-x)P = 0$; substituting in

$$0 = Pdp + Qdq + Xdx + Zdz + Ydy,$$

we obtain

$$0 = P(dp - q\,dx + q\,dq - x\,dq),$$

and therefore we may write as the intermediate integral

$$W = p + \tfrac{1}{2}q^2 - xq + a = 0,$$

To obtain the complete integral of this equation, we apply Charpit's method; we must obtain an integral of

$$\frac{dx}{-1} = \frac{dy}{-q+x} = \frac{dp}{-q} = \frac{dq}{0}.$$

One is given by $q = \beta$; and therefore

$$p = -a + \beta x - \tfrac{1}{2}\beta^2.$$

These values, substituted in

$$dz = p\,dx + q\,dy,$$

lead to the integral

$$z = \beta y + \tfrac{1}{2}\beta x (x - \beta) - ax - c,$$

which contains three arbitrary constants.

To obtain the modified integral (§ 280), we write this

$$z = f = -ax + \beta y + \tfrac{1}{2}\beta x (x - \beta) - c,$$

considering c as a function of a and β. Then we have

$$0 = \frac{df}{da} = -x - \frac{\partial c}{\partial a};$$

$$0 = \frac{df}{d\beta} = y + \tfrac{1}{2}x^2 - \beta x - \frac{\partial c}{\partial \beta};$$

$$p = -a + \beta x - \tfrac{1}{2}\beta^2; \quad r = \beta; \quad q = \beta; \quad s = 0 = t; \quad \frac{\partial^2 f}{\partial a^2} = 0 = \frac{\partial^2 f}{\partial a\,\partial\beta}; \quad \frac{\partial^2 f}{\partial\beta^2} = -x;$$

$$\frac{\partial f}{\partial c} = -1; \quad \frac{\partial^2 f}{\partial c^2} = 0; \quad \frac{\partial^2 f}{\partial a\,\partial c} = 0 = \frac{\partial^2 f}{\partial\beta\,\partial c}; \quad \frac{dp}{d\beta} = x - \beta; \quad \frac{dq}{d\beta} = 1 = -\frac{dp}{da}; \quad \frac{dq}{da} = 0.$$

Hence $R_1 = 0$; $T_1 = 1$; $S_1 = 0$; $V_1 = 0$; and the equation in f is

$$\frac{d^2 f}{d\beta^2} = 0,$$

or, on substitution in terms of c,

$$-x - \frac{\partial^2 c}{\partial\beta^2} = 0,$$

or finally

$$\frac{\partial^2 c}{\partial\beta^2} = \frac{\partial c}{\partial a}.$$

But, by § 270, an integral of this equation is

$$c = \int_{-\infty}^{\infty} e^{-\lambda^2} f(\beta + 2\lambda a^{\frac{1}{2}})\, d\lambda;$$

and therefore an integral of the original equation is given by the elimination of a and β between

$$z = \beta y - ax + \tfrac{1}{2}\beta x (x - \beta) - \int_{-\infty}^{\infty} e^{-\lambda^2} f(\beta + 2\lambda a^{\frac{1}{2}})\, d\lambda,$$

$$0 = x a^{\frac{1}{2}} + \int_{-\infty}^{\infty} \lambda e^{-\lambda^2} f'(\beta + 2\lambda a^{\frac{1}{2}})\, d\lambda,$$

$$0 = y + \tfrac{1}{2}x^2 - \beta x - \int_{-\infty}^{\infty} e^{-\lambda^2} f'(\beta + 2\lambda a^{\frac{1}{2}})\, d\lambda.$$

When the definite integral is integrated by parts, the second of these equations may be replaced by

$$0 = x - \int_{-\infty}^{\infty} e^{-\lambda^2} f'' \left(\beta + 2\lambda a^{\frac{1}{2}}\right) d\lambda.$$

Ex. 2. Integrate the equations:

(i) $r - t = \dfrac{2p}{x}$;

(ii) $x^4 r - 4x^2 qs + 4q^2 t + 2px^3 = 0$;

(iii) $(x+q)^2 r + 2(x+q)(y+p)s + (y+p)^2 t + 2(x+q)(y+p) = 0$;

(iv) $x^2 r + 2x^2 s + \left(x^2 - \dfrac{b^2}{x^2 q^2}\right) t = 2z$;

(v) $r + 2qs + (q^2 - x^2) t = q$;

(vi) $x^4 r - 4x^2 qs + 3qt + 2x^3 p = 0$;

(vii) $r + 2qs + q^2 t = b^2 t$;

(viii) $2ps + t - p = 0$.

(Ampère and Imschenetsky.)

A fuller discussion is contained in the valuable memoir by Imschenetsky, *Grunert's Archiv der Mathematik und Physik,* t. LIV.

MISCELLANEOUS EXAMPLES.

1. Prove that the integral of the equation

$$(x+y)(r-t) + 4p = 0,$$

as given by Monge's method, is

$$(x+y)z + e^{\frac{2y}{x+y}} F(x+y) = e^{\frac{2y}{x+y}} \int e^{-\frac{2y}{a}} f(2y-a)\, dy,$$

where $y+x$ is to be substituted for a after integration, and f and F denote arbitrary functions.

Hence solve the equation

$$(p+q)(r-t) = 4x(rt-s^2).$$

2. Solve by Monge's method the equations:

(i) $q(1+q)r + (p+q+2pq)s + p(1+p)t = 0$;

(ii) $(1+pq+q^2)r + s(q^2-p^2) - (1+pq+p^2)t = 0$;

(iii) $(r-t)xy - s(x^2-y^2) = qx - py$;

(iv) $x^2 r - y^2 t = xp - yq$;

    (v) $r - 2s + t = x + \phi(x+y)$;

    (vi) $(r - s)x = (t - s)y$;

    (vii) $x^2 r - y^2 t - 2xp + 2z = 0$;

    (viii) $(r - s)y + (s - t)x + q - p = 0$;

    (ix) $xr + (y - x)s - yt = q - p$.

3. Solve the equation $r + t = 2s$; and determine the arbitrary functions by the conditions, that $bz = y^2$ when $x = 0$ and $az = x^2$ when $y = 0$.

4. Integrate the equation

$$\frac{r}{x^2} - \frac{t}{y^2} = \frac{p}{x^3} - \frac{q}{y^3};$$

and obtain a first integral of the equation

$$rq\left(\frac{q}{z} - \frac{1}{y}\right) - 2s\frac{pq}{z} + tp\left(\frac{p}{z} - \frac{1}{x}\right) + rt - s^2 + \frac{pq}{xy}\left(1 - \frac{px + qy}{z}\right) = 0.$$

5. Investigate a solution of the equation

$$rt - s^2 = 0,$$

subject to the condition $q^2 = x^2(1 + p^2)$, in the form

$$z = ay + (a^2 - x^2)^{\frac{1}{2}} + a \log \frac{a - (a^2 - x^2)^{\frac{1}{2}}}{x}.$$

6. Integrate the equation

$$(1 + p^2)t - 2pqs + (1 + q^2)r = 0,$$

having given that $py - qx = 0$; and shew that a particular solution is

$$(x^2 + y^2)^{\frac{1}{2}} = c \cosh \frac{z}{c}.$$

7. Solve the equations:

    (i) $e^{2y}(r - p) = e^{2x}(t - q)$;   (ii) $qys = pyt + pq$;

    (iii) $xr + xys + yq = 0$;   (iv) $xr + 2ys + p = 4x$;

    (v) $2xr - 2t + 3p = 0$;   (vi) $x(r - a^2 t) = 2p$.

8. Prove that the only real solution of the simultaneous equations

$$\left. \begin{array}{c} \dfrac{\partial^2 u}{\partial x^2} + \dfrac{\partial^2 u}{\partial y^2} = 0 \\[2mm] \left(\dfrac{\partial u}{\partial x}\right)^2 + \left(\dfrac{\partial u}{\partial y}\right)^2 = 1 \end{array} \right\}$$

is

$$u = x \cos a + y \sin a + \beta.$$

9. Prove that the only real solutions which simultaneously satisfy the equations

$$r+t=2a \brace s^2-rt=b^2 \, ,$$

are comprised in

$$z=\tfrac{1}{2}x^2\,(a+c\cos a)+cxy\sin a+\tfrac{1}{2}y^2\,(a-c\cos a)+\beta x+\gamma y+\delta,$$

where $c^2=a^2+b^2$, and a, β, γ, δ, are arbitrary parameters.

10. Obtain an intermediate integral of

$$pqr=s\,(1+p^2)\,;$$

and shew that its general integral is obtained by eliminating a between the equations

$$z-\phi\,(a)-ax-(1+a^2)^{\frac{1}{2}}f\,(y)=0 \brace \phi'\,(a)+x+a\,(1+a^2)^{-\frac{1}{2}}f\,(y)=0 \, ,$$

where ϕ and f denote arbitrary functions.

(Serret, and Graindorge.)

11. Integrate the equations:

(i) $xp+yq+x^2r+\tfrac{5}{2}xys+y^2t=0\,;$

(ii) $(xp+yq)\,(rt-s^2)+q^2r-\tfrac{5}{2}pqs+p^2t=0\,;$

(iii) $(x^2-y^2)\,(t-r)+4\,(px+qy-z)=0.$

Also solve, by changing the independent variables to ξ and η, where $x^2=\xi\eta$ and $xy=\xi$,

$$x^2r-2xys+y^2t+2yq=0\,;$$

and, by changing the independent variables to ξ and η, where $x=e^{\xi+\eta}$ and $y=e^{\xi-\eta}$,

$$x^2r-y^2t=(xp-yq)f\,(xy).$$

12. Integrate the equations:

(i) $\dfrac{\partial^2 z}{\partial x^2}+\dfrac{2}{x}\dfrac{\partial z}{\partial x}=a^2\dfrac{\partial^2 z}{\partial y^2}\,;$ (ii) $\dfrac{\partial^2 z}{\partial y^2}=a^2\left(\dfrac{\partial^2 z}{\partial x^2}+\dfrac{2}{x}\dfrac{\partial z}{\partial x}-\dfrac{2}{x^2}z\right);$

(iii) $\dfrac{\partial^2 z}{\partial x^2}=\dfrac{a^2}{x^4}\dfrac{\partial^2 z}{\partial y^2}\,;$ (iv) $\dfrac{\partial^2 z}{\partial x^2}=\dfrac{2}{x^2}z-a\dfrac{\partial^2 z}{\partial x\,\partial y}\,;$

(v) $\dfrac{\partial^2 z}{\partial x\,\partial y}+\dfrac{1}{x+y}\left(\dfrac{\partial z}{\partial x}+\dfrac{\partial z}{\partial y}\right)-\dfrac{2}{(x+y)^2}z=0.$ (Gregory.)

13. Find the surface whose equation satisfies the equation

$$\frac{\partial^2 z}{\partial x\,\partial y}=0,$$

and whose trace on the plane of xy is the hyperbola $xy=a^2$.

14. Integrate the simultaneous equations:

(i)
$$\left.\begin{aligned}\frac{\partial u}{\partial x} &= \frac{\partial v}{\partial y}\\[1mm]\frac{\partial v}{\partial x} &= -\frac{\partial u}{\partial y}\end{aligned}\right\};$$

(ii)
$$\left.\begin{aligned}m\frac{\partial}{\partial x}\left(\frac{\partial a}{\partial x}+\frac{\partial \beta}{\partial y}\right)&=n\left(\frac{\partial^2 a}{\partial x^2}+\frac{\partial^2 a}{\partial y^2}\right)\\[1mm]m\frac{\partial}{\partial y}\left(\frac{\partial a}{\partial x}+\frac{\partial \beta}{\partial y}\right)&=n\left(\frac{\partial^2 \beta}{\partial x^2}+\frac{\partial^2 \beta}{\partial y^2}\right)\end{aligned}\right\}.$$

15. Shew that the simultaneous equations

$$rt+c\,(r+t)=0,\qquad pq+c'\,(py-qx)=0,$$

represent a series of coaxal paraboloids; and that they cut any fixed plane, perpendicular to the axis, in a series of similar conics the ratio of whose axes is

$$(c'-c)^{\frac{1}{2}}:(c'+c)^{\frac{1}{2}}.$$

16. Shew that the equation

$$Gs+Hp+K=0,$$

in which G, H, K, are functions of x, y, z, and q, has an intermediate integral, if

$$G\left(\frac{\partial H}{\partial x}-\frac{\partial K}{\partial z}\right)+H\left(\frac{\partial K}{\partial y}-\frac{\partial G}{\partial x}\right)+K\left(\frac{\partial G}{\partial z}-\frac{\partial H}{\partial y}\right)=0;$$

and obtain the integral.

Hence obtain the integral of the equation

$$\{(x+yz)\,s-ypq\}\,(x+y)=qy\,(1-z)$$

in the form

$$z=e^{-\int\frac{y\,dy}{(x+y)\phi\,(y)}}\left[\psi\,(x)-\int e^{\int\frac{y\,dy}{(x+y)\phi\,(y)}}\frac{x\,dy}{(x+y)\,\phi\,(y)}\right].$$

(Imschenetsky, and Graindorge.)

17. Obtain a solution of the equation

$$\frac{\partial^2 u}{\partial x^2}+\frac{\partial^2 u}{\partial y^2}+\frac{\partial^2 u}{\partial z^2}=0$$

in a series of ascending powers of x. (Lagrange.)

Solve the equation

$$a\frac{\partial^2 u}{\partial x^2}+2h\frac{\partial^2 u}{\partial x\,\partial y}+2g\frac{\partial^2 u}{\partial x\,\partial z}+b\frac{\partial^2 u}{\partial y^2}+2f\frac{\partial^2 u}{\partial y\,\partial z}+c\frac{\partial^2 u}{\partial z^2}=0,$$

where the coefficients are constants; and discuss the case in which the discriminant of the left-hand side is zero.

18. Verify that the partial differential equation

$$\frac{\partial^2 z}{\partial x^2}=a^2 y^{2b}\frac{\partial^2 z}{\partial y^2}$$

is integrable in finite terms, if $b\,(2i\pm1)=2i$, where i is a positive integer.

Solve also

$$\frac{\partial^2 u}{\partial x^2} - a^2 \frac{\partial^2 u}{\partial y^2} = \frac{i(i+1)}{x^2} u.$$

(Legendre.)

19. Shew that the complete integral of

$$\frac{1}{a^2} \frac{\partial^2 u}{\partial t^2} = \frac{\partial^2 u}{\partial r^2} + \frac{2}{r} \frac{\partial u}{\partial r} - \frac{n(n+1)u}{r^2}$$

(n being an integer) may be exhibited in the form

$$u = r^n \left(\frac{1}{r} \frac{\partial}{\partial r}\right)^n \frac{\phi(r+at) + \psi(r-at)}{r},$$

where ϕ and ψ are arbitrary functions; and obtain, in the form of a definite integral, the complete solution of

$$\frac{1}{a^2} \frac{\partial^2 u}{\partial t^2} = \frac{\partial^2 u}{\partial r^2} + \frac{1}{r} \frac{\partial u}{\partial r}.$$

20. Obtain as a definite integral the solution of

$$2 \frac{\partial^2 V}{\partial x \partial y} = \frac{1}{x-y} \left(\frac{\partial V}{\partial x} - \frac{\partial V}{\partial y}\right).$$

21. Obtain a solution of the equation

$$\frac{\partial U}{\partial t} = a^4 \frac{\partial^4 U}{\partial x^4}$$

in the form

$$\pi U = \int_{-\infty}^{\infty} \int_{-\infty}^{\infty} e^{-u^2 - v^2} \phi(x + 2^{\frac{3}{2}} a u v^{\frac{1}{2}} t^{\frac{1}{4}}) \, du \, dv.$$

22. Change the dependent variable from z to y in the equation

$$q(1+q)r - (p+q+2pq)s + p(1+p)t = 0;$$

and hence obtain the solution of the equation in the form

$$x + f(z) = F(x+y+z).$$

23. Shew that, if there be five functions z_1, z_2, z_3, z_4, z_5, each of which satisfies the equations

$$\left.\begin{array}{l} r = a_1 s + a_2 p + a_3 q + a_4 z \\ t = b_1 s + b_2 p + b_3 q + b_4 z \end{array}\right\},$$

where the a's and b's are functions of x and y alone, then between them there is a linear relation with constant coefficients of the form

$$C_1 z_1 + C_2 z_2 + C_3 z_3 + C_4 z_4 + C_5 z_5 = 0.$$

If, in addition, any four of them z_1, z_2, z_3, z_4, be such as to satisfy identically the equation

$$\begin{vmatrix} z_1, & z_2, & z_3, & z_4 \\ p_1, & p_2, & p_3, & p_4 \\ q_1, & q_2, & q_3, & q_4 \\ s_1, & s_2, & s_3, & s_4 \end{vmatrix} = 0,$$

then there is also a relation of the form

$$C_1 z_1 + C_2 z_2 + C_3 z_3 + C_4 z_4 = 0.$$

(Appell.)

24. Shew that the function $F(a, \beta, \gamma, \theta, \epsilon, x, y)$, given by the series

$$\Sigma\Sigma \frac{\Pi(a+m+n-1)}{\Pi(a-1)\,\Pi(m)\,\Pi(n)} \frac{\Pi(\beta+m-1)\,\Pi(\gamma+n-1)\,\Pi(\theta-1)\,\Pi(\epsilon-1)}{\Pi(\theta+m-1)\,\Pi(\epsilon+n-1)\,\Pi(\beta-1)\,\Pi(\gamma-1)} x^m y^n,$$

the summation extending for all integral values of m and n from zero to infinity, satisfies the two equations

$$(x-x^2)r - xys + \{\theta - (a+\beta+1)x\}p - \beta yq - a\beta z = 0,$$
$$(y-y^2)t - xys + \{\epsilon - (a+\gamma+1)y\}q - \gamma xp - a\gamma z = 0.$$

Hence shew that $F(a, \delta+c, -c, \theta, \epsilon, x, y)$ is a solution of

$$(x-x^2)r - 2xys + (y-y^2)t + \{\theta - (a+\delta+1)x\}p + \{\epsilon - (a+\delta+1)y\}q - a\delta z = 0,$$

c being an arbitrary constant.

(Appell.)

25. If there be three functions z_1, z_2, z_3, satisfying the equations

$$r = a_1 p + a_2 q + a_3 z,$$
$$s = b_1 p + b_2 q + b_3 z,$$
$$t = c_1 p + c_2 q + c_3 z,$$
$$z_1 (p_2 - q_3) + z_2 (p_3 - q_1) + z_3 (p_1 - q_2) = 0,$$

where the a's, b's, and c's, are functions of x and y, then there exists between the functions z_1, z_2, z_3, a linear relation with constant coefficients.

(Appell.)

26. Shew that the integral of the equation

$$s + xyp + kyz = 0$$

may, by differentiation, be connected with that of

$$s + xyp + (k+n)\,yz = 0,$$

k being a constant and n being an integer.

Hence solve the former equation in the case when k is a negative integer.

Obtain the solution when k is a positive integer.

(Tanner.)

27. Obtain the solution of

$$s = e^s$$

in the form

$$e^s = 2 \frac{\phi'(x)\,\psi'(y)}{\{\phi(x) + \psi(y)\}^2},$$

where ϕ and ψ are arbitrary functions.

(Liouville.)

Hence integrate $\qquad s = zp.$

(Tanner.)

Integrate also

$$s = \phi(x)\,\psi(y)\,e^s$$

in the form

$$e^s = \frac{2n^2\,\theta(x)\,\chi(y)}{\sin^2 n\,(F+f)},$$

where n is a constant, $F'(x) = \phi(x)\,\theta(x)$, and $f'(y) = \psi(y)\,\chi(y)$, and θ and χ denote arbitrary functions.

(R. Russell.)

28. Integrate by Ampère's method the equations:

(i) $\quad zs + \dfrac{z}{q^2} t + pq = 0;$

(ii) $\quad \dfrac{ax^2}{y^2} r + \dfrac{by^2}{x^2} t + (lx + my + nz)(rt - s^2)$

$$= (lx + my + nz)\left\{ 2\left(\frac{z}{xy} - \frac{p}{y} - \frac{q}{x}\right) + \left(\frac{z}{xy} - \frac{p}{y} - \frac{q}{x}\right)^2 \right\};$$

(iii) $\quad qr + (p + x)\,s + yt + y\,(rt - s^2) + q = 0.$

(Imschenetsky.)

GENERAL EXAMPLES

OF

DIFFERENTIAL EQUATIONS

GENERAL EXAMPLES.

1. Obtain the complete primitives and (where they exist) the singular solutions of the equations :

$$(i) \quad p^2 - y = 0;$$

$$(ii) \quad 3y(1 - 2p^2) = x(1 - 4p^3);$$

$$(iii) \quad p(ny - px) = c;$$

$$(iv) \quad y(1 + p^2) - 2xp(1 + p) = 2(1 - p);$$

$$(v) \quad y^2p^2 - 2xyp - x^2 + 2y^2 + a^2 = 0;$$

$$(vi) \quad y^2 - 2xy(p^2 + p + 1) + x^2p^2 = 0;$$

$$(vii) \quad xp^2 - 2yp + 3a = 0;$$

$$(viii) \quad xp^2 = (x - a)^2;$$

$$(ix) \quad y^2 - 2xyp + (x^2 - ay)p^2 = 0$$

$$(x) \quad p(y + xp) + c = 0;$$

$$(xi) \quad (xp - y)^3 = x^5(2y - xp);$$

$$(xii) \quad (xp - y)^2 = x - y;$$

$$(xiii) \quad y^2 = p^2(y - p);$$

$$(xiv) \quad (xp - y)(x^2 - y^2p) = cp;$$

$$(xv) \quad p^2x^2 + py(2x - y) + y^2 = 0;$$

$$(xvi) \quad p^2x^2 - 4pxy - 8px + 4y^2 + 8y = 0;$$

$$(xvii) \quad (x^2 + y^2)(1 + p^2) = a^2(px - y)^2;$$

$$(xviii) \quad p^2 = 2y^2x^3p + 4y^3x^2;$$

$$(xix) \quad y = xp^2 + p^3;$$

$$(xx) \quad y(1 + p^2) = 2xp;$$

$$(xxi) \quad x^2(p^2 + 1) = c(x + yp)^2;$$

$$(xxii) \quad (x^2 + y^2p^2)(y - xp) = xp;$$

$$(xxiii) \quad (x^2 - xy)p + 2x^2 + 3xy - y^2 = 0;$$

$$(xxiv) \quad y = x(p + p^2);$$

$$(xxv) \quad x(3y^2 - x)p = y(2x - y^2);$$

$$(xxvi) \quad x^4y^2(y - xp)^3 = p^2;$$

$$(xxvii) \quad p = 2x + x^3y - xy^2;$$

$$(xxviii) \quad (1 - x^2)p = xy - y^2;$$

(xxix) $2py = xy^2 + x^3$;

(xxx) $y = p + p^3$;

(xxxi) $x^4 (p + y^2) + a^2 = 0$;

(xxxii) $x^2 (p - ay - y^2) + 2 + ax = 0$;

(xxxiii) $x^2 (x - 2y) p = y^3 - 4xy^2 + 2x^3$;

(xxxiv) $(x^2 - y^2)(1 + p) = 2xy(1 - p)$;

(xxxv) $(xy + 4x) p = y^2 + 2x + 2y$;

(xxxvi) $p(x^2 + y^2 + a^2) + 2xy = 0$;

(xxxvii) $xp^2 + yp - y^4 = 0$;

(xxxviii) $p^2 y^2 - 6px^3 + 4x^2 y = 0$;

(xxxix) $(x^2 - x^2 y^2 - y^4) p^2 - 2xyp + y^2 = 0$;

(xl) $y^2 p^2 + 2xypa + (1 - a)(y^2 - b^2) + ax^2 = 0$;

(xli) $(ax + by)(1 + y^2)^{\frac{1}{2}} = (ay + bx)(1 + x^2)^{\frac{1}{2}} p$;

(xlii) $y^2 (p^2 + 1) - 4ayp - 4a(x - a) = 0$;

(xliii) $y(x - y - xp)^2 - p(x^2 - 2xy)(x - y - xp) + yp^2 = 0$;

(xliv) $(y - px)(py + y - px) = p$;

(xlv) $(x + y) p^2 + 2px = y$;

(xlvi) $2y = x(1 + p^2)$;

(xlvii) $xy(x^2 - y^2)(p^2 - 1) + (x^2 - y^2)^2 p + a^2 (xp + y)^2 = 0$;

(xlviii) $(a^2 - c^2) y^2 p^2 - 2c^2 xyp + a^2 y^2 - c^2 x^2 = a^2 c^2$;

(xlix) $(x^2 + y^2)(1 + p^2) = 4a^2 (xp - y)^2$;

(l) $9y^4 (x^2 - 1) p^2 - 6xy^5 p - 4x^2 = 0.$

2. Given the relation

$$y - ax = \tfrac{1}{2} x^2 (a - c)^2 - \tfrac{1}{4} b^2 (a - c)^4,$$

where c is the arbitrary parameter, shew how to find the corresponding differential equation. Find the curves which may or may not be singular solutions: and discuss the different cases.

3. Prove that, if the differential equation

$$Lp^2 + 2Mp + N = 0$$

has

$$Ac^2 + 2Bc + 1 = 0$$

for its primitive, then $LN - M^2 = 4(A - B^2) T^2$, where

$$T = \frac{\partial (A, B)}{\partial (x, y)} ;$$

and interpret the result geometrically.

4. Prove that the equation

$$(1 - x^4)^{-\frac{1}{2}} dx + (1 - y^4)^{\frac{1}{2}} dy = 0$$

is satisfied when

$$y = \left(\frac{1 - x^2}{1 + x^2} \right)^{\frac{1}{2}}.$$

5. Verify that a primitive of the equation

$$(1+x^3)^{-\frac{1}{2}} dx = (1+y^3)^{-\frac{1}{2}} dy$$

is

$$x^2y^2 + 2axy(x+y) + a^2(x-y)^2 - 4(x+y) + 4a = 0,$$

where a is an arbitrary constant.

6. Obtain the primitive of

$$(x-a)(x-b)\frac{dy}{dx} + y^2 + \lambda(y+x-a)(y+x-b) = 0$$

in the form

$$\frac{y+\lambda(y+x-a)}{y+\lambda(y+x-b)} = A\left(\frac{x-b}{x-a}\right)^\lambda;$$

and discuss in particular the cases when $\lambda = 0$, $\lambda = -1$.

7. Transform the equation

$$(x-a)(x-b)\frac{dy}{dx} + y^2 + (y+x-a)(y+x-b) = 0$$

to an equation of the second order in z by a substitution

$$y\frac{dz}{dx} = zf(x).$$

Hence integrate the equation: and discuss, in particular, the case in which $a = b$.

8. Shew that the equation $y = xp + f(p)$ is the only ordinary equation of the first order the primitive of which can be obtained by replacing p in the equation by an arbitrary constant.

9. Shew that, if the primitive of $f(x, y, p) = 0$ be $\phi(x, y, c) = 0$, then the primitive of $f(p, xp - y, x) = 0$ is given by the elimination of p between the equations

$$f(p, xp - y, x) = 0,$$
$$\phi(p, xp - y, c) = 0.$$

10. By reciprocation with regard to the parabola $y^2 = 2x$ or otherwise, shew how to deduce the complete primitive of the equation

$$f\left(\frac{y}{p} - x, \frac{1}{p}, \frac{1}{y}\right) = 0$$

from that of

$$f(x, y, p) = 0;$$

and extend the result to equations of the second order.

Solve the equation

$$\frac{d^2y}{dx^2} = (1+x^2)y.$$

11. Solve the equations:

(i) $\dfrac{d^2y}{dx^2} - 3\dfrac{dy}{dx} + 2y = \sinh 2x$;

(ii) $(1+x^2)\dfrac{d^2y}{dx^2} - x\dfrac{dy}{dx} + y = 0$;

(iii) $3x^3\dfrac{d^3y}{dx^3} + 2x^2\dfrac{d^2y}{dx^2} + x\dfrac{dy}{dx} - y = 3x^3 + x^2 + x$;

(iv) $\dfrac{d^6y}{dx^6} + 2\dfrac{d^3y}{dx^3} + y = xe^{-x}$;

(v) $x^2(x+y)\dfrac{d^2y}{dx^2} = \left(y - x\dfrac{dy}{dx}\right)^2$;

(vi) $\dfrac{d^3y}{dx^3} + \dfrac{6}{x}\dfrac{d^2y}{dx^2} + \dfrac{6}{x^2}\dfrac{dy}{dx} = 0$;

(vii) $\dfrac{d^2y}{dx^2} - x^2\dfrac{dy}{dx} - (x+1)^2 y = 0$;

(viii) $2x^2\dfrac{d^2y}{dx^2} - x\dfrac{dy}{dx} - 2y = \sin x$;

(ix) $\dfrac{d^2y}{dx^2} + x^2 y = \dfrac{2y}{x^2}$;

(x) $x^3\dfrac{d^2y}{dx^2} + 3x^2\dfrac{dy}{dx} + xy = 1$;

(xi) $x\dfrac{d^2y}{dx^2} + (2 + x - x^2)\dfrac{dy}{dx} - x(x+3) y = 0$;

(xii) $\dfrac{d^2y}{dx^2} - \cot x\dfrac{dy}{dx} + y\sin\ x = \sin^2 x\cos(\cos x)$;

(xiii) $4(1+x+x^2)^2\dfrac{d^2y}{dx^2} = 3y$;

(xiv) $(x^2+3x+4)\dfrac{d^2y}{dx^2} + (x^2+x+1)\dfrac{dy}{dx} - y(2x+3) = 0$;

(xv) $\dfrac{d^2y}{dx^2} - \dfrac{1}{x}\dfrac{dy}{dx} + \dfrac{4a^2}{x^6} y = 0$;

(xvi) $\dfrac{d^2y}{dx^2} + \dfrac{dy}{dx} + ye^{-2x} = 0$;

(xvii) $\dfrac{d^2y}{dx^2} + \dfrac{dy}{dx}\tan x - y\cos^2 x = 0$;

(xviii) $x^2\dfrac{d^2y}{dx^2} + x^3\dfrac{dy}{dx} + \tfrac{1}{4}(x^2+6)(x^2-4) y = x^2 e^{-\frac{1}{4}x^2}$;

(xix) $(x+1)^2(x^2+2x+3)\dfrac{d^2y}{dx^2} = 12y$;

(xx) $\dfrac{d^2y}{dx^2} = (1+x^2) y$;

(xxi) $(x^4+x^2)\dfrac{d^3y}{dx^3} + (6x^3+3x)\dfrac{d^2y}{dx^2} = 12xy$;

(xxii) $x\dfrac{d^2y}{dx^2} - (x^2-x)\dfrac{dy}{dx} + (x-1) y = e^{\frac{1}{2}x^2}$;

(xxiii) $\dfrac{d^2y}{dx^2} + 2\dfrac{dy}{dx}\tan x + 3y = \tan^2 x\sec x$;

(xxiv) $\dfrac{d^2y}{dx^2} + 2x\dfrac{dy}{dx} + x^2 y = 0$;

(xxv) $\dfrac{d^2y}{dx^2} + x^4\dfrac{dy}{dx} - x^3 y = \log x$;

(xxvi) $(4x^3 - g_2 x - g_3)\dfrac{d^2 y}{dx^2} + (6x^2 - \tfrac{1}{2}g_2)\dfrac{dy}{dx} + n^2 y = 0$;

(xxvii) $\tfrac{1}{2}x^2 (x-1)^2 \dfrac{d^2 y}{dx^2} + x(x^2-1)\dfrac{dy}{dx} + (1+x-x^2)y = 0$;

(xxviii) $x^2 \dfrac{d^2 y}{dx^2} + (2x - x^2)\dfrac{dy}{dx} + \{n(n+1)-x\}\,y = 0$;

(xxix) $x^2 \dfrac{d^2 y}{dx^2} + 2(x-1)\dfrac{dy}{dx} - n(n+1)\,y = 0$;

(xxx) $x\dfrac{d}{dx}\left(x\dfrac{dy}{dx} - y\right) - 2x\dfrac{dy}{dx} + 2y + x^2 y = 0$;

(xxxi) $(2x-1)\dfrac{d^2 y}{dx^2} - (3x-4)\dfrac{dy}{dx} + (x-3)\,y = 0$;

(xxxii) $x\dfrac{d^3 y}{dx^3} + 3\dfrac{d^2 y}{dx^2} - ax^2 y = 0$;

(xxxiii) $\dfrac{d^2 y}{dx^2} + \left(\dfrac{2}{x} - 3\right)\dfrac{dy}{dx} + \left(2 - \dfrac{3}{x}\right)y = 0$;

(xxxiv) $\dfrac{d^2 y}{dx^2} = \dfrac{1 - 3x}{(1-x)(1-2x)^2}\,y$;

(xxxv) $x^2(2-x)\dfrac{d^2 y}{dx^2} + 2x\dfrac{dy}{dx} - 2y = (x-1)^2$;

(xxxvi) $2x(x^2-1)\dfrac{d^2 y}{dx^2} - \dfrac{dy}{dx} = y\,(x^2-1)^{\frac{3}{2}}$;

(xxxvii) $\dfrac{d^2 y}{dx^2} - x^2(1+x)\dfrac{dy}{dx} + x(x^4-2)\,y = 0$;

(xxxviii) $x^2 \dfrac{d^2 y}{dx^2} + (2ax^2 + bx)\dfrac{dy}{dx} + (a^2 x^2 + abx + c)\,y = 0$;

(xxxix) $x^2 \dfrac{d^2 y}{dx^2} - 2x\dfrac{dy}{dx} + 2y = (x+1)^2$;

(xl) $\dfrac{d^2 y}{dx^2} + \dfrac{dy}{dx} = \dfrac{2}{x^2}\,y$;

(xli) $x^2(x^2-1)\dfrac{d^2 y}{dx^2} = (x^2-2)\left(x\dfrac{dy}{dx} - y\right)$;

(xlii) $x(1+x)^2 \dfrac{d^2 y}{dx^2} + x(1+x)\dfrac{dy}{dx} + y = x$.

12. The equation

$$x^3 \dfrac{d^3 y}{dx^3} - 4x^2 \dfrac{d^2 y}{dx^2} + x(8+x^2)\dfrac{dy}{dx} - 2(4+x^2)\,y = 0$$

has $x\sin x$ and $x\cos x$ for particular integrals: find the primitive.

13. Integrate completely the equation

$$\dfrac{d^3 y}{dx^3} + a^2 x\dfrac{d^2 y}{dx^2} + \tfrac{1}{3}a^4 x^2 \dfrac{dy}{dx} + \tfrac{1}{27}a^6 x^3 y = 0,$$

having given that $e^{-\frac{1}{6}a^2 x^2}$ is a solution.

14. Verify that the left-hand side of the equation

$$(\sin x - x \cos x) \frac{d^2 y}{dx^2} - x \sin x \frac{dy}{dx} + y \sin x = x$$

vanishes when $y = \sin x$: and obtain the primitive.

15. One integral of the equation

$$\frac{d^2 y}{dx^2} + \frac{dy}{dx} \tan x - \tfrac{1}{4} y \left(2 + 5 \tan^2 x\right) = 0$$

is given by

$$y^2 = \sec x;$$

obtain the primitive.

16. The complete primitive of $\dfrac{d^2 y}{dx^2} + P \dfrac{dy}{dx} + Qy = 0$ is of the form

$$y = A\phi(x) + B\psi(x);$$

shew that the additional term in the complete primitive of

$$\frac{d^2 y}{dx^2} + P \frac{dy}{dx} + Qy = F(x)$$

is given by

$$\int^x F(\xi) \frac{\phi(\xi)\psi(x) - \psi(\xi)\phi(x)}{\phi(\xi)\psi'(\xi) - \psi(\xi)\phi'(\xi)} \, d\xi.$$

17. Solve the equation

$$\frac{d^3 y}{dx^3} - c^2 \frac{dy}{dx} + c^2 \frac{a}{x} y = 0,$$

where a is an even positive integer: and prove that, when $a = 2$,

$$\frac{y}{x^2} = \int (A e^{cx} + B e^{-cx}) x^{-3} dx + C.$$

18. Prove that, if $y = \phi(x)$ is a solution of the equation

$$\frac{d^2 y}{dx^2} + \frac{a}{x} \frac{dy}{dx} + by = 0,$$

then $y = \dfrac{1}{x} \phi'(x)$ is a solution of the same equation with $a+2$ written for a.
Thence (or otherwise) obtain the integral of the equation when a is an even integer.

19. Prove that the transformation

$$y = \eta \xi^{-n}, \quad \xi = x - (x^2 - 1)^{\frac{1}{2}},$$

transforms Legendre's differential equation into a hypergeometric equation in which η is the dependent variable, and ξ^2 the independent variable.

20. Shew that, if

$$y^5 - 5y^3 + 5y - 4x + 2 = 0,$$

then

$$\frac{d^2 y}{dx^2} + \frac{2x-1}{2x(x-1)} \frac{dy}{dx} - \frac{y}{25x(x-1)} = 0.$$

Account for the difference between the degree of the algebraic equation and the order of the differential equation.

21. Prove that the equation

$$\frac{d^2y}{dx^2}+\phi(x)\frac{dy}{dx}+\psi(x)y=0$$

has a complete primitive of the form $y=Af\left(\dfrac{x}{m}\right)+Bf\left(\dfrac{x}{n}\right)$, where A and B are arbitrary constants, provided

$$y_1=\frac{m^2\psi(mx)-n^2\psi(nx)}{m\phi(mx)-n\phi(nx)}$$

is a solution of the equation

$$\frac{dy_1}{dx}+m\phi(mx)y_1-y_1^2=m^2\psi(mx),$$

and that then

$$\log f(x)+\int y_1 dx=0.$$

Shew that there is a failure when $m+n=0$.

Solve the equation

$$abx(a+b-x)\frac{d^2y}{dx^2}+\{(a+b)x^2-(a^2+b^2)x-ab(a+b)\}\frac{dy}{dx}-x^2y=0.$$

22. Prove that, for values of x between 0 and 1, the primitive of the equation

$$x(1-x)\frac{d^2y}{dx^2}+\tfrac{1}{2}(a+\beta+1)(1-2x)\frac{dy}{dx}-a\beta y=0$$

is $AF\{\tfrac{1}{2}a,\ \tfrac{1}{2}\beta,\ \tfrac{1}{2},\ (1-2x)^2\}+B(1-2x)F\{\tfrac{1}{2}a+\tfrac{1}{2},\ \tfrac{1}{2}\beta+\tfrac{1}{2},\ \tfrac{3}{2},\ (1-2x)^2\}$,

where A and B are arbitrary constants, and $F(a,b,c,x)$ denotes the hypergeometric series.

23. Shew that the function

$$Q_n(z)=\frac{\pi^{\frac{1}{2}}}{2^{n+1}}\frac{\Pi(n)}{\Pi(n+\frac{1}{2})}\frac{1}{z^{n+1}}F\left(\tfrac{1}{2}n+\tfrac{1}{2},\ \tfrac{1}{2}n+1,\ n+\tfrac{3}{2},\ \frac{1}{z^2}\right),$$

where $F(a,b,c,x)$ denotes the hypergeometric series formed with the elements $a,\ b,\ c,\ x$, satisfies Legendre's equation

$$(1-z^2)\frac{d^2y}{dz^2}-2z\frac{dy}{dz}+n(n+1)y=0.$$

Shew also that, when the real part of n is greater than -1,

$$Q_n(z)=\int_0^\infty\{z+(z^2-1)^{\frac{1}{2}}\cosh\theta\}^{-n-1}d\theta.$$

24. Investigate the differential equation which, with the usual notation of Bessel's functions, has its primitive of the form

$$y=\psi^m(x)[AJ_n\{\psi(x)\}+BY_n\{\psi(x)\}],$$

obtaining the result

$$\frac{d^2y}{dx^2}-\left\{\frac{\psi''}{\psi'}+(2m-1)\frac{\psi'}{\psi}\right\}\frac{dy}{dx}+(m^2-n^2+\psi^2)\left(\frac{\psi'}{\psi}\right)^2y=0.$$

Deduce the integrals of the equations

$$x\frac{d^2y}{dx^2}-(2n-1)\frac{dy}{dx}+xy=0,$$

$$x\frac{d^2y}{dx^2}+(n+1)\frac{dy}{dx}+\tfrac{1}{4}y=0.$$

25. Express the primitive of

$$\frac{dy}{dx} + y^2 + \frac{1}{x} = 0$$

in terms of Bessel's functions, of argument $2x^{\frac{1}{2}}$ and of order unity.

26. Prove that $y = \left(\frac{1}{x}\frac{d}{dx}\right)^m (A\cosh nx + B\sinh nx)$, where m is a positive integer, is the primitive of

$$\frac{d^2y}{dx^2} + \frac{2m}{x}\frac{dy}{dx} - n^2y = 0.$$

Shew also that any solution of

$$\frac{d^2y}{dx^2} + \frac{2}{x}\frac{dy}{dx} - n^2y = 0$$

is a solution of

$$\frac{d^4y}{dx^4} - 2\left(n^2 + \frac{6}{x^2}\right)\frac{d^2y}{dx^2} + n^2\left(n^2 + \frac{4}{x^2}\right)y = 0;$$

and obtain the primitive of the latter.

27. Shew that the complete primitive of the equation

$$\frac{d^2y}{dx^2} + y = \frac{n(n+1)}{x^2}y,$$

when n is an integer, can be exhibited in the form

$$y = x^{n+1}\{A\psi_n(x) + B\Psi_n(x)\},$$

where $\qquad \psi_n(x) = (-1)^n 1\,.\,3\ldots(2n+1)\left(\frac{1}{x}\frac{d}{dx}\right)^n \frac{\sin x}{x},$

and $\Psi_n(x)$ is the same function of $\dfrac{\cos x}{x}$.

28. In the differential equation $\dfrac{d^2y}{dx^2} = By$, the coefficient B is a polynomial in x. Shew that, if Q denote the operation of integrating from 0 to x, each of the series

$$1 + Q^2B + Q^2BQ^2B + \ldots, \quad x + Q^2Bx + Q^2BQ^2Bx + \ldots,$$

of which each term is obtained from its predecessor by multiplying by B and then integrating twice in succession from 0 to x, converges for all finite values of x and satisfies the equation.

29. Obtain a series in ascending powers of x to satisfy the equation

$$xy'' + (n+1)y' + y = 0.$$

If $\qquad\qquad \phi_n(x) = \sum_{r=0}^{\infty} \frac{(-1)^r x^r}{(n+r)!\,r!},$

and if u denote the continued fraction

$$\frac{1}{n+1} - \frac{x}{n+2} - \frac{x}{n+3} - \frac{x}{n+4} - \ldots,$$

prove that $\qquad\qquad \phi_n(x) = \frac{1}{n!}e^{-\int_0^x u\,dx}.$

30. Shew that, if

$$y = \frac{\sin(x-a)}{\sin x} e^{x \cot a},$$

where a is a constant, then

$$\frac{d^2 y}{dx^2} = \left(\frac{2}{\sin^2 x} + \cot^2 a\right) y;$$

and deduce the primitive of this equation, (i) in general, (ii) when $a = \frac{1}{2}\pi$.

31. Prove that, if n is not negative and m is a positive integer, the equation

$$(x^2 - 1)\frac{d^2 y}{dx^2} + (2n+2) x \frac{dy}{dx} = m(m+2n+1) y$$

has the two solutions

$$K_m(x) = (x^2 - 1)^{-n} \frac{d^m}{dx^m} \{(x^2 - 1)^{m+n}\},$$

$$L_m(x) = (x^2 - 1)^{-n} \int_{-1}^{1} \frac{(t^2 - 1)^n}{x - t} K_m(t)\, dt,$$

x being numerically greater than unity.

32. Prove that a solution of the equation

$$\frac{1}{y}\frac{d^2 y}{dx^2} = \frac{m(m-1)}{x^2} + b^2$$

is given by

$$y = x^m \int_0^{\pi} e^{bx \cos\theta} \sin^{2m-1}\theta\, d\theta.$$

33. Shew that

$$y = \int_0^{\pi} \cos(mx \cos\phi + nx \sin\phi + \theta)\, d\phi$$

satisfies the equation

$$\frac{d^2 y}{dx^2} + \frac{1}{x}\frac{dy}{dx} + (m^2 + n^2) y = -2\frac{n}{x}\sin\theta \cos mx.$$

34. Apply the method of solution by definite integrals to find the primitive of the equation

$$(1 - x^2)\frac{d^2 y}{dx^2} - 2x\frac{dy}{dx} - \frac{1}{4} y = 0,$$

for values of x satisfying the inequality

$$-1 < x < 1.$$

Prove that the equation is transformed into itself by the substitutions

$$(x-1)(x'-1) = 4, \quad y(x+1)^{\frac{1}{2}} = y'(x'+1)^{\frac{1}{2}};$$

and thence obtain the primitive for values of x satisfying

$$1 < x < \infty.$$

35. Obtain the primitive of the equation

$$\frac{d^3 y}{dx^3} + 2x\frac{dy}{dx} + y = 0$$

in the form

$$\int_0^{\infty} e^{-\frac{1}{6}t^6} \left\{ A e^{-t^2 x} + B e^{\frac{1}{2}t^2 x} \sin\left(\frac{\sqrt{3}}{2} t^2 x + C\right) \right\} dt,$$

where A, B, C, are arbitrary constants.

36. Prove that the equation

$$x\frac{d^3y}{dx^3}+(n+2p+3)\frac{d^2y}{dx^2}-x\frac{dy}{dx}-(n+1)y=0,$$

where n and p are real quantities greater than -1, is satisfied by

$$y=\int_0^\theta t^n(1-t^2)^p e^{-tx}dt,$$

for $\theta=1,-1,\infty$.

37. Shew that the primitive of the equation

$$x^2\frac{d^2y}{dx^2}-(m+n-1)x\frac{dy}{dx}+mny+a\left(x\frac{d^2y}{dx^2}-p\frac{dy}{dx}\right)=0,$$

where m, p, n, are positive integers in order of magnitude, is a polynomial in x

38. Shew that every solution of the equation

$$x^2(x+3)\frac{d^3y}{dx^3}-3x(x+2)\frac{d^2y}{dx^2}+6(x+1)\frac{dy}{dx}-6y=0$$

is a polynomial in x.

39. Shew that, if the equation

$$\frac{d^2y}{dx^2}+P\frac{dy}{dx}+Qy=0$$

has $$y=Au+Bv$$

for its primitive, u and v being polynomial functions of x, there are no squared factors in the denominator of P and no cubed factors in the denominator of Q.

40. Find the condition that the equation

$$\frac{d^2y}{dx^2}+\frac{1}{x}\frac{dy}{dx}-\left(1+\frac{a^2}{x^2}\right)y=0$$

should have one solution expressible in integral powers of x; and shew that, when the condition is satisfied, every other solution of the equation possesses a logarithmic infinity at the origin.

41. Prove that, if the equation

$$(Ax^2+Bx)\frac{d^2y}{dx^2}+(Dx+E)\frac{dy}{dx}+Fy=0$$

has a polynomial function of x as one solution, it is necessary (but not sufficient) that the quadratic $Ap(p-1)+Dp+F=0$ have a positive integer for a root.

What are the conditions that the complete primitive should be a polynomial function of x?

42. Solve the equation

$$x^2(x-1)^2\frac{d^2v}{dx^2}+\tfrac{3}{16}v=0;$$

and shew that x is a single-valued function of the quotient of two linearly independent solutions.

43. Prove that, if it is possible to determine constants a, h, b, so that $ay_1^2 + 2hy_1y_2 + by_2^2$ is constant, where y_1 and y_2 are linearly independent solutions of the equation

$$\frac{d^2y}{dx^2} + P\frac{dy}{dx} + Qy = 0,$$

then

$$\frac{dQ}{dx} + 2PQ = 0;$$

and solve the equation when this condition is satisfied.

Obtain also the respective conditions which must be satisfied by P and Q, when two linearly independent integrals y_1 and y_2 obey one of the relations:

(i) $\quad y_1y_2 + ay_1 + by_2 + c = 0$;

(ii) $\quad y_1^3 + y_2^3 = 1$.

44. A relation of the form

$$ay_1^2 + by_2^2 + cy_3^2 + 2fy_2y_3 + 2gy_3y_1 + 2hy_1y_2 = 0$$

subsists among the linearly independent integrals of the equation

$$\frac{d^3y}{dx^3} + P\frac{dy}{dx} + Qy = 0;$$

prove that

$$\frac{dP}{dx} = 2Q,$$

and solve the differential equation.

45. Shew that the differential equation of the third order, satisfied by the ratio of any solution of $\frac{d^2y}{dx^2} + Iy = 0$ to any solution of $\frac{d^2y}{dx^2} + Jy = 0$, is

$$\frac{d}{dx}\left\{\frac{\frac{d^2s}{dx^2} + (I-J)s}{\frac{ds}{dx}}\right\} - \frac{1}{2}\left\{\frac{\frac{d^2s}{dx^2} + (I-J)s}{\frac{ds}{dx}}\right\}^2 = 2J.$$

46. Having given one solution of the equation

$$\frac{d^2y}{dx^2} + P\frac{dy}{dx} + Qy = 0,$$

where P and Q are functions of x, find the complete primitive of the equation and also of

$$\frac{dy}{dx} = Q - Py + y^2.$$

Integrate the equation

$$\frac{dy}{dx} = x - 1 + xy + y^2.$$

47. Find the form of $f(y)$ in order that, by changing the independent variable from x to z, where $z = x - f(y)$, the equation

$$\frac{d^2y}{dx^2} - \frac{3}{y}\left(\frac{dy}{dx}\right)^2 + 3\frac{dy}{dx} - y = 0$$

may be reduced to a linear equation; and hence obtain the primitive.

Similarly solve the equation

$$\frac{dy}{dx}\left\{2(y^2 - x)\left(\frac{dy}{dx}\right)^2 + 4y\frac{dy}{dx} + 1\right\} + (x + y^2)\frac{d^2y}{dx^2} = 0.$$

48. Integrate the equation

$$\frac{d^2r}{d\theta^2} - \frac{3}{r}\left(\frac{dr}{d\theta}\right)^2 + \tfrac{1}{2}r = \frac{r^4}{c^3},$$

with the condition that $\frac{dr}{d\theta}$ vanishes when $2r = c$.

49. Prove that the differential equation

$$\left(\frac{d^2y}{dx^2}\right)^2 \frac{d^5y}{dx^5} - 5\frac{d^2y}{dx^2}\frac{d^3y}{dx^3}\frac{d^4y}{dx^4} + \tfrac{40}{9}\left(\frac{d^3y}{dx^3}\right)^3 = 0$$

has $\left(\frac{d^2y}{dx^2}\right)^n$ as an integrating factor, for each of two values of n. Hence integrate the equation completely.

50. Shew that the primitive of the equation

$$x^{\frac{3}{2}}\frac{d^2y}{dx^2} = f(yx^{-\frac{1}{2}})$$

is

$$\int \{a + \tfrac{1}{4}z^2 + 2\int f(z)\,dz\}^{-\frac{1}{2}}\,dz = b + \log x,$$

where

$$z = yx^{-\frac{1}{2}}.$$

Solve the equation

$$(a + 2bx + cx^2)^{\frac{3}{2}}\frac{d^2y}{dx^2} = f\{y\,(a + 2bx + cx^2)^{-\frac{1}{2}}\}$$

by using a similar substitution; and reduce the equation

$$\frac{d^2y}{dx^2} = (ax^2 + 2hxy + by^2 + 2gx + 2fy + c)^{-\frac{3}{2}}$$

to this form.

51. Integrate the equations :

(i) $\quad 1 + \left(\frac{dy}{dx}\right)^2 = c^2\frac{d^2y}{dx^2}$;

(ii) $\quad 9xy^2\frac{d^2y}{dx^2} + 2 = 0$;

(iii) $\quad \left\{x - \left(\frac{dy}{dx}\right)^2\right\}\frac{d^2y}{dx^2} = x^2 - \frac{dy}{dx}$;

(iv) $\quad x\frac{dy}{dx} - ny + my^2 = x^{2n}$;

(v) $\quad \frac{d^2y}{dx^2} + \left(\frac{dy}{dx}\right)^2 + \frac{4}{x}\frac{dy}{dx} + \frac{2}{x^2} = 0$;

(vi) $\quad \left\{1 + \left(\frac{dy}{dx}\right)^2\right\}\frac{dy}{dx}\frac{d^3y}{dx^3} = \left\{3\left(\frac{dy}{dx}\right)^2 - 1\right\}\left(\frac{d^2y}{dx^2}\right)^2$;

(vii) $\quad xy\frac{d^2y}{dx^2} - 2x\left(\frac{dy}{dx}\right)^2 + (y+1)\frac{dy}{dx} = 0$;

(viii) $\quad xy\frac{d^2y}{dx^2} - x\left(\frac{dy}{dx}\right)^2 + 2y\frac{dy}{dx} = n\frac{x+2}{(x+1)^2}y^3$;

(ix) $\quad \frac{d^2y}{dx^2} + \frac{x}{x^2-a^2}\frac{dy}{dx} = \frac{y}{y^2-a^2}\left(\frac{dy}{dx}\right)^2$.

52. Prove that the orthogonal trajectories of the curve

$$\left(r - \frac{k}{r^2}\right)\cos\theta = \text{constant}$$

are the curves

$$\left(r^2 + \frac{2k}{r}\right)\sin^2\theta = \text{constant};$$

and sketch roughly the two families of curves.

53. Prove that the orthogonal trajectory of the family of curves

$$r^{2n} = 2a^n r^n \cos n\theta + a^{2n} = c^{2n},$$

where c is the variable parameter, is

$$r^n \cos(n\theta + \gamma) = a^n \cos\gamma,$$

where γ is the variable parameter.

54. Prove that the system of curves, orthogonal to the family represented by the elimination of t between the equations

$$x = f(t, a), \quad y = \phi(t, a),$$

a being the parameter of the family, is given by substituting for a the complete integral of

$$\left(\frac{\partial f}{\partial t}\frac{\partial f}{\partial a} + \frac{\partial \phi}{\partial t}\frac{\partial \phi}{\partial a}\right)\frac{da}{dt} + \left(\frac{\partial f}{\partial t}\right)^2 + \left(\frac{\partial \phi}{\partial t}\right)^2 = 0.$$

Find the orthogonal system in the case

$$x = f(a) + a\cos t, \quad y = a\sin t.$$

55. Shew that, if $u + v\sqrt{-1}$ is a function of $x + y\sqrt{-1}$, the oblique trajectories of the system of curves $\phi(u, v) = c$ are given by

$$\frac{\partial\phi}{\partial u}(du - dv\tan a) + \frac{\partial\phi}{\partial v}(dv + du\tan a) = 0.$$

56. Find the differential equation of a curve cutting the coaxal circles

$$x^2 + y^2 + 2vx + c = 0,$$

where v is parametric, at a constant angle; and integrate the equation.

57. Prove that, if $\qquad f(x + iy) = u + iv,$

where i denotes $(-1)^{\frac{1}{2}}$, and u, v, are real, the curves

$$u = a, \quad v = \beta,$$

cut orthogonally.

Shew that,

(i)   when $f(u) = \log u$, the curves are concentric circles:

(ii)   when $f(u) = e^u$, the curves are catenaries of equal strength:

(iii)   when $f(u) = \cos^{-1} u$, the curves are confocal conics:

(iv)   when $f(u) = \tan^{-1} u$, the curves are stereographic projections of meridians and parallels of latitude on a sphere, which has the plane of x, y, for its equatorial plane.

58. Prove that the equation

$$y \frac{d^2y}{dx^2} - \frac{1}{2}\left(\frac{dy}{dx}\right)^2 - \frac{3}{2}\frac{x^2y}{1-x^3}\frac{dy}{dx} + \frac{3}{8}\frac{xy^2}{1-x^3} = 0$$

is satisfied by each of the quantities

$$(\alpha+x)^{\frac{1}{2}} + (1+\alpha x)^{\frac{1}{2}}, \quad (\alpha+x)^{\frac{1}{2}} - (1+\alpha x)^{\frac{1}{2}},$$

where α is a cube root of -1. Obtain the primitive of the equation.

59. Solve the equations

$$\left.\begin{array}{l} \dfrac{d^2y}{dx^2} + \dfrac{dz}{dx} - 2z = e^{2x} \\[2mm] \dfrac{dz}{dx} + 2\dfrac{dy}{dx} - 3y = 0 \end{array}\right\}.$$

60. Integrate the equations

$$\frac{dy}{dx} + yf'(x) - z\phi'(x) = 0,$$

$$\frac{dz}{dx} + y\phi'(x) + zf'(x) = 0,$$

f and ϕ being known functions of x.

61. Solve the equations

$$D(D-2)x - (D-1)y = 0,$$
$$(2D-1)x + D^2(D-1)y = t,$$

where D denotes d/dt.

62. Integrate the equations

$$\frac{dx}{x+y-z} = \frac{dy}{y+z-x} = \frac{dz}{z+x-y}.$$

63. Solve

$$\frac{dx}{x+y-xy^2} = \frac{dy}{x^2y-x-y} = \frac{dz}{z(y^2-x^2)}.$$

64. Illustrate the method of variation of parameters by obtaining an approximate solution of

$$\frac{dx}{dt} = -y + kx^2, \quad \frac{dy}{dt} = x - ky^2,$$

in circular functions, k being a small quantity whose square may be neglected.

Hence (or otherwise) shew how to integrate

$$a\frac{dx}{dt} + \beta\frac{dy}{dt} = a'x + \beta'y + x\sin t + y\cos t,$$

$$\beta\frac{dx}{dt} - a\frac{dy}{dt} = \beta'x - a'y + x\cos t - y\sin t.$$

65. Solve the equations

$$y = xp + p^2 - aq, \quad z = xq + pq,$$

where

$$p = \frac{dy}{dx}, \quad q = \frac{dz}{dx};$$

and shew that there are three distinct solutions.

66. Solve the system of equations

$$\frac{dx}{dt} = ax + hy + gz, \quad \frac{dy}{dt} = hx + by + fz, \quad \frac{dz}{dt} = gx + fy + cz.$$

Shew that, when

$$a - \frac{gh}{f} = b - \frac{hf}{g} = c - \frac{fg}{h} = \lambda,$$

the solution is

$$x = Ae^{\lambda t} + \frac{B}{f} e^{\mu t}, \quad y = A'e^{\lambda t} + \frac{B}{g} e^{\mu t}, \quad z = A''e^{\lambda t} + \frac{B}{h} e^{\mu t},$$

where

$$\frac{A}{f} + \frac{A'}{g} + \frac{A''}{h} = 0,$$

and

$$\mu = \lambda + \frac{gh}{f} + \frac{hf}{g} + \frac{fg}{h}.$$

67. Let P, Q, R, be any functions of three independent variables; and with a point A (x, y, z), associate the plane $(X - x) P + (Y - y) Q + (Z - z) R = 0$. Let a consecutive point $B (x + dx, y + dy, z + dz)$ be taken in the plane through A. Shew that there are two directions through A such that the intersection of the planes, associated with A and with B, is the direction AB.

Interpret the result when the condition of integrability for the equation

$$P dx + Q dy + R dz = 0$$

is satisfied; and find (i) when the two directions are perpendicular, (ii) if the two directions can be coincident.

68. A doubly infinite system of similar conics in parallel planes have their centres collinear and their corresponding axes parallel. Shew that they can be cut orthogonally by a family of surfaces only if the line of centres is perpendicular to their planes.

69. Find a solution of the equation

$$(x^2 + 2xy + y^2 - z^2) dx - (x^2 + 2xy + y^2 + z^2) dy + 2 (x + y) z dz = 0,$$

which is satisfied by $x = y = z = 1$.

70. Obtain integral equivalents of the equations:

(i) $2 (y + z) dx + (x + 3y + 2z) dy + (x + y) dz = 0;$

(ii) $(yz + z^2) dx - xz dy + xy dz = 0;$

(iii) $zy (1 + 4xz) dx - zx (1 + 2xz) dy - xy dz = 0;$

(iv) $y^2 dx + (3xy - 2z^2)\, dy - 2yz\, dz = 0$;

(v) $(2x + yz)\, y\, dx - x^2\, dy + (x + 2z)\, y^2\, dz = 0$;

(vi) $yz^2 (x^2 - yz)\, dx + zx^2 (y^2 - zx)\, dy + xy^2 (z^2 - xy)\, dz = 0$;

(vii) $(2yz + zx - z^2)\, dx - zx\, dy - (x^2 + xy - xz)\, dz = 0$;

(viii) $y^2 z\, (x \cos x - \sin x)\, dx + x^2 z\, (y \cos y - \sin y)\, dy$
$\qquad\qquad\qquad + xy\, (y \sin x + x \sin y + xy \cos z)\, dz = 0$;

(ix) $(2z^2 - xy + y^2)\, z\, dx + (2z^2 + x^2 - xy)\, z\, dy - (x + y)\, (xy + z^2)\, dz = 0$;

(x) $(y + z)\, (x^2 y^2 z^2 - 1)\, dx + x^2 y^2 z^2\, (xy + xz) \left(\dfrac{dy}{y} + \dfrac{dz}{z} \right) + x\, (dy + dz) = 0$;

(xi) $(y^2 z - y^3 + x^2 y)\, dx - (x^2 z + x^3 - xy^2)\, dy + (x^2 y - xy^2)\, dz = 0$;

(xii) $(y^2 z^2 - yz)\, dx - (xz - 1)\, dy - (xy - y^2)\, dz = 0$;

(xiii) $z\, (1 - z^2)\, dx + z\, dy - (x + y + xz^2)\, dz = 0$;

(xiv) $(3x^2 + 2xy - y^2 + z)\, dx + (x^2 - 2xy - 3y^2 + z)\, dy + (x + y)\, dz = 0$;

(xv) $(z^3 + zx^2 + 2yx^2 - y^2 z)\, dy + (y^3 + yx^2 + 2zx^2 - z^2 y)\, dz$
$\qquad\qquad\qquad\qquad\qquad = 2x\, (y^2 + z^2 + yz)\, dx.$

71. Find an integral equivalent of the equation
$$x\, dy\, dz\, (b - c) + y\, dz\, dx\, (c - a) + z\, dx\, dy\, (a - b) = 0.$$

72. Shew that the differential equation of a plane curve
$$f\, (x,\, y,\, z,\, y\, dz - z\, dy,\, z\, dx - x\, dz,\, x\, dy - y\, dx) = 0,$$
where x, y, z, are homogeneous point-coordinates, is transformable to
$$f\, (v\, dw - w\, dv,\, w\, du - u\, dw,\, u\, dv - v\, du,\, u,\, v,\, w) = 0$$
by the introduction of line-coordinates from the equation
$$ux + vy + wz = 0.$$
Obtain the primitive of the equation
$$ax^n\, (y\, dz - z\, dy) + by^n\, (z\, dx - x\, dz) + cz^n\, (x\, dy - y\, dx) = 0$$
in the form
$$\left. \begin{aligned} Aa + Bb + Cc &= 0 \\ Ax^{1-n} + By^{1-n} + Cz^{1-n} &= 0 \end{aligned} \right\} ;$$
and apply the above transformation to deduce the primitive of
$$ax\, (y\, dz - z\, dy)^n + by\, (z\, dx - x\, dz)^n + cz\, (x\, dy - y\, dx)^n = 0.$$

73. Integrate the partial differential equations:

(i) $p^3 = qz$;

(ii) $z = p^m q^m$;

(iii) $\dfrac{p^2}{y} - \dfrac{q^2}{x} = \dfrac{1}{z} \left(\dfrac{1}{x} + \dfrac{1}{y} \right)$;

(iv) $(pq - p)\, (1 + x^2) + qxy^2 = 0$;

(v) $3\, (p + q)^2 + (p - q)^2 z = 48$;

(vi) $zp - xq = y$;

(vii) $(p - q)(x + y) = z$;

(viii) $p^2 + q^2 = (x^2 + y^2)z$;

(ix) $(3x + y - z)p + (x + y - z)q = 2(z - y)$;

(x) $y(2z + x)p - z^2 q + yz = 0$;

(xi) $p^2 + q^2 - 2pq \tanh 2y = \operatorname{sech}^2 2y$;

(xii) $(x + y - z)(p - q) + a(px - qy + x - y) = 0$;

(xiii) $pq(x - y) + px - qy = (p - q)z$;

(xiv) $xyp + z^2 q + yz = 0$;

(xv) $q = \dfrac{1 + p^2}{1 + y^2} x + yp(z - px)^2$;

(xvi) $xyp - y(2x + y)q + 2xz = 0$;

(xvii) $px + qy = z(1 + pq)^{\frac{1}{2}}$;

(xviii) $z(p^2 - q^2) = x - y$;

(xix) $(2xy - x)p + 3xq + (2y - 1)z = 0$;

(xx) $qx + py = (p^2 - q^2)^n$;

(xxi) $(ax + hy + gz)p + (hx + by + fz)q = gx + fy + cz$;

(xxii) $xp + (ax + by)q = cx + ey + fz$;

(xxiii) $(p - q)x = y(z - px - qy)$;

(xxiv) $(px + qy)^2 - z(px + qy) + p^2 + q^2 = 0$;

(xxv) $(py + x)(qz + x) = (pz + x + y)(qy + z + y)$;

(xxvi) $aq(x + pz) + bp(y + qz) = 0$;

(xxvii) $(p - y)^2 + (q - x)^2 = 1$;

(xxviii) $xp - yq = xqf(z - px - qy)$;

(xxix) $x(y^n - z^n)p + y(z^n - x^n)q = z(x^n - y^n)$;

(xxx) $(z^2 + 1)yp^2 + xzpq = 4x^2 y$;

(xxxi) $(p^2 + q^2)^n (qx - py) = 1$;

(xxxii) $(xz + y^2)p + (yz - 2x^2)q + 2xy + z^2 = 0$;

(xxxiii) $px^2 + z = y^2 q^2$;

(xxxiv) $py + qx + pq = 0$;

(xxxv) $(p - q)(e^{x+y} + xyz) + (yq - xp)(z + xye^{-x-y}) + (x - y)(1 - z^2) = 0.$

74. Shew that any partial differential equation of the form

$$F\left(p,\ q,\ z-px-qy,\ \frac{y}{x}\right)=0$$

has a family of solutions in common with $z-px-qy=a$.

75. Obtain a complete integral of the equation

$$pxy\,(xy+2z^2)+z\,(qy-z)\,(yz-x^2)=0$$

in the form

$$\frac{x+a}{y}+\frac{y+b}{z}+\frac{z}{x}=0\ ;$$

and investigate the nature of the integral

$$\frac{x}{y}+\frac{y}{z}+\frac{z}{x}+\frac{2}{y^{\frac{1}{2}}z^{\frac{1}{2}}}=0.$$

Also shew that

$$(z-a)^2=4x+4by,$$

$$cy\,(z-a)=c^2x+y^2,$$

are integrals of the equation

$$xp^2+ypq=1\ ;$$

and indicate their relation to one another.

76. Obtain a complete integral of the equation

$$4z=p^2+q^2$$

in the form

$$z=f\,(x+\lambda y)\ ;$$

and derive from it the solutions

(i) $z=(x-a)^2+(y-b)^2,$

(ii) $(x^2+y^2+c^2-z)^2=4c^2\,(x^2+y^2).$

77. Prove that the general integral of

$$z=xp+yq+pq$$

can be obtained by eliminating t between the equations

$$x\frac{d}{dt}\left(\frac{p}{q}\right)-z\frac{d}{dt}\left(\frac{1}{q}\right)+\frac{dp}{dt}=0,$$

$$y\frac{d}{dt}\left(\frac{q}{p}\right)-z\frac{d}{dt}\left(\frac{1}{p}\right)+\frac{dq}{dt}=0,$$

when, in the latter, p and q are arbitrary functions of the variable t.

78. Solve the equation

$$b\,(bcy+axz)\,p+a\,(acx+byz)\,q=ab\,(z^2-c^2)\,;$$

and shew that the solution represents any surface generated by lines meeting two given lines.

79. Investigate the integral of

$$p^2+q+x+z=0,$$

containing the line $x=z$, $y=0$; and obtain it in the form

$$z=4e^{-2y}+(2x+4y-3)\,e^{-y}-x-1.$$

80. Find the surface, which satisfies the equation

$$(x+y)\,p+(x-y)\,q=z,$$

and passes through the curve $x^2+y^2=1$, $z=c$.

81. Determine the surface which satisfies the differential equation

$$(x^2-a^2)\,p+(xy-az\tan a)\,q=xz-ay\cot a,$$

and passes through the curve $z=0$, $x^2+y^2=a^2$.

82. Integrate the equation

$$(x^3+3xy^2)\,p+(y^3+3x^2y)\,q=2\,(x^2+y^2)\,z\,;$$

and find an integral which represents a surface passing through the circle

$$x^2+y^2=c^2,\quad z=a.$$

83. Obtain the primitives of the equation

$$\frac{x^2}{p^2}+\frac{y^2}{q^2}=z^2\,;$$

and illustrate the results by a geometrical interpretation.

Investigate the equations of surfaces possessing the property that the parallelepiped contained by the radius vector from the origin, a unit length along the perpendicular to the tangent plane, and a unit length along the axis of z, is of constant volume.

84. Find the differential equation of a surface, normals to which have intercepts between the coordinate planes of constant ratios $b-c:c-a:a-b$. Obtain a complete primitive in the form

$$\frac{x^2}{a+\lambda}+\frac{y^2}{b+\lambda}+\frac{z^2}{c+\lambda}+\mu=0\,;$$

and determine the nature of the solution

$$(b-c)^{\frac{1}{2}}x+(c-a)^{\frac{1}{2}}y+(a-b)^{\frac{1}{2}}z=0.$$

85. If the complex of lines $\phi\,(a,\,b,\,c,\,f,\,g,\,h)=0$ be the complex of tangents to a surface, the equation

$$\frac{\partial\phi}{\partial a}\frac{\partial\phi}{\partial f}+\frac{\partial\phi}{\partial b}\frac{\partial\phi}{\partial g}+\frac{\partial\phi}{\partial c}\frac{\partial\phi}{\partial h}=0$$

is satisfied. Is the condition sufficient, as well as necessary?

Transform this equation to

$$\frac{\partial b}{\partial g}+\frac{\partial b}{\partial a}\frac{\partial b}{\partial f}=0,$$

where the variables are now the quotients of $a,\,b,\,f,\,g,$ by h; and shew how the complex of tangent lines is contained in the general integral of this equation.

86. The square of the reciprocal of the perpendicular from the origin on the tangent plane of a surface is equal to $f\,(u)$, where u is the reciprocal of the radius vector. Obtain a primitive of the differential equation of the surface in polar coordinates in the form

$$\int\{f(u)-u^2\}^{-\frac{1}{2}}\,du=\{\phi+\sin^{-1}(\tan a\cot\theta)\}\sin a-\sin^{-1}(\sec a\cos\theta)+\beta.$$

87. Integrate the equations:

(i) $zx_1=p_1\,(p_1+p_2)+x_1p_2\,(p_3+x_3p_2)$;

(ii) $a^2\,(2z+x_1p_1)=zx_1p_1\,(x_3p_2+x_2p_3)$;

(iii) $(x_2+x_3+z)\,p_1+(x_3+x_1+z)\,p_2+(x_1+x_2+z)\,p_3=0.$

88. Shew that the equation

$$p_1{}^2+p_2{}^2+p_3{}^2=2\,(p_1x_1+p_2x_2+p_3x_3)$$

has the integrals

(i) $\dfrac{x_1{}^2}{z+a_1}+\dfrac{x_2{}^2}{z+a_2}+\dfrac{x_3{}^2}{z+a_3}=1,$

(ii) $(z-c)\,(\lambda^2+\mu^2+\nu^2)=(\lambda x_1+\mu x_2+\nu x_3)^2,$

(iii) $(z+x_3{}^2-2x_1{}^2-2x_2{}^2)^2+8x_3{}^2\,(x_1{}^2+x_2{}^2)=0$;

and shew how to derive (ii) and (iii) from (i), also (i) and (iii) from (ii).

89. Shew that the equations

$$(x-y)\frac{\partial f}{\partial x}-2\,(x-y)\frac{\partial f}{\partial y}+3\,(x+y+2z)\frac{\partial f}{\partial z}=0,$$

$$(y+z)\frac{\partial f}{\partial x}+2\,(2x-3y-z)\frac{\partial f}{\partial y}-3\,(2x+y+3z)\frac{\partial f}{\partial z}=0,$$

have a common solution. Find the solution; and shew that it may be determined by a single quadrature.

90. Obtain the most general integral common to the equations

$$\left(\frac{\partial u}{\partial x}\right)^2 + \left(\frac{\partial u}{\partial y}\right)^2 + \left(\frac{\partial u}{\partial z}\right)^2 = f(x^2 + y^2 + z^2),$$

$$\frac{1}{x}\frac{\partial u}{\partial x} = \frac{1}{y}\frac{\partial u}{\partial y}.$$

91. Find all the solutions common to the two equations

$$\left.\begin{array}{l} p_1 p_2 p_3 = p_4 \\ p_1 x_1 = p_2 x_2 + p_3 x_4 + p_4 x_3 \end{array}\right\}.$$

92. Shew that the most general function satisfying both the equations

$$-x_1 p_1 + x_2 p_2 + x_4 p_3 + x_3 p_4 = 0,$$

$$2(x_3 + x_4)p_2 + x_2(p_3 + p_4) = 0,$$

is a function of the two quantities

$$x_1{}^2\{x_2{}^2 - (x_3 + x_4)^2\}, \quad (x_3 - x_4) \div x_1.$$

93. Shew that, if

$$u = e^{a\left(\frac{d^2}{dx^2} + \frac{1}{x}\frac{d}{dx}\right)} e^{kx^2},$$

then

$$\frac{\partial u}{\partial a} = 4k^2 \frac{\partial u}{\partial k} + 4ku;$$

and thence deduce that

$$u(1 - 4ak) = e^{\frac{kx^2}{1 - 4ak}}.$$

94. Integrate the equations:

(i) $\quad P(Q+q)r + \{Q(Q+q) - P(P+p)\}s - Q(P+p)t = 0;$

where $P = \dfrac{y+qz}{yp-xq}, \quad Q = \dfrac{x+pz}{xq-yp};$

(ii) $x^2 r - y^2 t = xy;$

(iii) $x^4 r = a^2 t;$

(iv) $qr + ps + qt = 0;$

(v) $x^2 r - y^2 t - xp + yq = xy;$

(vi) $z(r - \lambda^2 t) = p^2 - \lambda^2 q^2;$

(vii) $(y-x)(q^2 r - 2pqs + p^2 t) = (p+q)^2 (p-q);$

(viii) $(y-x)\{pq(r-t) + (q^2 - p^2)s\} + 2pq(p+q) = 0;$

(ix) $xy^3 r + (a+b)x^2 y^2 s + abx^3 yt = y^3 p + abx^3 q;$

(x) $(qz+x)^2 \{(x+y)r - p^2(p+q)\} + (pz+y)^2\{(x+y)t - q^2(p+q)\}$
$\quad = 2(qz+x)(pz+y)\{(x+y)s - (1+pq)(p+q)\}.$

95. If z is any solution of the equation

$$\frac{\partial^2 z}{\partial x \partial y} + \frac{\kappa}{(x-y)^2} z = 0,$$

κ being a constant, then

$$\frac{\partial z}{\partial x} + \frac{\partial z}{\partial y}, \quad x\frac{\partial z}{\partial x} + y\frac{\partial z}{\partial y}, \quad x^2\frac{\partial z}{\partial x} + y^2\frac{\partial z}{\partial y},$$

are also solutions.

96. Solve the equation

$$\frac{\partial^2 z}{\partial x \partial y} = -xyz;$$

and verify that

$$z = \int_0^\pi \cos(xy \cos\phi)\,d\phi$$

is a solution.

97. Shew that an integral of the equation

$$\frac{\partial}{\partial x}\left(x\frac{\partial u}{\partial x}\right) + \frac{\partial}{\partial y}\left(x\frac{\partial u}{\partial y}\right) = 0$$

is given by

$$u = \int_0^\pi f(x\cos\phi + iy)\,d\phi,$$

where f is an arbitrary function; and that the real part of the integral

$$\int_0^\pi x\cos\phi f(x\cos\phi + iy)\,d\phi$$

satisfies the equation

$$\frac{\partial}{\partial x}\left(\frac{1}{x}\frac{\partial u}{\partial x}\right) + \frac{\partial}{\partial y}\left(\frac{1}{x}\frac{\partial u}{\partial y}\right) = 0.$$

98. Solve the equation

$$(pt - r)^2 = q^2 rt;$$

and shew that the common solution of the equations

$$r = q, \quad t = p,$$

is

$$z = A_0 + \sum_{\mu=1}^{\mu=3} A_\mu e^{x\omega^\mu + y\omega^{2\mu}},$$

where ω is a cube root of unity.

99. If p, q, r, are three arbitrary functions of u, subject to the condition

$$p^2 + q^2 + r^2 = 0;$$

if u be determined as a function of x, y, z, by the equation

$$au = xp + yq + zr,$$

where a is any constant; and if

$$v = F(u) + \frac{G(u)}{a - xp' - yq' - zr'},$$

where F and G are arbitrary functions; then

$$\frac{\partial^2 v}{\partial x^2} + \frac{\partial^2 v}{\partial y^2} + \frac{\partial^2 v}{\partial z^2} = 0.$$

100. Shew that the most general integral function of the nth degree in the variables, which satisfies the equation

$$\frac{\partial^2 u}{\partial x^2} + \frac{\partial^2 u}{\partial y^2} + \frac{\partial^2 u}{\partial z^2} = \nabla^2 u = 0,$$

is

$$u = V - \frac{r^2 \nabla^2 V}{2(2n-1)} + \frac{r^4 \nabla^4 V}{2 \cdot 4(2n-1)(2n-3)} - \cdots,$$

where V is an arbitrary homogeneous integral function of x, y, z, of degree n, and r^2 denotes $x^2 + y^2 + z^2$.

Shew also that the Particular Integral of the equation

$$\nabla^2 u = \phi,$$

where ϕ is a homogeneous integral function of x, y, z, of degree $n-2$, is

$$u = \frac{r^2 \phi}{2(2n-1)} - \frac{r^4 \nabla^2 \phi}{2 \cdot 4(2n-1)(2n-3)} + \cdots.$$

INDEX

A CATALOG OF SELECTED

DOVER BOOKS
IN SCIENCE AND MATHEMATICS

A CATALOG OF SELECTED
DOVER BOOKS
IN SCIENCE AND MATHEMATICS

QUALITATIVE THEORY OF DIFFERENTIAL EQUATIONS, V.V. Nemytskii and V.V. Stepanov. Classic graduate-level text by two prominent Soviet mathematicians covers classical differential equations as well as topological dynamics and ergodic theory. Bibliographies. 523pp. 5⅜ × 8½.   65954-2 Pa. $14.95

MATRICES AND LINEAR ALGEBRA, Hans Schneider and George Phillip Barker. Basic textbook covers theory of matrices and its applications to systems of linear equations and related topics such as determinants, eigenvalues and differential equations. Numerous exercises. 432pp. 5⅜ × 8½.   66014-1 Pa. $10.95

QUANTUM THEORY, David Bohm. This advanced undergraduate-level text presents the quantum theory in terms of qualitative and imaginative concepts, followed by specific applications worked out in mathematical detail. Preface. Index. 655pp. 5⅜ × 8½.   65969-0 Pa. $14.95

ATOMIC PHYSICS (8th edition), Max Born. Nobel laureate's lucid treatment of kinetic theory of gases, elementary particles, nuclear atom, wave-corpuscles, atomic structure and spectral lines, much more. Over 40 appendices, bibliography. 495pp. 5⅜ × 8½.   65984-4 Pa. $12.95

ELECTRONIC STRUCTURE AND THE PROPERTIES OF SOLIDS: The Physics of the Chemical Bond, Walter A. Harrison. Innovative text offers basic understanding of the electronic structure of covalent and ionic solids, simple metals, transition metals and their compounds. Problems. 1980 edition. 582pp. 6⅛ × 9¼.   66021-4 Pa. $16.95

BOUNDARY VALUE PROBLEMS OF HEAT CONDUCTION, M. Necati Özisik. Systematic, comprehensive treatment of modern mathematical methods of solving problems in heat conduction and diffusion. Numerous examples and problems. Selected references. Appendices. 505pp. 5⅜ × 8½.   65990-9 Pa. $12.95

A SHORT HISTORY OF CHEMISTRY (3rd edition), J.R. Partington. Classic exposition explores origins of chemistry, alchemy, early medical chemistry, nature of atmosphere, theory of valency, laws and structure of atomic theory, much more. 428pp. 5⅜ × 8½. (Available in U.S. only)   65977-1 Pa. $11.95

A HISTORY OF ASTRONOMY, A. Pannekoek. Well-balanced, carefully reasoned study covers such topics as Ptolemaic theory, work of Copernicus, Kepler, Newton, Eddington's work on stars, much more. Illustrated. References. 521pp. 5⅜ × 8½.   65994-1 Pa. $12.95

PRINCIPLES OF METEOROLOGICAL ANALYSIS, Walter J. Saucier. Highly respected, abundantly illustrated classic reviews atmospheric variables, hydrostatics, static stability, various analyses (scalar, cross-section, isobaric, isentropic, more). For intermediate meteorology students. 454pp. 6⅛ × 9¼. 65979-8 Pa. $14.95

RELATIVITY, THERMODYNAMICS AND COSMOLOGY, Richard C. Tolman. Landmark study extends thermodynamics to special, general relativity; also applications of relativistic mechanics, thermodynamics to cosmological models. 501pp. 5⅜ × 8½. 65383-8 Pa. $13.95

APPLIED ANALYSIS, Cornelius Lanczos. Classic work on analysis and design of finite processes for approximating solution of analytical problems. Algebraic equations, matrices, harmonic analysis, quadrature methods, much more. 559pp. 5⅜ × 8½. 65656-X Pa. $13.95

INTRODUCTION TO ANALYSIS, Maxwell Rosenlicht. Unusually clear, accessible coverage of set theory, real number system, metric spaces, continuous functions, Riemann integration, multiple integrals, more. Wide range of problems. Undergraduate level. Bibliography. 254pp. 5⅜ × 8½. 65038-3 Pa. $8.95

INTRODUCTION TO QUANTUM MECHANICS With Applications to Chemistry, Linus Pauling & E. Bright Wilson, Jr. Classic undergraduate text by Nobel Prize winner applies quantum mechanics to chemical and physical problems. Numerous tables and figures enhance the text. Chapter bibliographies. Appendices. Index. 468pp. 5⅜ × 8½. 64871-0 Pa. $12.95

ASYMPTOTIC EXPANSIONS OF INTEGRALS, Norman Bleistein & Richard A. Handelsman. Best introduction to important field with applications in a variety of scientific disciplines. New preface. Problems. Diagrams. Tables. Bibliography. Index. 448pp. 5⅜ × 8½. 65082-0 Pa. $12.95

MATHEMATICS APPLIED TO CONTINUUM MECHANICS, Lee A. Segel. Analyzes models of fluid flow and solid deformation. For upper-level math, science and engineering students. 608pp. 5⅜ × 8½. 65369-2 Pa. $14.95

ELEMENTS OF REAL ANALYSIS, David A. Sprecher. Classic text covers fundamental concepts, real number system, point sets, functions of a real variable, Fourier series, much more. Over 500 exercises. 352pp. 5⅜ × 8½. 65385-4 Pa. $11.95

PHYSICAL PRINCIPLES OF THE QUANTUM THEORY, Werner Heisenberg. Nóbel Laureate discusses quantum theory, uncertainty, wave mechanics, work of Dirac, Schroedinger, Compton, Wilson, Einstein, etc. 184pp. 5⅜ × 8½. 60113-7 Pa. $6.95

INTRODUCTORY REAL ANALYSIS, A.N. Kolmogorov, S.V. Fomin. Translated by Richard A. Silverman. Self-contained, evenly paced introduction to real and functional analysis. Some 350 problems. 403pp. 5⅜ × 8½. 61226-0 Pa. $10.95

PROBLEMS AND SOLUTIONS IN QUANTUM CHEMISTRY AND PHYSICS, Charles S. Johnson, Jr. and Lee G. Pedersen. Unusually varied problems, detailed solutions in coverage of quantum mechanics, wave mechanics, angular momentum, molecular spectroscopy, scattering theory, more. 280 problems plus 139 supplementary exercises. 430pp. 6½ × 9¼. 65236-X Pa. $13.95

ASYMPTOTIC METHODS IN ANALYSIS, N.G. de Bruijn. An inexpensive, comprehensive guide to asymptotic methods—the pioneering work that teaches by explaining worked examples in detail. Index. 224pp. 5⅜ × 8½.   64221-6 Pa. $7.95

OPTICAL RESONANCE AND TWO-LEVEL ATOMS, L. Allen and J.H. Eberly. Clear, comprehensive introduction to basic principles behind all quantum optical resonance phenomena. 53 illustrations. Preface. Index. 256pp. 5⅜ × 8½.
65533-4 Pa. $8.95

COMPLEX VARIABLES, Francis J. Flanigan. Unusual approach, delaying complex algebra till harmonic functions have been analyzed from real variable viewpoint. Includes problems with answers. 364pp. 5⅜ × 8½.   61388-7 Pa. $9.95

ATOMIC SPECTRA AND ATOMIC STRUCTURE, Gerhard Herzberg. One of best introductions; especially for specialist in other fields. Treatment is physical rather than mathematical. 80 illustrations. 257pp. 5⅜ × 8½.   60115-3 Pa. $6.95

APPLIED COMPLEX VARIABLES, John W. Dettman. Step-by-step coverage of fundamentals of analytic function theory—plus lucid exposition of five important applications: Potential Theory; Ordinary Differential Equations; Fourier Transforms; Laplace Transforms; Asymptotic Expansions. 66 figures. Exercises at chapter ends. 512pp. 5⅜ × 8½.   64670-X Pa. $12.95

ULTRASONIC ABSORPTION: An Introduction to the Theory of Sound Absorption and Dispersion in Gases, Liquids and Solids, A.B. Bhatia. Standard reference in the field provides a clear, systematically organized introductory review of fundamental concepts for advanced graduate students, research workers. Numerous diagrams. Bibliography. 440pp. 5⅜ × 8½.   64917-2 Pa. $11.95

UNBOUNDED LINEAR OPERATORS: Theory and Applications, Seymour Goldberg. Classic presents systematic treatment of the theory of unbounded linear operators in normed linear spaces with applications to differential equations. Bibliography. 199pp. 5⅜ × 8½.   64830-3 Pa. $7.95

LIGHT SCATTERING BY SMALL PARTICLES, H.C. van de Hulst. Comprehensive treatment including full range of useful approximation methods for researchers in chemistry, meteorology and astronomy. 44 illustrations. 470pp. 5⅜ × 8½.   64228-3 Pa. $11.95

CONFORMAL MAPPING ON RIEMANN SURFACES, Harvey Cohn. Lucid, insightful book presents ideal coverage of subject. 334 exercises make book perfect for self-study. 55 figures. 352pp. 5⅜ × 8¼.   64025-6 Pa. $11.95

OPTICKS, Sir Isaac Newton. Newton's own experiments with spectroscopy, colors, lenses, reflection, refraction, etc., in language the layman can follow. Foreword by Albert Einstein. 532pp. 5⅜ × 8½.   60205-2 Pa. $11.95

GENERALIZED INTEGRAL TRANSFORMATIONS, A.H. Zemanian. Graduate-level study of recent generalizations of the Laplace, Mellin, Hankel, K. Weierstrass, convolution and other simple transformations. Bibliography. 320pp. 5⅜ × 8½.   65375-7 Pa. $8.95

CATALOG OF DOVER BOOKS

CHALLENGING MATHEMATICAL PROBLEMS WITH ELEMENTARY SOLUTIONS, A.M. Yaglom and I.M. Yaglom. Over 170 challenging problems on probability theory, combinatorial analysis, points and lines, topology, convex polygons, many other topics. Solutions. Total of 445pp. 5⅜ × 8½. Two-vol. set.

Vol. I 65536-9 Pa. $7.95
Vol. II 65537-7 Pa. $7.95

FIFTY CHALLENGING PROBLEMS IN PROBABILITY WITH SOLUTIONS, Frederick Mosteller. Remarkable puzzlers, graded in difficulty, illustrate elementary and advanced aspects of probability. Detailed solutions. 88pp. 5⅜ × 8½.
65355-2 Pa. $4.95

EXPERIMENTS IN TOPOLOGY, Stephen Barr. Classic, lively explanation of one of the byways of mathematics. Klein bottles, Moebius strips, projective planes, map coloring, problem of the Koenigsberg bridges, much more, described with clarity and wit. 43 figures. 210pp. 5⅜ × 8½.
25933-1 Pa. $6.95

RELATIVITY IN ILLUSTRATIONS, Jacob T. Schwartz. Clear nontechnical treatment makes relativity more accessible than ever before. Over 60 drawings illustrate concepts more clearly than text alone. Only high school geometry needed. Bibliography. 128pp. 6⅛ × 9¼.
25965-X Pa. $7.95

AN INTRODUCTION TO ORDINARY DIFFERENTIAL EQUATIONS, Earl A. Coddington. A thorough and systematic first course in elementary differential equations for undergraduates in mathematics and science, with many exercises and problems (with answers). Index. 304pp. 5⅜ × 8½.
65942-9 Pa. $8.95

FOURIER SERIES AND ORTHOGONAL FUNCTIONS, Harry F. Davis. An incisive text combining theory and practical example to introduce Fourier series, orthogonal functions and applications of the Fourier method to boundary-value problems. 570 exercises. Answers and notes. 416pp. 5⅜ × 8½.
65973-9 Pa. $11.95

AN INTRODUCTION TO ALGEBRAIC STRUCTURES, Joseph Landin. Superb self-contained text covers "abstract algebra": sets and numbers, theory of groups, theory of rings, much more. Numerous well-chosen examples, exercises. 247pp. 5⅜ × 8½.
65940-2 Pa. $8.95

Prices subject to change without notice.
Available at your book dealer or write for free Mathematics and Science Catalog to Dept. GI, Dover Publications, Inc., 31 East 2nd St., Mineola, N.Y. 11501. Dover publishes more than 175 books each year on science, elementary and advanced mathematics, biology, music, art, literature, history, social sciences and other areas.